Biocontrol-Based Integrated Management of Oilseed Rape Pests

Ingrid H. Williams
Editor

Biocontrol-Based Integrated Management of Oilseed Rape Pests

Springer

Editor
Ingrid H. Williams
Institute of Agricultural
and Environmental Sciences
Estonian University of Life Sciences
Tartu 51014
Estonia
ingrid.williams@bbsrc.ac.uk

ISBN 978-90-481-3982-8 e-ISBN 978-90-481-3983-5
DOI 10.1007/978-90-481-3983-5
Springer Dordrecht Heidelberg London New York

Library of Congress Control Number: 2010925250

Cover design: eStudio Calamar S.L.

Printed on acid-free paper

Springer is part of Springer Science+Business Media (www.springer.com)

Preface

Oilseed rape is a major arable crop in both Europe and North America. It is particularly suited to the cooler climates of central and northern Europe, to the western provinces of Canada, and to the state of North Dakota in the USA. The area sown to oilseed rape exceeds 5 million hectares in the EU and 6 million hectares in Canada. Most of the European crop is autumn sown whereas most of the North American crop is spring sown. Forecasts predict a continuing increased demand for oilseed production worldwide.

In both Europe (Chapters 1 and 5) and North America (Chapter 6), growers must protect their oilseed rape crops from insect pests. The pest complex varies considerably on the two continents. Coleopterous pests predominate in both; a weevil species introduced from Europe has now become a serious pest in North America. Substantial yield losses can also occur through infestation by Diptera and also, in North America, by species of Hemiptera and Lepidoptera. Further, in Europe, the relative importance of various pests differs between spring (Chapter 5) and winter rape.

Crop protection against insect pests is still largely through the application of chemical insecticides. These continue to provide reliable and cost-effective control but cause concern because they can harm non-target organisms, such as parasitoids (Chapter 13) and bees (Chapter 14). More efficient targeting of insecticides in time and space can be achieved using economic thresholds, crop monitoring and computer-based decision support systems (Chapter 15). Crop management systems for the future, however, must combine sustainability with environmental acceptability to satisfy both social and economic demands; they should be high-yielding yet energy efficient, providing a good economic net return (Chapters 16 and 17). Consequently, there is now considerable emphasis on minimizing pesticide applications within integrated pest management systems and enhancing the use of natural biocontrol agents. This approach has received further impetus in Europe but the development of widespread resistance in the pollen beetle to pyrethroids, the main group of insecticides now used on the crop (Chapter 12).

The past decade has seen considerable progress in our knowledge of the natural enemies that contribute to biocontrol, particularly the parasitoids (Chapters 2 and 3), the ground beetles (Chapter 4) and the spiders (Chapter 10) and of how their distribution patterns, both within (Chapter 8) and without the crop (Chapter 9), and their behavioural ecology affects their ability to locate the crop (Chapter 7). Push-pull strategies are being developed that use host plant preferences and behavioural responses to semiochemicals to influence pest and natural enemy distributions on the crop. There is also potential for natural enemy conservation through modification of within-field crop husbandry practices, such as soil tillage (Chapter 11) as well as, on the landscape scale, through habitat and environmental manipulation to encourage vegetational diversity of the agroecosystem incorporating hedgerows, cover crops, flowering conservation headlands and field margins to provide refuge, food, over-wintering sites and alternative prey or hosts for natural enemies (Chapters 9, 10, and 17).

I thank the authors of the various chapters of this book for their expertise in collating the state-of-the-art knowledge that they have presented in their reviews. The book is intended to serve as a text for researchers, university teachers, graduate scientists, extension workers and growers involved in pest management. I hope it will play its part in furthering the development of integrated pest management systems that aim to incorporate biocontrol. I also thank the Estonian University of Life Sciences for financially supporting my contribution to the book, and Professor Anne Luik for her encouragement throughout.

Tartu, Estonia
July 2009

Ingrid H. Williams

Contents

Contributors

Lynda M. Castle Visual Communications Unit, Rothamsted Research, Harpenden, Hertfordshire AL5 2JQ, UK, lynda.castle@bbsrc.ac.uk

Samantha M. Cook Plant and Invertebrate Ecology Department, Rothamsted Research, Harpenden, Hertfordshire AL5 2JQ, UK, sam.cook@bbsrc.ac.uk

Lloyd M. Dosdall Department of Agricultural, Food and Nutritional Science, 4-10 Agriculture-Forestry Centre, University of Alberta, Edmonton, AB, Canada T6G 2P5, lloyd.dosdall@afhe.ualberta.ca

Thomas Drapela Department of Integrative Biology and Biodiversity Research, Institute of Zoology, University of Natural Resources and Applied Life Sciences Vienna, Vienna A-1180, Austria, thomas.drapela@boku.ac.at

Barbara Ekbom Department of Ecology, Swedish University of Agricultural Sciences, Uppsala, Sweden, barbara.ekbom@ekol.slu.se

Thomas Frank Department of Integrative Biology and Biodiversity Research, Institute of Zoology, University of Natural Resources and Applied Life Sciences, Vienna A-1180, Austria, thomas.frank@boku.ac.at

Andrew W. Ferguson Plant and Invertebrate Ecology Department, Rothamsted Research, Harpenden, Hertfordshire AL5 2JQ, UK, andrew.ferguson@bbsrc.ac.uk

Udo Heimbach Institute for Plant Protection in Field Crops and Grassland, Julius Kühn-Institute (JKI) Federal Research Centre for Cultivated Plants, Braunschweig 38104, Germany, udo.heimbach@jki.bund.de

Andreas Johnen proPlant GmbH, Münster 48147, Germany, a.johnen@proplant.de

Zdzisław Klukowski Department of Crop Protection, Wroclaw University of Environmental and Life Sciences, Wroclaw, Poland, zklukowski@up.wroc.pl

Reet Karise Institute of Agricultural and Environmental Sciences, Estonian University of Life Sciences, Tartu 51014, Estonia, reet.karise@emu.ee

Märt Kruus Institute of Agricultural and Environmental Sciences, Estonian University of Life Sciences, Tartu 51014, Estonia, mart.kruus@emu.ee

Anne Luik Institute of Agricultural and Environmental Sciences, Estonian University of Life Sciences, Tartu 51014, Estonia, anne.luik@emu.ee

Marika Mänd Institute of Agricultural and Environmental Sciences, Estonian University of Life Sciences, Tartu 51014, Estonia, marika.mand@emu.ee

Peter G. Mason Research Centre, 960 Carling Avenue, Agriculture and Agri-Food Canada, Ottawa, ON, K1A 0C6, Canada, peter.mason@agr.gc.ca

Dietmar Moser Department of Integrative Biology and Biodiversity Research, Institute of Zoology, University of Natural Resources and Applied Life Sciences Vienna, Vienna A-1180, Austria, dietmar.moser@boku.ac.at

Andreas Müller Institute for Plant Protection in Field Crops and Grassland, Julius Kühn-Institute (JKI) Federal Research Centre for Cultivated Plants, Braunschweig 38104, Germany, andreas.mueller@jki.bund.de

Christer Nilsson Swedish University of Agricultural Sciences, Alnarp SE-230 53, Sweden, NCNilsson@telia.com

Jean Roger-Estrade INRA-AgroParisTech, UMR211 Agronomie, Thiverval-Grignon, France, estrade@grignon.inra.fr

Adrien Rusch INRA-AgroParisTech, UMR211 Agronomie, Thiverval-Grignon, France, adrien.rusch@grignon.inra.fr

Jean Pierre Sarthou INRA, INPT-ENSAT, UMR1201 Dynamiques Forestières dans l'Espace Rural, Castanet-Tolosan F-31326, France, sarthou@ensat.fr

Matthew Skellern Plant and Invertebrate Ecology Department, Rothamsted Research, Harpenden, Hertfordshire AL5 2JQ, UK, matthew.skellern@bbsrc.ac.uk

Thomas Thieme BTL Biotest-Labor GmbH Sagerheide, Sagerheide 18184, Germany, tt@biotestlab.de

Carsten Thies Department of Crop Science, Agroecology Georg-August-University, Göttingen 37073, Germany, carsten.thies@agr.uni-goettingen.de

Teja Tscharntke Department of Crop Science, Agroecology Georg-August-University, Göttingen 37073, Germany, ttschar@gwdg.de

Bernd Ulber Department of Crop Sciences, Plant Pathology and Crop Protection Division, Georg-August University Göttingen, Göttingen D-37077, Germany, bulber@gwdg.de

Muriel Valantin-Morison INRA-AgroParisTech, UMR211 Agronomie, Thiverval-Grignon, France, morison@grignon.inra.fr

Eve Veromann Institute of Agricultural and Environmental Sciences, Estonian University of Life Sciences, Tartu 51014, Estonia, eve.veromann@emu.ee

Eneli Viik Institute of Agricultural and Environmental Sciences, Estonian University of Life Sciences, Tartu 51014, Estonia, eneli.viik@emu.ee

Douglas J. Warner Faculty of Engineering and Information Sciences, Agriculture and Environment Research Unit, Science and Technology Research Institute, University of Hertfordshire, Hatfield AL10 9AB, UK, d.j.warner@herts.ac.uk

Ingrid H. Williams Institute of Agricultural and Environmental Sciences, Estonian University of Life Sciences, Tartu 51014, Estonia, ingrid.williams@bbsrc.ac.uk

Johann G. Zaller Department of Integrative Biology and Biodiversity Research, Institute of Zoology, University of Natural Resources and Applied Life Sciences Vienna, Vienna A-1180, Austria, johann.zaller@boku.ac.at

Chapter 1
The Major Insect Pests of Oilseed Rape in Europe and Their Management: An Overview

Ingrid H. Williams

Abstract The oilseed rape crop in Europe is attacked by six major pests that often require control by growers to protect seed yield: the cabbage stem flea beetle, pollen beetle, cabbage seed weevil, cabbage stem weevil, rape stem weevil and brassica pod midge. These attack the crop successively at various growth stages and damage different parts of the plant. They are all widespread but their relative importance varies with country and year. Their control is still mainly through the application of chemical insecticides, often applied prophylactically. The pollen beetle has developed widespread resistance to pyrethroids, the main group of insecticides now used, increasing the urgency for alternative control strategies. The past decade has seen considerable progress in our knowledge of the parasitoids, predators and pathogens that contribute to biocontrol of the pests and of how to incorporate biocontrol into integrated pest management systems. More efficient targeting of insecticides in time and space can be achieved using economic thresholds, crop monitoring and computer-based decision support systems. Push-pull strategies are being developed that use host plant preferences and behavioural responses to semiochemicals to influence pest and natural enemy distributions on the crop. There is also potential for natural enemy conservation through modification of within-field crop husbandry practices as well as, on the landscape scale, through habitat and environmental manipulation to encourage vegetational diversity of the agroecosystem incorporating hedgerows, cover crops, flowering conservation headlands and field margins to provide refuge, food, overwintering sites and alternative prey or hosts for natural enemies.

1.1 Introduction

Oilseed rape is the major oilseed crop grown in northern and central Europe. In 2006, over 5.3 million ha were grown with a production of 15.5 million tonnes

---

I.H. Williams (✉)
Institute of Agricultural and Environmental Sciences, Estonian University of Life Sciences, Tartu 51014, Estonia
e-mail: ingrid.williams@bbsrc.ac.uk

I.H. Williams (ed.), *Biocontrol-Based Integrated Management of Oilseed Rape Pests*, DOI 10.1007/978-90-481-3983-5_1,

(Eurostat 2009). Major producers were Germany, France, Poland and the UK accounting for 74% of this area and 80% of production. Provisional statistics for 2007 indicate a 22% increase in area and a 15% increase in production over 2006. Forecasts predict a continuing increased demand for oilseed rape production in Europe. The crop has a high yield potential under efficient agronomic cultivation and is an important break crop in cereal rotations. It is valued both for human nutrition for its high-quality lipid acid composition and as a component of animal feeds. There is also strong incentive to increase production from the non-food sector, since the EU support for production for biodiesel.

The oilseed rape crop in Europe is dominated by *Brassica napus* spp. *oleifera*, with some *Brassica campestris* (turnip rape) grown in Scandinavia and the Baltic States. Most of the crop is winter sown; the proportion of spring sown rape increases in northern climates. The crop is mostly grown for its seed, which is crushed to extract the oil. The oil is used for cooking, as a biofuel and as a lubricant, as well as in the production of paints, soaps and plastics. Rape meal is used in animal feeds. Its agronomy and husbandry is outlined by Alford (2003a).

Integrated pest management (IPM) has the potential to improve the efficiency, profitability and environmental acceptability of crop production and, thereby, to contribute towards its sustainable production. Naturally-occurring agents of biological control, i.e., the parasitoids, predators and pathogens that attack the pests of oilseed rape, can provide economically viable control of some pests and reduce the need for insecticides.

The last decade has seen considerable advances in our knowledge of naturally-occurring biocontrol agents and how to incorporate them into IPM strategies. This knowledge base has been substantially added to by two EU programmes. The first (acronym: BORIS) was a 3-year Framework 4 Concerted Action (1997–2000) (CT-96-1314-Minimizing pesticide use and environmental impact by the development and promotion of bio-control strategies for oilseed rape pests) conducted under the FAIR programme. A consortium of participants from various European countries reviewed the natural enemies of oilseed rape insect pests; the outputs from this project were published as a monograph (Alford 2003b). The second project (acronym: MASTER) was a 4-year research project (2001–2005) (QLK5-CT-2001-01447 Integrated Pest Management Strategies incorporating biocontrol for European oilseed rape pests) co-funded by the EU Framework 5 Quality of Life and Management of Living Resources programme. This project had the following five objectives: 1. To determine the identity, status and potential of biocontrol agents for oilseed rape pests. 2. To develop economically-viable and environmentally-acceptable IPM Strategies for the crop. 3. To determine the socio-economic feasibility, importance and economic efficiency of the IPM strategies and constraints to their adoption. 4. To construct phenological models for major pests and their key biocontrol agents for integration into decision support systems. 5. To produce Technical Guidelines for farmers, advisors and policy makers on the IPM strategies.

This chapter identifies the economically-important pests of oilseed rape in Europe, outlines the damage they cause and reviews the advances made towards integrating biocontrol into the management of the insect pests of oilseed rape in Europe. It relies heavily on the outputs from the two EU-projects BORIS and MASTER as well as on research conducted by other European researchers who did not participate in these projects; many of the latter have contributed other chapters to this book.

1.2 Major Insect Pests and the Damage They Cause

The oilseed rape crop in Europe is attacked by a diversity of herbivores, including insects, nematodes, slugs and pigeons (Alford et al. 2003). Here we consider only the six major insect pests; these are widespread and abundant and cause sufficient economic damage in some years to require insecticide treatment by growers (Bromand 1990, Garbe et al. 2000).

A recent survey of winter oilseed rape growers in Germany, Poland, Sweden and the UK and of spring/turnip rape growers in Estonia and Finland, conducted as part of the EU-project MASTER showed that they applied control measures against a total of eight pests during the growing season 2002–2003 (Menzler-Hokkanen et al. 2006). Of these, the pollen beetle (*Meligethes aeneus* (Fabricius), Coleoptera: Nitidulidae) was deemed to require control by the majority of growers in each country. The cabbage stem flea beetle (*Psylliodes chrysocephala* (L.), Coleoptera: Chrysomelidae), cabbage seed weevil (*Ceutorhynchus obstrictus* (Marsham) syn. *C. assimilis* (Paykull), Coleoptera: Curculionidae) and the brassica pod midge (*Dasineura brassicae* Winnertz, Diptera: Cecidomyidae) were controlled on winter rape in Germany, Poland, Sweden and the UK (12–20% of growers). The stem weevils (the cabbage stem weevil, *Ceutorhynchus pallidactylus* (Marsham), syn. *C. quadridens* (Panzer) and, in some countries, the rape stem weevil, *Ceutorhynchus napi* Gyllenhal, both Coleoptera: Curculionidae), were controlled on winter rape in Germany, Poland and the UK (14–16% of growers) but not in Sweden. The cabbage stem weevil is present throughout Europe but the rape stem weevil is only present in central Europe. The cabbage root fly (*Delia radicum* L. Diptera: Anthomyidae) was controlled in winter rape in Germany only (12% of growers). Flea beetles (*Phyllotreta* spp., Coleoptera: Chrysomelidae) were controlled by about a third of growers of spring rape crops in Estonia and Finland.

This review focuses on the six most widespread major pests only, namely the cabbage stem flea beetle, the pollen beetle, the cabbage seed weevil, the pod midge, the cabbage stem weevil and the rape stem weevil. The less widely distributed cabbage root fly and the flea beetles, as well as other minor pests, of importance in some countries and seasons, including *Ceutorhynchus picitarsis* (the rape winter stem weevil), *Athalia rosae* (the turnip sawfly), *Brevicoryne brassicae* (cabbage aphid) and *Myzus persicae* (peach/potato aphid) are considered at greater length by Ekbom (Chapter 5, this volume).

1.2.1 Cabbage Stem Flea Beetle

1.2.1.1 Description

The cabbage stem flea beetle is 4–5 mm long, usually black with a blue-green metallic sheen (Alford 1999, Kirk 1992) (Fig. 1.1); a brown variant also occurs (Bonnemaison and Jourdheuil 1954). It has large hind femurs enabling it to jump. The antennae have 10 segments.

Fig. 1.1 Cabbage stem flea beetle, *Psylliodes chrysocephala* (Photo: Rothamsted Research)

1.2.1.2 Distribution

The cabbage stem flea beetle is the most widely distributed stem-mining pest of winter oilseed rape crops throughout regions of northern Europe with a maritime climate (Bromand 1990, Garbe et al. 2000); it is also recorded from the Middle East, Asia, North Africa Canada and the USA (Balachowsky 1963, Bonnemaison 1965, Cox 1998). Spring-sown crops are not infested.

1.2.1.3 Life Cycle

The cabbage stem flea beetle is univoltine. Adults migrate to emerging winter oilseed rape crops in early autumn (late August/early September); they require temperatures above 16°C for flight (Ebbe-Nyman 1952). Once in the crop their flight muscles atrophy (Ebbe-Nyman 1952, Bonnemaison 1965). The number of adults on the crop increases during the autumn, declines during the winter and few are found after April (Williams and Carden 1961). On arrival, the adults feed on the cotyledons and young leaves of the emerging crop. The ovaries of the females are immature at this stage but mature within about 2 weeks of feeding on the crop (Bonnemaison and Jourdheuil 1954, Williams and Carden 1961). Mating occurs soon after emergence and continues throughout the winter (Bonnemaison and Jourdheuil 1954).

Females lay their eggs in cracks in the soil close to rape plants or on the lower parts of newly-emerged rape plants (Sáringer 1984). The eggs are elongate-oval, pale orange in colour and 0.9–0.96 mm long and 0.4–0.43 mm wide (Cox 1998). Optimum conditions for egg-laying are high humidity and a temperature of 4–16°C (Bonnemaison and Jourdheuil 1954, Sáringer 1984). Thus most eggs are laid during the autumn. Egg-laying may continue throughout autumn and winter if the weather is mild (Ebbe-Nyman 1952, Alford 1979, Schulz 1985), or cease during cold periods resuming again in the early spring. Fecundity per female may be up to 1,000 eggs (Bonnemaison 1965, Sáringer 1984). Alford (1979) calculated that eggs require 240 accumulated day-degrees above 3.2°C to hatch.

Larvae are found in winter rape plants from September onwards (Bonnemaison and Jourdheuil 1954, Alford 1979). There are three larval instars. The larvae are up to 8 mm long, creamy-white with three pairs of legs, a black head and a black dorsal plate on the tail segment; the first and second instars are speckled with black dots (Ebbe-Nyman 1952, Kirk 1992). On hatching, the neonate larva can move up to 50 cm to find a host plant (Bonnemaison and Jourdheuil 1954); it then penetrates the upper surface of the petiole of a lower leaf, near the stem (Bonnemaison and Jourdheuil 1954, Queinnec 1967). The larvae first mine the older leaf petioles, then move to younger ones and later enter the stems and growing points feeding throughout the winter when the weather is mild (Ebbe-Nyman 1952, Bonnemaison and Jourdheuil 1954, Williams and Carden 1961, Schulz 1985, Nilsson 1990). At 4°C, larvae mature in 220 days (Bonnemaison and Jourdheuil 1954). From late winter onwards, mature larvae leave the plants to pupate in the soil, burrowing to a depth of 7–9 cm below the surface (Bonnemaison and Jourdheuil 1954, Williams and Carden 1961).

New generation adults emerge in late spring (late May-early July) after about 3 months of pupation (Williams and Carden 1961). They feed on the leaves, stems and pods of oilseed rape and other crucifers for a few weeks (Ebbe-Nyman 1952, Bonnemaison and Jourdheuil 1954, Alford 1979, Sáringer 1984) and then, in midsummer, enter a period of aestivation in sheltered areas such as hedgerows and woodlands creeping into crevices in vegetation (Bonnemaison and Jourdheuil 1954, Williams and Carden 1961). In the autumn, after 1–2 months of aestivation, they become active again and seek out cruciferous plants once more (Cox 1998).

1.2.1.4 Damage

In the autumn, feeding by adult cabbage stem flea beetles on the emerging winter rape crop leaves characteristic holes in the cotyledons and young leaves. Damage can be severe, particularly if the weather is dry and crop growth is slow, and may reduce plant density, but it is rarely necessary to apply insecticides to control adult feeding damage. Crops sown early in the autumn are usually more damaged than those sown later (Leach et al. 1994). Adult feeding is less important in the spring (Bonnemaison and Jourdheuil 1954).

It is the larvae of the cabbage stem flea beetle which cause most damage to rape crops. They invade the plants from October to March, tunneling first into the leaf petioles, where they can form extensive galleries, then down into the main stem and even the growing points. Plant density and plant architecture affects larval infestation; the number of larvae/m^2 has been found to decrease with decrease in plant density from 90 to 30 plants/m^2, while, in contrast, the number of larvae/plant increased at lower plant densities (Nuss and Ulber 2004). However, where a vigorous hybrid cultivar was grown, the larger petioles at lower plant densities provided sufficient food for larvae preventing their infestation of the terminal shoots. The tunnelling causes loss of plant vigour in the spring, lowering yields even when larval densities are low (Nilsson 1990). At high larval densities, plant growth may be severely affected with stem wilting, delayed flowering and even plant collapse, reducing winter survival and plant density (Bonnemaison and Jourdheuil 1954, Williams and Carden 1961, Graham and Alford 1981, Lane and Cooper 1989, Nilsson 1990, 2002, Winfield 1992). Yield loss due to the larvae can vary considerably from year to year; during 1982–1984, Nilsson (1990) estimated mean losses of 23–114 kg in oil yield per larva per plant (Nilsson 1990). Larval damage can also make the plants more susceptible to frost damage (Winfield 1992) and increase infestation by stem canker (*Leptosphaeria maculans* syn. *Phoma lingam*) (Schultz and Daebeler 1984).

1.2.2 Pollen Beetle

1.2.2.1 Description

Meligethes aeneus (Fabricius) is the dominant species of pollen beetle found on rape crops throughout Europe. It is a small black beetle (1.9 mm long and 1.3–1.5 mm wide) with foreshortened elytra (Fig. 1.2). The antennae have 11 segments with a compact 3-segemented club (Kirk-Spriggs 1991, 1996). Several other *Meligethes* spp. may also be found on rape crops in Europe (Nolte and Fritzsche 1952, Jurek 1972, Karltorp and Nilsson 1981), notably *Meligethes viridescens* (Fabricius) which may be abundant in some spring crops (Fritzsche 1957); it emerges later in the spring and requires higher temperatures for oviposition and development than *M. aeneus*.

Fig. 1.2 Pollen beetle, *Meligethes aeneus* (Photo: Rothamsted Research)

1.2.2.2 Distribution

Pollen beetles are generally abundant throughout Europe on both winter and spring rape crops. However, they are usually more abundant and cause more damage to spring rape and hence are a more serious pest in northern Europe where relatively more spring rape is grown, and the life-cycle of pest and crop are well synchronized as the beetle migrates to the crop at its susceptible green bud stage.

1.2.2.3 Life Cycle

Pollen beetles overwinter as adults in the soil, vegetation and leaf litter of field margins, woodlands and hedgerows, emerging in early spring (March to June) when temperatures exceed 10°C (Nilsson 1988a, Láska and Kocourek 1991). They are polyphagous feeders and, on emergence, feed on pollen from plants of many different families for a week or two (Free and Williams 1978). When temperatures exceed 12°C, they seek cruciferous plants, for mating and oviposition (Free and Williams 1978); they usually arrive on oilseed rape crops when they are at the green bud stage and feed on pollen in the buds and flowers. They lay their eggs (2–3 mm long) in buds at least 3 mm long, ovipositing through a hole they chew into the base of the bud (Scherney 1953, Nilsson 1988b, Ekbom and Borg 1996). The eggs are placed either beside the anthers or occasionally between the sepals and petals (Scherney 1953). Single females caged with buds have laid a mean of 246 eggs each with 2–3 eggs per bud (Scherney 1953), although up to 10 eggs per bud have been reported (Ekbom and Borg 1996). The eggs hatch after a few days and development from egg to adult takes about 1 month, depending on temperature.

There are two larval instars (Osborne 1964). Larvae are up to 5 mm long, creamy-white with a brown-black head and three pairs of legs. The first instar feeds on pollen within the bud for 5–10 days (Nilsson 1988c), the second feeds on pollen from open flowers, moving into younger flowers every few days (Williams and Free 1978). On maturity after about 2 weeks (Bromand 1983), larvae drop from the flowering canopy to the soil below and pupate just below soil surface. New generation beetles emerge during the summer and feed on pollen from plants of many families (Williams and Free 1978). Their ovaries do not mature and they do not mate before seeking overwintering sites between mid-July and late September (Scherney 1953, Fritzsche 1957, Blazejewska 1960).

1.2.2.4 Damage

Crops are most susceptible to yield-decreasing damage by the pollen beetle at the early bud stage and become less susceptible as plants develop (Williams and Free 1979, Tatchell 1983, Nilsson 1987, 1988d, 1994, Axelsen and Nielsen 1990). Winter rape crops may escape damage if they pass the susceptible bud stage before the main migration of beetles into the crops but backward winter rape crops and spring rape crops are often in the bud stage during crop colonization. Losses of 70% yield in spring crops not treated with insecticide have been reported (Nilsson 1987). Feeding by both adults and larvae cause bud abscission leaving podless stalks or results in distorted or weakened pods (Gould 1975, Free and Williams 1978, Nilsson 1988d); some of the latter may later be attacked by brassica pod midge.

Once the crop is in flower, pollen beetles cause little further damage to most crops as both adults and second instar larvae can then readily access the pollen on which they feed (Williams and Free 1978). However, where spring-sown composite hybrids are grown, the adults can cause yield-limiting damage to the crop; they concentrate on the male-fertile plants to feed on pollen, thereby reducing the amount of pollen available to pollinate the male-sterile plants (Cook et al. 1999, 2004). Large numbers can affect both seed set and yield.

Not all pollen beetle damage to buds results in yield loss as rape plants are capable of considerable compensatory growth in response to attack (Williams and Free 1979, Tatchell 1983, Lerin 1987, Axelsen and Nielsen 1990); feeding damage to the terminal raceme leads to increased production of new side racemes, pods and seeds per pod. Seed yield may be little affected if the damage occurs early although seeds from pods on these racemes are often smaller and contain less oil (Sylvén and Svensson 1976, Nilsson 1987, Axelsen and Nielsen 1990). Plants may even overcompensate and produce increased yield although late maturing pods may not give an evenly harvestable crop.

1.2.3 Cabbage Seed Weevil

1.2.3.1 Description

The cabbage seed weevil is 2–3 mm long with a pronounced rostrum and 7-segmented antennae. It is lead-grey in colour with grey-black legs and has two rows of whitish hairs between the longitudinal furrows on its elytra (Alford et al. 2003) (Fig. 1.3).

Fig. 1.3 Cabbage seed weevil, *Ceutorhynchus obstrictus* syn. *C. assimilis* (Photo: Rothamsted Research)

1.2.3.2 Distribution

The cabbage seed weevil is a widespread and major pest of both winter and spring rape crops throughout Europe and North America (see also Dosdall and Mason Chapter 6, this volume).

1.2.3.3 Life Cycle

The cabbage seed weevil is univoltine and oligophagous on *Brassica* species. Adults emerge from overwintering sites in spring (May–June); males usually emerge before females (Ankersmit 1956). They fly to cruciferous plants, including oilseed rape, when temperatures exceed their flight threshold temperature of 13–15°C (Free and Williams 1979b, Lerin 1991). Adults of both sexes usually migrate to winter rape during flowering. They feed on the buds, flowers, pods and stem tips of plants (Williams and Free 1978). Females are reproductively immature on emergence from hibernation, their ovaries maturing after about 2 weeks of feeding (McCaffrey et al. 1991).

The female lays her eggs, usually singly in pods through a hole she bores in the pod wall with her mouthparts (Heymons 1921, Doucette 1947, Skrocki 1972); small young pods (20–40 mm long) are preferred for oviposition (Free and Williams 1978). Caged females lay between 24 and 240 eggs (Bonnemaison 1957, Dmoch 1965, Lerin 1991). After oviposition, the female deposits a pheromone on the pod wall which deters further oviposition into the pod (Kozlowski et al. 1983, Ferguson and Williams 1991, Mudd et al. 1997, Ferguson et al. 1999a, b). The egg hatches 1–2 weeks later (Dmoch 1965).

There are three larval instars (Heymons 1921, Hoffmann 1951, Dmoch 1965). The larvae are up to 5 mm long, creamy-white and legless (Kirk 1992). The larva feeds on the developing seeds inside the pod for 2–3 weeks (but may up to 5 weeks depending on temperature), destroying about five seeds (Heymons 1921, Carlson et al. 1951, Bonnemaison 1957). When mature, it chews a hole through the pod wall, exits through it and drops to the ground. It burrows to a depth of up to 130 mm to pupate for 9–23 days (Heymons 1921, Wolf and Krausse 1921, Hanson et al. 1948, Bonnemaison 1957, Dmoch 1965).

New generation weevils emerge after 11–24 days, from late July onwards. They feed for a week or two on cruciferous plants (Bonnemaison 1965, Williams and Free 1978) and then seek overwintering sites in perennial vegetation and leaf litter of field margins and woodlands (Bonnemaison 1957, Dmoch 1965). The ovaries do not mature and they do not mate before winter diapause.

1.2.3.4 Damage

The cabbage seed weevil is more damaging to winter rape than to spring rape; the adults usually infest winter rape during flowering and spring rape before flowering (Free and Williams 1978, 1979b). The adults cause little direct damage to the crop although their feeding and egg-laying punctures in the pods may allow the brassica pod midge to oviposit into them and increase the likelihood of fungal attack by *Phoma lingam*.

Yield-limiting economic damage to the crop is through larval feeding in the pods. Each larva eats about five seeds in a pod, representing 8–15% of the seeds present (Bonnemaison 1957) and reducing yield of the pod by about 18% (Williams and Free 1978). Infestation of a winter rape crop by an average of one adult weevil per plant has been estimated to reduce crop seed yield by about 4%. Although the crop can compensate for some seed loss (Williams and Free 1979), late compensation can result in immature pods at harvest.

1.2.4 Brassica Pod Midge

1.2.4.1 Description

The brassica pod midge is a small fly. The male is 0.7–1.5 mm long with a yellow-grey abdomen (Fig. 1.4), whereas the female is 0.9–2.2 mm long with a pinkish-red abdomen (Fig. 1.5) (Sylvén 1949). The antennae are many-segmented with distinct whorls of hairs. The female has a long extendable ovipositor (Alford et al. 2003).

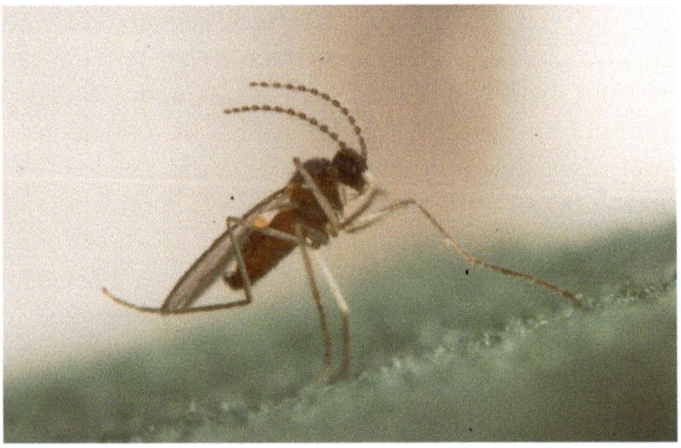

Fig. 1.4 Brassica pod midge, *Dasineura brassicae* (male) (Photo: Rothamsted Research)

Fig. 1.5 Brassica pod midge, *Dasineura brassicae* (female) (Photo: Rothamsted Research)

1.2.4.2 Distribution

The brassica pod midge is an important pest of both winter and spring rape crops throughout most of Europe (Bromand 1990).

1.2.4.3 Life Cycle

The brassica pod midge is multivoltine. It infests both winter rape, on which it usually has two generations (Williams et al. 1987a), as well as spring rape on which it has only one in UK (Williams et al. 1987b) but may have two in mainland Europe, for example in Poland (Z. Klukowski pers comm).

Adults emerge in the spring from mid-May to mid-July after overwintering as larvae in cocoons in fields that grew oilseed rape in previous years. Axelsen (1992a) calculated that 50% emerged after 141°C above 7.1°C after 1 January. The two sexes emerge concurrently and mate at the emergence site (Fröhlich 1956, Buhl 1960). Females dispense a sex pheromone to attract males (Williams and Martin 1986). The adults are short-lived (1–3 days) in the field, but can live for up to 30 days in the laboratory on moist sand at 10°C (Williams unpublished). The males die soon after mating (Buhl 1960) and only the mated females migrate to rape crops (Williams et al. 1987a, b).

The female midge lays her eggs in batches of 20–30 in pods, usually in pods damaged in some way (Sylvén 1949), as her ovipositor is thought not to be strong enough to pierce the pod wall of an undamaged pod (Ankersmit 1956). Cabbage seed weevil feeding, oviposition or larval exit punctures are most often used (Speyer 1921, Sylvén 1949, Ankersmit 1956, Skrocki 1979); Ferguson et al. (1995) found that the midge appears to prefer pods containing weevil eggs which their larvae may predate. The eggs hatch after 3–4 days (Czajkowska and Dmoch 1975).

Larvae are up to 2 mm long, whitish-yellow and legless (Kirk 1992). There are three larval instars. They feed on the inner pod wall for up to a month. Midge-infested pods split open prematurely, releasing the larvae, which drop to the ground and burrow beneath the soil surface to a depth of about 5 cm where they spin cocoons (Sylvén 1949). A proportion pupate immediately, emerging as new generation adults a few weeks later and infest the crop anew, while others enter winter diapause to emerge the following year or even up to 5 years later (Buhl 1960, Axelsen 1995). Later in the year, a greater proportion of larvae enter diapause (Buhl 1960, Czajkowska 1987a). Ploughing of fields where midge cocoons overwinter is probably an important mortality factor for this pest (Axelsen 1992b, 1995).

1.2.4.4 Damage

Brassica pod midge infestations are often most severe on crop headlands (Free and Williams 1979a, b, Ferguson et al. 2004). Infested pods become swollen and yellow, splitting prematurely and shedding larvae and seed. As much as 82% of seed weight can be lost from infested pods.

1.2.5 Cabbage Stem Weevil

1.2.5.1 Description

The cabbage stem weevil is 2.5–3.5 mm long. It is greyish-brown with brown-red legs and has scattered white scales over its body and fine hairs on the elytra, those concentrated centrally on the base of the elytra forming a white rectangular spot (Jourdheuil 1963, Kirk 1992) (Fig. 1.6).

Fig. 1.6 Cabbage stem weevil, *Ceutorhynchus pallidactylus* (Photo: Rothamsted Research)

1.2.5.2 Distribution

The cabbage stem weevil is a pest of both winter and spring rape throughout Europe (Bromand 1990). It has also been reported from Russia, North Africa and North America.

1.2.5.3 Life Cycle

The cabbage stem weevil is univoltine. Adults migrate to oilseed rape crops in the spring (March/June) and feed on the margins, veins and petioles of leaves and on young stems (Jourdheuil 1963).

The eggs are laid in small groups of two to eight on the underside of leaf petioles and sometimes in young stems. The egg is smooth, shiny and translucent, 0.7 mm long and 0.6 mm wide (Jourdheuil 1963, Alford 1999). Eggs hatch after 6–11 days.

The larvae are white with a creamish head, legless and up to 6 mm long (Jourdheuil 1963, Kirk 1992). There are three larval instars (Alford et al. 2003). The larvae feed for 3–6 weeks, first in the petioles and later in the stems and lateral shoots. On maturity in July and August, third instar larvae exit the plant and pupate in the soil a few centimetres below the surface (Jourdheuil 1963, Bonnemaison 1965). New generation adults emerge after about 3 weeks, from mid-July onwards, and feed on cruciferous plants, grazing on the undersides of leaf petioles and veins before entering hibernation. They overwinter outside the crop under plant debris, compost and bushes (Winfield 1961, Jourdheuil 1963).

1.2.5.4 Damage

Cabbage stem weevil larvae cause more damage to spring than to winter rape crops. Larval tunnelling within the petioles and later within the stems destroys the pith and causes loss of plant vigour, an increase in the number of podless stalks, reduced leaf area, earlier leaf abscission, delayed flowering and lodging of plants with consequent loss of seed yield (Graham and Gould 1980, Broschewitz and Daebeler 1987, Walczak et al. 1997, Kelm and Walczak 1998, Kelm and Klukowski 2000). Damage to the stems also facilitates infestation by the fungal diseases, *Phoma lingam* and *Botrytis cinerea* (Broschewitz et al. 1993, Krause et al. 2006).

1.2.6 Rape Stem Weevil

1.2.6.1 Description

The rape stem weevil is 3.2–4 mm long with a long rostrum. It is greyish with three rows of whitish hairs between the longitudinal furrows of the elytra (Alford et al. 2003) (Fig. 1.7).

Fig. 1.7 Rape stem weevil, *Ceutorhynchus napi* (Photo: Rothamsted Research)

1.2.6.2 Distribution

The rape stem weevil is a serious pest of winter oilseed rape in central continental Europe, especially in Austria, France, Germany, Poland and Switzerland. It is not a pest of winter rape in northern Europe or the UK, or a pest of spring-sown rape.

1.2.6.3 Life Cycle

Rape stem weevil adults migrate to oilseed rape in late winter/early spring (February/March). The female lays her eggs singly into the stem pith close to the growing tip of winter oilseed rape plants during stem elongation in late winter/early spring (Lerin 1993). Eggs hatch in 1–2 weeks (Büchi 1996).

The larvae are up to 8 mm long, creamish-white to yellowish in colour and with wart-like protrusions on each segment of the thorax and abdomen (Dosse 1952). They feed within the stem for 3–5 weeks before leaving the plant to pupate in the soil. New generation adults remain in their earthen chambers overwinter and emerge the following spring.

1.2.6.4 Damage

The deposition of rape stem weevil eggs in the stems of winter rape plants during March and April causes substantial deformation of stem growth with swelling, twisting, and longitudinal splitting (Le Pape and Bronner 1987, Lerin 1993). The larvae cause further damage as they tunnel into the pith to feed (Ballanger 1983, Büchi 1988, 1996, Debouzie and Ballanger 1993, Lerin 1993). Damage to the stems also predisposes them to infestation by fungal pathogens, particularly *Phoma lingam* (Broschewitz et al. 1993); the severity of *P. lingam* stem infection has been found to increase with increasing number of rape stem weevil oviposition punctures (Krause et al. 2006). Yield losses up to 50% have been reported (Alford et al. 2003).

1.3 Pest Management

1.3.1 Chemical Control

Management of insect pests in commercial crops of oilseed rape throughout Europe is still largely by the application of chemical insecticides. These are often applied routinely and prophylactically to the entire area of the crop, frequently without regard to pest incidence.

1.3.1.1 Insecticide Use

Insecticides may be applied to the oilseed rape crop several times during its life. Recent surveys of insecticide usage across Europe have revealed that crops of winter and spring rape commonly receive between one and four applications and that some receive more than five (Menzler-Hokkanen et al. 2006, Richardson 2008).

Seed is commonly supplied treated with insecticide to control the cabbage stem flea beetle. During the last two decades, the use of organochlorine insecticides has been withdrawn in most countries because of concerns about the safety of these insecticides to human operators and their harmful effects on non-target organisms. They were replaced by carbosulfan, isofenphos and imidacloprid plus beta-cyfluthrin (or clothianidin if cabbage root fly is a problem) applied as a seed dressing, or by pyrethroids applied as a spray. The latter is applied at early crop emergence when adult damage is first seen and, if necessary, again later in the autumn to prevent larval damage (cotyledon, two to four leaf stages), although many crops are sprayed prophylactically (Smith and Hewson 1984, Nilsson 1990, 2002).

Chemical control of spring and summer pests has, for the past 20 years or more, been largely by the application of synthetic pyrethroids (see also Thieme et al.

Chapter 12 this volume). Infestations of the cabbage stem weevil, the rape stem weevil and the pollen beetle are usually controlled in early spring before crop flowering, from the early stem elongation to yellow bud stages. A list of the pyrethroids, organophosphates, neonicotinoids and other active substances approved for use against the pollen beetle in Europe, is given in Richardson (2008). Pyrethroid application against cabbage seed weevil and brassica pod midge is commonly recommended at early to mid-flowering if economic thresholds have been breached (Ballanger 1997, Steinbach et al. 2007). Insecticides are relatively cheap and so are often applied prophylactically, especially in a tank-mix with a fungicide if disease protection is considered necessary.

Although insecticides give effective control of the pests, there is a need to develop alternative strategies for pest management (see also Ekbom Chapter 5, Nilsson Chapter 16, and Rusch et al., Chapter 17 this volume). Over-use of insecticides reduces the economic competitiveness of the crop and threatens biological diversity. Insecticides may kill the natural enemies of the pests, such as their parasitoids (see also Ulber et al. Chapter 13 this volume), which would otherwise be a natural resource of great potential benefit to the farmer and consumer. By killing natural enemies and thereby reducing biological control of the pests, pesticide applications must be increased further to achieve effective control (Pickett et al. 1995, Alford et al. 1996, Murchie et al. 1997b). Insecticides applied during flowering also kill bees which may be foraging on the crop (Mänd et al. Chapter 14 this volume).

1.3.1.2 Resistance to Insecticides

Insecticides can lose their effectiveness over pests. Three decades ago, Lakocy (1977) reported widespread cross-resistance in the pollen beetle to chlorinated hydrocarbon, organophosphorus and carbamate insecticides in Poland. More recent studies indicate that the long-term selective pressure exerted by the often-prophylactic high-frequency application of low-priced pyrethroids over the last two decades has caused the pollen beetle to develop widespread resistance to pyrethroids in many European countries and suspected developing resistance in others (reviewed by Thieme et al. Chapter 12 this volume). The ban on the use of organophosphates is recognized as a key contributing factor. It is estimated that two-thirds of the oilseed rape crop in Europe is now affected; in Germany alone, 200,000 ha were damaged in 2006, and 30,000 ha, with an estimated value of 22–25 million euros, were destroyed (Zlof 2008). To date, tests in France (Ballanger et al. 2003) and in Germany (Müller et al. 2008) have not yet detected positive pyrethroid resistance in any other major pests of oilseed rape.

Resistance management is essential to integrated pest management (Roush 1989, Denholm and Rowland 1992, Clarke et al. 1997, Leonard 1997). Measures are urgently needed to reduce the use of insecticides in oilseed rape, to preserve activity of the limited armoury of available insecticides and to minimize environmental pollution.

To manage the resistance problem in the pollen beetle a range of chemical control options and ecological approaches have been proposed (Zlof 2008).

Insecticides should not be used prophylactically but only when economic thresholds have been breached. At least two and preferably three to four modes of action to which pollen beetle is not resistant should be made available to growers, to be used as part of spatial rotation of cropping and IPM strategies. The insecticides should have minimal impact on beneficial organisms. Growers are being advised to consider the use of non-pyrethroid treatments, such as thiacloprid, to control the pollen beetle (Richardson 2008). The need to replace plant protection products withdrawn or to be withdrawn from the market by new EU Directives (http://www.europa.eu.int./comm/food/index_en.html) is recognised. Approaches to be used within IPM strategies include (i) the greater use of monitoring to determine pest abundance more accurately, (ii) the use of trap crops to attract and to concentrate pests thereby reducing pest density and damage to the main crop, and (iii) the promotion of conservation biological control with particular emphasis on minimum tillage and avoidance of insecticides during flowering (Cook and Denholm 2008).

1.3.2 Biological Control

Naturally-occurring agents of biological control, i.e., the parasitoids, predators and pathogens that attack the pests of oilseed rape, can provide economically-viable control of some pests and reduce the need for insecticides. In oilseed rape, biocontrol of the pests focuses on the conservation and enhancement of their natural enemies, rather than on the importation and establishment of exotic natural enemies or the inundative or inoculative releases of mass-reared natural enemies. This is considered to be the easiest and least costly method of biocontrol without harming and disturbing the natural ecosystem.

1.3.2.1 Parasitoids

Parasitoids exert substantial natural control on rape pest populations. The six major pests are reported to host to at least 80 species of parasitoid (Alford 2003b). Most are hymenopterous wasps, especially braconids, chalcids and ichneumonids, and most attack the larval stages. However, these groups can be difficult to identify correctly to species and many records in the literature are probably inaccurate and unreliable due to erroneous identification to species and recent revisions of the taxonomy of major groups.

Reviews of the literature and extensive sampling during the EU-funded project MASTER identified 12 species of parasitoid as key species attacking the six major pests of oilseed rape (Ulber et al. Chapter 2 this volume). They are widespread and abundant throughout most of Europe where their hosts occur. Percent parasitism of target pests varied between countries and years, commonly ranging between 20 and 50%. They are therefore of considerable potential economic importance for conservation biocontrol of these major pests. The new guide to the identification of adults of these key parasitoids by Ferguson et al. (Chapter 3 this volume) should greatly aid future research and the development of IPM strategies incorporating biocontrol.

Evidence is growing that parasitoids can provide effective biocontrol of rape pests. For example, in Finland and Sweden, 50% of all insecticides are targeted against the pollen beetle, yet simple management of its parasitoid *Phradis morionellus* can result in complete control of this pest (Hokkanen et al. 1988). In the UK, the pteromalid wasp, *Trichomalus perfectus*, has been shown to be a widely distributed and increasingly important enemy of the cabbage seed weevil and capable of exerting considerable natural control of this pest (Murchie et al. 1997b).

Like their hosts, the parasitoids migrate into the rape crop from their overwintering sites each spring. Landscape structure can affect parasitization rates; proximity of field margin strips and fallow can increase parasitization of the pollen beetle (Hokkanen 1991, Thies and Tscharntke 1999). The effect of semi-natural habitats on parasitism of pollen beetle populations is reviewed by Thies and Tscharntke (Chapter 9 this volume). Within-field the distribution of pest and its parasitoids should co-incide in time and space for effective biocontrol (Murchie et al. 1999, Ferguson et al. 2000, Williams et al. 2000); their spatio-temporal distributions and the factors that influence them are reviewed by Williams and Ferguson (Chapter 8 this volume).

Parasitoids find their hosts in response to physical and chemical cues emanating from their hosts and the infested host plant (see Williams and Cook Chapter 7 this volume). Plant structure can variously affect parasitization rates. For example, fewer rape stem weevil larvae are parasitized by *Tersilochus fulvipes* in stems that exceeded 8–9 mm in diameter (Ulber 2003), whereas parasitism of cabbage seed weevil larvae by *Trichomalus perfectus* is unrelated to infested pod position on plants or to host density (Ulber and Vidal 1998, Ferguson et al. 2000). Broad-spectrum insecticides applied to the crop, particularly during or after flowering, kill many parasitoids (Murchie et al. 1997b, Ulber et al. Chapter 13 this volume). Application of insecticides during spray windows determined by phenological models of pests and parasitoids can reduce parasitoid mortality (Johnen et al. Chapter 15 this volume). Many parasitoid species overwinter in the soil of the rape field. Post-harvest soil cultivations, particularly ploughing and rotary harrowing, reduce parasitoid survival, whereas minimal cultivation is less harmful (Nilsson 1985, Nilsson Chapter 11 this volume).

1.3.2.2 Predators

A wide range of generalist predators attacks the pests of oilseed rape (Büchs 2003a, Büchs and Alford 2003, Büchs et al. 2006); ground beetles (Coleoptera: Carabidae), rove beetles (Coleoptera: Staphylinidae) and spiders (Arachnida: Araneae) are the main epigaeic predators while long-legged flies (Diptera: Dolichopodidae) and dance flies (Diptera: Hybotidae) are the most abundant predators in the crop canopy. They can be present in large numbers throughout the year and can utilise alternative food, such as Collembola, to sustain their populations at times when rape pests are unavailable. Much of our knowledge of their predation potential comes from studies in cereals (e.g., Kromp 1999), sugar beet (e.g., Büchs et al. 1997) and cabbage (Hokkanen and Holopainen 1986). Information about their identity, status and importance in the oilseed rape crop has been sparse and their relative contribution to biocontrol in the rape crop uncertain unti recently (Büchs 2003a). Present knowledge of the role of predators in the biocontrol of minor pests in spring rape crops is reviewed by Ekbom (Chapter 5 this volume).

The ground beetles or carabids are the most important polyphagous predators within arable cropping systems in Europe (Kromp 1999), but until recently, there has been little information on their activities in the oilseed rape crop and assessments of their importance have varied. The pests are most vulnerable to predation by carabids as eggs or young larvae in the soil or when, as mature larvae, they drop to the ground to pupate. Early reports indicated that the carabid *Clivina fossor* can cause larval mortality of the pollen beetle and the cabbage seed weevil of 60–65% (Bonnemaison 1957, Scherney 1959). Basedow (1973) attributed 39 and 65% of larval mortality of the pollen beetle and the brassica pod midge, respectively, to epigaeic predators. In contrast, Hokkanen et al. (1988) in Finland attributed only a 3% reduction of pollen beetle larvae and pupae on spring rape to carabids and other epigaeic predators. Similarly, Nilsson and Andreasson (1987) considered that carabids were of little importance as predators of pollen beetle larvae in Sweden.

More recent studies have contributed substantially to our knowledge of the carabid complex in oilseed rape crops, particularly in Germany and the UK, and of their potential role in biocontrol (Büchs and Nuss 2000, Warner et al. 2000, 2003). Information collated during the EU-project MASTER identified 15 carabid species that were dominant or subdominant in the Estonia, Germany, Poland, Sweden, and the UK but that the community composition varied between country, between winter and spring crops, and with crop husbandry (Büchs et al. 2006). Ten species were sufficiently widespread and dominant to be considered key predators; their life cycles, habitat preferences, mobility and feeding habits are reviewed by Williams et al. (Chapter 4 this volume).

The species composition of rove beetles and of spiders in oilseed rape fields in Germany has been studied by Krooss (1996) and by Harenberg (1997), respectively, but there is little information about their impact on the pests. As part of the EU project MASTER, Felsmann and Büchs (2006a, b) found that staphylinid larvae were usually more numerous where winter rape was grown in an integrated system (with reduced tillage and no insecticides) than in a standardised system (ploughed and with insecticides) and that, in most years, they were both spatially and temporally coincident with pollen beetle larvae. The dolichopids, hybotids and the web-forming spider *Theridion impressum* are probable predators of the brassica pod midge (Büchs 2003a); adult midge were found in ca. 38% of *T. impressum* webs in 2004 as part of the MASTER project (Felsmann et al. 2006). The responses of the major pests and the wolf spider, *Pardosa agrestis*, in relation to site and landscape factors are presented by Frank et al. (Chapter 10 this volume).

1.3.2.3 Pathogens

Pathogenic organisms with most potential for the control of oilseed rape pests include entomopathogenic fungi, entomopathogenic nematodes and protozoa (Hokkanen et al. 2003).

Agricultural soils are generally poor in entomopathogenic fungi although these may be abundant in nearby uncultivated land (Vänninen et al. 1989, Zec-Vojinovic et al. 2006a). Two species of fungus, *Beauveria bassiana* and *Metarhizium anisopliae*, show potential for biocontrol. *Beauveria bassiana* has been shown to reduce overwintering survival of the pollen beetle (Hokkanen 1993). *Metarhizium anisopliae,* applied as a soil inoculum or as an oil-spray formulation in semi-field studies, has given good mortality of the cabbage stem flea beetle, the pollen beetle and the cabbage seed weevil (Vänninen et al. 2000, Husberg and Hokkanen 2001). Husberg and Hokkanen (2001), however, found that some parasitoids were also susceptible to the fungus. Generally, parasitoids (Husberg and Hokkanen 2001), bees (Butt et al. 1994), and carabids (Goettel 1994) appear to be less susceptible to *M. anisopliae* than the pest species. Honey bees have been used to disseminate *M. anisopliae* to the flowering canopy of oilseed rape to infect both pollen beetles and cabbage seed weevils (Butt et al. 1998, Carreck et al. 2006, see also Mänd et al. Chapter 14 this volume). In the future, there is potential to employ entomopathogenic fungi, particularly *M. anisopliae*, as bio-insecticides to act synergistically with natural enemies in an integrated pest management strategy.

Species of the entomopathogenic nematode genera *Steinernema* and *Heterorhabditis* have been effective against pollen beetle larvae in the laboratory and in the field (Ehlers and Hokkanen 1996). As part of the EU-project MASTER, in a collaborative trial in six partner countries, a suspension of the nematode *Steinernema feltiae* was applied in water to 1–2 m^2 plots at the rate of 1 million infective juveniles/m^2. Considerable reduction (42–72%) in the mean numbers of cabbage stem flea beetle, pollen beetle, cabbage seed weevil and rape stem weevil that emerged from treated plots was achieved (Hokkanen et al. 2006). Delivery sytstems for the inundation/inoculation of entomopathogenic nematodes for control of oilseed rape pests are being investigated (Zec-Vojinovic et al. 2006b).

The protozoan *Nosema meligethi* can have a major impact on overwintering mortality of the pollen beetle and appears to be specific to *Meligethes* spp. (Lipa and Hokkanen 1995). Although common in 'wild' populations of the pollen beetle, it is rare in crops but can be artificially augmented; its distribution appears to be limited to Eastern Europe, Finland, Sweden, Denmark and Austria (Lipa and Hokkanen 1992).

Current knowledge of the role of pathogens in the control of pests of spring rape is reviewed by Ekbom (Chapter 5 this volume).

1.3.3 Integrated Pest Management

Integrated pest management (IPM) has the potential to improve the efficiency, profitability and environmental acceptability of crop production and, thereby, to contribute towards its sustainable production (see also Ekbom Chapter 5, Nilsson Chapter 16, and Rusch et al. Chapter 17 this volume). Although still in its infancy in the oilseed rape crop, encouraging results have being obtained for integrated management of the major insect pests.

Guidelines for Integrated Production of Arable Crops (including oilseed rape) in Europe published in 1997 by the International Organisation for Biological and Integrated Control of Noxious Animals and Plants (IOBC) advocated the use of biological, bio-technical, physical or agronomic methods rather than chemical methods of plant protection. However, then, just over a decade ago, the only options specified for oilseed rape pest control were the use of economic thresholds to determine the need to apply pesticides and the establishment of flowering border strips to attract natural enemies of pests (IOBC 1997).

The past decade has seen considerable research progress towards the development of IPM strategies for the oilseed rape crop. Research has highlighted the potential for more precise targeting of insecticides to the crop in time and space, for minimising the need for insecticides by encouraging the conservation of naturally-occurring biocontrol agents and for utilising semiochemicals to manipulate the distributions on the crop, as for example, in the push-pull strategy.

1.3.3.1 Targeting Insecticides

Decision-making in integrated pest management involves determining the need to treat pests with insecticides, if their numbers exceed that which can be tolerated on economic grounds. A recent survey of 1,005 oilseed rape growers in Europe, as part of the MASTER project, revealed that they use a range of decision-making criteria to determine the need for pest control on their crops: seeing the pest on the crop (70% of growers), economic thresholds (47%), consultant recommendation (24%), plant growth stage (16%), advice from neighbours (16%), computer-based decision support systems (3%) and spraying by date (2%) (Menzler-Hokkanen et al. 2006).

More efficient targeting of insecticides in time and space can be achieved through the greater use of crop monitoring, economic thresholds, and computer-based decision support systems. These help determine more precisely the need for insecticide treatment and to define spray windows that target the pests while conserving their parasitoids. In this way, they can help integrate chemical and biological approaches to the management of oilseed rape pests in an economically-viable and environmentally-more-acceptable manner.

Economic Thresholds

The decision to apply an insecticide against an insect pest must be made on economic grounds if it is to meet the primary goals of integrated pest management for economic and environmental sustainability. Pest populations on a crop need to be reduced only when their densities cause a level of injury and consequent economic losses equal to or greater than the cost of their control (Stern et al. 1959, Stern 1973, Pedigo et al. 1986, Higley and Pedigo 1993).

The economic injury level (EIL) concept (Stern et al. 1959) offers a practical approach to decide on the need for crop protection against pests. It can be defined as the pest density on the crop (e.g., number of pest insects per plant) at which the cost of damage caused (price of yield loss) equals the cost of the control measures to be

Table 1.1 Control thresholds recommended for cabbage stem flea beetle (*Psylliodes chrysocephala*) in winter rape in various European countries. The lower threshold is used for backward and thin crops

Country	Source	Adults/crop unit	Larvae/plant
Austria	BFL (2000)	2–3/m^2	–
France	Pilorgé et al. (1997)	3 in 10 plants with feeding punctures	2–3
Germany	Garbe et al. (1996)	10% leaf area eaten	3–5
Poland	Anon. (2008)	1/m plant row	–
Sweden	Twengström (2000)	1–2/m plant row	–
Switzerland	Anon. (1997)	–	>2
UK	Oakley (2003)	–	3–5 (=30–60% leaf petioles with feeding scars)

applied: EIL = C (cost of control)/P (market value of the crop) D (yield loss per pest) (Pedigo 1999, Thacker 2002). The EIL changes as parameters in the equation change: they are subject to market prices, both for the crop and for the insecticides applied.

The economic threshold (sometimes also called the action or control threshold) is defined as the pest population level at which control measures should be implemented to prevent populations reaching the EIL. Economic thresholds are used to determine the need to control many of the major pests of oilseed rape in Europe although few have been developed experimentally or validated (Garbe et al. 1996, Tables 1.1, 1.2, and 1.3). They are based on pest biology, the damage that the pests cause to the crop, their effects on yield and their response to insecticides. They may be expressed as the number of pest insects per plant/plants, per length of row, per unit area of crop, per water trap or as the amount of damage (such as feeding punctures or scars) seen on plants; they vary considerably from country to country. Climate and crop husbandry practices, such as the cultivar grown, the plant density established and the fertilizer applied can all influence the thresholds. Reliable methods of crop monitoring are essential if thresholds are to be used to determine the need for crop protection.

Economic thresholds need revision as the economics of crop production, crop husbandry practices, and pest susceptibility to insecticides change. A recent EPPO meeting to discuss the development of widespread resistance of the pollen beetle to pyrethroids throughout Europe recommended that the economic thresholds for the pollen beetle be reviewed and that the influence of soil, cultivar, growth stage and climatic conditions on the compensatory ability of the crop be established (Zlof 2008). There is probably a need to revisit the thresholds for the other major pests too as many of them were established several decades ago.

Further development of the EIL concept is needed to take account of biological relationships between pests and tritrophic interactions (Poston et al. 1983). The presence of natural enemies providing biocontrol may increase the pest populations

Table 1.2 Control thresholds (number of beetles/plant) recommended for pollen beetle (*Meligethes aeneus*) on winter and spring rape crops in various European countries. Growth stages (GS) according to Meier (2001)

		Winter rape				Spring rape
Country	Source	GS 50–51	GS 52–53	GS 55–59	GS 60–61	GS 51–53
Austria	BFL (2000)	–	4–5	–	–	–
Czech Republic	Richardson (2008)	1–3	–	–	3–5	1–3
Denmark	Nielsen (2000)	3	6	–	–	1–3
Estonia	Kaarli (2004)	1–2	–	–	4	1–2
Finland	Hokkanen et al. (1988)	–	–	–	–	1–3
France	Pilorgé et al. (1997)	3–4 (1[a])	–	7–8 (2–3)	–	–
Germany	Garbe et al. (1996)	3–4 (1–2[a])	7–8 (3–4[a])	>8 (>4[a])	–	1–3
Hungary	Richardson (2008)	4–6	–	–	–	–
Ireland	Richardson (2008)	15 (5[a])	–	–	–	3
Latvia	Richardson (2008)	1–3	–	3–5	–	1–3
Luxembourg	Richardson (2008)	4–6	–	–	–	–
Netherlands	Richardson (2008)	4–6	–	–	–	5
Norway	Richardson (2008)	2	–	–	–	2
Poland	Anon. 2008)	1	3–5	3–5	–	1–5
Slovenia	Richardson (2008)	3	–	–	–	–
Sweden	Twengström (2000)	2–3	3–4	5–6	–	0.5–2
Switzerland	Anon. (1997)	5	–	3–5	–	–
UK	Oakley (2003)	–	15 (5[a])	–	–	3

GS 50 = flower buds present, still enclosed by leaves, GS 51 = flower buds visible from above 'green bud', GS 52 = flower buds free, level with youngest leaves, GS 53 = flower buds raised above the youngest leaves, GS 59 = first petals visible, flower buds still closed 'yellow bud', GS 61 = 10% of flowers on main raceme open, main raceme elongating
[a]Recommended on backward crops.

that a crop can tolerate before it sustains economic damage, but their presence is rarely considered when recommending economic thresholds or assessing the need for insecticide application.

Cabbage Stem Flea Beetle

In the autumn, early treatment of cabbage stem flea beetle is based on the number of adults per unit of crop (per plant, row length or crop area) whereas a later treatment is based on the number of larvae per plant; the larval threshold of five per plant may be reduced to three per plant for thin of backward crops (Table 1.1, Lane and Walters

Table 1.3 Control thresholds recommended for ceutorhynchid weevils on winter rape in various European countries

Country	Source	Cabbage seed weevil *C. assimilis*	Cabbage stem weevil *C. pallidactylus*	Rape stem weevil *C. napi*
		Weevils/plant	Weevils/trap in 3 days or weevils/*n* plants	Weevils/trap in 3 days or weevils/*n* plants
Austria	BFL (2000)	0.5–1	. . . . 10[a]. . . .	
Denmark	Nielsen (2000)	6	–	NP
Estonia	Kaarli (2004)	1–2 on 10% plants	1/6 plants	NP
Finland	–	–	–	NP
France	Pilorgé et al. (1997)	0.5	–	–
Germany	Steinbach et al. (2007)	0.5–1	30	10
Poland	Anon. (2000) and Z. Klukowski (pers comm)	4/25 plants	20 (or 6/25 plants)	10 (or 2–4/25 plants)
Sweden	Twengström (2000)	0.5–1	–	NP
Switzerland	Büchi (1988)	0.5–1	–	1 puncture/plant or 40–60% plants with punctures
UK	Oakley (2003)	2	–	NP

NP, pest not present.
[a]Stem weevils not distinguished.

1993). In the spring, the larval threshold may be increased as the plants are more advanced and better able to tolerate larval damage, for example, from five per plant to five to ten per plant (Lane and Walters 1993). No relationship has been found between adult numbers caught in yellow water traps and subsequent larval densities (Hossfeld 1993). However, a survey, involving 49 locations in northern Germany, showed that a threshold of 50 adults, caught over a 3-week period during migration in autumn per standard yellow water trap placed on the ground, can be used for risk assessment. Beetle numbers below this threshold never resulted in larval populations of more than three to five larvae per plant.

Pollen Beetle

The control thresholds used for control of pollen beetle in winter and spring rape crops in several countries in Europe are given in Table 1.2. Thresholds for winter rape vary from one to 15 per plant, those for the more susceptible spring rape crop are lower, from 0.5 to 5 per plant with advisors in many countries recommending a range of values increasing in number as bud growth advances and the damage caused per beetle lessens. Where spring-sown composite hybrids are grown, pollen beetles may remove so much pollen (Cook et al. 1999, 2004) that seed yield is reduced and consequently the threshold for composite hybrid crops is lower than that for conventional and modern hybrid crops. The condition of the crop and the cultivar grown can also affect the threshold; for example, in the UK, where the pollen beetle is considered to do less damage to the crop than in continental Europe, the threshold is 15 pollen beetles per winter rape plant on conventional cultivars, five per plant on delayed crops (e.g., those damaged by pigeons), and only one per five plants on composite hybrid cultivars, in which pollen beetle feeding on pollen can reduce pollination and pod set. The greater edge distribution of the pollen beetle during crop colonization (Free and Williams 1979a, Frearson et al. 2005) is recognized in Austria, where the threshold is four to five beetles per main raceme at the crop edge but one to three in the crop centre (Richardson 2008).

Cabbage Seed Weevil

The control threshold for the cabbage seed weevil varies between 0.5 and 1 weevil per plant in Austria, France, Germany, Sweden and Switzerland to six per plant in Denmark (Table 1.3). Interactions with other pests and the method of insecticide application can also affect the threshold. The lower thresholds may be used on farms where the brassica pod midge is also a problem (Garbe et al. 1996, Lane and Gladders 2000). For example, in the UK, the threshold for cabbage seed weevil alone is two per plant but when the pod midge is also present it is one per plant. When insecticide is applied together with a fungicide in a tank-mix, the effective threshold is also one per plant (Lane and Walters 1993).

Stem Weevils

Control thresholds for the cabbage stem weevil and the rape stem weevil on winter rape have been established in only a few countries (Table 1.3). In Austria, Germany and Poland they are monitored by means of standard yellow water traps (Fig. 1.8) and treatment is recommended when the threshold number has been caught over

Fig. 1.8 Standard yellow water trap in oilseed rape crop canopy as used for monitoring insect pests (Photo: Ingrid Williams)

3 days. The cabbage stem weevil is not considered to be a pest in Switzerland and the rape stem weevil is not present in the UK or northern Europe. In France, no threshold exists for the latter pest but treatment is advised within a week after the first adults are seen (Pilorgé et al. 1997).

Crop Monitoring

Crop monitoring to determine the level of infestation by a pest is an essential aspect of integrated pest management. It can inform when the control threshold for a pest has been reached and help assess the effectiveness of any control measure that has been applied. Monitoring should take account of the existence of clusters and gaps in the spatial distributions of the major pests on the crop (Williams and Ferguson Chapter 8 this volume); it should not be restricted to the edge of the crop where pest densities are often greater than in the crop centre, leading to overestimation of pest abundance.

In the oilseed rape crop, monitoring can be 'active' or 'passive'. Active monitoring involves visiting the crop and looking for and counting the pest. Adult pests may be counted on a certain number of whole plants or parts of plants (Cooper and Lane 1991), or larvae may be counted within stems or leaf petioles. Passive monitoring involves placing insect traps in the crop, examining their contents at intervals and recording the presence and abundance of pests caught. Yellow water traps, sticky traps, omni-directional flight traps or traps baited with pheromones or plant volatiles, such as isothiocyanates, have all been used for this purpose. Techniques for trapping and monitoring the pests of oilseed rape as well as their parasitoids have been reviewed by Williams et al. (2003). Trap catch data can help define the time of pest emergence and/or migration to the crop and be used to aid the timing and targeting of insecticide application (Williams 1989, Smart et al. 1993, 1996, 1997).

Computer-Based Decision Support

Computerised models and decision support systems (DSSs) provide a means of collating and synthesizing interrelated data into formats that can be used for decision-making on crop management options by farmers and their advisors (Knight 1997). Modeling approaches range from simple regression models to simulation models and expert systems. Models for tactical decision-making require a detailed understanding of the biology of the interacting crop plant/pest/parasitoid complex as modified by climatic, edaphic and agronomic factors. The best models are interactive, capable of incorporating recent advances in knowledge and of automatically importing relevant information, such as weather data. They integrate information on the whole pest complex on the crop and have regard for the effects on non-target organisms, such as parasitoids, when recommending chemical treatments, thereby integrating both chemical and biological control of the pests.

Data from crop monitoring over several years have been used to develop accumulated day-degree models to forecast the date of emergence and migration of some of the pests of oilseed rape to enable insecticide to be applied more accurately and effectively. Models have been published for the cabbage stem flea beetle (Alford 1979) and the brassica pod midge (Axelsen 1992b). Weather-based phenological models for all six of the major pests are incorporated into the computer-based decision support system proPlant (Johnen and Meier 2000, Walters et al. 2003). The system predicts potential pest infestation based on weather parameters and provides site-specific crop protection treatment decisions, a selection of suitable pesticides, computes optimum dates and rates of application and evaluates the efficacy of past applications. During the EU-project MASTER, these models were validated in several EU countries (Johnen et al. Chapter 15 this volume).

Spray Windows to Conserve Parasitoids

Where pest numbers exceed their economic thresholds, treatment with insecticide may be deemed essential, yet consideration of pest and parasitoid phenology can help maximise pest control while minimising harm to their parasitoids. Spray windows defined by weather-based phenological models and diel periodicities of pests and their parasitoids have potential to help farmers and advisors in the future to decide when to apply insecticide to control the insect pests at times when they will have minimum effects on parasitoids while at the same time reducing use of insecticide and increasing reliance on biocontrol.

Studies of the phenology of the cabbage seed weevil and its key parasitoid *T. perfectus* have shown that the pest usually starts to arrive on the crop before its parasitoid (Murchie et al. 1997b). This provides an opportunity for the temporal targeting of pyrethroids before the main migration period of the parasitoids. Avoidance of treatments post-flowering, when the parasitoids are most active, conserves natural populations of *T. perfectus* and has been shown to be effective and economically-viable as a method for controlling cabbage seed weevil populations in UK (Alford et al. 1996).

During the EU-funded project MASTER, phenological models were constructed for key parasitoids of the six major pests of winter rape and integrated into the proPlant decision support system (Johnen et al. Chapter 15 this volume). The effectiveness of different insecticide application dates, as recommended by the proPlant phenological models now need to be checked in field trials.

Knowledge of the diel periodicities of pests and parasitoids within the crop may help define times of day when parasitoids are less vulnerable to insecticides than pests. Ferguson et al. (2010) found that, during flowering, peak flight activity of the pollen beetle and its parasitoid *P. interstitialis* and of *T. obscurator*, the key parasitoid of the cabbage stem weevil, was around midday and that few insects were caught before 10.00 h. Further work on pest and parasitoid behaviour and vulnerability within the crop canopy is needed to determine whether application of insecticide in early morning (or late evening) would help conserve parasitoids without compromising control of the pests.

1.3.3.2 Push-Pull Strategy

'Push-pull' strategies in integrated pest management exploit behaviour-modifying stimuli, particularly those associated with host plant location, to manipulate the distribution of pests and their natural enemies on a crop (Cook et al. 2007a). Attractive stimuli (e.g., preferred host plants, aggregation or sex pheromones, visual, oviposition or gustatory stimulants) are used to 'pull' the pests to a trap crop while repellent and/or deterrent stimuli (e.g., less-preferred host plants, antifeedants, oviposition-deterring, epideictic or alarm pheromones) are used to 'push' pests from the protected crop while concentrating natural enemies on it for biocontrol. Development of such strategies requires a good understanding of the environmental and behavioural factors that influence the distribution of the insects on the crop and of the visual and volatile cues that the pests and their bio-control agents use in resource location so that this behaviour can be manipulated or disrupted.

Insect Distributions

Information on the spatio-temporal distributions of pests and their natural enemies on an arable crop are essential for any IPM strategy and particularly so for a push-pull strategy using a trap crop, where the aim is to alter the distribution of the insects on the crop. Any heterogeneity of insect distributions will also have implications for sampling, decision-making, and crop yields. Further, it may allow temporal and/or spatial targeting of insecticides, thereby enhancing their efficiency while conserving natural enemies. Until recently, knowledge of pest and natural enemy distributions on oilseed rape was poor. Insect sampling along line transects, and from discrete points at the edges and the centres of crops during pest immigration into and infestation of the crop, indicated that, on most crops, distributions were not uniform and that edges were often more heavily infested (e.g., Free and Williams 1979a).

Good coincidence in time and space between pests and their natural enemies is essential for effective biocontrol. During the last decade, the sampling of insects from the nodes of a rectangular grid across the crop, their mapping and then analyses using spatial statistics, has provided a much more detailed and informative picture of the pattern of crop colonization by both pests and their natural enemies. Insect distributions on the crop are non-uniform, with differing patterns of aggregation, often edge-distributed and may be affected by many crop and environmental factors. Research on the within-field spatio-temporal relationships between pests and their parasitoids and predators, together with implications for biocontrol-base IPM are reviewed by Williams and Ferguson (Chapter 8 this volume) and Williams et al. (Chapter 4 this volume), respectively.

Precision farming is the process of adjusting husbandry practices within a field to measured spatial variability but at present, it is not feasible for the grower to measure patterns of crop infestation by pests or their natural enemies because of the lack of appropriate sensors (Sylvester-Bradley et al. 1999). Consequently, it is difficult to know how to target crop protection measures to areas of high pest infestation, except by the use of trap crops or headland applications. However, in the future, advances in our knowledge of the environmental factors and behavioural responses determining the spatio-temporal distributions of the pests may lead to the development of IPM incorporating spatially targeted treatments to areas of the main crop.

Semiochemicals

Many pests of oilseed rape and some of their natural enemies have been shown to respond to semiochemicals, both plant- and insect-derived. These have potential for use in the push-pull strategy to alter the distributions of the insects. Plant-derived attractants with potential include more attractive species or cultivars of host plants (see below under trap crops). The pests (Bartlet 1996, Walczak et al. 1998, Cook et al. 2002) and some of their parasitoids (Murchie et al. 1997a, Dmoch 1998) are attracted to volatile chemicals released by the plant, especially glucosinolate metabolites, such as isothiocyanates and nitriles. Crucifer species and cultivars differ in their emission of these attractants. Pests utilize these attractants to locate the oilseed rape crop and parasitoids use them to locate the habitat of their hosts (reviewed by Williams and Cook Chapter 7 this volume). Insect-derived attractants with potential for use in manipulating insects are the sex and/or aggregation pheromones. Female brassica pod midge produce a sex pheromone from glands on the ovipositor to attract males for mating (Williams and Martin 1986, Isidoro et al. 1992). Evans and Bergeron (1994) found evidence for an aggregation pheromone in the cabbage seed weevil.

Repellents are compounds, either natural or synthetic, which cause avoidance behaviour in insects. In the push-pull strategy, less attractive host plant species or cultivars, such as those emitting lower quantities of attractant volatiles (see below under trap crops) can also be used to deter infestation. Insect-derived repellents with potential include the oviposition-deterring pheromone of the cabbage seed weevil. The female weevil uses this pheromone to regulate the distribution of her eggs amongst the pods of oilseed rape. It is produced by the abdominal seventh urotergite (Mudd et al. 1997, Ferguson et al. 1999a, b) and brushed onto the pod after oviposition (Ferguson and Williams 1991). If it could be identified, synthesised and applied to the crop there is clear potential for deterring egg-laying by this pest.

To date, none of the pheromones produced by the major pests of oilseed rape have been chemically identified or synthesised so the push-pull strategies currently under development utilise plant-derived semiochemical stimuli to manipulate pest and natural enemy distributions on the crop.

Trap Crops

Traps crops of the more attractive turnip rape (*Brassica rapa*) have been used for many years to lure pests away from the main oilseed rape (*B. napus*) crop, thereby lowering damage to it (Hokkanen 1991).

To control pollen beetle attacking spring rape in Finland, Hokkanen et al. (1986), developed a system of surrounding the crop with an early-flowering trap crop (ca. 10% by area) of turnip rape or winter rape. These preferentially attracted the beetles and were treated with insecticide whenever pest thresholds were exceeded. This system resulted in less use of insecticide and the conservation of parasitoids on the main crop.

Büchi (1990), in Switzerland, sowed crops with a mixture of 98% winter oilseed rape and 2% turnip rape (*Brassica rapa*). He found that the turnip rape was more heavily infested than the oilseed rape by the cabbage stem flea beetle, the rape stem weevil, the cabbage stem weevil and the pollen beetle but not by the cabbage seed weevil or the brassica pod midge. Büchi (1995) established perimeter strips with a mixture of turnip rape (*Brassica rapa*; 2–15%) and winter rape around crops of winter rape. These strips were not successful in controlling the rape stem weevil, but concentrated 28–80% of total pollen beetle during early immigration. To avoid migration of the beetles to the main crop, farmers were recommended to treat the perimeter with insecticide when a threshold of one beetle per plant was reached. Treatment of strips only also contributed towards enhancement of parasitoids.

More recently, Cook et al. (2002, 2007b) in the UK have further investigated the potential and mechanisms of action of turnip rape as a trap crop to protect oilseed rape from infestation by the pollen beetle.

The Integrated Crop Management system tested within the EU project MASTER (see Nilsson Chapter 16 this volume) also used turnip rape in the seed admixture as an internal trap crop to reduce damage to the main cultivar (Büchi 1990, 1995, Nilsson 2004).

Host Plant Resistance

Recent advances in plant breeding and genetic engineering hold promise for improving the resistance of oilseed rape to pest attack and thereby of also reducing the need for insecticide application.

One approach involves altering the glucosinolate content of the rape plant (Mithen 1994). Glucosinolates and their catabolites are important cues to host selection by crucifer-specialist insect pests, aiding both orientation to and recognition of the host plant (for a review, see Williams and Cook Chapter 7 this volume). Two strategies to minimize pest attack have been proposed (Bartlet et al. 1999). The first involves production of rape lines with low constitutive, but high induced glucosinolate levels. These would be more difficult for crucifer-specialist insect pests to locate when the plant is not damaged, but would be protected from generalist feeders and pathogens once damaged. The second involves production of rape lines with a high proportion of glucosinolate types that do not catabolise to isothiocyanates, the chemicals that attract the pests. The overall glucosinolate concentration of the plant would be maintained as protection from other herbivores and disease, but the plants would be less attractive to the crucifer-specialist pests. The implications for parasitoid foraging would also need to be considered.

Oilseed rape is also being genetically modified for insect resistance by introducing Bt toxin and proteinase inhibitors. Bt toxin is a natural toxin produced by the soil-dwelling bacterium *Bacillus thuringiensis*. Different strains of Bt produce different toxins, usually specific to a particular order of insects. None of the major pests of oilseed rape in Europe have yet been found to be susceptible to Bt toxins, but Hokkanen and Wearing (1996) postulated that suitable strains might be found to control them. Proteinase inhibitors inhibit the activity of proteinase enzymes that split proteins into peptides. Genes coding for proteinase inhibitors are being introduced into oilseed rape to target the digestive enzymes of insect pests with the aim of causing starvation and death on ingestion (e.g., Girard et al. 1998)

1.3.4 Integrated Crop Management

The concept of integrated crop management (ICM) involves planning a system for the management of the whole farm, including cropped and non-cropped areas, to enhance biodiversity and maximise natural regulatory mechanisms by means of on-farm husbandry practices and habitat management of landscape structures. There are many definitions of ICM. The International Organisation for Biological Control defines ICM as '... using natural resources and regulating mechanisms to replace polluting inputs and to secure sustainable farming' (El Titi et al. 1993). The Integrated Crop Production Alliance of the UK defines ICM as '*A whole farm policy aiming to provide the basis for efficient and profitable production which is economically viable and environmentally responsible. It integrates beneficial natural processes into modern farming practices using advanced technology and aims to minimise environmental risks while conserving, enhancing and recreating that which is of environmental importance*'.

Rusch et al. (Chapter 17 this volume) stress that the development of such strategies requires a thorough understanding of agroecosystem functioning. They review the effects of different elements, from the field scale to the landscape scale that are known to enhance biological control in agroecosystems and limit pest damage with particular reference to oilseed rape. They present a methodological framework to design and assess sustainable cropping systems, with a particular emphasis on complementariness between models, systemic trials and more analytical approaches.

1.3.4.1 Crop Husbandry

ICM encompasses IPM, but extends the principles to the control of weeds and diseases on the cropped areas of the farm. Conventional crop husbandry practices (Arthur et al. 1999, Christen et al. 1999) with potential for modification within an ICM system for winter oilseed rape were identified during the EU-funded Concerted Action BORIS by Alford et al. (2000). These include crop rotation, tillage, sow dates and rates, row spacing, cultivar choice, nutrient use, timing, rate and targeting of pesticide inputs, choice of insecticide and use of economic thresholds. The effects of many of these remain to be tested (Williams et al. 2002), but some are of particular importance to the management of pests and are reviewed by Rusch et al. (Chapter 17 this volume).

Winter rape is usually grown within a cereal rotation, following winter barley or set aside and preceding winter wheat. The INTEX project, started in 1990 at three different locations in Germany differing in soil type, compared the efficiency and productivity of oilseed rape crop rotations with different levels of extensification and a long-term fallow area (Gerowitt and Wildenhayn 1997). Reduced management (without insecticides and with reduced nitrogen fertilizer) resulted in greater pest abundance but also higher activity density of predators (Büchs et al. 1997, Büchs 2003b).

Each pest must migrate to colonise the crop in the autumn or spring. Crop rotation may reduce or delay pest infestation by making it more difficult for the pests to locate the crop. However, the coleopteran pests are strong fliers, and even weak fliers, such as the brassica pod midge, are probably carried long distances by wind. Little is known about the distances that natural enemies move to the crop, but the use of microsatellite markers to investigate population genetics of parasitoid populations may soon provide new information on this. Siting the crop so that it is bordered by wind breaks or other barriers or modifiers to flight may alter the migration or distribution of pests on the crop. Ploughing after harvest may kill some overwintering pests, e.g., the brassica pod midge, or developing pupae, e.g., of the cabbage seed weevil. Sowdate and seedbed conditions are also important to pest damage. Early drilling and establishment promotes strong autumn growth and ability of plants to withstand pest injury. Early flowering minimises damage by the pollen beetle.

Minimal (non-inversion) tillage throughout the rotation favours not only epigeal predators, such as carabid beetles, but also those parasitoids that overwinter in the soil (Nilsson 1985, Hokkanen 1989, Nilsson Chapter 11 this volume). The use of economic thresholds and decision support systems based on pest forecasting and population modelling for pest management rather than prophylactic treatment has already been considered (see above). Selection of pesticides with maximum selectivity and minimum environmental impact and the use of reduced application rates may encourage natural enemies. Further refinement and development of novel strategies, like the push-pull strategies incorporating semiochemicals (see above) also have scope for further minimising pesticide use.

One objective of the EU project MASTER was to design and test an ICM System for growing winter rape that aimed to be energy-efficient and high-yielding, giving a good economic net return with high natural control (see Nilsson Chapter 16 this volume). Two systems were compared in joint experiments in five countries. The Standard Farming System aimed to depict a practical, modern way of growing rape. The ICM System was designed to be more sustainable and resource-efficient and to enhance biological control of pests, by reduced tillage with no ploughing, use of a seed mix to provide an internal trap crop to reduce pest damage, plant density to increase parasitisation, and insecticide application using control thresholds or unsprayed plots.

1.3.4.2 Landscape Management

Landscape context influences the structure of the pest and parasitoid community within a crop. There is considerable interest in the integration of agroecosystem diversification into ICM strategies for arable crops including oilseed rape. Natural enemy conservation through habitat and environmental manipulation aims to encourage vegetational diversity of the agroecosystem incorporating hedgerows, cover crops, flowering conservation headlands and field margins to provide refuge, food, overwintering sites and alternative prey or hosts for natural enemies. The value of trap crops has been discussed above. Connectivity between these habitats is also recognised as beneficial.

Several studies have investigated the value of field margins, particularly to the pollen beetle and its natural enemies. Hausamman (1996) and Lethmayer et al. (1997) found that sown weed strips adjacent to or in the middle of rape crops did not increase damage by the beetle. Although new generation pollen beetles used flowering white mustard (*Sinapis alba*) strips temporarily for feeding, they did not use them for overwintering. Büchi (2002) investigated the mortality of pollen beetle larvae within rape crops 3 and 30 m from two types of compensation areas, wild flower strips and extensive meadows. When adjacent to wild flower strips, parasitization of beetle larvae by *Phradis* sp. was increased at 3 m, but not at 30 m into the crop. However, estimated mortality of beetle larvae due to parasitization did not exceed 10%, whereas that from predation was 55%. No consistent relationships, however, could be found between predation and distance from the compensation areas. Thies and Tscharntke (1999) found that parasitization rates of the pollen beetle in a rape crop were greater (58%) where the crop was adjacent to 6-year-old field margins than when adjacent to 1-year-old margins (24%). They also analysed the trophic interactions of the beetle and its parasitoids on rape crops in 15 agricultural landscapes differing in structural complexity and interannual changes in rape crop area (see Thies and Tscharntke Chapter 9 this volume). They found that beetle damage decreased and parasitism rates increased and as the percentage of semi-natural habitats in the landscape increased. In addition, parasitism increased following contraction of the rape crop area in the landscape and decreased with its expansion. They concluded that complex landscapes with good connectivity of uncultivated perennial habitats enhance populations of parasitoids and can contribute to the reduction of pest populations below the economic threshold.

Habitat manipulation requires knowledge of crop-pest-natural enemy interactions and of their spatio-temporal dynamics on the landscape scale. There is little understanding currently of the key factors that cause natural enemies to aggregate in uncultivated areas and to migrate into cropped areas. Push-pull strategies that encourage such aggregation and migration have potential for enhancing the effectiveness of pest control.

1.4 Conclusions

The insect community on the oilseed rape crop is complex comprising a diversity of pest and beneficial species. Sustainable, environmentally less harmful approaches to pest control require a holistic and integrated approach to farm and crop management. Many aspects of crop husbandry have potential for modification within IPM systems: crop rotation, tillage, sow dates and rates, row spacing, cultivar choice and nutrient use. Although pests on commercial crops are still largely controlled by application of chemical insecticides, there is good potential for reducing the amount applied through crop monitoring, adherence to economic thresholds and use of weather-based decision support systems to determine the need for and the most effective time for application. A greater understanding of insect distributions on the crop should allow these to be targeted more appropriately. Underpinning strategic research to develop push-pull and biocontrol strategies for the crop and, in the future, genetic engineering offers potential for producing cultivars more resistant to insect attack.

Acknowledgements Writing of this chapter was supported by the Estonian Targeting Finance Project SF0170057s09.

References

Alford DV (1979) Observations on the cabbage stem flea beetle, *Psylliodes chrysocephala*, on winter oilseed rape in Cambridgeshire. *Ann Appl Biol* 93: 117–123.

Alford DV (1999) *A textbook of agricultural entomology*. Blackwell, Oxford, UK.

Alford DV (2003a) The oilseed rape crop. In: Alford DV (ed.) *Biocontrol of oilseed rape pests*. Blackwell, Oxford, UK.

Alford DV (ed.) (2003b) *Biocontrol of oilseed rape pests*. Blackwell, Oxford, UK.

Alford DV, Ballanger Y, Büchi R, Büchs W, Ekbom B, Hansen LM, Hokkanen HMT, Kromp B, Nilsson C, Ulber B, Walters KFA, Williams IH, Young JEB (2000) Final report CT-96-1314. 'Minimising pesticide use and environmental impact by the development and promotion of bio-control strategies for oilseed rape pests.'

Alford DV, Nilsson C, Ulber B (2003) Insect pests of oilseed rape crops. In: Alford DV (ed.) *Biocontrol of oilseed rape pests*. Blackwell, Oxford, UK.

Alford DV, Walters KFA, Williams IH, Murchie AK (1996) A commercially viable low-cost strategy for the management of seed weevil populations on winter oilseed rape in the UK. *Proc British Crop Prot Conf, Pests and Diseases, Brighton,* pp 609–614.

Ankersmit GW (1956) De levenswijze en de bestrijding can de koolzaad snuitkever (*Ceutorhynchus assimilis* Payk.) en de koolzadgalmug (*Dasyneura brassicae* Winn.). *Verslagen van landbouwkundige onderzoekingen der Rijklandbouw-proefstations* 132: 1–59.

Anon (1997) Krankheiten und Schädlinge im Raps. *UFA Revue*: 7–8.

Anon (2000) Zalecenia Ochrony Roslin na lata 2000/1, dotyczace zwalczania choröb, szkodników oraz chwastów róslin uprawynch. *Rosliny Rolnicze*, Czesc II, Instytut Ochrony Roslin, Poznan, Poland.

Anon (2008) Zalecenia Ochrony Roślin na lata 2008/9, dotyczące zwalczania choröb, szkodników oraz chwastów roślin uprawynch. *Rośliny Rolnicze*, Czesc II, Instytut Ochrony Roślin, Poznań, Poland.

Arthur E, Dalla Costa L, Dominguez J, Garbe V, Meakin P, Messean A, Meynard J-M, Pouzet A (1999) Presentation of some results of the Concerted Action on the management of oilseed crops in the European Union. *Oléagineux Corps Gras Lipides* 6: 1–21.

Axelsen J (1992a) The development time of the pod gall midge, *Dasyneura brassicae* Winn. (Dipt., Cecidomyiidae). *J Appl Entomol* 114: 263–267.

Axelsen JA (1992b) The population dynamics and mortalities of the pod gall midge (*Dasyneura brassicae* Winn.) (Dipt., Cecidomyiidae) in winter rape and spring rape (*Brassica napus* L.) in Denmark. *J Appl Entomol* 114: 463–471.

Axelsen JA (1995) The winter mortality and emergence time of *Dasineura brassicae* in Denmark. *IOBC/wprs Bull* 18(4): 81–87.

Axelsen J, Nielsen PS (1990) Compensation in spring sown oilseed rape after attack by pollen beetles (*Meligethes aeneus* F.). *Tidsskrift Planteavl* 94: 195–199.

Balachowsky AS (ed.) (1963) *Entomologie appliquée à l'agriculture*. I. Coléopterès. Masson, Paris, France.

Ballanger Y (1983) Nuisibilité du charancon de le tige du colza. *Phytoma* 384: 35–37.

Ballanger Y (1997) Insectes des siliques du colza d'hiver. *Phytoma* 491: 43–47.

Ballanger Y, Détourné D, Delorme R, Pinochet X (2003) Difficulties to control pollen beetle (*Meligethes aeneus* F.) in France revealed by unusual high level infestations in winter rape fields. *Proc GCIRC 11th Int Rapeseed Cong, Copenhagen, 6–10 July 2003*: 1048–1050.

Bartlet E (1996) Chemical cues to host-plant selection by insect pests of oilseed rape. *Agric Zool Rev* 7: 89–116.

Bartlet E, Williams IH, Pickett JA (1999) The ideal glucosinolate profile for pest resistance in oilseed rape. *IOBC/wprs Bull* 22: 13–17.

Basedow T (1973) Der Einfluss epigäischer Raubarthropoden auf die Abundanz phytophager Insekten in der Agrarlandschaft. *Pedobiologia* 13: 410–422.

BFL (Bundesamt und Forschungszentrum für Landwirtschaft) (2000) *Richlinien für die pflanzenschutzarbeit.* BMLF, Vienna.

Blazejewska A (1960) Studia nad biologia slodyszka rzepakowego. (Studies on biology of *Meligethes aeneus* Fabr.). *Studia Societatis Torunensis Torun – Polonia. Sectio E (Zoologia)* 4: 1–70.

Bonnemaison L (1957) Le charançon des siliques (*Ceuthorrhynchus assimilis* Payk.) Biologie et méthodes de lutte. *Ann Epiphyties* 8: 387–543.

Bonnemaison L (1965) Insect pests of crucifers and their control. *Annu Rev Entomol* 10: 233–256.

Bonnemaison L, Jourdheuil P (1954) L'altise d'hiver du colza (*Psylliodes chrysocephala* L.). *Ann Epiphyties* 5: 1345–1524.

Bromand B (1983) Possibility of continuous rearing of *Meligethes aeneus* Fabr. (Col.). *J Appl Entomol* 96: 419–422.

Bromand B (1990) Diversities in oilseed rape growing in the Western Palearctic region. *IOBC/wprs Bull* 13(4): 7–31.

Broschewitz B, Daebeler F (1987) Beitrag zur Biologie und Schadwirkung des Gefleckten Kohltriebrüsslers (*C. quadridens* Panz.) an winterraps. *Nachrichtenblatt Pflanzenschutz der DDR* 41: 34–37.

Broschewitz B, Steinbach P, Goltermann S (1993) Einfluss stengelbewohnender tierischer Schaderreger auf den Befall von Winterraps mit *Phoma lingam* und *Botrytis cinerea. Gesunde Pflanzen* 45: 106–110.

Büchi R (1988) Neue Bekämpfungsschwelle für den Rapsstengelrüssler *Ceutorhynchus napi* Gyll. *Mitt Scweiz Landwirtschaft* 35: 110–117.

Büchi R (1990) Investigations on the use of turnip rape as trap plant to control oilseed rape pests. *IOBC/wprs Bull* 13(4): 32–39.

Büchi R (1995) Combination of trap plants (*Brassica rapa* var. *silvestris*) and insecticide use to control rape pests. *IOBC/wprs Bull* 18(4): 102–121.

Büchi R (1996) Eiablage des Rapsstengelrüsslers *Ceutorhynchus napi* Gyll. In: Abhängigkeit von der Stengellänge bei verschiedenen Rapssorten. *Anzeiger Schädlingskunde* 69: 136–139.

Büchi R (2002) Mortality of pollen beetle (*Meligethes* spp.) larvae due to predators and parasitoids in rape fields and the effect of conservation strips. *Agric Ecosyst Environ* 90: 255–263.

Büchs W (2003a) Predators as biocontrol agents of oilseed rape pests. In: Alford DV (ed.) *Biocontrol of oilseed rape pests.* Blackwell, Oxford, UK.

Büchs W (2003b) Impact of on-farm landscape structures and farming systems on predators. In: Alford DV (ed.) *Biocontrol of oilseed rape pests.* Blackwell, Oxford, UK.

Büchs W, Alford DV (2003) Predators of oilseed rape pests. In: Alford DV (ed.) *Biocontrol of oilseed rape pests.* Blackwell, Oxford, UK.

Büchs W, Felsman DS, Klukowski Z, Luik A, Nilsson C, Ahmed B, Ulber B, Williams IH (2006a) Key predator species in oilseed rape crops – a review of literature and MASTER results. *CD Proc Int Symp Integrated Pest Management in Oilseed Rape, 3–5 April 2006, Goettingen, Germany.*

Büchs W, Harenberg A, Zimmermann J (1997) The invertebrate ecology of farmland as a mirror of the intensity of the impact of man? An approach to interpreting results of field experiments carried out in different crop management intensities of a sugar beet and an oil seed rape rotation including set-aside. *Biol Agric Hort* 15: 83–107.

Büchs W, Nuss H (2000) First steps to assess the importance of epigaeic active polyphagous predators on oilseed rape insect pests with soil pupating larvae. *IOBC/wprs Bull* 23(6): 151–163.

Buhl C (1960) Beobachtungen über vermehrtes schadauftreten der Kohlschotenmücke (*Dasyneura brassicae* Winn.) an Raps und Rübsen in Schleswig-Holstein. *Nachr Deutschen Pflanzenschutz Stuttgart* 12: 1–6.

Butt TM, Carreck NL, Ibrahim L, Williams IH (1998) Honey bee mediated infection of pollen beetle (*Meligethes* spp.) by the insect-pathogenic fungus, *Metarhizium anisopliae*. *Biocontrol Sci Techn* 8: 533–538.

Butt TM, Ibrahim I, Ball B, Clark S (1994) Pathogenicity of the entomogenous fungi *Metarhizium anisopliae* and *Beauveria bassiana* against crucifer pests and the honey bee. *Biocontrol Sci Techn* 4: 207–214.

Carlson EC, Lange WH, Sciaroni RM (1951) Distribution and control of the cabbage seed pod weevil in California. *J Econ Entomol* 44: 958–966.

Carreck NL, Butt TM, Clark SJ, Ibrahim L, Isger EA, Pell JK, Williams IH (2006) Honey bees can disseminate a microbial control agent to more than one inflorescence pest of oilseed rape. *Biocontrol Sci Techn* 17(1–2): 179–191.

Christen L, Evans E, Nilsson C, Haldrup C (1999) Oilseed rape cropping systems in NW Europe. *10th Int Rapeseed Cong, 26–29 September 1999, Canberra, Australia.*

Clarke JH, Clark WS, Hancock M (1997) Strategies for the prevention of pesticide resistance in the UK – lessons for and from the use of herbicides, fungicides and insecticides. *Pestic Sci* 51: 391–397.

Cook SM, Denholm I (2008) Ecological approaches to the control of pollen beetles in oilseed rape. *EPPO Bull* 38: 110–113.

Cook SM, Khan ZR, Pickett JA (2007a) The use of push-pull strategies in integrated pest management. *Annu Rev Entomol* 52: 375–400.

Cook S, Murray DA, Williams IH (1999) Pollen beetle, *Meligethes aeneus* Fabricius, incidence in the composite hybrid winter oilseed rape, Synergy. *10th Int.Rapeseed Cong 26–29 September 1999, Canberra, Australia.*

Cook SM, Murray DA, Williams IH (2004) Do pollen beetles need pollen? The effect of pollen on oviposition, survival and development of a flower-feeding herbivore. *Ecol Entomol* 29: 164–173.

Cook SM, Rasmussen HB, Birkett MA, Murray DA, Pye BJ, Watts NP, Williams IH (2007b) Behavioural and chemical ecology underlying the success of turnip rape (*Brassica rapa*) trap crops in protecting oilseed rape (*Brassica napus*) from the pollen beetle (*Meligethes aeneus*). *Arthropod-Plant Inte* 1: 57–67.

Cook SM, Smart LE, Potting RJP, Bartlet E, Martin JL, Murray DA, Watts NP, Williams IH (2002) Turnip rape (*Brassica rapa*) as a trap crop to protect oilseed rape (*Brassica napus*) from infestation by insect pest: Potential and mechanism of action. *Proc British Crop Protection Conf, Pests and Diseases, Brighton, November 2002,* pp 569–74.

Cooper DA, Lane A (1991) Monitoring of oilseed rape pests in England and Wales. *IOBC/wprs Bull* 14(6): 5–13.

Cox ML (1998) The genus *Psylliodes* Latreille (Chrysomelidae: Alticinae) in the UK. *Coleopterist* 2: 33–65.

Czajkowska M (1987a) Badania nad pryszczarkiem kapustnikeiem (*Dasyneura brassicae* Winn.) I. Biologia i ekologia. *Roczniki nauk Rolniczych, Seria E* 8: 115–131.

Czajkowska M, Dmoch J (1975) Badania nad pryszczarkiem kapusnikiem (*Dasyneura brassicae* Winn.). I. Obserwacje nad biologią i ekologią szkodnika na rzepaku ozimym. *Roczniki nauk Rolniczych, Seria E* 5: 87–98.

Debouzie D, Ballanger Y (1993) Dynamics of a *Ceutorhynchus napi* population in winter rape fields. *Acta Oecol* 14: 603–618.

Denholm I, Rowland M (1992) Tactics for managing pesticide resistance in arthropods: Theory and practice. *Annu Rev Entomol* 37: 91–112.

Dmoch J (1965) The dynamics of a population of the cabbage seedpod weevil (*Ceuthorrhynchus assimilis* Payk.) and the development of winter rape. *Ekologia Polska Seria A* 8: 249–287.

Dmoch J (1998) Kairomones and searching behaviour of *Trichomalus perfectus* Walker. *IOBC/wprs Bull* 21(5): 171–176.

Dosse G (1952) Zur biologie und Morphologie des scwarzen Triebrüsslers *Ceuthorrhynchus picitarsis* Gyll., mit differentialdiagnostischen Angaben zur Unterscheidung der Larven von *Ceuthorrhynchus napi* Gyll. *C. quadridens* Panz und C. Picitarsis Gyll. *Z Angew Entomol* 34: 303–312.

Doucette CF (1947) Host plants of the cabbage seedpod weevil. *J Econ Entomol* 40: 838–840.

Ebbe-Nyman E (1952) Rapsjordoppon *Psylliodes chrysocephala* L-bidragtill kannedam um des biologi ock hekampning. *Statens Vaxskydelanstalt Meddelande* 63: 1–103.

Ehlers RU, Hokkanen HMT (1996) Insect biocontrol with non-endemic entomopathogenic nematodes (*Steinernema* and *Heterorhabditis* spp.): Conclusions and recommendations of a combined OECD and COST workshop on scientific and regulatory policy issues. *Biocontrol Sci Techn* 6: 295–302.

Ekbom B, Borg A (1996) Pollen beetle (*Meligethes aeneus*) oviposition and feeding preference on different host plant species. *Ent Exp Appl* 78: 291–299.

El Titi A, Boller EF, Gendrier JP (eds.) (1993) Integrated production – principles and technical guidelines. *IOBC/wprs Bulletin* 16(1): 1–97.

Eurostat (2009) http://epp.eurostat.ec.europa.eu

Evans KA, Bergeron J (1994) The behavioural and electrophysiological response of cabbage seed weevils (*Ceutorhynchus assimilis*) to conspecific odour. *J Chem Ecol* 20: 979–989.

Felsmann DS, Büchs W (2006a) Predation rates and assessment of staphylinid beetle larvae from oilseed rape flower stands and their role in the regulation of pollen beetle larvae. *CD Proc Int Symp Integrated Pest Management in Oilseed Rape, 3–5 April 2006, Goettingen, Germany.*

Felsmann DS, Büchs W (2006b) Assessment of staphylinid beetle larvae from oilseed rape flower stands and their influence on pollen beetle larvae. *IOBC/wprs Bull* 29(7): 161–163.

Felsmann DS, Schlein O, Büchs W (2006) Management-related web densities, predation rates and prey composition of spiders (Linyphiidae and *Theridium impressum*) in oilseed rape fields. *CD Proc Int Symp Integrated Pest Management in Oilseed Rape, 3–5 April 2006, Goettingen, Germany.*

Ferguson AW, Campbell JM, Warner DJ, Watts NP, Schmidt JEU, Williams IH (2004) Phenology and spatial distributions of *Dasineura brassicae* and its parasitoids in a crop of winter oilseed rape: Implications for integrated pest management. *IOBC/wprs Bull* 27(10): 243–252.

Ferguson AW, Holdgate R, Mason NS, Clark SJ, Williams IH (2010) Phenologies and diel periodicities of within-crop flight by pests and parasitoids in oilseed rape. *IOBC/wprs Bull* (in press).

Ferguson AW, Kenward M, Williams IH, Clark SJ, Kelm M, Dudzic A (1995) Interactions between the cabbage seed weevil (*Ceuthorhynchus assimilis* Payk.) and the brassica pod midge (*Dasineura brassicae* Winn.) infesting oilseed rape pods. *Proc 9th Int Rapeseed Cong, Cambridge, UK, 4–7 July 1995*: 679–681.

Ferguson AW, Klukowski Z, Walczak B, Perry JN, Mugglestone MA, Clark SJ, Williams IH (2000) The spatio-temporal distribution of adult *Ceutorhynchus assimilis* in a crop of winter oilseed rape in relation to the distribution of their larvae and that of the parasitoid *Trichomalus perfectus*. *Entomol Exp Appl* 95: 161–171.

Ferguson AW, Solinas M, Ziesmann J, Isidoro N, Williams IH, Scubla P, Mudd A, Clark SJ, Wadhams LJ (1999a) Identification of the gland secreting oviposition-deterring pheromone in the cabbage seed weevil, *Ceutorhynchus assimilis* Payk. (Coleoptera: Curculionidae) and the mechanism of pheromone deposition. *J Insect Physiol* 45: 687–699.

Ferguson AW, Williams IH (1991) Deposition and longevity of oviposition-deterring pheromone in the cabbage seed weevil. *Physiol Entomol* 16: 27–33.

Ferguson AW, Ziesmann J, Blight MM, Williams IH, Wadhams LJ, Clark S, Woodcock CM, Mudd A (1999b) Perception of oviposition-deterring pheromone by cabbage seed weevil (*Ceutorhynchus assimilis*). *J Chem Ecol* 25: 1655–1670.

Frearson DJT, Ferguson AW, Campbell JM, Williams IH (2005) The spatial dynamics of pollen beetles in relation to inflorescence growth stage of oilseed rape: Implications for trap crop strategies. *Entomol Exp Appl* 116: 21–29.

Free JB, Williams IH (1978) The responses of the pollen beetle, *Meligethes aeneus* and the seed weevil *Ceuthorhynchus assimilis* to oil-seed rape *Brassica napus*, and other plants. *J Appl Ecol* 15: 761–774.

Free JB, Williams IH (1979a) The distribution of insect pests on crops of oil-seed rape (*Brassica napus* L.) and the damage they cause. *J Agr Sci* 93: 139–149.

Free JB, Williams IH (1979b) The infestation of crops of oilseed rape (*Brassica napus* L.) by insect pests. *J Agr Sci* 92: 203–218.

Fritzsche R (1957) Zur biologie und ökologie der räpschadlinge aus der gattung *Meligethes*. *Z Angew Entomol* 40: 220–280.

Fröhlich G (1956) Zur Biologie und Bekämpfung der Kohlschoten-Gallmücke (*Dasyneura brassicae* Winn.). *Nachr Deutschen Pflanzenschutz* 10: 12–128.

Garbe V, Broschewitz B, Erichsen E, Hossfeld R, Lauenstein G, Steinbach G, Ulber B, Zellner M (1996) Schadensschwellen bei Rapsschädlingen. Instrumente einer wirtschaftlichen Winterrapsproduktion. *Raps* 14: 58–63.

Garbe V, Gladders P, Lane A (2000) Report of the Concerted Action 'Research for the adaptation of oilseed crops to the new requirements of the common agricultural policy: Crop competetivity, seed quality, environment'. *IOBC/wprs Bulletin* 23(6): 173–178.

Gerowitt B, Wildenhayn M (eds.) (1997) *Őkologische und őkonomische auswirkungen von extensivivierungsmassnahmen im ackerbaau ergebnisse des gőttinger INTEX-projektes 1990–1994*. University of Gőttingen, Germany.

Girard C, Le Metayer M, Zaccomer B, Bartlet E, Williams I, Bonade-Bottino M, Pham-Delegue M, Jouanin L (1998) Growth stimulation of beetle larvae reared on a transgenic oilseed rape expressing a cysteine proteinase inhibitor. *J Insect Physiol* 44: 263–270.

Goettel M (1994) Host range and specificity in relation to safety of exotic fungi. *6th Int Colloquium on Invertebrate Pathology and Microbial Control, 27th Ann Meeting Soc Invertebrate Pathology, Montpelier, France*, pp 325–329.

Gould HJ (1975) Surveys of pest incidence on oil-seed rape in south central England. *Ann Appl Biol* 79: 19–26.

Graham CW, Alford DV (1981) The distribution and importance of cabbage stem flea beetle (*Psylliodes chrysocephala* L.) on winter oilseed rape in England. *Plant Pathol* 30: 141–145.

Graham CW, Gould HJ (1980) Cabbage stem weevil (*Ceutorhynchus quadridens*) on spring oilseed rape in southern England and its control. *Ann Appl Biol* 95: 1–10.

Hanson AJ, Carlson EC, Breakey EP, Webster RL (1948) *Biology of the cabbage seedpod weevil in NorthWest Washington*. State Coll. Wash, Agr, Exp. Sta. Bull. 498.

Harenberg A (1997) *Auswirkungen extensiv geführter Anbausysteme in verschiedenen Fruchtfolgen (Raps-, Zuckerrübenfruchfolge) und einer selbsbegrünenden Dauerbrache auf spinnen (Arachnida: Araneae)*. PhD thesis, Technical University of Braunschweig, Braunschweig, Germany.

Hausamman A (1996) Strip-management in rape crop: Is winter rape endangered by negative impacts of sown weed strips? *J Appl Entomol* 120: 505–512.

Heymons R (1921) Mitteilungen über den Rapsrüssler, *Ceutorrhynchus assimilis* Payk und Seinen Parasiten *Trichomalus fasciatus* Thoms. *Z Angew Entomol* 8: 93–111.

Higley LG, Pedigo LP (1993) Economic injury level concepts and their use in sustaining environmental quality. *Agr Ecosyst Environ* 46: 233–243.

Hoffmann A (1951) Contribution a l'étude de *Ceutorrhynchus assimilis* Payk. (le Ceuthorrhynque des siliques). *Revue Path* 30: 238–252.

Hokkanen HMT (1989) Biological and agrotechnical control of the rape blossom beetle *Meligethes aeneus* (Coleoptera, Nitidulidae). *Acta Entomol Fenn* 53: 25–29.
Hokkanen HMT (1991) Trap cropping in pest management. *Ann Rev Entomol* 36: 119–138.
Hokkanen HMT (1993) Overwintering survival and spring emergence in *Meligethes aeneus*: Effects of body weight, crowding, and soil treatment with *Beauveria bassiana*. *Ent Exp Appl* 67: 241–246.
Hokkanen H, Granlund H, Husberg G-B, Markkula M (1986) Trap crops used successfully to control *Meligethes aeneus*. *Ann Entomol Fenn* 52: 115–120.
Hokkanen H, Holopainen JK (1986) Carabid species and activity densities in biologically and conventionally managed cabbage fields. *J Appl Entomol* 102: 353–363.
Hokkanen H, Husberg GB, Söderblom M (1988) Natural enemy conservation for the integrated control of the rape blossom beetle *Meligethes aeneus* F. *Agriculturae Fenniae* 27: 281–294.
Hokkanen HMT, Menzler-Hokkanen I, Butt TM (2003) Pathogens of oilseed rape pests. In: Alford DV (ed.) *Biocontrol of oilseed rape pests*. Blackwell, Oxford, UK.
Hokkanen HMT, Wearing CH (1996) Assessing the risk of pest resistance evolution to *Bacillus thuringiensis* engineered into crop plants: A case study of oilseed rape. *Field Crops Res* 45: 171–179.
Hokkanen HMT, Zec-Vojinovic M, Husberg G-B, Menzler-Hokkanen I, Büchs W, Klukowski Z, Luik A, Nilsson C, Ulber B, Williams IH (2006) Effectiveness of entomopathogenic nematodes in the control of oilseed rape pests. *CD Proc Int Symp Integrated Pest Management in Oilseed Rape, 3–5 April 2006, Goettingen, Germany*.
Hossfeld R (1993) Die Gelbschale als Entscheidungshilfe bei der Bekämpfung des Rapserdflohs (*Psylliodes chrysocephala* L.). *Gesunde Pflanzen* 45: 291–295.
Husberg GB, Hokkanen HMT (2001) Effects of *Metarhizium anisopliae* on the pollen beetle *Meligethes aeneus* and its parasitoids *Phradis morionellus* and *Diospilus capito*. *Biocontrol* 46: 261–273.
IOBC International Organization for Biological Control (1997) Guidelines for integrated production of arable crops in Europe. *IOBC/wprs Bull* 20(5): 5–17.
Isidoro N, Williams IH, Solinas M, Martin A (1992) Mating behaviour and identification of the female sex pheromone gland in the brassica pod midge (*Dasineura brassicae* Winn.: Cecidomyidae, Diptera). *Boll Ist Entomol Univ Bologna* 47: 27–48.
Johnen A, Meier H (2000) A weather-based decision support system for managing oilseed rape pests. *Proc BCPC Conf Pests Diseases* 2: 793–800.
Jourdheuil P (1963) Tribu des Ceuthorrhynchini. In: Balachowsky AS (ed.) *Entomologie appliqué à l' agriculture. I. Coléopterès*. Masson, Paris, France.
Jurek M (1972) *Meligethes* spp. On winter rape *–Brassic napus* L. var.oleifera Metz. F. biennis Thel. *Polskie pismo Entomol* 42: 483–490.
Kaarli K (2004) *Olikultuuride kasvataja käsiraamat*. EMI, Saku, Estonia.
Karltorp M, Nilsson C (1981) Rapsbagger I mellansvenska våroljeväxtodlingar. *Växtskyddnotiser* 45: 146–154.
Kelm M, Klukowski Z (2000) The effect of stem weevil (*Ceutorhynchus pallidactylus* Marsh.) infestation on oilseed rape yield. *IOBC/wprs Bull* 23(6): 125–130.
Kelm M, Walczak B (1998) The relationship between the stem weevil (*Ceutorhynchus pallidactylus* Marsh.) injury and losses of the flower buds. *IOBC/wprs Bull* 21(5): 147–151.
Kirk WDJ (1992) *Insects on cabbages and oilseed rape*. Richmond, Slough, UK.
Kirk-Spriggs AH (1991) Nitidulidae. In: Cooter J, Cribb GEPW (eds.) *A coleopterist's handbook*. The Amateur Entomologist's Society, Feltham, London.
Kirk-Spriggs AH (1996) *Pollen beetles*. Coleoptera: Katerietidae and Nitidulidae: Meligethinae. The Royal Entomological Society of London, UK.
Knight JD (1997) The role of decision support systems in integrated crop protection. *Agr Ecosyst Environ* 64: 157–163.
Kozlowski MW, Lux S, Dmoch J (1983) Oviposition behaviour and pod marking in the cabbage seed weevil, *Ceutorhynchus assimilis*. *Entomol Exp Appl* 34: 277–282.

Krause U, Koopmann B, Ulber B (2006) Impact of rape stem weevil, *Ceutorhynchus napi*, on the early stem infection of oilseed rape by *Phoma lingam*. *IOBC/wprs Bull* 29(7): 323–328.
Kromp B (1999) Carabid beetles in sustainable agriculture: A review on pest control efficiency, cultivation impacts and enhancement. *Agric Ecoyst Environ* 74: 187–228.
Krooss S (1996) *Staphyliniden in abgestuft extensiv bewirtschafteten Agrarökosystemen: Populationsdynamik, Gemeinschaftsstruktur und Bedeutung als Prädatoren*. PhD thesis, University of Göttingen, Göttingen, Germany.
Lakocy A (1977) The influence of some biological and ecological factors on the development of resistance to insecticides and on the course of chemical control of *Meligethes aeneus* F. in the voivodships of Poznan and Wroclaw. *Prace Naukowe Instytutu Ochrony Roslin* 19: 123–181.
Lane A, Cooper D (1989) Importance and control of insect pests of oilseed rape. *Aspects Appl Biol* 23: 269–276.
Lane A, Gladders P (2000) Pests and diseases of oilseeds, brassica seed crops and field beans. In: Alford DV (ed.) *Pest and disease management handbook*. Blackwell, Oxford, UK.
Lane A, Walters KFA (1993) Recent incidence and cost effective control of pests of oilseed rape in England and Wales. *IOBC/wprs Bull* 16(9): 185–192.
Láska P, Kocourek F (1991) Monitoring of flight activity in some crucifer-feeding pests by means of yellow water traps. *Acta Entomol Bohemos* 88: 25–32.
Le Pape H, Bronner R (1987) The effects of *Ceuthorrhynchus napi* (Curculionidae, Coleoptera) on stem tissues of *Brassica napus var. oleifera*. *Proc 6th Int Symp Insect-Plant relationships*: 207–212.
Leach JE, Darby RJ, Williams IH, Fitt BDL, Rawlinson CJ (1994) Factors affecting growth and yield of winter oilseed rape (*Brassica napus*), 1985–1989. *J Agr Sci* 122: 405–413.
Leonard PK (1997) There has never been a better time or a greater need for resistance management. *Pestic Sci* 51: 387–390.
Lerin J (1987) Compensation in winter rape following simulated pollen beetle damage *IOBC/wprs Bull* 10(4): 57–63.
Lerin J (1991) Influence of host plant phenology on the reproduction of the rape weevil *Ceuthorynchus assimilis* Payk. *J Appl Entomol* 111: 303–310.
Lerin J (1993) Influence of growth rate on the splitting of the stem after an attack of *Ceutorhynchus napi* Gyll. *IOBC/wprs Bull* 16(9): 160–163.
Lethmayer C, Nentwig W, Frank T (1997) Effects of weed strips on the occurrence of noxious coleopteran species (Nitidulidae, Chrysomelidae, Curculionidae). *Z Pflanzenkrankh Pflanzenschutz* 104: 75–92.
Lipa JJ, Hokkanen HMT (1992) *Nosema meligethi* I. & R. (Microsporida) in populations of *Meligethes* spp. in Europe. *Biocontrol Sci Technol* 2: 19–125.
Lipa JJ, Hokkanen HMT (1995) Occurrence and dynamics of *Nosema meligethi* (Microsporida) in populations of *Meligethes aeneus* (Coleoptera, Nitidulidae) in Finland. *Entomol Fenn* 6: 11–18.
McCaffrey JP, Harmon BL, Ni XZ (1991) Field studies on the reproductive development of the cabbage seedpod weevil. *Proc Rapseed Cong Saskatchewan* 2: 506–511.
Meier U (ed.) (2001) Growth stages of mono- and dicotyledonous plants. http://www.bba.de/veroeff/bbch/bbcheng.
Menzler-Hokkanen I, Hokkanen HMT, Büchs W, Klukowski Z, Luik A, Nilsson C, Ulber B, Williams IH (2006) Insect problems in European oilseed rape cultivation, and how to deal with them: The OSR farmers' perspective. *IOBC/wprs Bull* 29(7): 91–94.
Mithen RF (1994) Application of biotechnology to the improvement of Brassica oilseed crops. *Aspects Appl Biol* 39: 23–34.
Mudd A, Ferguson AW, Blight MM, Williams IH, Scubla P, Solinas M, Clark SJ (1997) Extraction, isolation and composition of the oviposition-deterring secretion of the cabbage seed weevil *Ceutorhynchus assimilis* Payk. *J Chem Ecol* 23: 2227–2240.
Müller A, Heimbach U, Thieme T (2008) Pyrethroid sensitivity monitoring in Germany of oilseed rape pest insects other than pollen beetle. *EPPO Bull* 38: 85–90.

Murchie AK, Smart LE, Williams IH (1997a) Responses of *Dasineura brassicae* Winn. (Diptera: Cecidomyiidae) and its parasitoids *Platygaster subuliformis* Kieffer (Hymenoptera: Platygastridae) and *Omphale clypealis* Thomson (Hymenoptera: Eulophidae) to traps, baited with organic isothiocyanates, in the field. *J Chem Ecol* 23: 917–926.

Murchie AK, Williams IH, Alford DV (1997b) Effects of commercial insecticide treatments to winter oilseed rape on parasitism of *Ceutorhynchus assimilis* Paykull (Coleoptera: Curculionidae) by *Trichomalus perfectus* (Walker) (Hymenoptera: Pteromalidae). *Crop Protect* 16: 199–202.

Murchie AK, Williams IH, Perry J (1999) Edge distributions of *Ceutorhynchus assimilis* (Paykull) and its parasitoid *Trichomalus perfectus* (Walker) in a crop of winter oilseed rape (*Brassica napus* L.). *Biocontrol* 44: 379–390.

Nielsen GC (2000) *Vejledende bekaempelsestaerskler for skadedyr 2001*. Landskontoret Planteavl, Arhus, Denmark.

Nilsson C (1985) Impact of ploughing on emergence of pollen beetle parasitoids after hibernation. *J Appl Entomol* 100: 302–308.

Nilsson C (1987) Yield losses in summer rape caused by pollen beetles (*Meligethes aeneus*). *Swed J Agr Res* 17: 105–111.

Nilsson C (1988a) The pollen beetle (*Meligethes aeneus*) in winter and spring rape at Alnarp 1976–1978. I. Migration and sex ratio. *Växtskyddnotiser* 52: 134–138.

Nilsson C (1988b) The pollen beetle (*Meligethes aeneus*) in winter and spring rape at Alnarp 1976–1978. II. Oviposition. *Växtskyddnotiser* 52: 139–144.

Nilsson C (1988c) The pollen beetle (*Meligethes aeneus*) in winter and spring rape at Alnarp 1976–1978. III. Mortality. *Växtskyddnotiser* 52: 145–150.

Nilsson C (1988d) Pollen beetles (*Meligethes aeneus* F.) and flowering in rape. *Swed J Agr Res* 1: 113–118.

Nilsson C (1990) Yield losses in winter rape caused by cabbage stem flea beetle larvae (*Psylliodes chrysocephala*). *IOBC/wprs Bull* 13(4): 53–55.

Nilsson C (1994) Pollen beetles (*Meligethes* spp.) in oilseed rape crop (*Brassica napus* L.): Biological interactions and crop losses. *Department of Plant Protection Sciences, SLU Dissertations 1: 39 pp.*

Nilsson C (2002) Strategies for the control of cabbage stem flea beetle on winter rape in Sweden. *IOBC/wprs Bull* 25(2): 133–139.

Nilsson C (2004) Trap plants to avoid insecticide application against pollen beetles in oilseed rape. *IOBC/wprs* 27(10): 215–221.

Nilsson C, Andreasson B (1987) Parasitoids and predators attacking pollen beetles (*Meligethes aeneus* F.) in spring and winter rape in southern Sweden. *IOBC/wprs Bull* 10(4): 64–73.

Nolte HW, Fritzsche R (1952) Untersuchungen über das Vorkommen verschiedener *Meligethes* – Arten auf Raps. *Beiträge Entomol* 2: 434–448.

Nuss H, Ulber B (2004) Effect of sowing density of oilseed rape on the abundance and within-plant distribution of cabbage stem flea beetle, *Psylliodes chrysocephala*. *IOBC/wprs Bull* 27(10): 223.

Oakley J (2003) *Pest management in cereals and oilseed rape, a guide*. HGCA, London, UK.

Osborne P (1964) Morphology of the immature stages of *Meligethes aeneus* (F.) and *M. viridescens* (F.) (Coleoptera, Nitidulidae). *Bull Entomol Res* 55: 747–759.

Pedigo LP (1999) *Entomology and pest management*. Upper Saddle River, New Jersey, Prentice Hall.

Pedigo LP, Hutchins SH, Higley LG (1986) Economic injury levels in theory and practice. *Annu Rev Entomol* 31: 341–368.

Pickett JA, Butt TM, Doughty KJ, Wallsgrove RM, Williams IH (1995) Minimising pesticide input in oilseed rape by exploiting natural regulatory processes. *Proc 9th Int Rapeseed Cong Cambridge, UK, 4–7 July 1995*, pp 565–571.

Pilorgé E, Maisonneuve C, Ballanger Y (1997) *Les ravageurs du colza d'hiver*. Cetiom, France.

Poston FL, Pedigo LP, Welch SM (1983) Economic injury levels: Reality and practicality. *Bull Ent Soc Amer* 29: 49–53.

Queinnec Y (1967) Etude des factuers psycho-physiologiques pemettant la decouverte de la plant hôte par les larves neonates de l'altise du colza, *Psylliodes chrysocephala*. *Ann Epiphyties* 18: 28–74.

Richardson DM (2008) Summary of findings from a participant country pollen beetle questionnaire. *EPPO Bull* 38: 68–72.

Roush RT (1989) Designing resistance management programs: How can you choose? *Pestic Sci* 26: 423–441.

Sáringer G (1984) Summer diapause of the cabbage stem flea beetle, *Psylliodes chrysocephala*. *J Appl Entomol* 98: 50–54.

Scherney FV (1953) Zur biology der an Raps vorkommenden Meligethesarten. *Z Pflanzenbau und Pflanzenschutz* 4: 154–176.

Scherney F (1959) *Unsere laufkäfer, ihre biologie und wirtschaftliche bedeutung*. Verlag A Ziemsen, Wittenberg.

Schultz RR, Daebeler F (1984) Zum Schaden durch den Rapserdfloh. (*Psylliodes chrysocephala* L.) insbesondere seiner Imagines. *Nachr Bl Pflanzenschutz DDr* 38: 113–115.

Schulz RR (1985) Untersuchungen zur Vermehrungsrate des Rapsserdflohs (*Psylliodes chrysocephala* L.). *Arch Phytopathol Pfl* 21: 305–311.

Skrocki C (1972) Biologia chowacza podobnika (*Ceuthorrhynchus assimilis* Payk.). *Roczniki Nauk Rolniczych seria E* 9: 149–157.

Skrocki C (1979) Zależność składania jaj pryszczarka kapustnika (*Dasyneura brassicae* Winn.) of występowania chowacza podobnika (*Ceuthorrhynchus assimilis* Payk.) na rzepaku ozimym. *Roczniki Nauk Rolniczych Seria E* 9: 149–157.

Smart LE, Blight MM, Hick AJ (1993) Development of a monitoring system for the cabbage seed weevil and the pollen beetle. *IOBC/wprs Bull* 16(10): 351–354.

Smart LE, Blight MM, Hick AJ (1997) Effect of visual cues and a mixture of isothiocyanates on trap capture of cabbage seed weevil (*Ceutorhynchus assimilis* (Paykull) (Coleoptera: Curculionidae)). *J Chem Ecol* 23: 889–902.

Smart LE, Blight MM, Lane A (1996) Development of a monitoring system for spring and summer pests of oilseed rape. *Brighton Crop Prot Conf Pests Dis* 1: 167–172.

Smith DM, Hewson RT (1984) Control of the cabbage stem flea beetle and rape winter stem weevil on oilseed rape with deltamethrin. *Proc British Crop Protection Conf, Brighton*: 755–760.

Speyer W (1921) Beiträge zur biologie der kohschotenmücke (*Dasyneura brassicae* Winn.). *Mitt Biol Reichsanst Land- und Forstwirtsch* 21: 208–217.

Steinbach P, Gerowitt B, von Tiedemann A, Ulber B (2007) Integrierter Pflanzenschutz im Betrieb – Krankheits – und Schädlingskontrolle, Prognosen und Konzepte. In: Christen O Frauen W (eds.) *Winterraps DLG*. Verlags GmbH, Frankfurt/Main.

Stern VM (1973) Economic thresholds. *Annu Rev Entomol* 18: 259–280.

Stern VM, Smith RF, van den Bosch R, Hagen KS (1959) The integrated control concept. *Hilgardia* 29: 81–101.

Sylvén E (1949) Skidgallmyggan, *Dasyneura brassicae* Winn. *Statens Växtskyddanstalt Mededeling* 54: 1–120.

Sylvén E, Svensson G (1976) Effect of yield of damage caused by *Meligethes aeneus* F. (Col.) to winter rape, as indicated by cage experiments. *Ann Agr Fenn* 15: 24–33.

Sylvester-Bradley R, Lord E, Sparkes DL, Scott RK, Wiltshire JJJ, Orson J (1999) An analysis of the potential of precision farming in Northern Europe. *Soil Use Manag* 15: 1–8.

Tatchell GM (1983) Compensation in spring sown oil seed rape (*Brassica napus* L.) plants in response to injury to their flower buds and pods. *J Agr Sci* 101: 565–573.

Thacker JRM (2002) *An introduction to arthropod pest control*. Cambridge University Press, Cambridge, UK.

Thies C, Tscharntke T (1999) Landscape structure and biological control in agroecosystems. *Science* 285: 293–295.

Twengström E (ed.) (2000) Behovsanpassad bekämpning av skadegörare i jordbruksgrödor. Faktablad om växtskydd 1. *J Swedish University of Agr Sci*, 61 pp.

Ulber B (2003) Parasitoids of stem weevils. In: Alford DV (ed.) *Biocontrol of oilseed rape pests*. Blackwell, Oxford, UK.

Ulber B, Vidal S (1998) Influence of host density and host distribution on parasitism of *Ceutorhynchus assimilis* by *Trichomalus perfectus*. *IOBC/wprs Bull* 21(5): 185–195.

Vänninen I, Husberg G-B, Hokkanen HMT (1989) Occurrence of entomopathogenic fungi and entomoparasitic nematodes in cultivated soils in Finland. *Acta Entomol Fenn* 53: 65–71.

Vänninen I, Tyni-Juslin J, Hokkanen H (2000) Persistence of augmented *Metarhizium anisopliae* and *Beauveria bassiana* in Finnish agricultural soils. *BioControl* 45: 201–222.

Walczak B, Kelm M, Karp R (1997) Wpływ uszkodzeń spowodowanych przez chowacze łodygowe na jakość nasion rzepaku. *Prog Plant Protect* 37(2): 22–23.

Walczak B, Kelm M, Klukowski Z, Smart LE, Ferguson AW, Williams IH (1998) The effect of trap design and 2-phenylethyl isothiocyanate on catches of stem weevils (*Ceuthorynchus pallidactylus* Marsh. and *C. napi* Gyll.) in winter oilseed rape. *IOBC/wprs Bull* 21(5): 141–146.

Walters KFA, Young JEB, Kromp B, Cox PD (2003) Management of oilseed rape pests. In: Alford DV (ed.) *Biocontrol of oilseed rape pests*. Blackwell, Oxford, UK.

Warner DJ, Allen-Williams LJ, Ferguson AW, Williams IH (2000) Pest-predator spatial relationships in winter rape: Implications for Integrated Crop Management. *Pest Manage Sci* 56: 977–982.

Warner DJ, Allen-Williams LJ, Warrington S, Ferguson AW, Williams IH (2003) Mapping, characterisation, and comparison of the spatio-temporal distributions of cabbage stem flea beetle (*Psylliodes chrysocephala*), carabids, and Collembola in a crop of winter oilseed rape (*Brassica napus*). *Entomol Exp Appl* 109: 225–234.

Williams IH (1989) Pheromone monitoring of brassica pod midge (*Dasineura brassicae* Winn.). *Proc 7th Int Rapeseed Cong, Poznan, Poland, 11–14 May 1987*: 1074–1076.

Williams IH, Büchi R, Ulber B (2003) Sampling, trapping and rearing oilseed rape pests and their parasitoids. In: Alford DV (ed.) *Biocontrol of oilseed rape pests*. Blackwell, Oxford, UK.

Williams IH, Büchs W, Hokkanen H, Menzler-Hokkanen I, Johnen A, Klukowski Z, Luik A, Nilsson C, Ulber B (2002) MASTER: Management strategies for European rape pests – a new EU project. *Proc British Crop Protection Conf, Pests and Diseases, Brighton, November 2002*, pp 641–646.

Williams JJW, Carden PW (1961) Cabbage stem flea beetle in East Anglia. *Plant Pathol* 10: 85–95.

Williams IH, Free JB (1978) The feeding and mating behaviour of pollen beetles (*Meligethes aeneus* Fab.) and seed weevils (*Ceuthorhynchus assimilis* Payk.) on oil-seed rape (*Brassica napus* L.). *J Agr Sci* 91: 453–459.

Williams IH, Free JB (1979) Compensation of oil-seed rape (*Brassica napus* L.) plants after damage to their buds and pods. *J Agr Sci* 92: 53–59.

Williams IH, Martin AP (1986) Evidence for a female sex pheromone in the brassica pod midge *Dasineura brassicae*. *Physiol Entomol* 11: 353–356.

Williams IH, Martin AP, Kelm M (1987a) The phenology of the emergence of brassica pod midge (*Dasineura brassicae* Winn.) and its infestation of winter oil-seed rape (*Brassica napus* L.). *J Agr Sci* 108: 579–589.

Williams IH, Martin AP, Kelm M (1987b) The phenology of the emergence of brassica pod midge (*Dasineura brassicae* Winn.) and its infestation of spring oil-seed rape (*Brassica napus* L.). *J Agr Sci* 109: 309–314.

Williams IH, Murchie AK, Ferguson AW, Klukowski Z, Perry JN, Walczak B (2000) Within-field distributions of the seed weevil *Ceutorhynchus assimilis* Paykull and its parasitoid *Trichomalus perfectus* (Walker) on winter oilseed rape. *IOBC/wprs Bull* 23(6): 111–118.

Winfield AL (1961) Observations on the biology and control of the cabbage stem weevil, *Ceutorhynchus quadridens* (Panz.) on Trowse mustard (*Brassica juncea*). *B Entomol Res* 52: 589–600.

Winfield AL (1992) Management of oilseed rape pests in Europe. *Agr Zool Rev* 5: 51–95.

Wolf M, Krausse A (1921) Beiträge zur Kenntnis der Biologie von Oelfruchtschädlingen, insbesondere über der anteil der von *Ceuthorrhynchus assimilis* Payk. Verursachten, fälschlich den *Meligethes aeneus* R. ZugeschriebenenSchäden. *Arch Natur Gesch* 91: 1–45.

Zec-Vojinovic M, Hokkanen HMT, Büchs W, Klukowski Z, Luik A, Nilsson C, Ulber B, Williams IH (2006a) Natural occurrence of pathogens of oilseed rape pests in agricultural fields in Europe. *CD Proc Int Symp Integrated Pest Management in Oilseed Rape, 3–5 April 2006, Goettingen, Germany.*

Zec-Vojinovic M, Menzler-Hokkanen I, Hokkanen HMT (2006b) Application strategies for entomopathogenic nematodes in the control of oilseed rape pests. *CD Proc Int Symp Integrated Pest Management in Oilseed Rape, 3–5 April 2006, Goettingen, Germany.*

Zlof V (2008) Recommendations and conclusions of the Ad hoc EPPO Workshop on insecticide resistance of *Meligethes* spp. (pollen beetle) on oilseed rape. *EPPO Bull* 38: 65–67.

Chapter 2
Parasitoids of Oilseed Rape Pests in Europe: Key Species for Conservation Biocontrol

Bernd Ulber, Ingrid H. Williams, Zdzislaw Klukowski, Anne Luik, and Christer Nilsson

Abstract The six most important pests of oilseed rape are host to at least 80 species of parasitoid, mostly parasitic Hymenoptera, particularly braconids, chalcids and ichneumonids. Most of them attack the egg or larval stages of their hosts. Based on reviews of the literature and extensive sampling programmes during the EU project MASTER (QLK5-CT-2001-01447), 12 species have been identified as the key parasitoid species of these pests in winter oilseed rape, and, with little divergence, also in spring rape in nearly all European countries where their hosts occur. Some key species have been recorded for the first time in individual partner countries. They are sufficiently widespread and abundant across Europe to be of potential economic importance for conservation biological control of the target pests. Their incidence and abundance in European countries were associated with the occurrence of their hosts, thereby indicating close host-parasitoid-relationships.

New information on the identity, biology, phenology, distribution and impact of key parasitoid species in Europe was obtained by strategic research of the MASTER project. The level of parasitism of target pests was determined from samples of numerous field experiments and commercial crops of oilseed rape by dissection of larvae and by rearing adult parasitoids from their hosts. Percent parasitism of target pests varied between countries and years, commonly ranging between 20 and 50%, occasionally exceeding 80%.

2.1 Introduction

Parasitoids of various hymenopteran families form a substantial part of the natural enemy complex of the insect pests of oilseed rape (*Brassica napus* L.) and related species in Europe. Published literature on these parasitoids was first collated by participants of the EU-funded project BORIS, and published as detailed reviews by European authorities in Alford (2003).

B. Ulber (✉)
Department of Crop Sciences, Plant Pathology and Crop Protection Division, Georg-August University Göttingen, Göttingen, D-37077, Germany
e-mail: bulber@gwdg.de

I.H. Williams (ed.), *Biocontrol-Based Integrated Management of Oilseed Rape Pests*, DOI 10.1007/978-90-481-3983-5_2, © Springer Science+Business Media B.V. 2010

Here we review the identity, status and potential of parasitoids for conservation biocontrol of the pests. The information presented was largely obtained through collaborative research in six European countries (Estonia, Finland, Germany, Poland, Sweden and the UK) within the EU-funded project MASTER (Management Strategies for European Rape pests, QLK5-CT-2001-01447) during 2002–2006 (Williams et al. 2005, Williams 2006a). The project focussed on the six most widespread and economically-important insect pests of winter rape, namely the pollen beetle, *Meligethes aeneus* (Fabricius), the cabbage seed weevil, *Ceutorhynchus obstrictus* (Marsham) syn. *C. assimilis* (Paykull), the brassica pod midge, *Dasineura brassicae* (Winnertz), the cabbage stem flea beetle, *Psylliodes chrysocephala* (Linnaeus), the cabbage stem weevil, *Ceutorhynchus pallidactylus* (Marsham), and the rape stem weevil, *Ceutorhynchus napi* (Gyllenhal). It identified 11 key species of parasitoid from the parasitoid complex that were both abundant and widespread on winter rape crops throughout Europe and consequently had most potential for conservation biocontrol of these pests on winter rape (Table 2.1). Although the focus of the MASTER project was winter rape, some observations on these target pests and their parasitoids were also made in spring rape, and a further parasitoid species was identified as a key species primarily on spring

Table 2.1 Key larval parasitoids of the six most important pests of oilseed rape found during research of the EU-project 'MASTER' in the UK, Sweden (SE), Estonia (EE), Poland (PL) and Germany (DE)

Pest	Parasitoid	UK	SE	EE	PL	DE
Pollen beetle (*Meligethes aeneus*)	*Phradis interstitialis*	*	*	**	*	*
	Phradis morionellus	*	*	**	*	*
	Tersilochus heterocerus	*	*	**	*	*
	Diospilus capito	*	*	**	*	*
Cabbage seed weevil (*Ceutorhynchus obstrictus* syn. *C. assimilis*)	*Trichomalus perfectus*	*	*	**	*	*
	Mesopolobus morys	*	*	**	*	*
	Stenomalina gracilis	*	*	**	*	*
Brassica pod midge (*Dasineura brassicae*)	*Omphale clypealis*	*	*	–[1)]	–	*
	Platygaster subuliformis	*	**	*[1)]	*	**
Cabbage stem flea beetle (*Psylliodes chrysocephala*)	*Tersilochus microgaster*	**	**	–[1)]	**	*
Cabbage stem weevil (*Ceutorhynchus pallidactylus*)	*Tersilochus obscurator*	*	*	**	*	*
Rape stem weevil (*Ceutorhynchus napi*)	*Tersilochus fulvipes*	–[1)]	–[1)]	–[1)]	*	*

* = Parasitoid present in country.
** = Parasitoid associated with this pest for 1st time by MASTER.
– = Parasitoid not present in the country.
[1)] = Pest not present in the country.

rape. The importance of parasitoids for biocontrol of the pests of spring rape, particularly *Phyllotreta* flea beetles, is further reviewed by Ekbom (Chapter 5 this volume).

Pest and parasitoid activity in crops of oilseed rape (mostly winter rape but some spring rape too) was monitored for 4 years (2002–2006) in Estonia, Germany, Poland, Sweden and the UK, using yellow water traps placed and maintained at canopy level in the crop. Traps were mostly emptied three times each week. The datasets provided information on the phenology of pest and parasitoid migration into crops and their activity densities within them (see also Johnen et al. Chapter 15 this volume). Data on levels of parasitism found on both commercial crops and on unsprayed experimental plots in Estonia, Germany, Poland, Sweden and the UK between 1995 and 2005 were also collated and compared. Samples of pest larvae were collected either from plant samples or in water-filled trays below the crop canopy as they dropped to the ground to pupate. Standardized methods were used for determining levels of larval parasitism (Williams et al. 2003). Percentage endoparasitism of pollen beetle, stem weevil and cabbage stem flea beetle larvae was assessed either by dissection of host larvae and/or by rearing adult parasitoids from them (see also Klingenberg and Ulber 1994, Barari et al. 2004). Percentage ectoparasitism of cabbage seed weevil larvae was assessed by examining host larvae in pods. Percentage parasitism of brassica pod midge larvae was not determined.

Hymenopteran parasitoids are difficult to identify to species and taxonomic keys and literature on the different taxa are widely dispersed. To aid their identification, a practical, simple to use guide was produced for use by MASTER project partners (see Ferguson et al. Chapter 3 this volume). This collates essential information on the taxonomic characters of the key parasitoids from the literature, adding information obtained during the examination of thousands of parasitoid specimens collected and examined during the course of the MASTER project. It is liberally illustrated to highlight characteristic features of each key species enabling them to be distinguished from similar species by a non-specialist.

2.2 Parasitoids of the Pollen Beetle (*Meligethes aeneus*)

2.2.1 Identity of Species

The eggs or larvae of the pollen beetle in Europe are parasitized by at least nine species of hymenopteran endoparasitoid: four species of ichneumonid, three braconid, one encyrtid and one proctotrupid (Nilsson 2003, Table 2.2). Of these, *Phradis interstitialis*, *Phradis morionellus* (Fig. 2.1) and *Tersilochus heterocerus* (Fig. 2.2) are the most widespread and abundant and were identified by research within the MASTER project as the key larval parasitoids of this pest, particularly on winter rape (Table 2.1) (Williams et al. 2005, Ulber et al. 2006b). In addition, the braconid endoparasitoid, *Diospilus capito*, was found to parasitize pollen beetle

Table 2.2 Systematic list and classification of the parasitoids of the pollen beetle (*Meligethes aeneus*) reported in Europe

Order HYMENOPTERA
Superfamily ICHNEUMONOIDEA
Family ICHNEUMONIDAE (Subfamily TERSILOCHINAE)
Aneuclis Förster
– *incidens* (Thomson)
Phradis Holmgren
– *interstitialis* (Thomson)
– *morionellus* (Holmgren)
Tersilochus Holmgren (= *Thersilochus* Holmgren)
– *heterocerus* Thomson
Family BRACONIDAE (Subfamily HELCONINAE)
Blacus Nees
– *nigricornis* Haeselbarth
Diospilus Haliday
– *capito* (Nees)
Eubazus Nees (= *Calyptus* Haliday)
– *sigalphoides* (Marshall)
Superfamily CHALCIDOIDEA
Family ENCYRTIDAE
Cerchysiella Girault (= *Zeteticontus* Silvestri)
– *planiscutellum* (Mercet)
Superfamily PROCTOTRUPOIDEA
Family PROCTOTRUPIDAE
Brachyserphus Hellén
– *parvulus* (Nees)

larvae on winter rape but, more frequently, to be both abundant and widespread on spring rape. The other five parasitoids listed in Table 2.2 are of minor importance; they have been observed only occasionally with low levels of parasitism of pollen beetle larvae (Nilsson 2003). No parasitoids of the adult stage of the pollen beetle are known (Nilsson 2003).

Fig. 2.1 *Phradis morionellus*, a key parasitoid of the pollen beetle (Photo: Rothamsted Research)

Fig. 2.2 *Tersilochus heterocerus*, a key parasitoid of the pollen beetle (Photo: Rothamsted Research)

2.2.2 Distribution of Species

The key species, *P. interstitialis, P. morionellus* and *T. heterocerus*, are widely distributed throughout Europe wherever oilseed rape is grown (Nilsson 2003), including all countries contributing to the MASTER project (Table 2.3). Their occurrence and relative abundance is affected by the climate, the type of rape grown in the area and how it is cultivated (see also Nilsson Chapter 11 this volume). In central Europe and the UK, the most abundant species on winter rape are *P. interstitialis* and *T. heterocerus* (Wyrostkiewicz and Blazejewska 1985, Klingenberg and

Table 2.3 Occurrence of parasitoids of the pollen beetle (*Meligethes aeneus*) in UK, Sweden, Estonia, Poland, and Germany

	UK		Sweden		Estonia		Poland		Germany	
Family/species	L	R	L	R	L	R	L	R	L	R
Family ICHNEUMONIDAE										
Aneuclis incidens			*				*		*	
Phradis interstitialis	*	*	*		*	*	*	*	*	*
Phradis morionellus	*	*	*	*	*	*	*	*	*	*
Tersilochus heterocerus	*	*	*	*	*	*	*	*	*	*
Family BRACONIDAE										
Blacus sp.	*	*	*			*	*			
Blacus nigricornis			*			*	*		*	*
Diospilus capito	*		*	*	*	*	*	*	*	*
Eubazus sigalphoides							*			
Family ENCYRTIDAE										
Cerchysiella planiscutellum	*									
Family PROCTOTRUPIDAE										
Brachyserphus parvulus	*	*	*			*			*	

L = reported in the literature; R = reared from the host during the MASTER project or, in Estonia, caught in yellow water traps.

Ulber 1994, Büchi 2002, Kraus and Kromp 2002, Nilsson 2003, Ferguson et al. 2003) whereas in northern Europe (Estonia, Finland and Central Sweden) where more spring rape is grown, *P. morionellus* is often the more abundant (Hokkanen 1989, Billqvist and Ekbom 2001a, Nilsson 2003, Jönsson et al. 2004, Hokkanen 2006, Veromann et al. 2006b, d). In Estonia, although all four key parasitoids have been caught from spring rape, only *Phradis morionellus* has so far been caught from winter rape (Veromann et al. 2006c).

The braconid *D. capito* is also widely distributed throughout Europe wherever oilseed rape is grown and has also been reported from all MASTER countries (Nilsson 2003, Williams et al. 2005, Table 2.3). However, like *P. morionellus*, it is a more common parasitoid of pollen beetle larvae in northern Europe (Estonia, Finland, Sweden), particularly on spring rape (Nilsson and Andreasson 1987, Billqvist and Ekbom 2001b, Nilsson 2003, Veromann et al. 2006c, Hokkanen 2008). Populations on winter rape are generally low (Nilsson 2003). In Estonia, however, *D. capito* is a major parasitoid of pollen beetle larvae on both winter and spring rape (Luik et al. 2006); this may be due to the delayed phenology of host larvae on winter crops in this country. Numbers of *D. capito* caught in yellow water traps in Estonia increased with expansion of the area grown to winter rape (Veromann et al. 2006a).

Parasitism of pollen beetle larvae by *Aneuclis incidens*, *Blacus nigricornis*, *Brachyserphus parvulus*, *Cerchysiella planiscutellum* and *Eubazus sigalphoides* has been recorded infrequently and from various European countries; few specimens are generally found although *B. parvulus* and *B. nigricornis* can be common in some crops (Nilsson 2003).

The within-field spatio-temporal distributions of the pollen beetle and its key parasitoids are reviewed by Williams and Ferguson (Chapter 8 this volume).

2.2.3 Life Histories of Key Species

The life histories of the three key ichneumonid parasitoid species attacking pollen beetle larvae, namely *P. interstitialis, P. morionellus* and *T. heterocerus*, have been studied in detail by Jourdheuil (1960), Osborne (1960) and Nilsson (1994, 1997), and are reviewed by Nilsson (2003). They are all univoltine, koinobiont endoparasitoids. They overwinter as diapausing adults within their pupal cocoons in the soil of fields that have just grown oilseed rape. Overwintering mortality can be high and adversely affected by soil tillage (Nilsson 1985, 1989; Nilsson Chapter 11 this volume). The time of emergence and migration to new crops of oilseed rape the following spring varies between species, regions and years being dependent on weather parameters, particularly temperature and sunshine (see Johnen et al. Chapter 15 this volume). Adults of *P. interstitialis* often emerge 1–2 weeks earlier than the other two species, in early to mid-April, and may be found in rape crops already at the bud stage (Ulber and Nitzsche 2006, Williams 2006b). Female *P. interstitialis* prefer to oviposit through the bud walls into the eggs and first-instar larvae of their hosts (Nilsson 2003). Adults of *P. morionellus* and *T. heterocerus* commonly colonize the crop at the beginning of flowering, i.e., towards the end of April or early May in Germany and UK (Ulber and Nitzsche 2006, Williams 2006b). They oviposit into small larvae within buds and large second instar host larvae in open flowers, respectively (Nilsson 2003). Female parasitoids are attracted by volatiles emitted by oilseed rape (Jönsson et al. 2005, Williams et al. 2007; Williams and Cook Chapter 7 this volume). Following eclosion, the parasitoid remains in its first instar until the full-grown host larva drops to the ground to pupate. There, the parasitoid completes its larval development in a few days and finally kills the prepupal stage of its host. Pupation of the parasitoid larva occurs within the earthen cocoon of its host. Adults then diapause in their silken cocoons and emerge from the soil the following spring.

The braconid *D. capito* is a multivoltine koinobiont larval endoparasitoid with two to three generations per year in Northern Europe (Billqvist and Ekbom 2001b, Nilsson 2003). Host alternation between the pollen beetle and other beetle species was suggested by Meuche (1940) and Börner et al. (1942), but, in France, no other host of *D. capito* was found (Jourdheuil 1960). Adult *D. capito* often first appear in low numbers towards the end of flowering of winter rape, but are more numerous during flowering of spring rape (Börner et al. 1921, Kaufmann 1923, Miczulski 1967). Females oviposit in both first and second instar host larvae, in buds as well as flowers (Börner and Blunck 1920, Osborne 1960). New generation adults emerge from the soil approximately 10 days after migration of their host larvae into the soil to pupate. Few adults are thought to survive winter (Nilsson 2003).

In a recent survey (2007) at various locations in Germany, parasitism of pollen beetle larvae by *P. interstitialis* and *T. heterocerus* was observed from mid April to the end of June, while parasitism by *D. capito* was found only on spring rape from the end of May to mid August (Krueger and Ulber unpublished).

2.2.4 Percentage Parasitism

Parasitism of pollen beetle larvae can be a major factor for the population dynamics of this pest. Levels of parasitism exceeding 50% have been reported recently from several European countries, e.g., Austria (Kromp and Kraus 2006), Finland (Hokkanen 2006), Germany (Nitzsche 1998), Sweden (Nilsson 1989), Switzerland (Büchi 2002) and the UK (Williams 2006b).

Data collated during the MASTER project showed that parasitism of pollen beetle larvae from unsprayed crops of winter rape under various growing conditions for the years 1995–2005 (Ulber et al. 2006b) was often high, up to 97%, with average levels in Germany, Poland, Sweden and the UK within the range 25–50% (Fig. 2.3). In Estonia, percentage parasitism was lower (3–18%). In spring rape, similar high levels of parasitism as in winter rape were observed in Sweden and the UK (Fig. 2.4), but, in contrast, in Estonia and in Germany they were lower, between 0 and 16%. The relative abundance of the key parasitoids varied between countries. *Tersilochus heterocerus* and *P. interstitialis* predominated in Germany, Poland and the UK, while *P. morionellus* and *D. capito* were more common in Estonia and in Central Sweden.

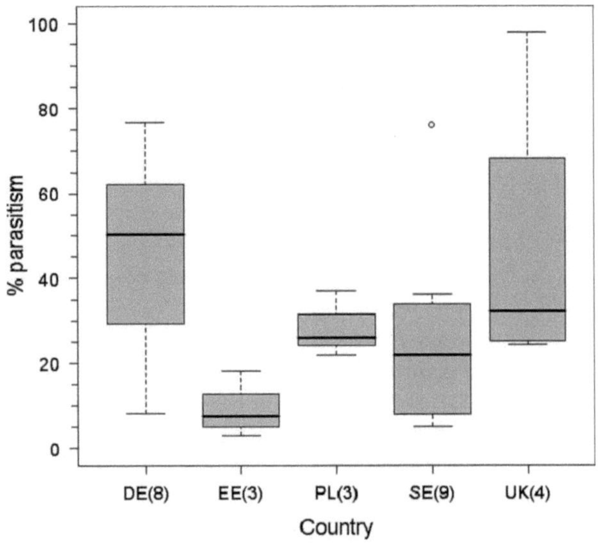

Fig. 2.3 Percentage larval parasitism of the pollen beetle, *Meligethes aeneus*, in crops of winter oilseed rape in Germany (DE), Estonia (EE), Poland (PL), Sweden (SE) and the UK in 1995–2005. In parentheses: number of oilseed rape crops examined

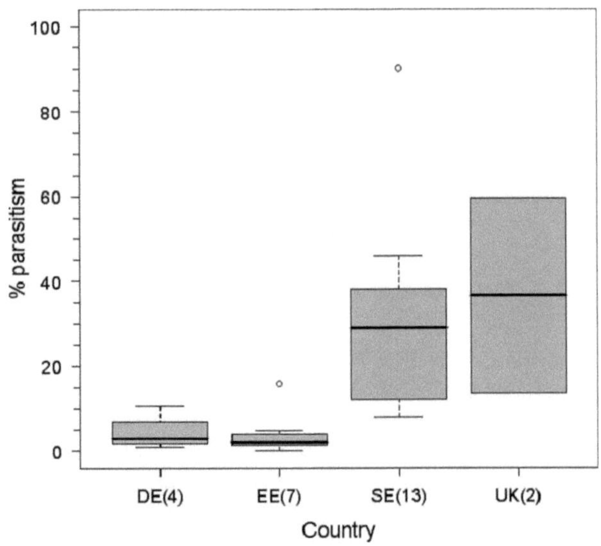

Fig. 2.4 Percentage larval parasitism of the pollen beetle, *Meligethes aeneus*, in crops of spring oilseed rape in Germany (DE), Estonia (EE), Sweden (SE) and the UK in 1995–2005. In parentheses: number of oilseed rape crops examined

Hokkanen (2008) studied parasitism of pollen beetle larvae by *P. morionellus* on spring rape crops in 13 regions of Finland from 1985 to 1995. The percentage of parasitism in each region weighted by the area of rape grown in the region was used as a measure for the proportion of pollen beetles removed from the new generation; it ranged from 8% in 1988 to 49% in 1987, with average levels between 20 and 40% in other years. By comparison with damage severity levels by the pollen beetle in these regions, Hokkanen concluded that parasitoids were able to significantly lower beetle abundance when 30–40% larval parasitism was exceeded.

Superparasitism, that is, more than one parasitoid egg or larva per pollen beetle larva, is common with *T. heterocerus* but not with *P. interstialis* (Nitzsche 1998). It was observed regularly even at parasitisation rates as low as 4% (Ulber unpublished). When the overall level of parasitism was very high, e.g., at 97% in the UK (Williams 2006b), the level of superparasitism was as high as 95%. Further, multiparasitism, that is, host larvae with more than one species of parasitoid, occurs frequently with both *T. heterocerus* and *Phradis* spp. but only one parasitoid develops to adult within each larva (Nitzsche 1998); thus parasitoid species are essentially competitors. Female *T. heterocerus* do not discriminate between host larvae that are already parasitized, either by conspecifics or by *Phradis* spp., and non-parasitized host larvae (Nitzsche 1998).

The braconid *D. capito* parasitised pollen beetle larvae on winter rape only occasionally during the MASTER project, but was more frequently found on spring rape crops in Estonia, Finland, Germany, Sweden and the UK. Other studies have found levels of pollen beetle larval parasitism of 8–29% on white mustard and spring rape in Sweden (Billqvist and Ekbom 2001b), 5–12% on spring rape in Finland (Hokkanen 1989) and 3–16% on spring rape in Germany (Krueger and Ulber unpublished).

2.3 Parasitoids of Cabbage Seed Weevil (*Ceutorhynchus obstrictus* syn. *C. assimilis*)

2.3.1 Identity of Species

The cabbage seed weevil is host to at least 31 species of parasitoid (Table 2.4), mostly larval ectoparasitoids, of which three pteromalids, *Trichomalus perfectus* (Fig. 2.5), *Stenomalina gracilis* (Fig. 2.6) and *Mesopolobus morys* (Fig. 2.7) dominate. Where ectoparasitoids from seed weevil larvae have been reared to adults (e.g., Laborius 1972, Murchie 1996, Ulber and Vidal 1998, Kevväi et al. 2006), *T. perfectus* has usually been the predominant species, followed by *M. morys* and then *S. gracilis*. *Mesopolobus morys* may be relatively more important on spring than on winter rape (Murchie 1996). These three species were identified as key species for biocontrol in Europe by the MASTER project (Table 2.1, Ulber et al. 2006b); other larval parasitoids appear to be insufficiently widespread or abundant to contribute much to biocontrol of this pest. The adult weevil is parasitized by the braconid *Microctonus melanopus*, a species which can be abundant locally (Bonnemaison 1957, Jourdheuil 1960). Mymarids are known to attack the eggs, but also appear to be of negligible importance for biocontrol (Williams 2003a).

Table 2.4 Systematic list and classification of the parasitoids of cabbage seed weevil (*Ceutorhynchus obstrictus* syn. *C. assimilis*) reported in Europe

Order HYMENOPTERA
Superfamily ICHNEUMONOIDEA
Family ICHNEUMONIDAE (Subfamily TERSILOCHINAE)
Aneuclis Förster
– *melanaria* (Holmgren) (= *diversus* Szépligeti; = *petiolaris* Szépligeti)
Tersilochus Holmgren
– sp.
Family BRACONIDAE (Subfamily DORYCTINAE)
Bracon Fabricius
– *fulvipes* Nees
– sp.
– *variator* Nees (= *discoideus* Wesmael [Note 1]; = *maculiger* Wesmael)
Family BRACONIDAE (Subfamily CHELONINAE)
Sigalphus Latreille
– *obscurellus* Nees
Family BRACONIDAE (Subfamily HELCONINAE)
Diospilus Haliday
– *morosus* Reinhardt
– *oleraceus* Haliday
Taphaeus Wesmael
– *affinis* Wesmael
Family BRACONIDAE (Subfamily EUPHORINAE)
Microctonus Wesmael
– *melanopus* Ruthe [also cited as *Perilitus melanopus*]
– cf. *deceptor* Wesmael
Superfamily CHALCIDOIDEA
Family EURYTOMIDAE
Eurytoma Illiger
– *aciculata* (Ratzeburg) [hyperparasitoid]
– *curculionum* Mayr
– sp.
Family PTEROMALIDAE (Subfamily PTEROMALINAE)
Anisopteromalus Ruschka
– *calandrae* (Howard)
Chlorocytus Graham
– *diversus* (Walker)
Habrocytus Thomson
– *dispar* (Curtis)
– *semotus* (Walker)
Mesopolobus Westwood = *Amblymerus* Walker; = *Eutelus* Walker;
= *Xenocrepis* (Förster)
– *mediterraneus* (Mayr) [hyperparasitoid]
– *morys* (Walker) (= *pura* Mayr)
Stenomalina Ghesquière
– *gracilis* (Walker) = *muscarum* (Linnaeus)
Trichomalus Thomson
– *perfectus* (Walker) = *decisus* (Walker); = *decorus* (Walker);
= *laevinucha* (Thomson)
– sp.
Zatropis Crawford
– sp.

Table 2.4 (continued)

Family EUPELMIDAE (Subfamily EUPELMINAE)
Eupelmus Dalman (= *Eupelmella* Masi; = *Macroneura* Walker)
– *vesicularis* (Retzius)
Family EULOPHIDAE
Eulophus Müller
– sp.
Necremnus Thomson
– *tidius* (Walker) (= *duplicatus* Gahan)
Tetrastichus Haliday
–*galectobus* (Ratzeburg)
Family MYMARIDAE (Subfamily MYMARINAE)
Anaphes Haliday
– *fuscipennis* Haliday
Mymar Curtis
– *autumnalis* (Förster)
Patasson Walker
– *brachygaster* Debauche
– *declinata* (Soyka)

NB Some authorities treat *Bracon discoideus* Wesmael as a separate species.

Fig. 2.5 *Trichomalus perfectus*, a key parasitoid of the cabbage seed weevil (Photo: Rothamsted Research)

Fig. 2.6 *Stenomalina gracilis*, a key parasitoid of the cabbage seed weevil (Photo: Rothamsted Research)

Fig. 2.7 *Mesopolobus morys*, a key parasitoid of the cabbage seed weevil (Photo: Rothamsted Research)

2.3.2 Distribution of Species

The three key parasitoid species, *T. perfectus*, *S. gracilis* and *M. morys*, are widely distributed throughout Europe (Williams 2003a). They were found in all five countries monitored during the MASTER project (Table 2.5). Other species of parasitoid have been infrequently reported from one or more countries but are not widespread (Williams 2003a, Table 2.5).

Literature on the within-field spatio-temporal distributions of the cabbage seed weevil and its key parasitoids is reviewed by Williams and Ferguson (Chapter 8 this volume).

2.3.3 Life Histories of Key Species

The three key pteromalid species attacking cabbage seed weevil larvae are thought to have similar life-histories, although only *T. perfectus* has been studied in detail (Dmoch and Klimek 1975, Murchie 1996, for a review see Williams 2003a).

Trichomalus perfectus is a univoltine ectoparasitoid with peaks of abundance on crops of oilseed rape 2–4 weeks after immigration of its host. More detailed information on its immigration phenology in relation to weather parameters is presented by Johnen et al. (Chapter 15 this volume). On locating a seed weevil larva within a pod, the female penetrates the pod with her ovipositor and lays a single egg on its surface. The parasitoid is a solitary idiobiont, so the host larva is immobilised on parasitisation and gradually discolours. The parasitoid egg hatches in 1–4 days and the larva feeds externally from its host for 7–10 days, eventually consuming it completely, except for its head capsule and skin. It pupates alongside its host's remains without forming a cocoon; the pupal stage lasts 8–15 days. The adult chews a small hole in the pod wall through which it exits the pod. New generation adults mate at emergence and can be found on the crop until harvest time. Only females are thought to overwinter, probably in evergreen foliage and other sheltered places. In addition

Table 2.5 Occurrence of parasitoids of the cabbage seed weevil (*Ceutorhynchus obstrictus* syn. *C. assimilis*) in the UK, Sweden, Estonia, Poland, and Germany

	UK		Sweden		Estonia		Poland		Germany	
Family/species	L	R	L	R	L	R	L	R	L	R
Family BRACONIDAE										
Bracon variator							*			
Sigalphus obscurellus	*									
Diospilus oleraceus	*						*		*	
Microctonus sp.	*	*								
Microctonus melanopus									*	
Microctonus cf. deceptor							*			
Family EURYTOMIDAE										
Eurytoma aciculata							*			
Eurytoma curculionum		*					*	*	*	*
Family PTEROMALIDAE										
Chlorocytus diversus	*									
Habrocytus sp.							*	*		
Habrocytus semotus	*									
Mesopolobus morys	*	*	*	*	*	*	*	*	*	*
Stenomalina gracilis	*	*	*	*	*	*	*	*	*	*
Trichomalus perfectus	*	*	*	*	*	*	*	*	*	*
Family EUPELMIDAE										
Eupelmus vesicularis							*	*		
Family EULOPHIDAE										
Necremnus sp.	*								*	
Necremnus tidius							*			
Tetrastichus galectobus	*									
Family MYMARIDAE										
Anaphes sp.						*	*			
Anaphes fuscipennis							*			
Mymar autumnalis									*	
Patasson sp.							*			
Patasson brachygaster	*									

L = species or genus reported in literature; R = species or genus reared from host.

to killing the larvae by parasitisation, *T. perfectus* can cause substantial mortality of host larvae by host-feeding.

2.3.4 Percentage Parasitism

Variable levels of parasitism of seed weevil larvae have been reported in the literature and they can be substantial (Williams 2003a) e.g., exceeding 50% in Germany (Nissen 1997), in Switzerland (Linz 1991, Büchi 1993) and the UK (Murchie 1996), thereby contributing to biocontrol of this pest.

Parasitism of seed weevil larvae from unsprayed crops of winter rape under various growing conditions in four European countries during the MASTER project

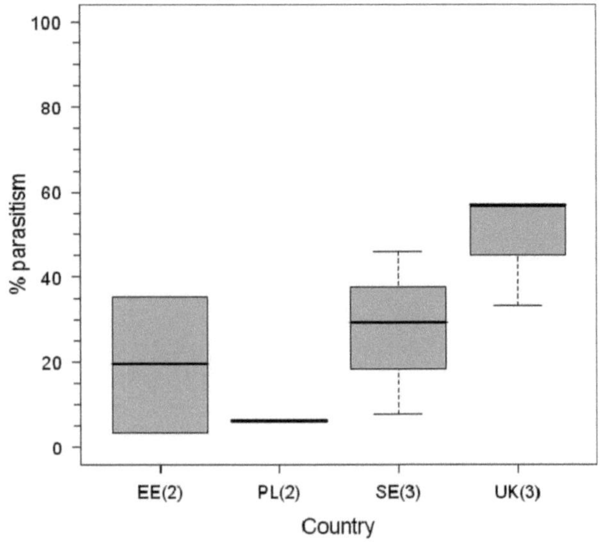

Fig. 2.8 Percentage larval parasitism of the cabbage seed weevil, *Ceutorhynchus obstrictus* syn. *C. assimilis*, in crops of winter oilseed rape, in Estonia (EE), Poland (PL), Sweden (SE) and the UK, in 1995–2005. In parentheses: number of oilseed rape crops examined

(Ulber et al. 2006b) ranged from 3–35% in Estonia to 33–57% in the UK (Fig. 2.8). In six of the 10 crops studied, percentage parasitism exceeded 30%. However, in two crops studied in Poland, parasitism was only 6%. In all four countries, *T. perfectus, M. morys* and *S. gracilis* were the only species of parasitoid found to attack seed weevil larvae during the MASTER project.

2.4 Parasitoids of the Brassica Pod Midge (*Dasineura brassicae*)

2.4.1 Identity of Species

The brassica pod midge is reported in the literature to be host to at least 31 species of parasitoid, all attacking the egg and larval stages (Williams and Walton 1990, Williams 2003b, Table 2.6). Of these, the platygastrid *Platygaster subuliformis* (Fig. 2.9) and the eulophid *Omphale clypealis* (Fig. 2.10) have been recorded most commonly in Europe and were identified as key parasitoid species during the MASTER project (Table 2.1). No parasitoids have been reported to attack the adults (Williams 2003b).

Platygaster subuliformis appears to be the most widespread parasitoid of the brassica pod midge in Europe (Williams 2003b, Ulber et al. 2006b). The species is described by Murchie et al. (1999); they found it to be the most important parasitoid of the larvae in the UK, and a new species record for the country. During the MASTER project it was also found to be the dominant species parasitizing brassica pod midge in Germany, Sweden and Poland (Ulber et al. 2006b). Identification of *Platygaster* specimens to species is difficult; *P. subuliformis* can be easily confused

Table 2.6 Systematic list and classification of the parasitoids of brassica pod midge (*Dasineura brassicae*) reported in Europe

Order HYMENOPTERA
Superfamily CHALCIDOIDEA
  Family EURYTOMIDAE
    Eurytoma Illiger
    – *aciculata* Ratzeburg
    – *dentata* Mayr
  Family TORYMIDAE
    Pseudotorymus Masi
    – *napi* Amerling & Kirchner (= *brassicae* Ruschka)
  Family EULOPHIDAE
    Aprostocetus Westwood
    – *epicharmus* (Walker) (= *variegatus* Szelényi)
    Necremnus Thomson
    – *leucarthros* (Nees)
    Neochrysocharis Kurdjumov
    – sp.
    Omphale Haliday (= *Secodes* Förster)
    – *clypealis* (Thomson)
    – *coilus* (Walker)
    Sigmophora Rondani
    – *brevicornis* (Panzer)
Superfamily PLATYGASTROIDEA
  Family PLATYGASTRIDAE
    Amblyaspis Förster
    – sp.
    Inostemma Haliday
    – *boscii* (Jurine)
    – *walkeri* Kieffer
    – nr *reticulatum* (Szelényi)
    Isocybus Förster
    – *thomsoni* Kieffer
    Piestopleura Förster
    – sp.
    Platygaster Latreille (= *Prosactogaster* Kieffer)
    – *boscii* Nees
    – *gladiator* Zetterstedt
    – *iolas* Walker
    – *munita* Walker
    – *niger* Nees
    – *nitida* (Thomson)
    – *oebalus* Walker
    – *subuliformis* (Kieffer)
    – *tisias* Walker
    Synopeas Förster
    – nr *lugubris* Thomson
    – sp.
Superfamily CERAPHRONOIDEA
  Family CERAPHRONIDAE
    Aphanogmus Thomson
    – *abdominalis* (Thomson)
    – *tenuicornis* Thomson

Table 2.6 (continued)

Ceraphron Jurine = *Calliceras* Nees
– *longipennis* (Kieffer) (= *insularis* (Kieffer))
– *pallipes* Thomson
– *serraticornis* Kieffer
– *xanthosoma* (Kieffer)
Family MEGASPILIDAE
Conostigmus Dahlborn
– *rufescens* Kieffer

Fig. 2.9 *Platygaster subuliformis*, a key parasitoid of the brassica pod midge (Photo: Rothamsted Research)

Fig. 2.10 *Omphale clypealis*, a key parasitoid of the brassica pod midge (Photo: Rothamsted Research)

with *P. minuta, P. gladiator, P. oebalus and P. tisias* (Murchie et al. 1999). Therefore, former records of other species of *Platygaster* and *Prosactogaster* in the literature (e.g., Laborius 1972) may be misidentifications and may refer to *P. subuliformis* as well.

A study of the parasitoid complex attacking brassica pod midge in the UK recorded only *P. subuliformis* and *O. clypealis* from winter rape but a further two species (an *Aphanogmus* sp. and a *Ceraphron* sp.) from spring rape (Murchie 1996).

Platygaster subuliformis was the more abundant on winter rape whereas *O. clypealis* was the more abundant on spring rape.

2.4.2 Distribution of Key Species

Platygaster spp. and *O. clypealis* are both widespread in distribution throughout northern and central Europe and probably occur almost everywhere that their host species exists whereas all other species have been reported from only a few countries and are infrequently encountered (Williams 2003b, Table 2.7). The key species

Table 2.7 Occurrence of parasitoids of the brassica pod midge (*Dasineura brassicae*) in the UK, Sweden, Estonia, Poland, and Germany

	UK		Sweden		Estonia		Poland		Germany	
Family/species	L	R	L	R	L	R	L	R	L	R
Family EURYTOMIDAE										
Eurytoma aciculata							*	*		
Eurytoma dentata							*			
Family TORYMIDAE										
Pseudotorymus napi							*	*	*	
Family EULOPHIDAE										
Aprostocetus epicharmus	*						*	*		
Necremnus leucarthros							*	*	*	
Neochrysocharis sp.							*			
Omphale clypealis	*	*	*	*		*			*	
Omphale coilus							*	*		
Sigmophora brevicornis									*	
Family PLATYGASTRIDAE										
Inostemma sp.			*						*	
Inostemma boscii									*	
Inostemma walkeri							*	*	*	
Piestopleura sp.									*	
Platygaster sp.			*	*		*	*	*	*	
Platygaster iolas							*	*		
Platygaster nitida							*	*		
Platygaster oebalus								*	*	
Platygaster subuliformis	*	*	*	*		*		*		*
Synopeas sp.			*	*						
Family CERAPHRONIDAE										
Aphanogmus sp.		*	*							
Aphanogmus abdominalis							*	*	*	
Ceraphron sp.		*	*	*						
Ceraphron pallipes									*	
Ceraphron serraticornis									*	
Ceraphron xanthosoma									*	
Family MEGASPILIDAE										
Conostigmus rufescens									*	

L = species or genus reported in literature; R = species or genus reared from host, or, in Estonia, caught in yellow water traps.

P. subuliformis appears to be the most widely distributed and abundant; it occurred in all countries participating in the MASTER project (Table 2.7). The key species *O. clypealis* was found in all MASTER countries except Poland. Surprisingly, both species were caught in yellow water traps in crops of oilseed rape in Estonia, although their host, the brassica pod midge, has not been recorded there (Veromann et al. 2006c).

The within-field, spatio-temporal distributions of the brassica pod midge and its parasitoids is reviewed by Williams and Ferguson (Chapter 8 this volume).

2.4.3 Life Histories of Key Species

Literature on the life histories and biology of both key species of parasitoid attacking the brassica pod midge is reviewed by Williams (2003b). Information about their responses to host plant volatiles is presented in Williams and Cook (Chapter 7 this volume).

Platygaster subuliformis is an egg-larval endoparasitoid (Murchie et al. 1999). Like its host it is probably multivoltine, although it may have fewer generations per year than its host as it takes longer to emerge pre-diapause. Emergence in the UK occurs during the first half of May with peak abundance of adults co-inciding with peak availability of host larvae (Ferguson et al. 2004). Each female parasitizes several host eggs within an infested pod, laying a single egg in each. The parasitoid is a koinobiont; its egg develops only after its host is nearly full-grown and at the prepupal and pupal stage of development within the host's larval skin. Part of the population emerges the same season, part remains in diapause in the soil inside host cocoons. Mating occurs soon after emergence and the mated females then migrate to rape crops. Further information on the phenology of its migration to winter rape in relation to weather parameters is presented in Johnen et al. (Chapter 15 this volume).

Omphale clypealis is a larval endoparasitoid. Like its host, it is probably multivoltine but its biology is poorly known. It overwinters within the cocoons of its host and emerges over a prolonged period during the spring and summer, starting about a month later than its host (Ferguson et al. 2004). Its sex ratio is strongly female-biased (Murchie 1996). Peak abundance of the parasitoid has been found to co-incide with that of its host. The females oviposit into mature host larvae through the pod wall and the parasitoid larva feeds within its host during its larval and pupal stages, completing its development after the mature host larva has dropped to the soil to pupate.

2.4.4 Percentage Parasitism

The few assessments of the levels of parasitism in the brassica pod midge are difficult to compare because of the multivoltine life-histories of both pest and key parasitoids and the ability of the pest to diapause for several years. However, several studies suggest that although percentage parasitism is variable it can also be substantial in some years. Thus, Murchie (1996) found that, in the UK, *P. subuliformis* emerged pre-diapause from 3 to 13% and from 0 to 18% of host larvae collected at weekly intervals (for 4 weeks) from two crops of winter rape and from 4 to 67% and from 27 to 74% of larvae, post-diapause. Ten percent of midge cocoons from a spring rape crop were parasitized. Ferguson et al. (2004) found that, in winter rape in 1999, only 7% of first generation midge larvae, which dropped to the ground to pupate, gave rise to adult insects (midge and parasitoids) that same year, and that of these 42% were parasitoids, mostly *O. clypealis.* Only 0.2% of both generations of midge larvae emerged as adults the following year, of which 49% were parasitoids, with similar numbers of *O. clypealis* and *P. subuliformis.*

In recent years, the infestation levels of oilseed rape by the brassica pod midge have been low and parasitism levels of pod midge larvae were not estimated during the MASTER project (Ulber et al. 2006b).

2.5 Parasitoids of Stem-Mining Pests

2.5.1 Identity of Species

2.5.1.1 Parasitoids of the Cabbage Stem Flea Beetle (*Psylliodes chrysocephala*)

Three ichneumonid, two braconid and one pteromalid parasitoid species have been reared from the larvae of the cabbage stem flea beetle and one braconid from the adult in Europe (Table 2.8).

Table 2.8 Systematic list and classification of the parasitoids of the cabbage stem flea beetle (*Psylliodes chrysocephala*) reported in Europe

Order HYMENOPTERA
Superfamily ICHNEUMONOIDEA
  Family ICHNEUMONIDAE
    Aneuclis Förster
    – *melanaria* (Holmgren) (= *diversus* Szépligeti; = *petiolaris* Szépligeti) [Note 1]
    Tersilochus Holmgren (= *Thersilochus* Holmgren)
    – *microgaster* (Szépligeti) [Note 2]
  Family BRACONIDAE
    Diospilus Haliday
    – *morosus* Reinhardt
    – *oleraceus* Haliday
    Microctonus Wesmael
    – *melanopus* Ruthe [Note 3]
Superfamily CHALCIDOIDEA
  Family PTEROMALIDAE
    Trichomalus Thomson
    – *lucidus* (Walker)
    – nr *lucidus* (Walker)

Note 1. Also cited in the literature under genera *Isugurus*, *Perilitus* and *Thersilochus.*
Note 2. Also cited in the literature as *Isurgus microgaster* Szépligeti.
Note 3. Also cited in the literature as *Perilitus melanopus* Ruthe.

Earlier studies from France (Jourdheuil 1960), Czech Republic (Šedivý 1983) and Germany (Dosse 1961, Lehmann 1965), reported that *Tersilochus tripartitus* Brischke (syn. *Tersilochus melanogaster* Thomson) was an abundant larval parasitoid of the cabbage stem flea beetle. However, since 1990, *T. tripartitus* has never been detected, and *Tersilochus microgaster* (Szépligeti) has been reported to be the most abundant and frequently occurring parasitoid of this pest in Europe (Klingenberg and Ulber 1994, Nitzsche 1998, Barari et al. 2004, Ulber and Wedemeyer 2004). Because Horstmann (pers comm) recently found that no host is known for *T. tripartitus,* the earlier reports apparently resulted from erroneous identification of *T. microgaster*. In the MASTER project, *T. microgaster* was identified as the only key larval parasitoid species for the cabbage stem flea beetle (Table 2.1). All other larval parasitoid species appear to be of minor importance (Ulber and Williams 2003).

The braconid *Microctonus melanopus* is the only species known to attack adult cabbage stem flea beetles but information on the status, importance and biology of this species is sparse (Jourdheuil 1960, Ulber and Williams 2003).

2.5.1.2 Parasitoids of the Cabbage Stem Weevil (*Ceutorhynchus pallidactylus*)

The larva of the cabbage stem weevil is host to three known parasitoid species (Table 2.9). The most abundant and widespread species is *Tersilochus obscurator*; it is the only one identified as a key species for biocontrol by the MASTER project (Table 2.1). Various species reported in the literature like *Thersilochus tripartitus* Brischke spp. *obscurator* Aubert (Aubert and Jourdheuil 1958) have proved to be synonyms of *T. obscurator* (Horstmann 1971, 1981). *Stibeutes curvispina* has been reported only from Germany; it parasitises the larvae or prepupae within the soil

Table 2.9 Systematic list and classification of the parasitoids of the cabbage stem weevil (*Ceutorhynchus pallidactylus*) reported in Europe

Order HYMENOPTERA
- Family ICHNEUMONIDAE
  - Subfamily PHYGADEUONTINAE
    - *Stibeutes* Förster
      - – *curvispina* (Thomson)
  - Subfamily TERSILOCHINAE
    - *Tersilochus* Holmgren (= *Thersilochus* Holmgren)
      - – *obscurator* Aubert [Note 1]
- Family BRACONIDAE
    - *Microctonus* Wesmael
      - – *melanopus* Ruthe [Note 2]
- Family PTEROMALIDAE
    - *Trichomalus* Thomson
      - – *lucidus* (Walker)

Note 1. Also cited in the literature as *Thersilochus tripartitus* Brischke spp. *obscurator* Aubert.
Note 2. Also cited in the literature as *Perilitus melanopus* Ruthe.

(Nissen 1997). The multivoltine ectoparasitoid *T. lucidus* was reared from larvae of cabbage stem weevil in Poland and Germany during the MASTER project (Ulber et al. 2006b).

The braconid, *M. melanopus*, attacks the adults (Table 2.9); this species is a non-specialist and attacks the adults of the cabbage stem flea beetle, cabbage seed weevil and the rape winter stem weevil, *Ceutorhynchus picitarsis* as well (Jourdheuil 1960).

2.5.1.3 Parasitoids of the Rape Stem Weevil (*Ceutorhynchus napi*)

Only two species of parasitoid are known to parasitise the larvae of the rape stem weevil: the ichneumonid *Tersilochus fulvipes* (Jourdheuil 1960, Šedivý 1983, Ulber 2000, 2003) and the pteromalid ectoparasitoid *Stenomalina gracilis* (Table 2.10). The former is abundant and widespread and considered a key species for biocontrol (Table 2.1). The latter has been reared from rape stem weevil larvae in Poland (Klukowski and Kelm 2000); it is a key parasitoid of cabbage seed weevil larvae (Table 2.1).

Table 2.10 Systematic list and classification of the parasitoids of the cabbage stem weevil (*C. napi*) reported in Europe

Order HYMENOPTERA
- Family ICHNEUMONIDAE
  - Subfamily TERSILOCHINAE
    - *Tersilochus* Holmgren (= *Thersilochus* Holmgren)
    - – *fulvipes* (Gravenhorst) (= gibbus Holmgren) [Note 1]
- Family PTEROMALIDAE
    - *Stenomalina* Ghesquière
    - – *gracilis* (Walker) (*muscarum* misidentification)

Note 1. Also cited in the literature as *Porizon fulvipes* (Gravenhorst) and as *Thersilochus fulvipes* (Gravenhorst) ssp. *gallicator* Aubert.

2.5.2 Distribution of Species

Tersilochus microgaster is the most widely distributed parasitoid of cabbage stem flea beetle in Europe. It was reared from this host for the first time in Germany (Klingenberg and Ulber 1994) and has also been identified, during the MASTER project from UK (Barari et al. 2005), Sweden and Poland (Table 2.11). It has not been reported from Estonia and Finland where its host, the cabbage stem flea beetle, is not present (Veromann et al. 2006a).

Other parasitoid species attacking cabbage stem flea beetle are less widespread on oilseed rape crops. Although *Aneuclis melanaria* has been reported from many European countries (Horstmann 1971, 1981): France (Aubert and Jourdheuil 1958, Jourdheuil 1960), Czech Republic (Šedivý 1983) and Germany (Ulber and Wedemeyer 2004), it was not found on either winter nor spring rape crops in UK, Sweden, Estonia or Poland during the MASTER project. *Cremastus carinifer*, has been reported from Germany (Meuche 1940) and France (Bonnemaison and

Table 2.11 Occurrence of parasitoids of cabbage stem flea beetle (*Psylliodes chrysocephala*) in UK, Sweden, Estonia, Poland, and Germany

	UK		Sweden		Estonia		Poland		Germany	
Family/species	L	R	L	R	L	R	L	R	L	R
Family ICHNEUMONIDAE										
Aneuclis melanaria										*
Cremastus carinifer									*	
Tersilochus microgaster	*	*	*	*			*			*
Family BRACONIDAE										
Diospilus oleraceus							*		*	
Diospilus morosus							*		*	
Microctonus sp.		*					*			
Family PTEROMALIDAE										
Trichomalus lucidus							*		*	*

L = species or genus reported in literature; R = species or genus reared from host. Pest not present in Estonia

Table 2.12 Occurrence of parasitoids of the cabbage stem weevil (*Ceutorhynchus pallidactylus*) in UK, Sweden, Estonia, Poland, and Germany

	UK		Sweden		Estonia		Poland		Germany	
Family/species	L	R	L	R	L	R	L	R	L	R
Family ICHNEUMONIDAE										
Tersilochus obscurator	*	*	*	*		*		*	*	*
Family BRACONIDAE										
Microctonus melanopus									*	
Family PTEROMALIDAE										
Trichomalus lucidus	*							*		*

L = species or genus reported in literature; R = species or genus reared from host. Pest not present in Estonia.

Jourdheuil 1954); however, the identification in France was later revised to *Aneuclis melanaria* (Jourdheuil 1960). *Diospilus morosus* and *D. oleraceus* have been reported from France (Jourdheuil 1960) and *D. morosus* also from Germany (Godan 1950). *Trichomalus lucidus* has been reared from the larvae of the cabbage stem flea beetle in Germany (Nissen 1997, Ulber and Wedemeyer 2004) and the UK (DV Alford *unpublished*); it has also been found to parasitise larvae of cabbage stem weevil in Germany (Ulber and Wedemeyer 2004).

Tersilochus obscurator, the key larval parasitoid of the cabbage stem weevil, has been reared from host larvae in Germany, Poland and Sweden, and now, for the first time during the MASTER project from Estonia and the UK (Table 2.12). In addition, it has been reported in the literature from Ireland, France, Switzerland, Austria, Czech Republic and Hungary (Jourdheuil 1960, Horstmann 1981, Šedivý 1983, Büchi 1995, Kraus and Kromp 2002).

Table 2.13 Occurrence of parasitoids of the rape stem weevil (*Ceutorhynchus napi*) in UK, Sweden, Estonia, Poland, and Germany

	UK		Sweden		Estonia		Poland		Germany	
Family/species	L	R	L	R	L	R	L	R	L	R
Family ICHNEUMONIDAE										
Tersilochus fulvipes								*	*	*
Family PTEROMALIDAE										
Stenomalina gracilis	*		*		*		*	*	*	

L = species or genus reported in literature; R = species or genus reared from host. Pest not present in the UK or Estonia.
NB *S. gracilis* is also a parasitoid of the cabbage seed weevil.

Tersilochus fulvipes, the key larval parasitoid of rape stem weevil, has been reported from most countries where its host occurs, including Austria (Kraus and Kromp 2002), the Czech Republic (Šedivý 1983), France (Jourdheuil 1960), Germany (Ulber 2000, 2003), Hungary (Horstmann 1981), Poland (Klukowski pers comm) and Switzerland (Günthardt 1949) and was reared from this host in Poland and Germany during the MASTER project (Table 2.13).

The within-field, spatio-temporal distributions of the stem-mining pests and their parasitoids is reviewed by Williams and Ferguson (Chapter 8 this volume).

2.5.3 Life Histories of Key Species

The biology of the key parasitoids of the cabbage stem flea beetle, the cabbage stem weevil and the rape stem weevil, namely *T. microgaster, T. obscurator* and *T. fulvipes*, respectively, have been studied extensively by Jourdheuil (1960). They are all univoltine, koinobiont, solitary endoparasitoids of the larvae and have similar life histories.

Adults overwinter in the fields where they have developed in their hosts on oilseed rape. According to the phenologies of the respective host larvae, they emerge from soil in early or late spring and migrate to the new oilseed rape crops in succession (Ferguson et al. 2006, Ulber and Nitzsche 2006). Further information on how weather parameters affect the emergence and migration of *T. microgaster* is presented by Johnen et al. (Chapter 15 this volume). Female parasitoids often show temporal synchrony with the vulnerable instars of their hosts in the crop. Crop location is aided by chemical cues emitted by infested host plants (see Williams and Cook, Chapter 7 this volume). The phenologies of emergence and the immigration of adult parasitoids into new oilseed crops was monitored by emergence traps, yellow water traps and Malaise traps in Germany and the UK (Ulber and Nitzsche 2006, Ferguson et al. 2006). Peak emergence of overwintering adult *T. microgaster* was observed in early March to April (Ulber and Wedemeyer 2004, Ferguson et al. 2006). Female parasitoids colonize new oilseed rape crops from March to May, indicating a high level of synchrony between immigration of parasitoids and the appearance of larval instars of the cabbage stem flea beetle within plants. First individuals of *T. obscurator* and *T. fulvipes* emerge in April and colonize new crops of oilseed rape simultaneously or a few days later, usually shortly before or at the beginning of flowering. Peak activity occurs in April/May when the crop is at full flowering and declines at the end of flowering (Jourdheuil 1960, Lehmann 1965, Nitzsche 1998, Ulber and Nitzsche 2006).

Female parasitoids forage on the rape plants, with antennation of the stem surface and ovipositor probing close to infested parts of the stem, suggesting that host microhabitat location and host recognition is assisted by contact chemosensory cues originating from the host plant or host (Ulber 2003). Females oviposit through the tissue of petioles or stems into host larvae while these are mining within the pith. After hatching, the parasitoid larva remains in its first instar within the host which apparently is not affected by parasitism (Jourdheuil 1960). However, after the mature host larva has migrated to the soil for pupation, the parasitoid larva develops rapidly and finally kills the host prepupa. The mature parasitoid larva spins a silken cocoon and pupates within the earthen cocoon prepared by the host. Adult parasitoids hatch in late summer and overwinter in diapause within the pupal cocoon in the soil. There is no information on alternative hosts for these *Tersilochus* species. Under laboratory conditions, the average longevity of newly-emerged females of *T. fulvipes* and *T. obscurator*, provided with rape flowers and water, was 53 and 58 days, respectively (Nitzsche 1998).

2.5.4 Percentage Parasitism

2.5.4.1 Cabbage Stem Flea Beetle

In all countries, the only parasitoid species found to attack the cabbage stem flea beetle during the MASTER project, with very few exceptions, was *T. microgaster*. Assessments of parasitism by *T. microgaster* were conducted in four countries. The level of parasitism of larvae was variable (Fig. 2.11), ranging between 0 and 57% in Germany and Sweden, below 6% in two crops in Poland, and 11% in one crop in UK. In earlier studies from France and Germany, parasitisation rates of larvae by *T. tripartitus* (probably syn. with *T. microgaster* – see above) ranged from 30 to 61% and from 3 to 27%, respectively (Aubert and Jourdheuil 1958, Jourdheuil 1960, Dosse 1961). In Germany, in 2001, 2002 and 2003, at peak abundance of host larvae in the first decade of May, the field parasitism levels were 25% (n = 280), 44% (n = 792) and 23% (n = 127), respectively (Ulber and Wedemeyer 2004). There was no positive relationship between the abundance of host larvae per plant and the level of parasitism. While in 2000/2001 and 2001/2002 high numbers of host larvae were present within rape plants throughout the winter, in 2002/2003 the number of larvae started to increase only from the middle of March onwards. This might have affected the spatial-temporal coincidence between parasitoid and host populations resulting in different levels of parasitism.

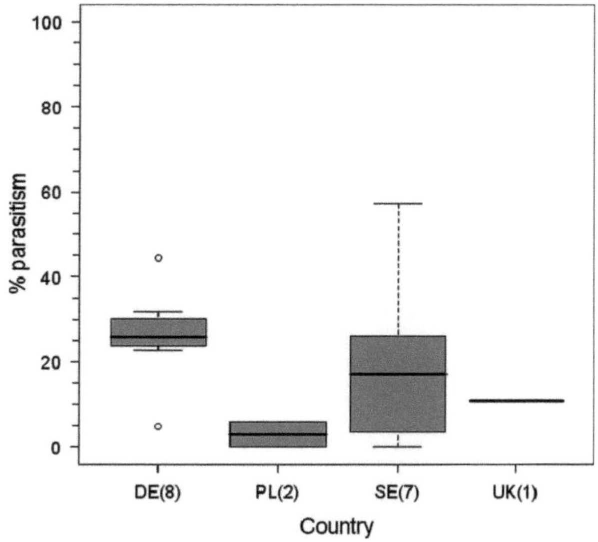

Fig. 2.11 Percentage larval parasitism of the cabbage stem flea beetle, *Psylliodes chrysocephala*, in crops of winter oilseed rape in Germany (DE), Poland (PL), Sweden (SE) and the UK in 1995–2005. In parentheses: number of oilseed rape crops examined

Superparasitism by *T. microgaster* occurs regularly, with up to nine encapsulated eggs and/or larvae of *T. microgaster* per individual host larva in 2002 in Germany (Ulber and Wedemeyer 2004). Superparasitism of parasitised larvae increased from 41 to 83% between 15 April and 22 May.

Parasitism levels by other species of parasitoid are negligible. *Aneuclis melanaria* parasitized only between 0.2 and 1.5% larvae in 1953, 1954, and 1955 in France (Jourdheuil 1960). In Germany, only 2–5% larvae were found parasitized by this species in the autumn of 1999, with no parasitism in the following years, despite high levels of larval infestation and extensive dissections and rearings of larvae (Ulber and Wedemeyer 2004). Parasitism by *Diospilus* spp. has also been reported to be low; this has been attributed to insufficient synchrony of the autumn generation of *D. morosus* and the host larvae (Jourdheuil 1960). In the studies conducted during the MASTER project from 2002 to 2005, no parasitism of cabbage stem flea beetle larvae by *Diospilus* spp. was found, even at higher host densities. *Trichomalus lucidus* (one female only) was reared from a total of 260 larvae sampled in May 2003 from a crop of oilseed rape at Goettingen/Germany (Ulber and Wedemeyer 2004), and two were reared from larvae in northern Germany (Nissen 1997).

2.5.4.2 Cabbage Stem Weevil

The parasitism of cabbage stem weevil larvae was determined from unsprayed crops of oilseed rape under various growing conditions in five European countries during the MASTER project (Ulber et al. 2006b). The level of parasitism ranged from 10 to 57%, with average levels in Germany, Poland and the UK at ca. 20% and in

Fig. 2.12 Percentage larval parasitism of cabbage stem weevil, *Ceutorhynchus pallidactylus*, in crops of winter oilseed rape in Germany (DE), Poland (PL), Sweden (SE) and the UK in 1995–2005. In parentheses: number of oilseed rape crops examined

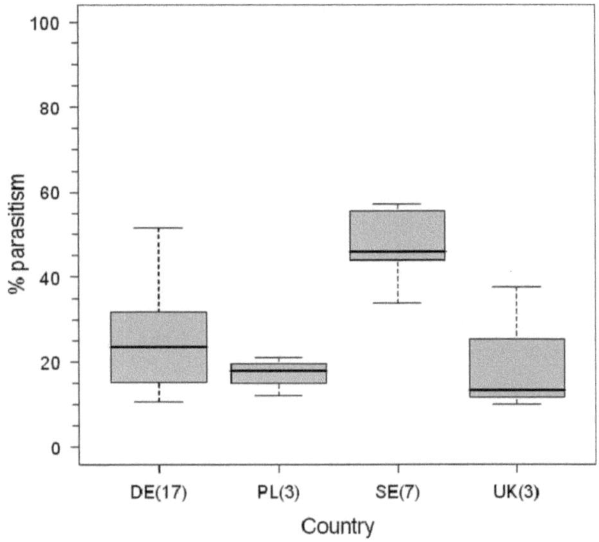

Sweden at ca. 50% (Fig. 2.12). With only very few exceptions, *T. obscurator* was the predominant parasitoid of the cabbage stem weevil.

As with the cabbage stem flea beetle, superparasitism of the cabbage stem weevil by *T. obscurator* was observed regularly in many crops of oilseed rape; for example in all crops sampled during 2003–2005 in the UK, with levels up to 39% of parasitized hosts (Williams unpublished). In Northern Germany, superparasitism was particularly evident at high levels of parasitism of host larvae, however, the level was not analysed in greater detail (Nissen 1997); encapsulation of parasitoid eggs and larvae within host larvae was also found.

2.5.4.3 Rape Stem Weevil

Levels of parasitism of the rape stem weevil determined in Germany and Poland during the project MASTER were considerably lower than those recorded in earlier studies from Austria, Czech Republic and France (Ulber 2003, Ulber et al. 2006b). In Germany and in Poland, parasitism in the majority of crops ranged between 2 and 14%, with a peak level of 21% in Germany (Fig. 2.13). *Tersilochus fulvipes* was the only parasitoid species identified from all samples. As the rape stem weevil is distributed only in Central Europe, there is no data on parasitism of this pest from the UK and Northern European countries.

Plant density affects plant architecture as well as microclimate and was found to affect percent parasitism of rape stem weevil larvae; it was higher in the lower sections of the main stems of plants when sowing densities were high (74 seeds/m^2) than when they were lower (25, 37 or 49 seeds/m^2) (Fischer and Ulber 2006). Presumably the thinner stems of high density plantings allow greater access of

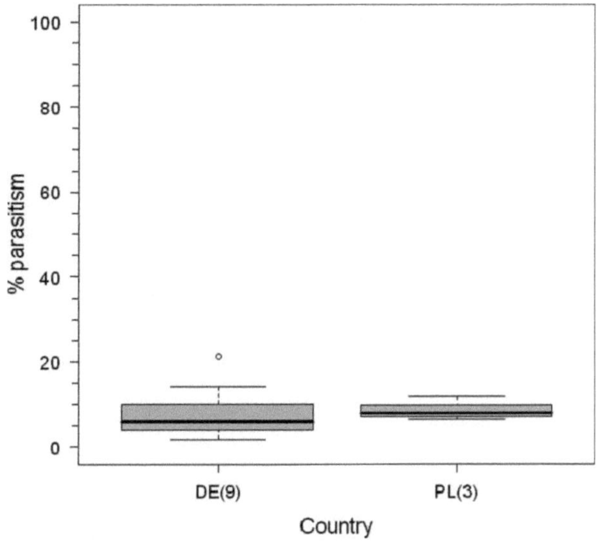

Fig. 2.13 Percentage larval parasitism of rape stem weevil, *Ceutorhynchus napi*, in crops of winter oilseed rape in Germany (DE) and Poland (PL) in 1995–2005. In parentheses: number of oilseed rape crops examined

parasitoids to their host larvae within the stems. However, at very low plant density (10 plants/m^2), a greater proportion of host larvae was parasitized within lateral branches and the level of parasitism was higher than at high plant density (70 plants/m^2) (Neumann and Ulber 2006). As the ovipositor length of *T. fulvipes* females is only 4.2 mm, thick stems can provide structural refuges for rape stem weevil larvae (Ulber 2003). Further, the species and cultivars of the *Brassica* host plant have significant effects on larval parasitism of rape stem weevil (Ulber et al. 2006a).

2.6 Conclusions and Implications for Biocontrol-Based IPM in Oilseed Rape

At least 80 species of hymenopteran species are known to parasitise the six economically most important pests of oilseed rape but only 12 of these were identified by the EU-funded project MASTER as sufficiently widespread and abundant across Europe to be of potential economic importance for biocontrol of these six pests. Most of the 12 parasitoid species were recorded from all five project partner countries where their host species is present.

The emergence and seasonal activity periods of the key parasitoids within crops of oilseed rape are closely synchronized with the phenologies of the pre-imaginal life stages of the target host populations. Immigration of parasitoids usually starts shortly after the beginning of host oviposition or hatching of host larvae on plants.

In most European countries, the level of parasitism of target pests is high, frequently ranging between 10 and 50%, with parasitism of cabbage seed weevil and pollen beetle in Sweden, Germany and the UK occasionally exceeding 70 and 90%, respectively. However, the level of parasitism of the six most damaging pests of oilseed rape varies between years and countries, and in some seasons the abundance of pest populations is too low for reliable estimations of percentage parasitism. Percentage parasitism of target pests frequently exceeds the threshold of 30% below which biological control has rarely been found to be successful (Hawkins and Cornell 1994). The most important consequence of parasitism is direct or later mortality of pest larvae, leading to reductions in adult pest populations for the following year. Thus, the results obtained during the MASTER project provide further evidence that the key parasitoids have potential to significantly reduce pest populations, in many years keeping pest densities below thresholds of economic damage, thereby exerting an important role for the natural regulation of pests.

Acknowledgements The project MASTER (QLK5-CT-2001-01447) was co-funded by the EU under its Framework 5 Quality of Life and Management of Living Resources programme. We thank all scientists and students who contributed to the project, particularly the following: Britt Ahman, Hassan Barari, Wolfgang Büchs, Suzanne Clark, Daniela Felsmann, Andrew Ferguson, Pierre Fernandez, Kirsa Fischer, Dave Frearson, Ross Holdgate, Heikki Hokkanen, Andreas Johnen, Mariusc Kaczmarzyk, Reelika Kevvai, Beate Klander, Mart Kruus, Neil Mason, Ingeborg Menzler-Hokkanen, Helene Nuss, Ross Piper, Oliver Schlein, Emma Smith, Tiiu Tarang, Eve Veromann, Nigel Watts and Rainer Wedemeyer. Writing of this chapter was also supported by the Estonian Targeting Finance Projects SF0172655s04 and SF0170057s09.

References

Alford DV (ed.) (2003) *Biocontrol of oilseed rape pests*. Blackwell, Oxford, UK.

Aubert JF, Jourdheuil P (1958) Nouvelle description et biologie de quelques Ichneumonides appartenant aux genres *Aneuclis* Först., *Isurgus* Först. et *Thersilochus* Holm. *Rev Path Veg Ent Agr* 37: 175–193.

Barari H, Cook SM, Clark SJ, Williams IH (2005) Effects of a turnip rape (*Brassica rapa*) trap crop on stem-mining pests and their parasitoids in winter oilseed rape (*Brassica napus*). *BioControl* 50: 69–86.

Barari H, Cook SM, Williams IH (2004) Rearing and identification of the larval parasitoids of *Psylliodes chrysocephala* and *Ceutorhynchus pallidactylus* from field-collected specimens. *IOBC/wprs Bull* 27(10): 263–272.

Billqvist A, Ekbom B (2001a) The influence of host plant species on parasitism of pollen beetles (*Meligethes* spp.) by *Phradis morionellus*. *Entomol Exp Appl* 98: 41–47.

Billqvist A, Ekbom B (2001b) Effect of host plant species on the interaction between the parasitic wasp *Diospilus capito* and pollen beetles (*Meligethes* spp.). *Agr Forest Entomol* 3: 147–152.

Bonnemaison L (1957) Le charançon des siliques (*Ceuthorrhynchus assimilis* Payk.). *Ann Épiphyties* 8: 387–543.

Bonnemaison L, Jourdheuil P (1954) L'altise d'hiver du Colza (*Psylliodes chrysocephala* L.). *Ann Épiphyties* 4: 345–524.

Börner C, Blunck H (1920) Zur Lebensgeschichte des Rapsglanzkäfers. *Mitt Biol Reichsanst Land- und Forstwirtsch, Berlin-Dahlem* 18: 91–109.

Börner C, Blunck H, Speyer W (1942) Wirtswechsel der Schlupfwespe *Diospilus capito* zwischen den Larven von Rapsglanzkäfer und Kohlblattrüßler. *Z PflKrankh PflSchutz* 52: 107–113.

Börner C, Blunck H, Speyer W, Dampf A (1921) Beiträge zur Kenntnis vom Massenwechsel (Gradation) schädlicher Insekten. *Arb Biol Reichsanst Land- und Forstwirtsch Berlin* 10: 405–466.

Büchi R (1993) Monitoring of parasitoids of the cabbage seed weevil, *Ceutorhynchus assimilis* during 1990 and 1991 in Switzerland. *IOBC/wprs Bull* 16(9): 145–149.

Büchi R (1995) Natürliche Gegenspieler von Rapsschädlingen. *Innovation* 3: 21–24.

Büchi R (2002) Mortality of pollen beetle (*Meligethes* spp.) larvae due to predators and parasitoids in rape fields and the effect of conservation strips. *Agr Ecosyst Environ* 90: 255–263.

Dmoch J, Klimek G (1975) Badania nad pasożytami chowacza podobnika (*Ceutorrynchus assimilis* Payk.). III. Obserwacje nad biologią *Trichomalus perfectus* (Walker). *Roczniki nauk Rolniczych Seria E* 5: 125–136.

Dosse G (1961) *Thersilochus melanogaster* Thoms. als Larvenparasit des Rapserdflohs *Psylliodes chrysocephala* L. *Z PflKrankh PflSchutz* 68: 575–580.

Ferguson AW, Barari H, Warner DH, Campbell JM, Smith ET, Williams IH (2006) Distributions and interactions of the stem miners, *Psylliodes chrysocephala* (L.) and *Ceutorhynchus pallidactylus* (Marsham), and their parasitoids in a crop of winter oilseed rape (*Brassica napus* L.). *Entomol Exp Appl* 119: 81–92.

Ferguson AW, Campbell JM, Warner DJ, Watts NP, Schmidt JEU, Williams IH (2003) Spatio-temporal distributions of *Meligethes aeneus* and its parasitoids in an oilseed rape crop and their significance for crop protection. *Proc 11th Int Rapeseed Cong, 6–10 July 2003, Copenhagen, Denmark,* 1057–1059.

Ferguson AW, Campbell JM, Warner DJ, Watts NP, Schmidt JEU, Williams IH (2004) Phenology and spatial distributions of *Dasineura brassicae* and its parasitoids in a crop of winter oilseed rape: Implications for integrated pest management. *IOBC/wprs Bull* 27(10): 243–252.

Fischer K, Ulber B (2006) Larval parasitism of *Ceutorhynchus napi* Gyll. and *Ceutorhynchus pallidactylus* (Mrsh.) in plots of different crop density of oilseed rape. *IOBC/wprs Bull* 29(7): 201.

Godan D (1950) Parasitierung von Rapserdflohlarven. *Anz Schädlingskunde* 23: 150.

Günthardt E (1949) Beiträge zur Lebensweise und Bekämpfung von *Ceuthorrhynchus quadridens* Panz. und *Ceuthorrhynchus napi* Gyll., mit Beobachtungen an weiteren Kohl- und Rapsschädlingen. *Mitt Schweiz Entomol Ges* 22: 44–591.

Hawkins BA, Cornell HV (1994) Maximum parasitism rates and successful biological control. *Science* 266: 1886.

Hokkanen HMT (1989) Biological and agrotechnical control of the rape blossom beetle *Meligethes aeneus* (Col., Nitidulidae). *Acta Entomol Fenn* 53: 25–29.

Hokkanen HMT (2006) *Phradis morionellus* on *Meligethes aeneus*: Long-term patterns of parasitism and impact on pollen beetle populations in Finland. *IOBC/wprs Bull* 29(7): 187–191.

Hokkanen HMT (2008) Biological control methods of pest insects in oilseed rape. *EPPO Bull* 38: 104–109.

Horstmann K (1971) Revision der europäischen Tersilochinen I (Hymenoptera, Ichneumonidae). *Veröffentlichungen der Zoologischen Staatssammlung München* 15: 47–138.

Horstmann K (1981) Revision der europäischen Tersilochinae II (Hymenoptera, Ichneumonidae). *Spixiana* 4: 1–76.

Jönsson M, Lindkvist A, Anderson P (2005) Behavioural responses in three ichneumonid pollen beetle parasitoids to volatiles emitted from different phenological stages of oilseed rape. *Entomol Exp Appl* 115: 363–369.

Jönsson M, Nilsson C, Anderson P (2004) Occurrence of pollen beetle parasitoids in the south of Sweden. *IOBC/wprs Bull* 27(10): 239–242.

Jourdheuil P (1960) Influence de quelques facteurs écologiques sur les fluctations de population d'une biocénose parasitaire: Étude relative à quelque Hyménoptères (Ophioninae, Diospilinae, Euphorinae) parasites de divers Coléoptères inféodès aux crucifères. *Ann Épiphyties* 11: 445–658.

Kaufmann O (1923) Beobachtungen und Versuche zur Frage der Überwinterung und Parasitierung von Ölfruchtschädlingen aus den Gattungen *Meligethes*, *Phyllotreta*, *Psylliodes* und *Ceutorrhynchus*. *Arb Biol Reichsanst Land- und Forstwirtsch Berlin* 12: 109–169.

Kevväi R, Veromann E, Luik A, Saarnit M (2006) Cabbage seed weevil (*Ceutorhynchus assimilis* Payk.) and its parasitoids in oilseed rape crops in Estonia. *Agron Res* 4(Special Issue): 227–230.

Klingenberg A, Ulber B (1994) Untersuchungen zum Auftreten der Tersilochinae (Hym., Ichneumonidae) als Larvalparasitoide einiger Rapsschädlinge im Raum Göttingen 1990 und 1991 und zu deren Schlupfabundanz nach unterschiedlicher Bodenbearbeitung. *J Appl Entomol* 117: 287–299.

Klukowski Z, Kelm M (2000) *Stenomalina gracilis* (Walker), a new parasitoid reared from *Ceutorhynchus napi* Gyll. in Poland. *IOBC/wprs Bull* 23(6): 135–138.

Kraus P, Kromp B (2002) Parasitization rates of the oilseed rape pests *Ceutorhynchus napi*, *Ceutorhynchus pallidactylus* (Coleoptera, Curculionidae) and *Meligethes aeneus* (Coleoptera, Nitidulidae) by ichneumonids in several localities of eastern Austria. *IOBC/wprs Bull* 25(2): 117–122.

Kromp B, Kraus P (2006) Levels of parasitism by ichneumonid wasps in various oilseed rape pests in eastern Austria. *Proc Symp Integrated Pest Management in Oilseed Rape, 3–5 April 2006, Göttingen, Germany.*

Laborius A (1972) Untersuchungen über die Parasitierung des Kohlschotenrüsslers (*Ceuthorrhynchus assimilis* Payk.) und der Kohlschotengallmücke (*Dasyneura brassicae* Winn.) in Schleswig-Holstein. *Z Angew Entomol* 72: 14–31.

Lehmann W (1965) Einfluss chemischer Bekämpfungsmassnahmen auf einige Rapsschädlinge und ihre Parasiten. II. Knospen- und Stengelschädlinge. *Arch Pflanzenschutz* 1: 209–219.

Linz B (1991) Hymenopteren als Parasitoide des Rapsglanzkäfers (*Meligethes* spp., Nitidulidae) und des Kohlschotenrüsslers (*Ceutorhynchus assimilis*, Curculionidae) auf Raps und Rübsen. *Mitt Dtsch Ges Allg Angew Ent* 8: 90–92.

Luik A, Veromann E, Kevväi R, Kruus M (2006) A comparison of the pests, parasitoids and predators on winter and spring oilseed rape crops in Estonia. *Proc Symp Integrated Pest Management in Oilseed Rape, 3–5 April 2006, Göttingen, Germany.*

Meuche A (1940) Untersuchungen am Rapserdfloh (*Psylliodes chrysocephala* L.) in Ostholstein. *Z Angew Entomol* 27: 464–495.

Miczulski B (1967) Błonkówki (*Hymenoptera*) w biocenozie upraw rzepaku. Część zesc III. Meczelkowate (Braconidae) i mszycarzowate (Aphidiidae). *Polskie Pismo Entomol* 37: 167–191.

Murchie AK (1996) *Parasitoids of cabbage seed weevil and brassica pod midge in oilseed rape.* PhD thesis, University of Keele, UK.

Murchie A, Polaszek KA, Williams IH (1999) *Platygaster subuliformis* (Kieffer) (Hym., Platygastridae) new to Britain, an egg-larval parasitoid of the brassica pod midge *Dasineura brassicae Winnertz* (Dipt., Cecidomyidae). *Entomol Monthly Magazine* 135: 217–222.

Neumann N, Ulber B (2006) Adult activity and larval abundance of stem weevils and their parasitoids at different crop densities of oilseed rape. *IOBC/wprs Bull* 29(7): 193–199.

Nilsson C (1985) Impact of ploughing on emergence of pollen beetle parasitoids after hibernation. *Z Angew Entomol* 100: 302–308.

Nilsson C (1989) The pollen beetle (*Meligethes aeneus* F.) in winter and spring rape at Alnarp 1976–1978. III. Mortality factors. *Växtskyddsnotiser* 52: 145–150.

Nilsson C (1994) Pollen beetles (*Meligethes* spp.) in oil seed rape crops (*Brassica napus* L.): Biological interactions and crop losses. *Dept Plant Protection Sciences, SLU. Dissertations* 1: 39p.

Nilsson C (2003) Parasitoids of the pollen beetles. In: Alford DV (ed.) *Biocontrol of oilseed rape pests*. Blackwell, Oxford, UK.

Nilsson C, Andreasson B (1987) Parasitoids and predators attacking pollen beetles (*Meligethes aeneus* F.) in spring and winter rape in southern Sweden. *IOBC/wprs Bull* 10(4): 64–73.

Nissen U (1997) *Ökologische Studien zum Auftreten von Schadinsekten und ihren Parasitoiden an Winterraps norddeutscher Anbaugebiete*. PhD thesis Christian-Albrechts-University of Kiel, Kiel, Germany.

Nitzsche O (1998) *Auftreten und Effizienz von Parasitoiden als natürliche Gegenspieler von Schadinsekten im Winterraps unter besonderer Berücksichtigung unterschiedlicher Bodenbearbeitungsmassnahmen nach Winterraps*. PhD thesis Georg-August-University of Göttingen, Göttingen, Germany.

Osborne P (1960) Observations on the natural enemies of *Meligethes aeneus* (F.) and *M. viridescens* (F.) (Col. Nitidulidae). *Parasitology* 50: 91–110.

Šedivý J (1983) Tersilochinae as parasitoids of insect pests of winter rape (Hymenoptera: Ichneumonidae). *Contributions Amer Entomol Inst* 20: 266–276.

Ulber B (2000) Bibliography of parasitoid species and levels of parasitism of rape stem weevil *Ceutorhynchus napi* Gyll. and cabbage stem weevil *Ceutorhynchus pallidactylus* (Mrsh.). *IOBC/wprs Bull* 23(6): 131–134.

Ulber B (2003) Parasitoids of ceutorhynchid stem weevils. In: Alford DV (ed.) *Biocontrol of oilseed rape pests*. Blackwell, Oxford, UK.

Ulber B, Eickermann M, Mennerich D (2006a) The parasitism of ceutorhynchid stem weevils on *Brassica* host plants. *Proc Symp on Integrated Pest Management in Oilseed Rape, 3–5 April 2006, Göttingen, Germany*.

Ulber B, Fischer K, Klukowski Z, Luik A, Veromann E, Nilsson C, Ahman B, Williams IH, Ferguson AW, Piper R, Barari H (2006b) Identity of parasitoids and their potential for biocontrol of oilseed rape pests in Europe. *Proc Symp on Integrated Pest Management in Oilseed Rape, 3–5 April 2006, Göttingen, Germany*.

Ulber B, Nitzsche O (2006) Phenology of parasitoids (Hym., Ichneumonidae-Tersilochinae) of oilseed rape pests in northern Germany in 1995–1997. *IOBC/wprs Bull* 29(7): 173–179.

Ulber B, Vidal S (1998) Influence of host density and host distribution on parasitism of *Ceutorhynchus assimilis* by *Trichomalus perfectus*. *IOBC/wprs Bull* 21(5): 185–195.

Ulber B, Wedemeyer R (2004) Incidence of larval parasitism of *Psylliodes chrysocephala* within oilseed rape crops in Germany. *IOBC/wprs Bull* 27(10): 275–280.

Ulber B, Williams IH (2003) Parasitoids of flea beetles. In: Alford DV (ed.) *Biocontrol of oilseed rape pests*. Blackwell, Oxford, UK.

Veromann E, Luik A, Kevväi R (2006a) Oilseed rape pests and their parasitoids in Estonia. *IOBC/wprs Bull* 29(7): 165–172.

Veromann E, Luik A, Kevväi R (2006b) Dynamics of parasitoids in oilseed rape crops in Estonia. *Proc Symp on Integrated Pest Management in Oilseed Rape, 3–5 April 2006, Göttingen, Germany*.

Veromann E, Luik A, Metspalu L, Williams I (2006c) Key pests and their parasitoids on spring and winter oilseed rape in Estonia. *Entomol Fenn* 17: 400–404.

Veromann E, Tarang T, Kevväi R, Luik A, Williams I (2006d) Insect pests and their natural enemies on spring oilseed rape in Estonia: Impact of cropping systems. *Agr Food Sci* 15: 61–72.

Williams IH (2003a) Parasitoids of the cabbage seed weevil. In: Alford DV (ed.) *Biocontrol of oilseed rape pests*. Blackwell, Oxford, UK.

Williams IH (2003b) Parasitoids of the brassica pod midge. In: Alford DV (ed.) *Biocontrol of oilseed rape pests*. Blackwell, Oxford, UK.

Williams IH (2006a) Integrating biological control within IPM for winter oilseed rape in Europe: An overview of the MASTER project. *Proc Symp on Integrated Pest Management in Oilseed Rape, 3–5 April 2006, Göttingen, Germany*.

Williams IH (2006b) Integrating parasitoids into management of pollen beetle on oilseed rape. *Agron Res* 4(Special Issue): 465–470.

Williams IH, Buechi R, Ulber B (2003) Sampling, trapping and rearing oilseed rape pests and their parasitoids. In: Alford DV (ed.) *Biocontrol of oilseed rape pests*. Blackwell, Oxford, UK.

Williams IH, Büchs W, Hokkanen H, Menzler-Hokkanen I, Johnen A, Klukowski Z, Luik A, Nilsson C, Ulber B (2005) MASTER – Integrating biological control within IPM for winter oilseed rape across Europe. *Proc BCPC Int Cong, Crop Science & Technology, Glasgow, 31 October–2 November 2005,* 1: 301–308.

Williams IH, Frearson DJT, Barari H, McCartney A (2007) First field evidence that parasitoids use upwind anemotaxis for host-habitat location. *Entomol Exp Appl* 123: 299–307.
Williams IH, Walton M (1990) A bibliography of the parasitoids of the brassica pod midge (*Dasineura brassicae* Winn.). *IOBC/wprs Bull* 13(4): 46–52.
Wyrostkiewicz K, Blazejewska A (1985) *Isurgus heterocerus* Thoms. *I. morionellus* Holmgr. (Hym: Ichneumonidae) parazytoidy larw slodyszka rzepakowego – *Meligethes aeneus* F. (Col.: Nitidulidae). *Polskie Pismo Entomol* 55: 391–404.

Chapter 3
Key Parasitoids of the Pests of Oilseed Rape in Europe: A Guide to Their Identification

Andrew W. Ferguson, Ingrid H. Williams, Lynda M. Castle, and Matthew Skellern

Abstract The six major pests of oilseed rape, namely the cabbage stem flea beetle, the cabbage stem weevil, the rape stem weevil, the pollen beetle, the cabbage seed weevil and the brassica pod midge, are reported to host at least 80 species of hymenopteran parasitoids. Of these, 12 key species are widespread and abundant on oilseed rape crops throughout Europe; these species are important for conservation biocontrol of the pests on winter rape. This guide aims to help in their identification. It collates information from previously published keys, other literature and recent extensive examination of specimens and highlights selected key features of relevant taxa from superfamily to species. It is liberally illustrated with figures, newly-drafted and redrawn and/or modified from published literature. A glossary of terms is provided.

3.1 Introduction

This illustrated guide is intended as an aid to the identification of the species of hymenopteran parasitoid most important for conservation biocontrol of the pests of oilseed rape in Europe.

The six major pests of oilseed rape, namely the cabbage stem flea beetle, the cabbage stem weevil, the rape stem weevil, the pollen beetle, the cabbage seed weevil and the brassica pod midge (Table 3.1) are host to at least 80 species of hymenopteran parasitoid from 15 different families. A systematic list and classification of all species is given in Ulber et al. (Chapter 2 this volume); their life-histories, status, and importance are reviewed in Alford (2003). The majority belong to six hymenopteran families: the Ichneumonidae (12 species), the Braconidae (14 species), the Pteromalidae (11 species), the Eulophidae (10 species), the Platygastridae (17 species) and the Proctotrupidae (1 species).

A.W. Ferguson (✉)
Plant and Invertebrate Ecology Department, Rothamsted Research, Harpenden, AL5 2JQ, Hertfordshire, UK
e-mail: andrew.ferguson@bbsrc.ac.uk

I.H. Williams (ed.), *Biocontrol-Based Integrated Management of Oilseed Rape Pests*,
DOI 10.1007/978-90-481-3983-5_3,

Table 3.1 Key parasitoids of the six major pests of oilseed rape in Europe

Host	Parasitoid	Family
Cabbage stem flea beetle, *Psylliodes chrysocephala* (Linnaeus)	*Tersilochus microgaster* (Szépligeti)	Ichneumonidae
Cabbage stem weevil, *Ceutorhynchus pallidactylus* (Marsham)	*Tersilochus obscurator* Aubert	Ichneumonidae
Rape stem weevil, *Ceutorhynchus napi* Gyllenhal	*Tersilochus fulvipes* (Gravenhorst)	Ichneumonidae
Pollen beetle, *Meligethes aeneus* (Fabricius)	*Phradis interstitialis* (Thomson)	Ichneumonidae
	Phradis morionellus (Holmgren)	Ichneumonidae
	Tersilochus heterocerus Thomson	Ichneumonidae
	Diospilus capito (Nees)	Ichneumonidae
Cabbage seed weevil, *Ceutorhynchus obstrictus* (Marsham) syn. *C. assimilis* (Paykull)	*Trichomalus perfectus* (Walker)	Pteromalidae
	Stenomalina gracilis (Walker)	Pteromalidae
	Mesopolobus morys (Walker)	Pteromalidae
Brassica pod midge, *Dasineura brassicae* (Winnertz)	*Platygaster subuliformis* (Kieffer)	Platygastridae
	Omphale clypealis (Thomson)	Eulophidae

Twelve species of parasitoid are sufficiently widespread and abundant on oilseed rape crops throughout Europe to be of key economic importance for conservation biocontrol of pests (Ulber et al. Chapter 2 thisvolume, Table 3.1). They vary in importance with country and season but frequently exceed 50% host parasitism. All are larval endoparasitoids, except those of the cabbage seed weevil, which are larval ectoparasitic.

Hymenopteran parasitoids are difficult to identify to species. Taxonomic literature is widely dispersed, few voucher specimens are readily available and several genera have been recently revised. Consequently, many species records, particularly those in the older literature, are suspect due to possible erroneous identification (Alford 2003).

This guide collates selected information from many different sources, but in particular from the published keys and other publications by Delucci and Graham (1956), Graham (1959, 1963, 1969), Rosen (1964), Askew (1968), Horstmann (1971, 1981), Medvedev (1978), Vlug (1985), Tobias et al. (1986), Bouček and Rasplus (1991), Achterberg (1993), Achterberg and Quicke (2000), Goulet and Huber (1993), Murchie et al. (1999), Noyes (2000), Noyes et al. (2000), Fitton et al. (2000), Vidal (2003), Barari et al. (2005), and Gibson et al. (2005). Additional expertise and information was obtained by examination of specimens (over 39,000) of the key parasitoid species collected from crops of oilseed rape during 2001–2005 in Estonia, Germany, Sweden, Poland and the UK, during the course of the EU-funded project MASTER (QLK5-CT-2001-01447) and through consultation with taxonomic authorities (see Acknowledgements). Voucher specimens of the key species are deposited in collections of partner organisations of the MASTER project (see www.rothamsted.bbsrc.ac.uk/pie/master/master.htm). The guide is liberally

Table 3.2 Systematic list and classification of species of the family Ichneumonidae reported to be parasitic on pests of oilseed rape in Europe

	Host(s)
ORDER HYMENOPTERA	
Superfamily ICHNEUMONOIDEA	
Family ICHNEUMONIDAE	
Subfamily Phygadeuontinae	
Stibeutes Förster	
– *curvispina* (Thomson)	Rape stem weevil
Subfamily Tersilochinae	
Aneuclis Förster	
– *incidens* (Thomson)[†]	Pollen beetle
– *melanaria* (Holmgren)[†]	Cabbage stem flea beetle; cabbage seed weevil
(= *diversus* Szépligeti)	
(= *petiolaris* Szépligeti)	
Phradis Holmgren	
– *interstitialis* (Thomson)[*,†]	Pollen beetle
– *morionellus* (Holmgren)[*,†]	Pollen beetle
Tersilochus Holmgren (= *Tersilochus* Holmgren)	
(= *Thersilochus* Holmgren)	
– *fulvipes* (Gravenhorst)[a,*,†]	Rape stem weevil
(= *gibbus* Holmgren)	
– *heterocerus* Thomson[*,†]	Pollen beetle
– *microgaster* (Szépligeti)[b,*,†]	Cabbage stem flea beetle
– *obscurator* Aubert[c,*,†]	Cabbage stem weevil
– *stenocari* (Gregor)[†]	Rape winter stem weevil (*Ceutorhynchus picitarsis*)
– *triangularis* (Gravenhorst)[†]	[not parasitic on rape pests]
– *tripartitus* Brischke[d,†]	Cabbage stem flea beetle

Key species are marked*. Species included in this guide are marked[†].
[a]Also cited in the literature as *Porizon fulvipes* (Gravenhorst) and as *Thersilochus fulvipes* (Gravenhorst) ssp. *gallicator* Aubert.
[b]Also cited in the literature as *Isurgus microgaster* Szépligeti.
[c]Also cited in the literature as *Thersilochus tripartitus* Brischke spp. *obscurator* Aubert.
[d]Possibly a misidentification of *Tersilochus microgaster* (Szépligeti)[†].

illustrated with figures, many new and, where indicated, redrawn and/or modified from the above-listed keys. Examination of vouched specimens of species in conjunction with this guide will aid correct identification. Additional characters to aid identification will be found in the above-mentioned keys.

The guide is arranged in order of the families: Ichneumonidae and Braconidae, the Pteromalidae, the Eulophidae, the Platygastridae and the Proctotrupidae. It lists the key species that are associated with rape pests in each of these families and presents the key characters of the main taxa to which the key species belong, from superfamily to genus (Tables 3.2, 3.3, 3.4, 3.5, and 3.6). The guide is not intended to be a comprehensive taxonomic key but seeks to highlight features that will help to distinguish the key species from others that may be reared from the pests or may be found in oilseed rape crops.

Table 3.3 Systematic list and classification of species of the family Braconidae reported to be parasitic on pests of oilseed rape in Europe

	Host(s)
ORDER HYMENOPTERA	
Superfamily ICHNEUMONOIDEA	
Family BRACONIDAE	
Subfamily Cheloninae	
Sigalphus Latreille	
– *obscurellus* Nees	Rape winter stem weevil; cabbage seed weevil
Subfamily Doryctinae	
Bracon Fabricius	
– *fulvipes* Nees	Cabbage seed weevil
– *variator* Nees	Cabbage seed weevil
(= *discoideus* Wesmael)	
(= *maculiger* Wesmael)	
Subfamily Euphorinae	
Microctonus Wesmael	
– *areolatus* Thomson	Cabbage flea beetle (*Phyllotreta nemorum*)
– cf. *deceptor* Wesmael	Cabbage seed weevil
– *melanopus* Ruthe[a]	Cabbage stem flea beetle; rape winter stem weevil; cabbage seed weevil
– *vittatae* Muesbeck	Cabbage flea beetle (*Phyllotreta nemorum*)
Townesilitus Haeselbarth and Loan	
– *bicolor* (Wesmael)	Cabbage flea beetle (*Phyllotreta nemorum*)
Subfamily Helconinae	
Blacus Nees	
– *nigricornis* Haeselbarth†	Pollen beetle
Diospilus Haliday	
– *capito* (Nees)*,†	Pollen beetle
– *morosus* Reinhardt†	Cabbage stem flea beetle; cabbage seed weevil
– *oleraceus* Haliday†	Cabbage stem flea beetle; rape winter stem weevil; cabbage seed weevil
Eubazus Nees (= *Calyptus* Haliday)	
– *sigalphoides* (Marshall)	Pollen beetle
Taphaeus Wesmael	
– *affinis* Wesmael	Cabbage seed weevil
– *tidius* (Walker)	Cabbage seed weevil

Key species are marked*. Species included in this guide are marked†.
[a]Also cited in the literature as *Perilitus melanopus* Ruthe.

3.2 Key Characters of the Order Hymenoptera, Suborder Apocrita and Superfamilies Ichneumonoidea, the Chalcidoidea, the Platygastroidea and the Proctotrupoidea

The most distinctive and constant feature of the order Hymenoptera is the fusion of the first abdominal segment (the propodeum) with the thorax to form the mesosoma. All other abdominal segments form the metasoma (Fig. 3.1).

Table 3.4 Systematic list and classification of species of the family Pteromalidae reported to be parasitic on pests of oilseed rape in Europe

	Host(s)
ORDER HYMENOPTERA	
Superfamily CHALCIDOIDEA	
Family PTEROMALIDAE	
Subfamily Pteromalinae	
Anisopteromalus Ruschka	
– *calandrae* (Howard)	Cabbage seed weevil
Chlorocytus Graham	
– *diversus* (Walker)	Cabbage seed weevil
Habrocytus Thomson	
– *dispar* (Curtis)	Cabbage seed weevil
– *semotus* (Walker)	Cabbage seed weevil
Mesopolobus Westwood (= *Amblymerus* Walker) (= *Eutelus* Walker) (= *Xenocrepis* Förster)	
– *morys* (Walker)*,† (= *pura* Mayr)	Cabbage seed weevil
Stenomalina Ghesquière	
– *gracilis* (Walker)[a],*,†	Cabbage seed weevil Rape stem weevil
Trichomalus Thomson	
– *lucidus* (Walker)†	Cabbage stem weevil
– *perfectus* (Walker)*,† (= *decisus* Walker) (= *decorus* (Walker)) (= *laevinucha* (Thomson))	Cabbage seed weevil
Zatropis Crawford	
– sp.	Cabbage seed weevil

Key species are marked*. Species included in this guide are marked†.
[a]Misidentified as *S. muscarum.*

The Hymenoptera are divided into two suborders, the Symphyta (sawflies) and the Apocrita, which comprises the majority of species. In the Apocrita, the first (sometimes the first and second) segment(s) of the metasoma are constricted to form the petiole, a narrow 'waist' or stalk joining the gaster (remaining segments of the metasoma) with the propodeum (Fig. 3.1).

The Apocrita are divided into two main groups, the Aculeata (ants, bees and wasps) and the Parasitica, small wasps whose larvae are usually ecto- or endoparasitic on other insects. The species of Parasitica which are parasitoids of the larvae of the major pests of oilseed rape belong to the four superfamilies: the Ichneumonoidea, the Chalcidoidea, the Platygastroidea and the Proctotrupoidea.

3.2.1 Key Characters of the Superfamily Ichneumonoidea

The superfamily Ichneumonoidea comprises two families: the Ichneumonidae and the Braconidae.

Table 3.5 Systematic list and classification of species of the family Eulophidae reported to be parasitic on pests of oilseed rape in Europe

	Host(s)
ORDER HYMENOPTERA	
Superfamily CHALCIDOIDEA	
Family EULOPHIDAE	
Subfamily Entodoninae	
Neochrysocharis Kurdjumov	
– sp.†	Brassica pod midge
Omphale Haliday (= Secodes Förster)	
– *clypealis* (Thomson)*,†	Brassica pod midge
– *coilus* (Walker)	Brassica pod midge
Subfamily Eulophinae	
Eulophus Müller	
– sp.	Cabbage seed weevil
Subfamily Tetrastichinae	
Aprostocetus Westwood	
– *epicharmus* (Walker)† (= *variegatus* Szelényi)	Brassica pod midge
Necremnus Thomson	
– *tidius* (Walker) (= *duplicatus* Gahan)	Cabbage seed weevil
– *leucarthros* (Nees)	Brassica pod midge
Sigmophora Rondan	
– *brevicornis* (Panzer)	Brassica pod midge
Tetrastichus Haliday	
– *galectobus* (Ratzeburg)	Cabbage seed weevil

Key species is marked*. Species included in this guide are marked†.

1. Basal (head) end of metasoma constricted to form the petiole (Fig. 3.1).
2. Forewing with no costal cell and at least one closed cell which may be open at its basal end or very narrow (Fig. 3.2, not Fig. 3.3).
3. Antennae with 14 or more segments (including scape, pedicel and flagellum) (as in Fig. 3.14).

3.2.2 Key Characters of the Superfamily Chalcidoidea

The superfamily Chalcidoidea is a diverse group with a wide range of hosts. It is the most important group of parasitic Hymenoptera in applied biocontrol (Noyes 1985). It comprises 20 families, including the Pteromalidae and the Eulophidae.

1. Body commonly metallic, often strongly so.
2. Length usually $\leq$ 5 mm, but some > 20 mm.
3. Forewings with no cells enclosed by tubular veins (as in Fig. 3.26).
4. Antennae almost always < 15 segments (including scape, pedicel and flagellum), rarely > 13. Scape elongate and elbowed in appearance, as in ants. Flagellum, especially in females, differentiated into funicle and clava. Clava conspicuously larger than any preceding segment and composed of two or more fused segments. Longitudinal sensilla present on at least one flagellar segment, with

their distal apices free, separated from the cuticle (unique to Chalcidoidea) (as in Fig. 3.27).

5. Prepectus present and separating tegula and pronotum so that they do not touch (as in Fig. 3.28a).

Table 3.6 Systematic list and classification of species of the family Platygastridae reported to be parasitic on pests of oilseed rape in Europe

ORDER HYMENOPTERA
Superfamily PLATYGASTROIDEA
Family PLATYGASTRIDAE
Amblyaspis Förster
– sp.
Inostemma Haliday
– *boscii* (Jurine)
– *walkeri* Kieffer
– nr. *reticulatum* (Szelényi)
Isocybus Förster
– *thomsoni* Kieffer
Piestopleura Förster
– sp.
Platygaster Latreille (= *Prosactogaster* Kieffer)
– *boscii* Nees
– *gladiator* Zetterstedt
– *iolas* Walker†
– *munita* Walker
– *niger* Nees
– *nitida* (Thomson)
– *oebalus* Walker†
– *subuliformis* (Kieffer)*,†
– *tisias* Walker†
Synopeas Förster
– nr. *lugubris* Thomson
– sp.

All are parasitoids of brassica pod midge. Key species is marked*. Species included in this guide are marked†.

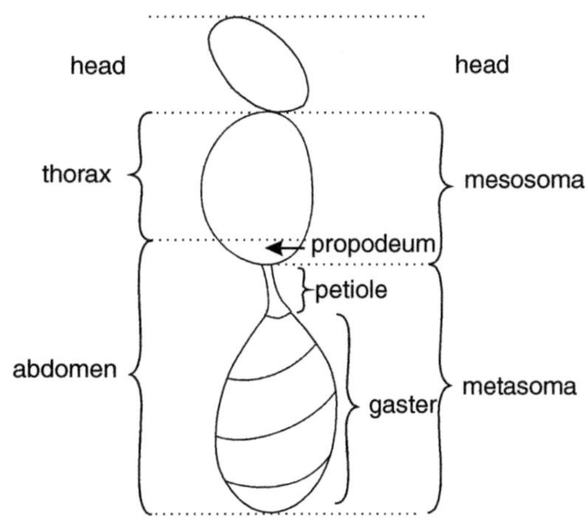

Fig. 3.1 Morphological divisions of the body of the suborder Apocrita (order Hymenoptera)

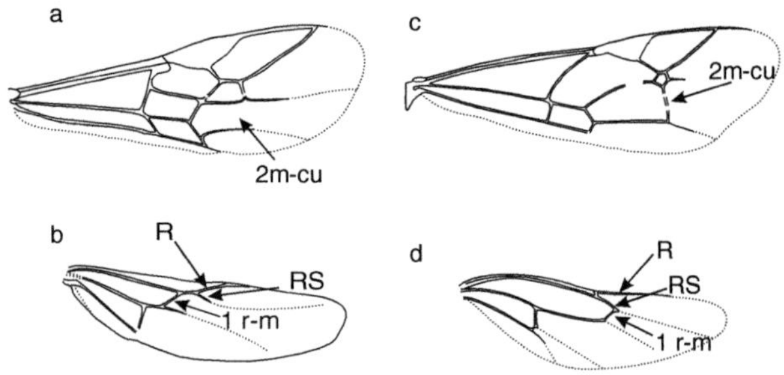

Fig. 3.2 Forewings and hindwings of (**a**) and (**b**) Braconidae, and (**c**) and (**d**) Ichneumonidae (redrawn after van Achterberg and Quicke 2000). Letters indicate the names of veins. Vein 2m-cu is absent in the forewing of Braconidae

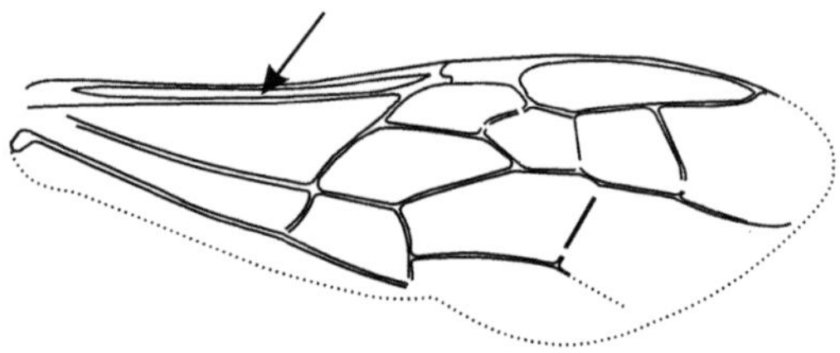

Fig. 3.3 A hymenopteran forewing with costal cell (*arrowed*) (redrawn after Goulet and Huber 1993)

3.2.3 Key Characters of the Superfamily Platygastroidea

The superfamily Platygastroidea comprises two families, the Scelionidae and the Platygastridae.

1. Body rarely metallic
2. Small (forewings 0.5–6.0 mm)
3. Forewing without cells enclosed by tubular veins, many species with no wing veins (as in Fig. 3.49).
4. Ovipositor weakly sclerotised and completely retracted within the metasoma when not in use.
5. Gaster <seven visible tergites. Moderately well sclerotised and dorso-ventrally compressed, sometimes with morphological adaptations (horns, sacks, humps, elongation) to accommodate the length of the ovipositor.
6. Antennae $\leq$12 segments.

3.2.4 Key Characters of the Superfamily Proctotrupoidea

The superfamily Proctotrupoidea is a diverse group including nine families. Most species belong to the families Diapriidae and Proctotrupidae.

1. Non-metallic.
2. Morphologically diverse.

3. Forewing usually with a closed costal cell (as in Fig. 3.54).
4. Strongly sclerotised.

3.3 Parasitoids of the Families Ichneumonidae and Braconidae

The family Ichneumonidae has 39 subfamilies, two of which include parasitoids of rape pests: the Phygadeuontinae and the Tersilochinae. The Phygadeuontinae is a little studied group and there are no good keys to their identification (Horstmann *pers. comm*); it includes only one species reported to attack rape pests. The Tersilochinae includes 12 species (of which six are key species) reported to attack rape pests, all in the genera *Aneuclis*, *Phradis* and *Tersilochus*. They are all small, univoltine, koinobiont larval endoparasitoids of Coleoptera (Table 3.2).

The family Braconidae also includes several species reported to attack oilseed rape pests (Table 3.3). Only one, *D. capito*, is sufficiently widespread and abundant to be considered a key species for biocontrol.

3.3.1 Key Characters of the Family Braconidae

1. Forewing vein 2m-cu absent (Fig. 3.2a). Hindwing vein 1 r-m (also known as rs-m) branches before veins R and RS divide (Fig. 3.2b).
2. Second and third metasomal tergites fused, with two pairs of spiracles (Fig. 3.4a).

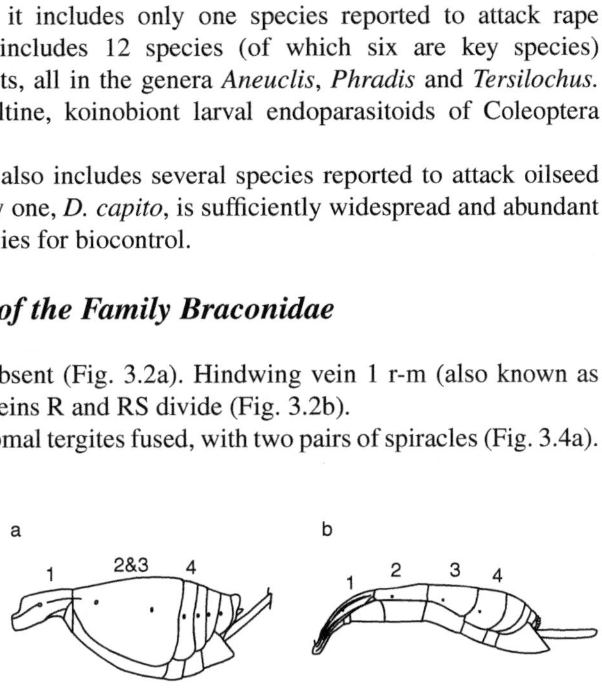

Fig. 3.4 Metasoma of (**a**) Braconidae, and (**b**) Ichneumonidae (redrawn after Goulet and Huber 1993)

Key characters of *Blacus nigricornis*

1. General appearance as in Fig. 3.5.
2. Antennae with 17 (sometimes 18 in males) segments. First two segments (scape and pedicel) rounded and bulbous. Segments after ninth or tenth abruptly shorter (Fig. 3.5).
3. Wing venation with forewing cell 2cu open (Fig. 3.5).
4. Ovipositor long (ca. as long as the hind wing), and gently curved downwards, curve more pronounced towards the tip (Fig. 3.5).
5. Maxillary palps with 6 segments, the fourth longer than the rest; labial palps with three segments.
6. Mesosoma and first metasomal tergite coarsely granulated.

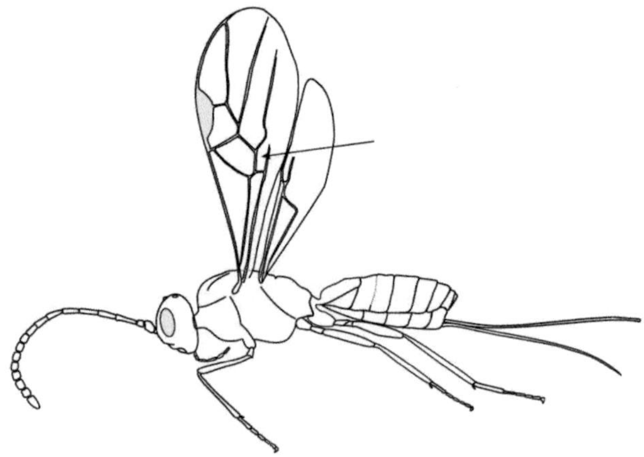

Fig. 3.5 *Blacus nigricornis* ♀. Forewing cell 2cu *arrowed*

Key to *Diospilus spp.*

1. General appearance as in Fig. 3.6.
2. First metasomal tergite slender, subparallel in basal half, largely smooth, may be sculptured at edges (Fig. 3.7a and b). Marginal cell of forewing normal (Fig. 3.8a), not short (Fig. 3.8b)... *D. capito*

Fig. 3.6 *Diospilus capito* ♀

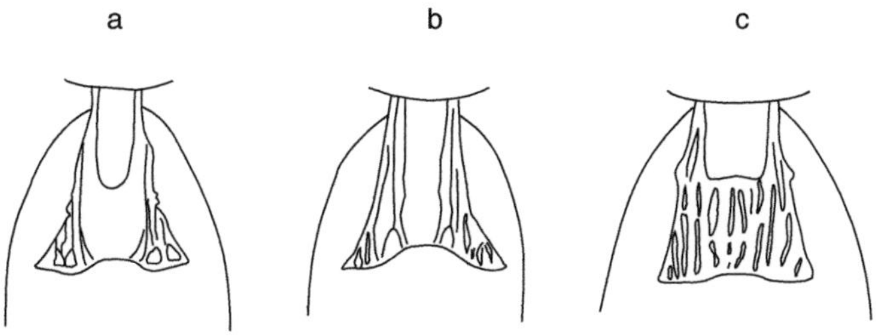

Fig. 3.7 First metasomal tergites of (**a**) and (**b**) two specimens of *Diospilus capito*, (**c**) *Diospilus oleraceus* and *Diospilus morosus*

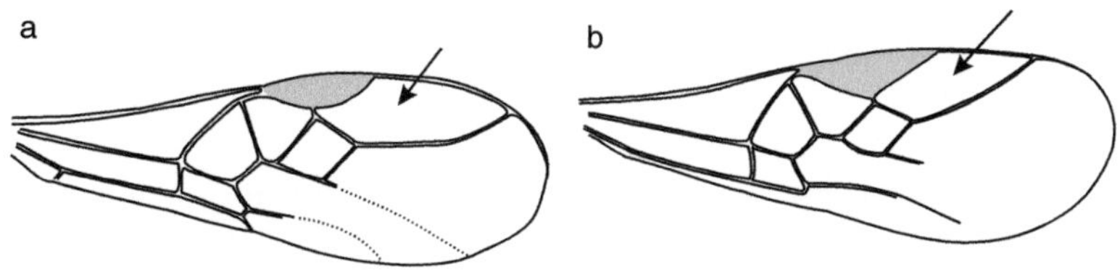

Fig. 3.8 Forewings of (**a**) *Diospilus capito* and *Diospilus oleraceus* and (**b**) *Diospilus morosus* (redrawn after Tobias et al. 1986). Marginal cell *arrowed*

– First metasomal tergite distinctly sculptured, robust and gradually widening towards the apex (Fig. 3.7c)..3

3. Marginal cell of forewing normal (Fig. 3.8a)............................... *D. oleraceus*

– Marginal cell of forewing short (Fig. 3.8b)................................. *D. morosus*

3.3.2 Key Characters of the Family Ichneumonidae

1. Forewing vein 2m-cu present (Fig. 3.2c). Hindwing vein 1 r-m (also known as rs-m) joins vein RS apically to the division of veins R and RS (Fig. 3.3d).
2. Second and third metasomal tergites usually separate and articulated, as indicated by the single pair of spiracles on each tergite (Fig. 3.4b).

3.3.2.1 Key Characters of the Subfamily Tersilochinae

1. General appearance as Fig. 3.9b.
2. First metasomal tergite (on petiole) with spiracles on its posterior half in most genera, including all species parasitic on rape pests (Fig. 3.10a, b), not at, or in front of, the mid-point (not Fig. 3.10c, d).
3. Forewing vein 2m-cu with single fenestra, not two separate fenestra. Pterostigma short and broad. Forewing areolet open (Fig. 3.11a, not b).

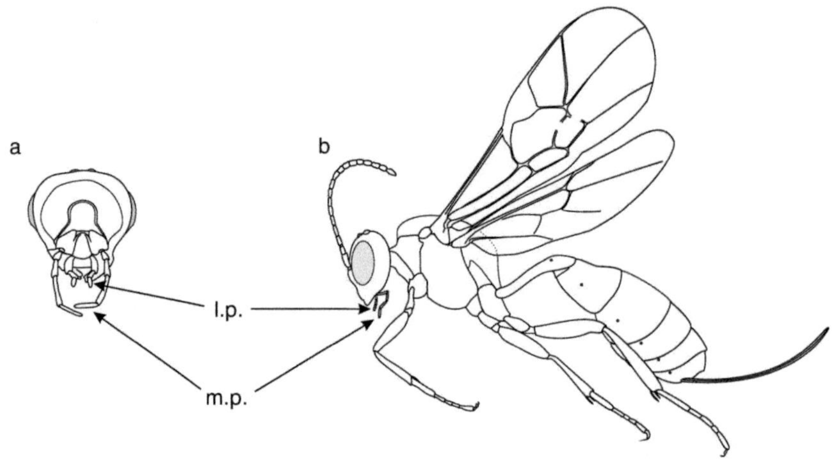

Fig. 3.9 Maxillary (m.p.) and labial palps (l.p.) on (**a**) head, rear view, (**b**) a tersilochine (redrawn after Goulet and Huber 1993)

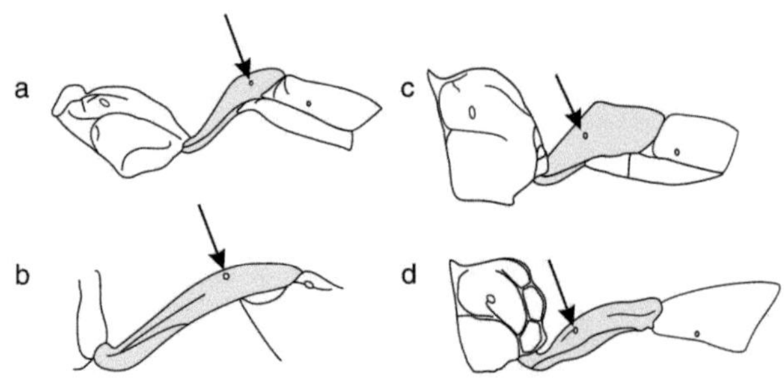

Fig. 3.10 First metasomal tergites (in *grey*): (**a**) and (**b**) as in Tersilochinae; (**c**) and (**d**) not as Tersilochinae (redrawn after Goulet and Huber 1993). *Arrows* indicate spiracles

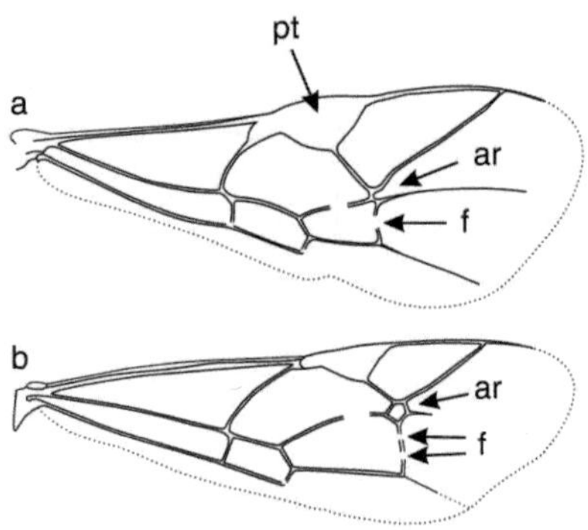

Fig. 3.11 Forewings of Ichneumonidae: (**a**) Tersilochinae, (**b**) not Tersilochinae (redrawn after Goulet and Huber 1993). ar = areolet. f = fenestra. pt = pterostigma

4. Maxillary palps (outer pair around mouth) with four segments, labial palps (inner pair around mouth) with three segments (Fig. 3.9).
5. Lower margin of clypeus with a single comb-like row of regularly-spaced and parallel setae (Fig. 3.12a, not b).

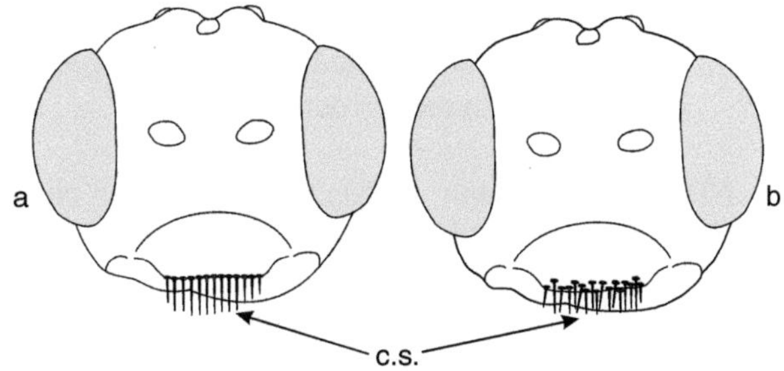

Fig. 3.12 Clypeal setae (c.s.) of (**a**) Tersilochinae, (**b**) other subfamilies (redrawn after Fitton et al. 2000)

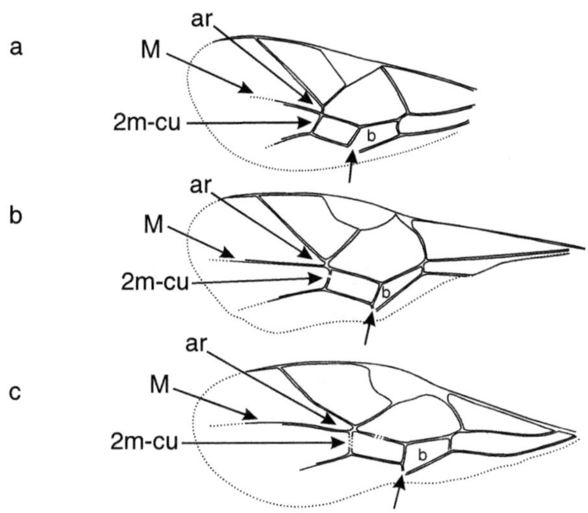

Fig. 3.13 Forewings in (**a**) *Aneuclis* (redrawn after Horstmann 1971), (**b**) *Phradis*, and (**c**) *Tersilochus*. *Arrows* to brachial cells (**b**) indicate their openings. ar = areolet

Key to the Genera *Aneuclis*, *Phradis* and *Tersilochus*

1. Forewing brachial cell wide open (Fig. 3.13a). Forewing vein 2m-cu at least partly pigmented. Ovipositor with a simple curve, not sinuous apically. Head usually granulated...genus *Aneuclis*

 – Forewing brachial cell closed or nearly so (Fig. 3.13b, c). Maxillary palps clearly shorter than height of head (Fig. 3.9b). Forewing vein 2m-cu joins vein M at or after junction of veins forming the areolet, at most only slightly before it (Fig. 3.13b, c)...2

2. Forewing vein 2m-cu leaves vein M at or slightly before (slightly basal to) the junction of veins forming the areolet (Fig. 3.13b).......................genus *Phradis*

 – Forewing vein 2m-cu joins vein M after (apical to) the junction of veins forming the areolet (Fig. 3.13c). Surface of head and thorax completely, or almost completely, dull and granulated. Path of the sternaulus across the mesopleuron (see Fig. 3.24) is indicated by a series of pits or wrinkles, or by more coarse granulation than the surrounding surface.............................. genus *Tersilochus*

NB see Horstmann (1981) for more key characters separating the genus *Tersilochus* from 11 other genera.

Key to *Aneuclis* spp.

The five species of the genus *Aneuclis* are difficult to separate. This key separates only *A. incidens* and *A. melanaria*, the two species that parasitise rape pests.

Key characters of *Aneuclis incidens* ♀♀

1. Antennae with 16 segments.
2. Forewing vein 2m-cu usually joins vein M at the junction of veins forming the areolet (Fig. 3.13a), but rarely it joins vein M before (basal to) or after (apical to) this junction.
3. Sternaulus (see Fig. 3.24) not clearly defined or indicated only by coarser granulation along its path across the mesopleuron.

Key characters of *Aneuclis melanaria* ♀♀

1. Antennae with 18 segments.
2. Forewing vein 2m-cu always joins vein M clearly after (apical to) the junction of veins forming the areolet.
3. Path of sternaulus across the mesopleuron (see Fig. 3.24) delineated by wrinkles

Key to *Phradis* spp.

There are 12 species in the genus *Phradis* but this key separates only *P. morionellus* and *P. interstitialis*, both key parasitoids of the pollen beetle.

Key characters of *Phradis morionellus* ♀♂ (see Fig. 2.1).

1. Antennae with 15–16 (sometimes 17) segments; basal segments of flagellum elongate (Fig. 3.14).
2. Ovipositor incised (notched) dorsally just before tip (Fig. 3.15).
3. Thyridiae approximately triangular and not longer than wide (Figs. 3.16 and 3.17).

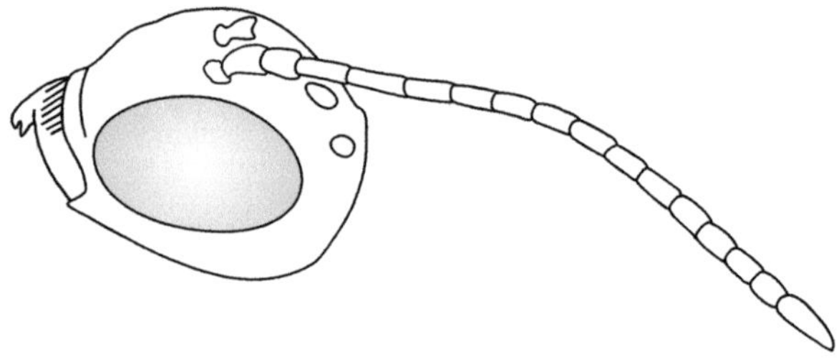

Fig. 3.14 Antenna of *Phradis morionellus*

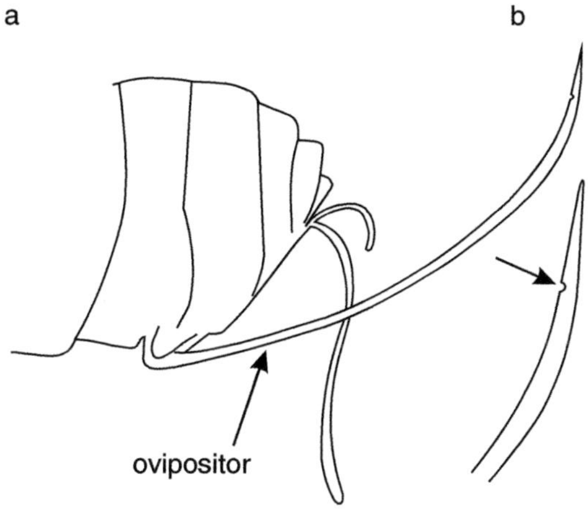

Fig. 3.15 *Phradis morionellus* (**a**) tip of abdomen, and (**b**) tip of ovipositor. Incision *arrowed*

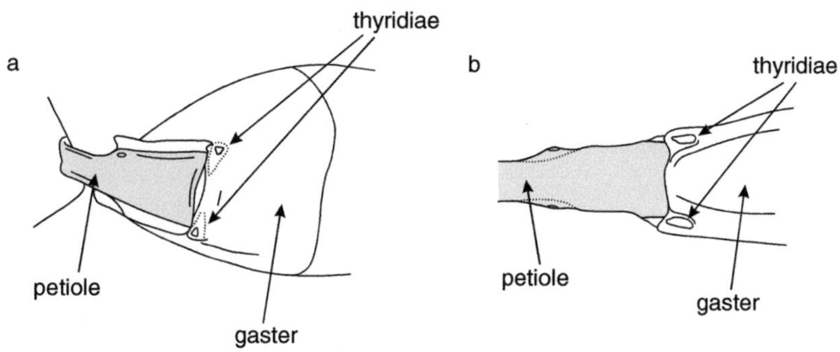

Fig. 3.16 Petiole and gaster of (**a**) *Phradis morionellus*, and (**b**) *Phradis interstitialis*, indicating thyridiae

Fig. 3.17 Shape of the thyridiae in *Phradis morionellus*, indicating their length and width

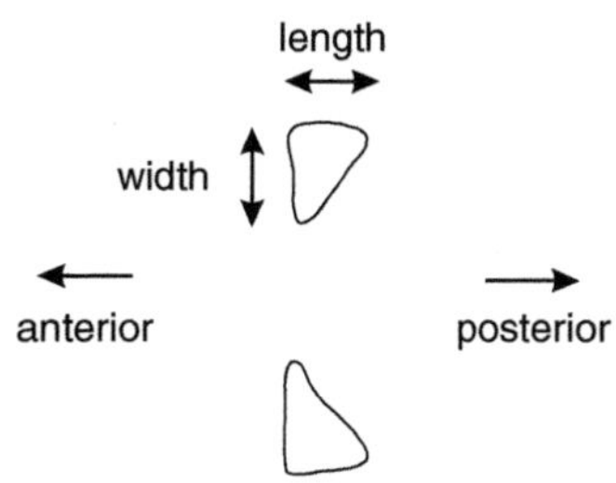

Key characters of *Phradis interstitialis* ♀♂

1. Antennae with 17–18 segments (up to 20 in males).
2. Ovipositor slender, with an upward curve that increases towards the tip and not incised dorsally just before tip (Fig. 3.18, not Fig. 3.15)
3. Thyridiae oval and 1.5–2 times as long as wide (Fig. 3.16b)

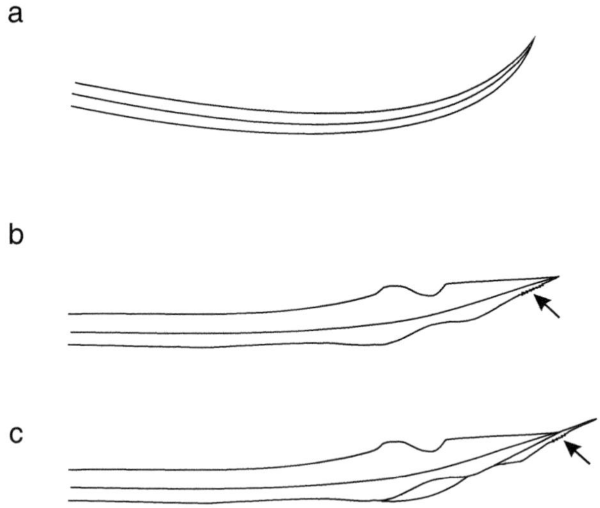

Fig. 3.18 Ovipositor tips of (**a**) *Phradis interstitialis*, and (**b**) and (**c**) *Tersilochus heterocerus*. In (**c**) one ventral component of the ovipositor is extended as when sawing through host cuticle. *Arrows* indicate finely serrated areas

Key to *Tersilochus* spp.

Eight species of the genus *Tersilochus* are reported to attack coleopteran pests of rape; four are key species. Species can be hard to separate. Males of groups *jocator* or *obliquus* cannot be determined to group or to species. Tersilochines of the group *obliquus* are not known to be parasitoids of oilseed rape pests.

1. Ovipositor tip evenly incised dorsally and lacking fine teeth ventrally. Antennae with 24 segments..*Tersilochus stenocari* ♀

   – Ovipositor tip toothed dorsally and with fine or shallow teeth ventrally.. 2

2. Fourth antennal segment shorter than third and fifth segments (Fig. 3.19). Antennae with 16 (sometimes 15 or 17) segments. Ovipositor slightly and smoothly curved upwards, dorsally clearly toothed, ventrally shallowly toothed

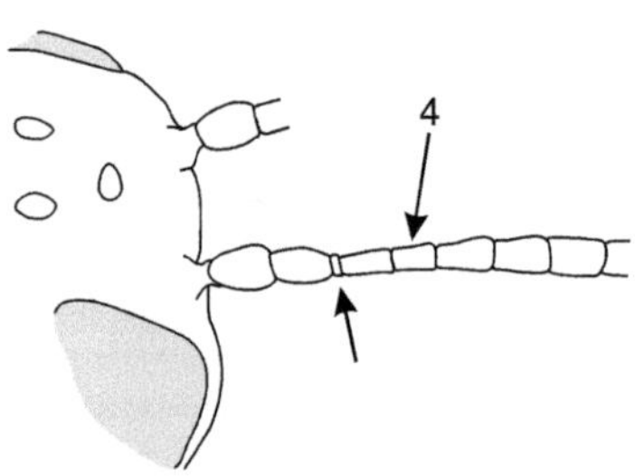

Fig. 3.19 Antenna (proximal segments) of *Tersilochus heterocerus* indicating segment 4. *Arrowed* annulus is not a segment

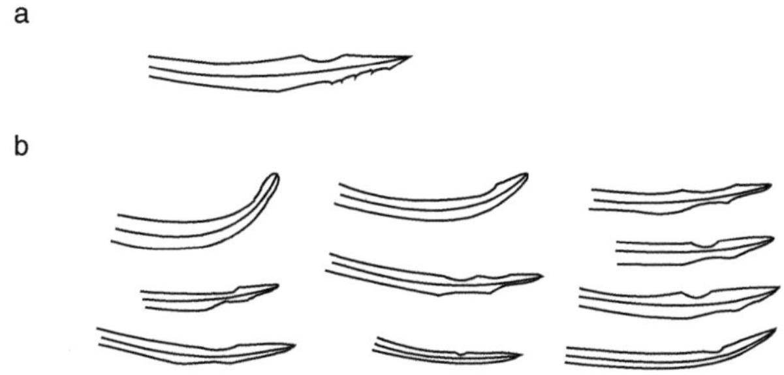

Fig. 3.20 Ovipositor tips of (**a**) *Tersilochus* spp. group *jocator*, and (**b**) *Tersilochus* spp. group *obliquus* (redrawn after Horstmann 1971)

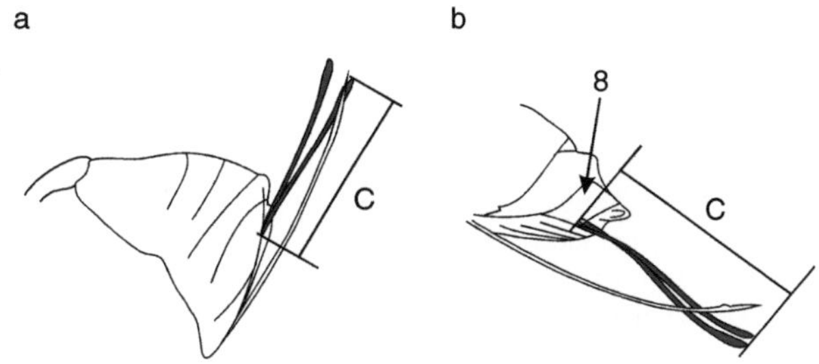

Fig. 3.21 (**a**) and (**b**) ovipositor sheaths of *Tersilochus obscurator* indicating measurement of their length, C. Proximal end of the sheath may be concealed behind the eighth metasomal tergite (*arrowed*) as in (**b**)

and finely serrated close to tip (Fig. 3.18b, c)....... *Tersilochus heterocerus* ♀♂ (see also Fig. 2.2).

– Length of antennal segments decreasing from the third segment onwards.. 3

3. Ovipositor tip with two dorsal teeth and fine teeth ventrally (Fig. 3.20a). Antennae with 21–27 segments .... *Tersilochus* spp. group *jocator* ♀.... 4

– Ovipositor tip convex dorsally; ovipositor variably and shallowly incised distally, both dorsally and sometimes ventrally (Fig. 3.20b). Antennae 19–24 segments .... *Tersilochus* spp. group *obliquus* ♀

4. Length of ovipositor sheaths ≤ length of first metasomal tergite (Figs. 3.21 and 3.22) .... *Tersilochus triangularis* ♀
NB *Tersilochus triangularis* is not known as a parasitoid of rape pests.

– Length of ovipositor sheaths > length of first metasomal tergite 5

5. Ovipositor sheath: first metasomal tergite ratio ('sheath ratio') ≥ 2.0 (Figs. 3.21 and 3.22). Sternaulus weakly defined by a line of pits spanning 25–50% of the mesopleuron (Figs. 3.23 and 3.24), centrally or forward of its centre, and not

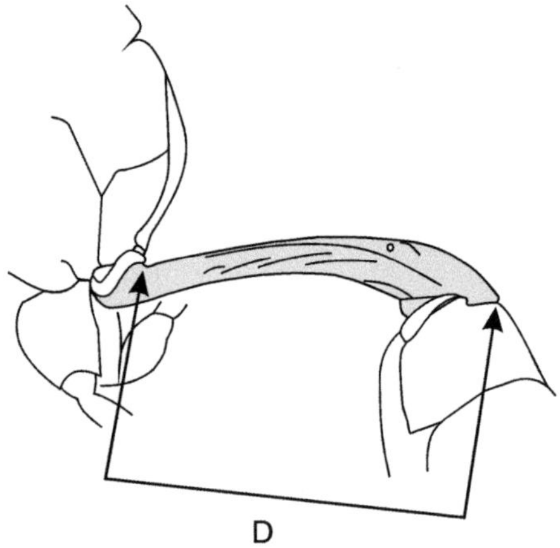

Fig. 3.22 First metasomal segment (= petiole; in *grey*) of *Tersilochus obscurator* indicating measurement of the length, D, of its tergite

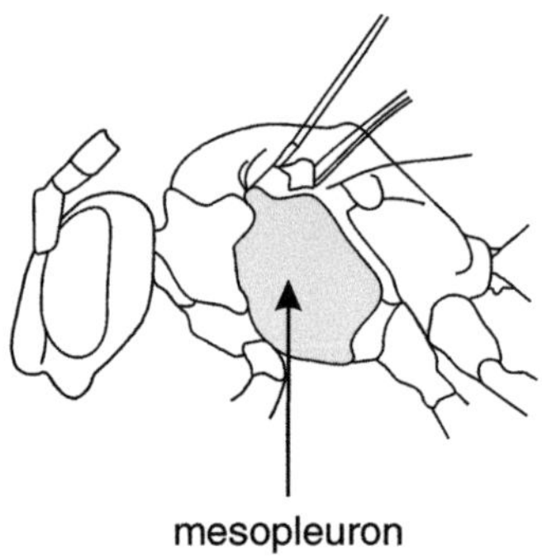

Fig. 3.23 Head and thorax of an hymenopteran indicating the mesopleuron (redrawn after Goulet and Huber 1993)

reaching its anterior nor posterior margins. Antennae with 25 or 26 segments .... *Tersilochus fulvipes* ♀

– Ovipositor sheath: first metasomal tergite ratio ('sheath ratio') ≤ 1.9 (Figs. 3.21 and 3.22).. 6

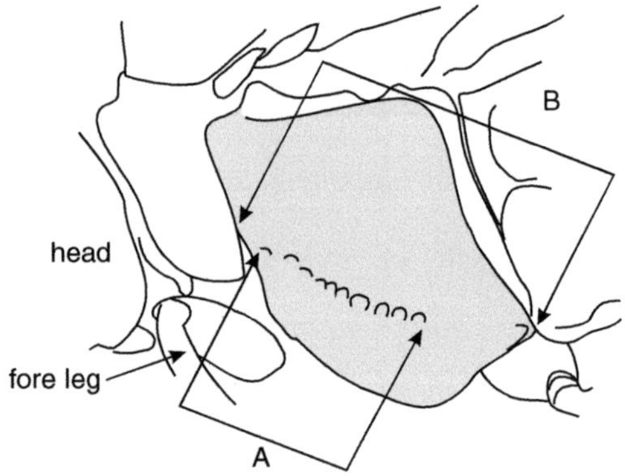

Fig. 3.24 Mesopleuron (in *grey*) of *Tersilochus microgaster* indicating measurement of sternaulus (*B*) and of the line of pits along it (*A*)

6. Ovipositor sheath: first metasomal tergite ratio 1.1–1.5 (Figs. 3.22 and 3.23). Sternaulus clearly defined by a line of pits spanning ≥50% of its path across the mesopleuron (Fig. 3.24)..7

   – Ovipositor sheath: first metasomal tergite ratio 1.4–1.9 (average 1.6) (Figs. 3.21 and 3.22). Sternaulus usually only weakly defined by a line of pits spanning 0–70% (average 40%) of its path across the mesopleuron as measured in Fig. 3.24, the pits often not reaching the anterior margin of the mesopleuron, never reaching its posterior margin. Antennae with 21–26 segments, commonly 23.. *Tersilochus obscurator* ♀

7. Ovipositor sheath: first metasomal tergite ratio 1.1–1.5 (average 1.3) (Figs. 3.21 and 3.22). Sternaulus clearly defined by a line of pits spanning 50–85% (average 70%) of its path across the mesopleuron, sometimes reaching its anterior margin, never reaching its posterior margin (Fig. 3.24). Petiole, measured in the last third of the distance from the mesosoma to the spiracles (Fig. 3.25a, b), broader than high in transverse section, flattened dorsally and with longitudinal ridges (carinae) laterally (Fig. 3.25c). Antennae with 21–23 segments... *Tersilochus microgaster* ♀

   – Sternaulus with line of pits almost across entire mesopleuron Petiole, measured in the last third of the distance from the mesosoma to the spiracles (Fig. 3.25a, b), round in cross-section with only weak longitudinal ridges (carinae) laterally (Fig. 3.25d). Antennae ≤ 24 segments.. *Tersilochus tripartitus* ♀

NB Reports in the literature that *Tersilochus tripartitus* is a parasitoid of the cabbage stem flea beetle may have resulted from erroneous identification of *T. microgaster* (Ulber et al. Chapter 2 this volume).

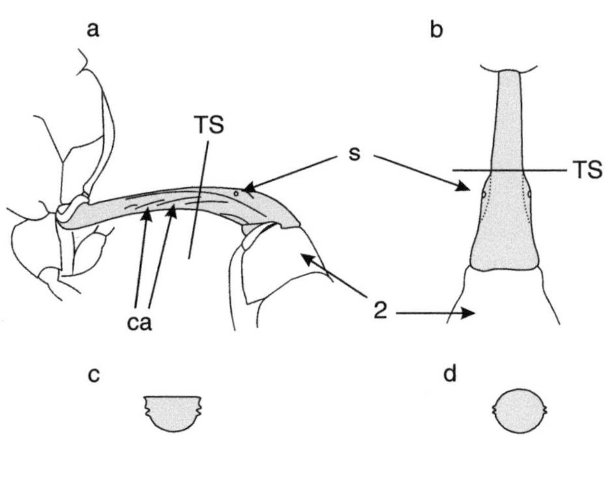

Fig. 3.25 Petioles (in *grey*) of a tersilochine in (**a**) lateral view, and (**b**) dorsal view, to indicate location (TS) of (**c**) and (**d**), transverse sections of (**c**) *Tersilochus microgaster*, and (**d**) *Tersilochus tripartitus*. ca = carinae. s = spiracle. 2 = second metasomal tergite

3.4 Parasitoids of the Family Pteromalidae

The family Pteromalidae is large and varied and species can be hard to separate. It includes nine species known to attack rape pests, including the three key species *Mesopolobus morys*, *Stenomalina gracilis* and *Trichomalus perfectus* (Table 3.4). All three are larval ectoparasitoids of weevils.

3.4.1 Key Characters of the Pteromalidae

1. Head and body metallic colour.
2. Head + body length 2.5–4 mm.
3. Forewings with no cells enclosed by tubular veins, veins represented by creases or lines of hairs. Forewing membrane clear ('hyaline'), not shaded or reticulate. Anterior margin of forewing without conspicuous long dark bristles. Hind-wing normal, not long and stalked (Fig. 3.26b, not c).
4. Antennae with 13 segments, differentiated into scape, pedicel and flagellum. Scape elongate, giving an elbowed appearance as in ants. Flagellar segments differentiated into two or three small anelli, five or six segments in the funicle and three distal segments fused to form the clava. Longitudinal sensilla present on at least one flagellar segment of the antennae with their distal apices free, separated from the cuticle (Fig. 3.27).
5. Mesosoma with prepectus present and clearly defined, separating tegula and pronotum so they do not touch. Prepectus not obviously smaller than tegula in side view. Mesopleuron divided into two parts, the mesepisternum and mesepimeron. Scutellum not conspicuously hairy (Fig. 3.28).

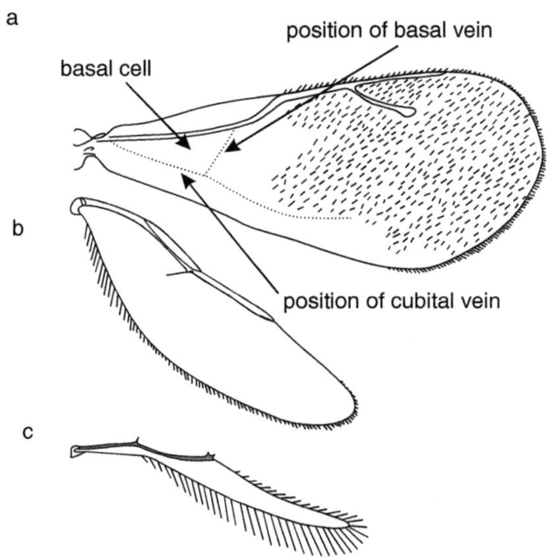

Fig. 3.26 (**a**) Forewing, and (**b**) hind wing of Pteromalidae (modified after Bouček and Rasplus 1991). (**c**) stalked hind wing not found in Pteromalidae (redrawn after Goulet and Huber 1993)

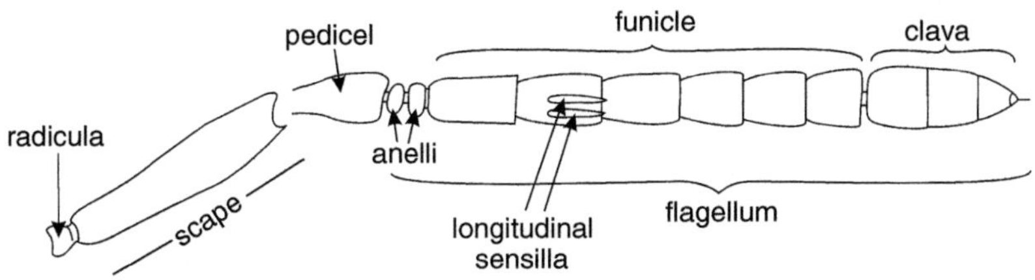

Fig. 3.27 Antenna of Pteromalidae (redrawn after Graham 1969)

Fig. 3.28 Mesosoma (lateral view) of Pteromalidae (redrawn after Bouček and Rasplus 1991)

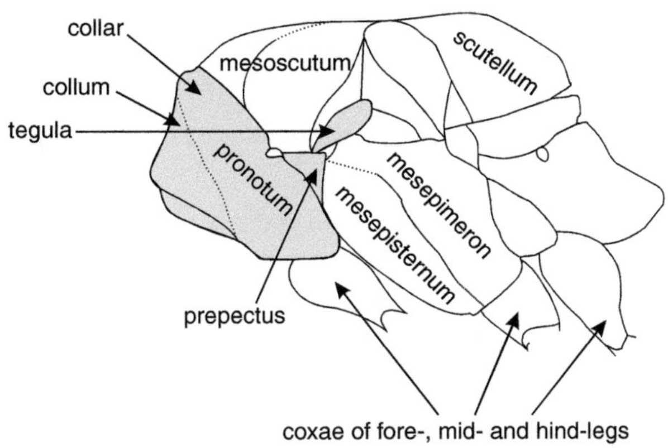

Fig. 3.29 Metasoma (ventral view) of Pteromalidae (redrawn after Bouček and Rasplus 1991)

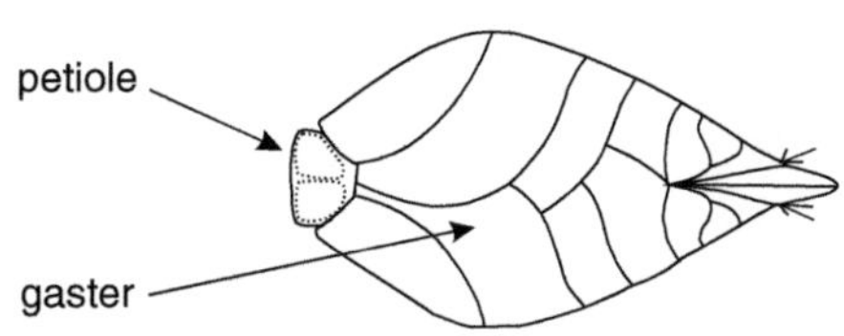

6. Gaster constricted at its junction with the petiole. Petiole small (Fig. 3.29).
7. Legs all with five tarsal segments.

3.4.2 *Key to* Mesopolobus morys, Stenomalina gracilis, Trichomalus perfectus *and* Trichomalus lucidus

1. Antennae with three anelli and five funicular segments (Fig. 3.30).
   .... could be *Mesopolobus morys*
   .... see key characters of *M. morys* below.

   – Antennae with two anelli and six funicular segments (Fig. 3.30b, c)
   2

2. Anterior margin of clypeus with a central tooth flanked by two slightly shorter teeth (Fig. 3.31b). Base of central tooth forms a slight vertical ridge in the clypeus.
   ....could be *Stenomalina gracilis*
   .... see key characters of *S. gracilis* below.

   – Anterior margin of clypeus with no central tooth but is shallowly notched, giving a wavy appearance with two shallow teeth (Fig. 3.31c).. could be *Trichomalus* spp.
   .... see key characters of *Trichomalus perfectus* and *T. lucidus* below.

Key characters of *Mesopolobus morys*

1. Antennae with three anelli and five funicular segments (as in Fig. 3.30a)
2. Mesoscutum not conspicuously or densely hairy (Fig. 3.28).
3. Clypeal anterior margin with no teeth but truncate with small notches either side. Sculpture on clypeus granulated (with faint striations radiating from anterior margin) and virtually indistinguishable from sculpture on frons (Fig. 3.31a).
4. Dorsal anterior margin of pronotal collar (Fig. 3.28) rounded, its edge not sharply defined and angular, so separation between pronotal collar and collum less distinct than in *T. perfectus* and *S. gracilis* (Fig. 3.32a, not b or c).
5. Basal cell of forewing (Fig. 3.26) bare, with no hairs within it and few or none outlining it (Fig. 3.33a, not b or c).

Description of *Mesopolobus morys*

Overall rather squat in appearance. Colour when fresh: head and all body strongly iridescent metallic black/green; legs pale yellow except for the proximal 75% of femora and final tarsi, which are very dark brown. Antennae rather club-like, i.e., flagellum gets wider up to the clava. The first segment of the clava is the widest antennal segment. Funicular segments not longer than broad (Fig. 3.30a) (see also Fig. 2.7).

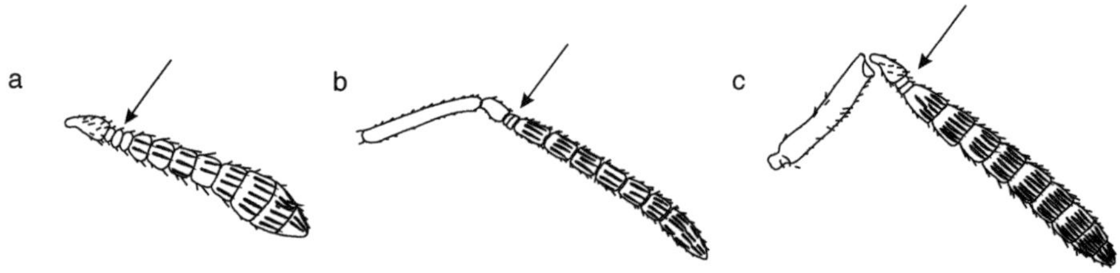

Fig. 3.30 Antennae of (**a**) *Mesopolobus teliformis* (redrawn after Graham 1969), (**b**) *Stenomalina gracilis* and (**c**) *Trichomalus perfectus* (redrawn after Graham 1969). Anelli *arrowed*

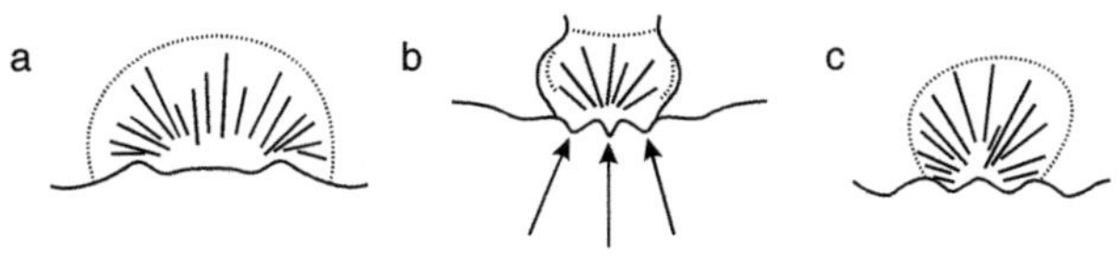

Fig. 3.31 Clypeal margin of (**a**) *Mesopolobus morys*, (**b**) *Stenomalina gracilis* (modified after Bouček and Rasplus 1991), and (**c**) *Trichomalus perfectus*. Teeth *arrowed*

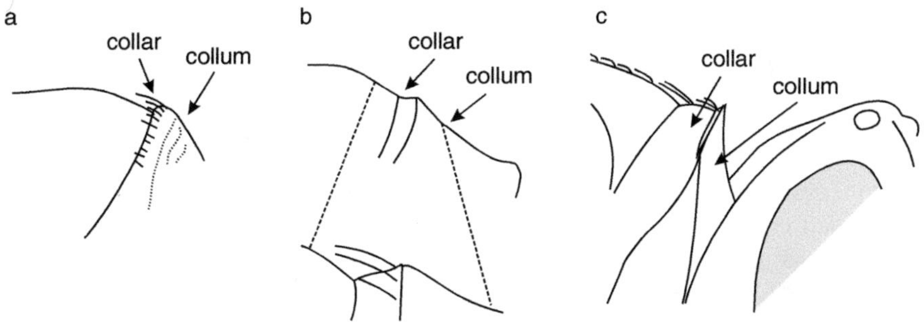

Fig. 3.32 Pronotal collar of (**a**) *Mesopolobus morys*, (**b**) *Stenomalina gracilis*, and (**c**) *Trichomalus perfectus*

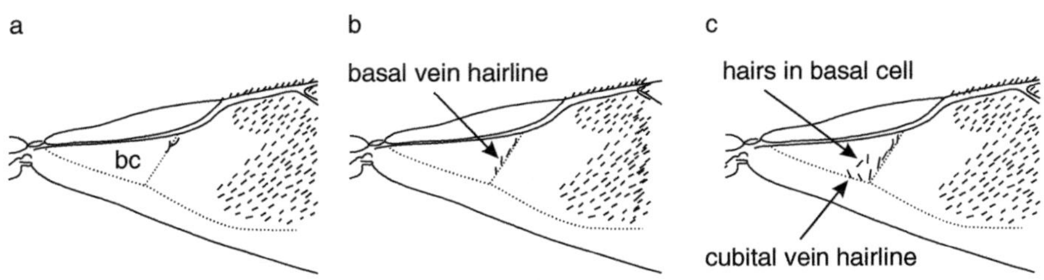

Fig. 3.33 Forewing basal cell (bc) hairs as in (**a**) *Mesopolobus morys*, (**b**) *Stenomalina gracilis*, and *Trichomalus perfectus* (redrawn after Bouček and Rasplus 1991)

Key characters of *Stenomalina gracilis*

1. Antennae slender with two anelli and six funicular segments (Fig. 3.30b).
2. Clypeus with central tooth flanked by two slightly shorter teeth on anterior margin (Fig. 3.31b). Base of central tooth forms a slight vertical ridge on clypeus. Sculpture of clypeus merges with reticulate sculpture of frons with clear striations radiating from anterior margin of clypeus.
3. Pronotal collum (Fig. 3.28) dorsally concave in profile and forward-extended. Dorsal anterior margin of pronotal collar upwardly produced and angular but less so than in *T. perfectus* (Fig. 3.32b, not c).
4. Basal cell of forewing (Fig. 3.26a) with hairs on basal hairline but no hairs within cell or on cubital vein (Fig. 3.33b, not c).

Description of *Stenomalina gracilis*

Overall long and slender in appearance. Colour when fresh: head and mesosoma metallic colour ranging from bronze-red with a green iridescence to black-green with a strong green iridescence, the abdomen being less iridescent and more bronze. Wet specimens appear less green. Tibiae, fibiae and tarsi all straw-yellow except for final tarsal segment which is dark brown. Antennae slender, basally barely wider than scape, at least the first two funicular segments longer than broad (see also Fig. 2.6).

Key characters shared by *Trichomalus perfectus* and *T. lucidus*

1. Antennae with two anelli and six funicular segments (Fig. 3.30c)
2. Anterior margin of clypeus with a wavy appearance comprised of two shallow teeth, one each side of a central shallow notch (Fig. 3.31c)
3. Pronotal collar smooth at its dorsal anterior margin with a raised and sharply-defined angular edge (Fig. 3.32c, not b).
4. Basal cell of forewing (Fig. 3.26a) with no hairs or few hairs (ca. one to three) within it. Basal hairline with hairs throughout its length. Variable numbers (usually none to four) of hairs on cubital vein, sometimes cubital vein hairy throughout (Fig. 3.33c).
5. Gaster first tergite conspicuously hairy laterally (not with only a few hairs as in *Pteromalus* spp.) (Fig. 3.34a, b).
6. Hind coxa (dorsal surface) hairy basally (nearest body) (Fig. 3.34). Dorsal hairs not restricted to distal (far) end of hind coxa as in *Pteromalus* spp.

Key characters distinguishing *Trichomalus perfectus* and *T. lucidus*

– Hind coxa basally with dorsal hairs curved and dense giving a 'furry' appearance (Fig. 3.34). Forewing marginal vein: stigmal vein ratio 1.4–1.6 (for vein nomenclature see Fig. 3.36). Propodeum with hairs arising from an area lateral to the plical carina leaving a bare area on the lateral surface of the propodeum distinctly anterior to the nucha and approximately square (Fig. 3.35a) .... *Trichomalus perfectus*

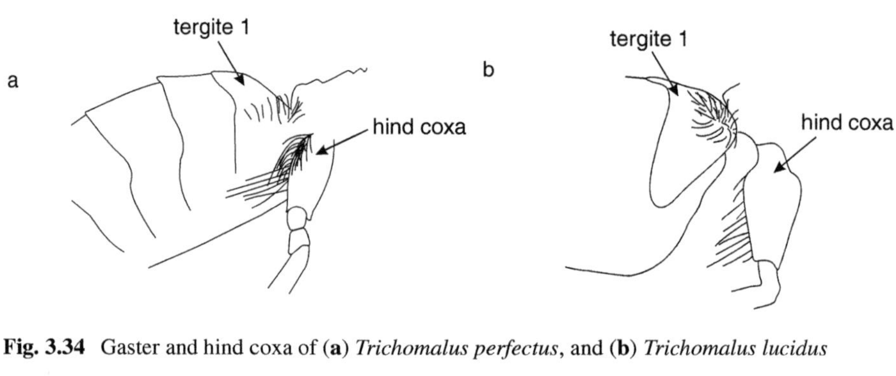

Fig. 3.34 Gaster and hind coxa of (**a**) *Trichomalus perfectus*, and (**b**) *Trichomalus lucidus*

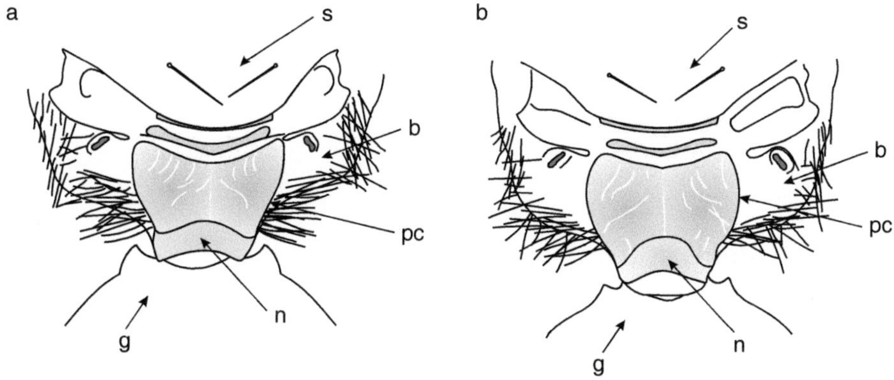

Fig. 3.35 Propodeum of (**a**) *Trichomalus perfectus*, and (**b**) *Trichomalus lucidus* (drawn after Gibson et al. 2005). b = bare area. g = gaster. n = nucha. pc = plical carina. s = scutellum

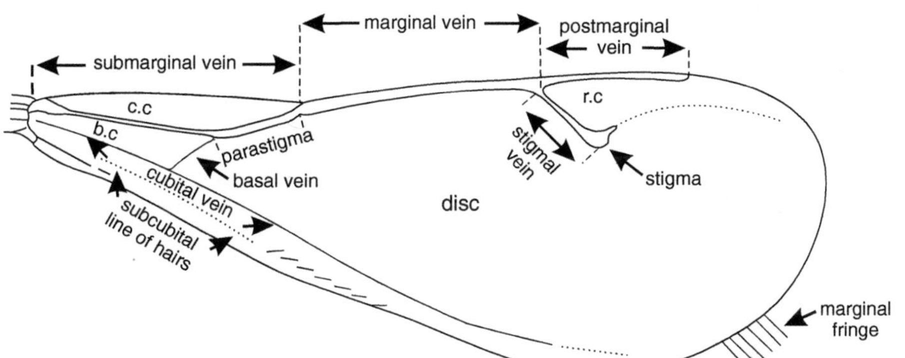

Fig. 3.36 Forewing of Eulophidae (redrawn after Graham 1959). b.c. = basal cell. c.c. = costal cell. r.c. = radial cell

– Hind coxae with dorsal proximal hairs spiny and sparse. Forewing marginal vein: stigmal vein ratio 1.65–1.85 (for vein nomenclature see Fig. 3.36). Propodeum with hairs arising from an area extending laterally and diagonally forwards from a point on the plical carina close to the nucha, leaving a bare area on the lateral surface of the propodeum that extends backwards in an acute angle towards the plical carina – nuchal juncture (Fig. 3.35b).................................. *Trichomalus lucidus*

Description of *Trichomalus perfectus*

Overall appearance stocky and powerful. Head and body metallic bronze-red with some green iridescence, most obvious on head and mesosoma. Legs pale brown to pale yellow. Hind coxa (dorsal surface) densely hairy proximally. Clypeus with strong striations radiating from its anterior margin and radiating striations continue in the reticulate sculpture of the frons (Fig. 3.31c) (See also Fig. 2.5).

Description of *Trichomalus lucidus*

Overall appearance similar to *T. perfectus*. Head and body iridescent dark copper-green. Legs yellow to orange. Hind coxa (dorsal surface) clearly but sparsely hairy proximally.

3.5 Parasitoids of the Family Eulophidae

The Eulophidae comprises four subfamilies: the Entodoninae, the Tetrastichinae, the Eulophinae and the Euderinae, the first three of which include species reported to attack brassica pod midge and cabbage seed weevil on oilseed rape (Table 3.5). The Entodoninae includes *Omphale clypealis*, a key parasitoid of brassica pod midge. The Tetrastichinae are mostly endoparasitoids of the eggs, larvae and pupae of Diptera, Hymenoptera or Lepidoptera, although some are ectoparasitoids or hyperparasitoids. This is a guide to females only.

3.5.1 Key Characters of the Family Eulophidae

1. Body almost always at least partly metallic-coloured.
2. Forewing with typical chalcid forewing venation. Forewing membrane not reticulate. Marginal vein distinct and several times longer than broad (Fig. 3.36).
3. Hindwing not long and stalk-like (not Fig. 3.26c).
4. Mesosoma with prepectus as big, or bigger, than the tegula in side view (Fig. 3.37). Mesoscotum with notauli distinctly curved when complete (Fig. 3.38a).

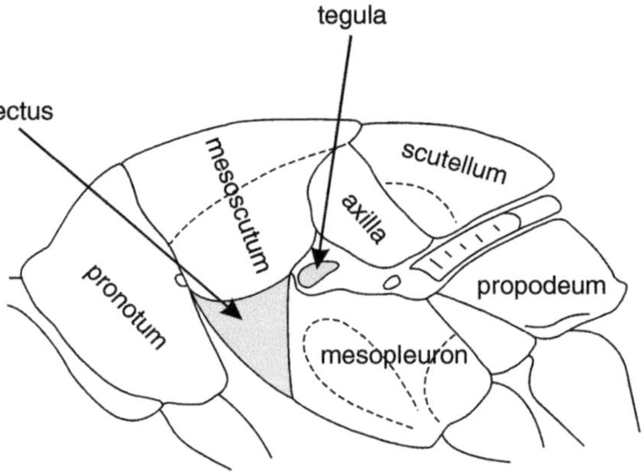

Fig. 3.37 Mesosoma (lateral view) of Eulophidae (redrawn after Graham 1959)

5. Forelegs and middle legs similar in size. Coxae of hind legs approximately cylindrical not flattened (Fig. 3.39, not Fig. 3.40). Hind femurs not swollen (not Fig. 3.41). Hind tibia without darker bristles arranged in a conspicuous pattern. Foretibial spur straight. Tarsi of all legs with four segments (Fig. 3.42).
6. Gaster distinctly constricted at junction with propodeum.
7. Antennae with five or fewer funicle segments (Fig. 3.43a, b).

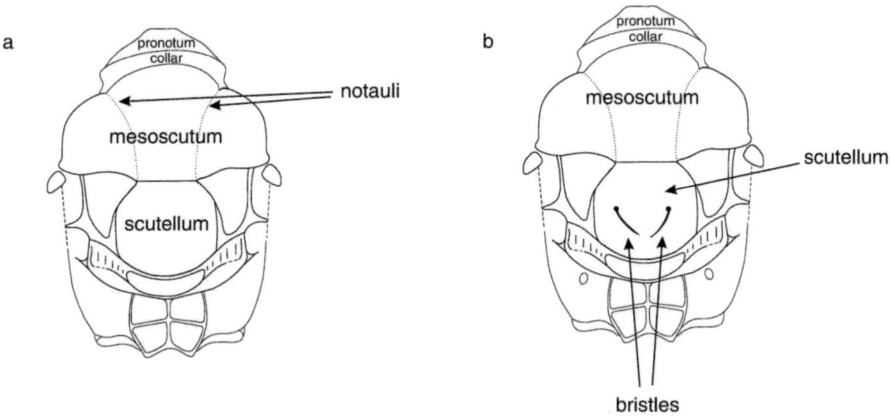

Fig. 3.38 Mesosoma (dorsal views) of (**a**) Eulophidae, and (**b**) Entedoninae (modified after Graham 1959)

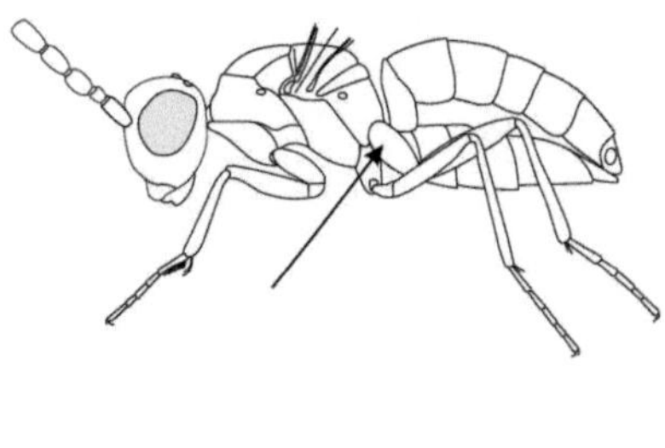

Fig. 3.39 Chalcidoid with cylindrical hind coxa (*arrowed*) (redrawn after Goulet and Huber 1993)

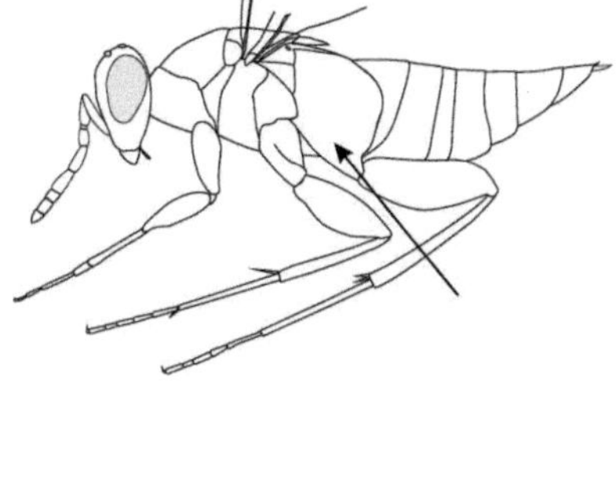

Fig. 3.40 Chalcidoid with flattened hind coxa (*arrowed*) (redrawn after Goulet and Huber 1993)

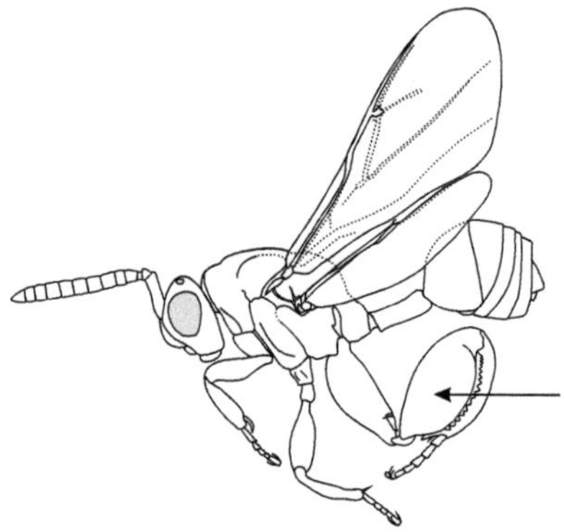

Fig. 3.41 Chalcidoid with swollen hind femur (*arrowed*) (redrawn after Goulet and Huber 1993)

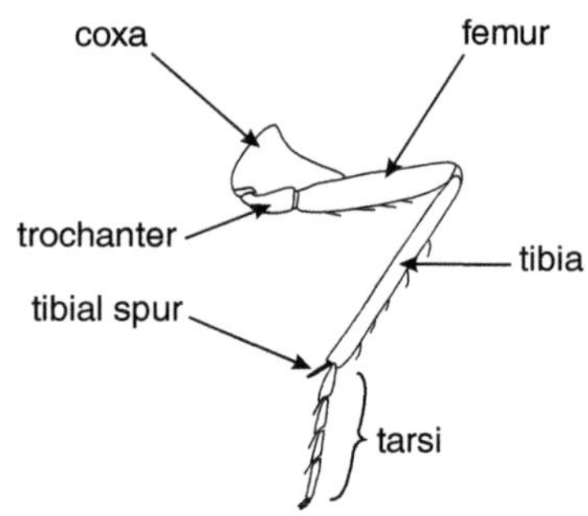

Fig. 3.42 Mid leg of *Omphale clypealis*

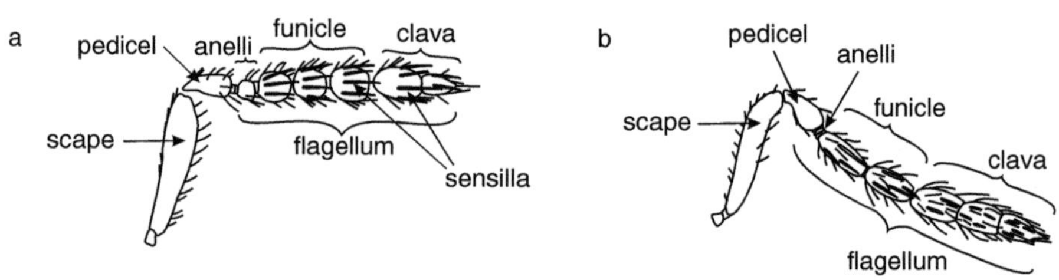

Fig. 3.43 Antenna of (**a**) Eulophidae (redrawn after Graham 1959), and (**b**) *Omphale clypealis*

3.5.1.1 Key Characters of Subfamilies Entodoninae and Tetrastichinae

1. Wing size normal.
2. Scutellum with two bristles near its middle and without a pair of longitudinal grooved lines either side of midline (Fig. 3.38b) .... Entedoninae

– Scutellum with four bristles and a pair of longitudinal grooved lines either side of midline. Postmarginal vein of forewing (Fig. 3.36) absent or rudimentary.. Tetrastichinae

Entedoninae: Characters distinguishing *Neochrysocharis* spp. and *Omphale clypealis*

1. Clypeus same colour as rest of head. Antennae clearly with two funicle segments and three closely-fused clava segments.............................. *Neochrysocharis* sp.

– Clypeus (Fig. 3.44) pale lemon-yellow. Antenna with two funicle segments (Fig. 3.43b) and first of the three claval segments less tightly fused to second segment than the second is to the third.............................. *Omphale clypealis*

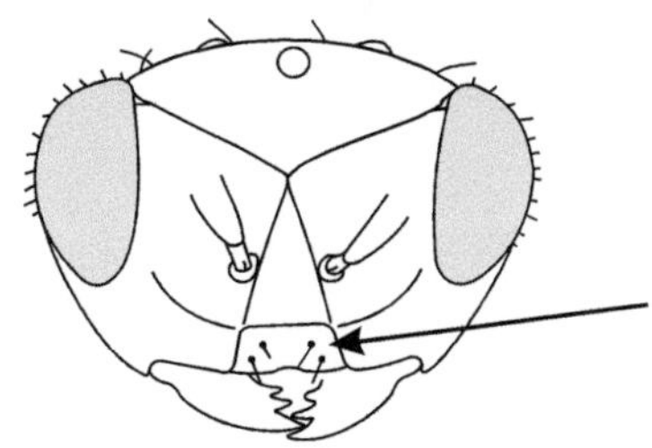

Fig. 3.44 Head of *Omphale clypealis*. Clypeus *arrowed*

Description of Omphale clypealis (after Graham 1963)

Body mainly green to blue-green (less so when wet). Gaster ovate, as long or slightly longer than head plus thorax. Clypeus entirely yellow, almost flat, shape as Fig. 3.44. Lower part of face with reticulate sculpture. Antennal flagellum black with two-segmented funicle and well-defined, three-segmented clava that is slightly broader than the funicle. Forewing with few, if any, hairs in the radial cell, the stigma rhomboidal and with the post-marginal vein slightly shorter than the stigmal vein (See also Fig. 2.10).

Tetrastichinae: Characters distinguishing *Aprostocetus epicharmus* from other Tetrastichinae attacking rape pests

1. Top of head with no ridge behind the lateral ocelli.
2. Mesosoma and metasoma weakly tinted with olive-blue and sometimes with much yellow (NB other species of Tetrastichinae are also strongly yellow-coloured).

3.6 Parasitoids of the Family Platygastridae

The family Platygastridae comprises about 1,000 known species, all endoparasitoids, mostly of Diptera, particularly the Cecidomyiidae. More than a third belong to the genus *Platygaster* (Vlug 1995) which includes *Platygaster subuliformis*, a key parasitoid of the brassica pod midge (Table 3.6). This is a guide to females only.

3.6.1 Key Characters of the Genus Platygaster

1. General appearance (of ♂) as in Fig. 3.45.
2. Forewings and hindwings with no veins (Fig. 3.45).
3. Antenna with 10 segments: scape, pedicel and eight flagellar segments (Fig. 3.46).
4. Petiole simple, not with a forward-extending cornutus as in *Inostemma* spp (Fig. 3.47).
5. Scutellum dome-shaped with a rounded posterior edge (Figs. 3.45 and 3.48a), not elongated into a backward-directed spine as in *Synopeas* spp (Fig. 3.48b).

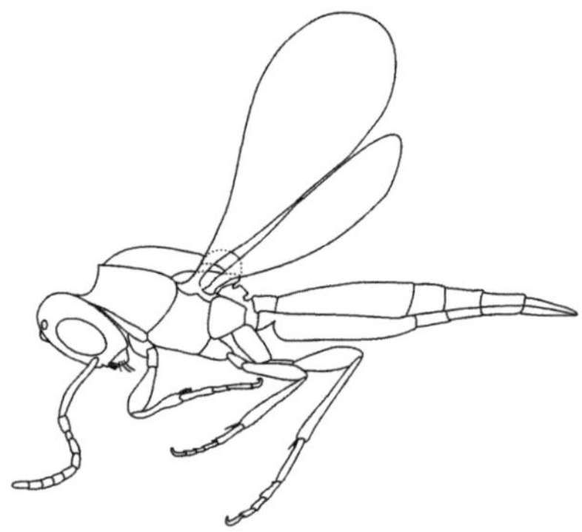

Fig. 3.45 Typical platygastrid (redrawn after Goulet and Huber 1993)

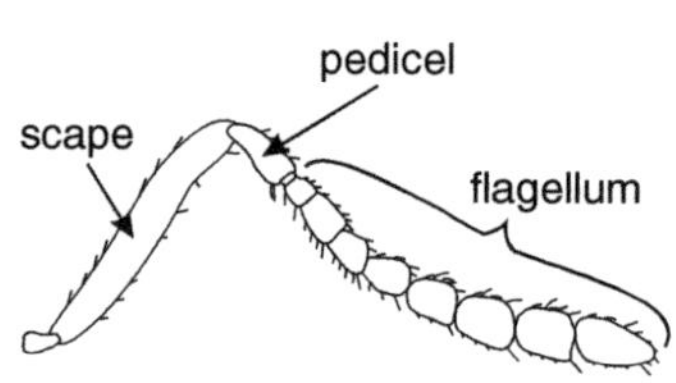

Fig. 3.46 Antenna of *Platygaster subuliformis* ♀ (redrawn after Murchie et al. 1999)

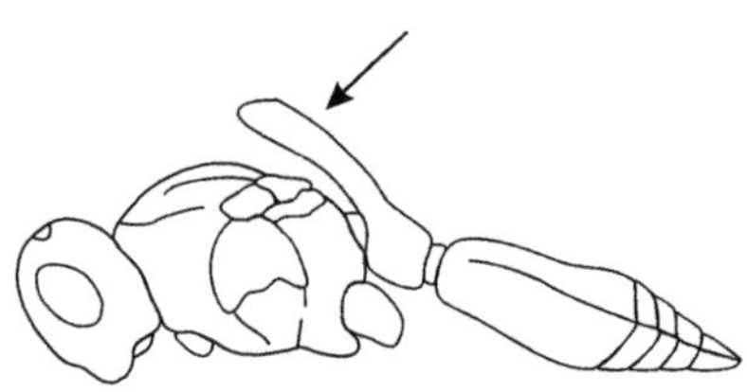

Fig. 3.47 *Inostemma boscii* (redrawn after Medvedev 1978). Cornutus *arrowed*

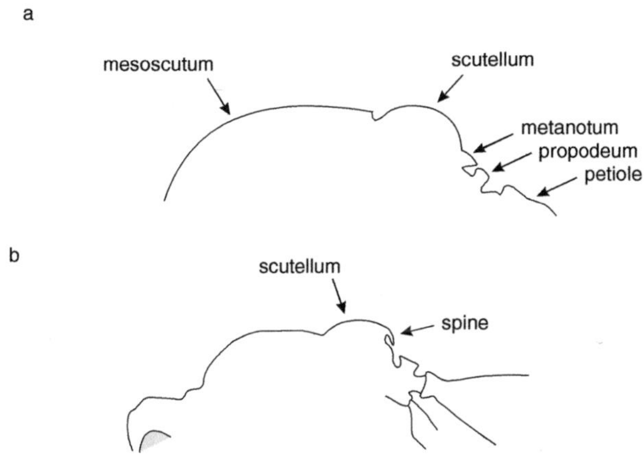

Fig. 3.48 Mesosoma (dorsal profiles) of (**a**) *Platygaster subuliformis* ♀ with dome-shaped scutellum and no spine, (**b**) a platygastrid with a backward-directed spine

Four other species of *Platygaster* could be confused with *Platygaster subuliformis*: *P. oebalus* (especially close), *P. tisias, P. iolas* and *P. munita.* The first three are reported to parasitise the brassica pod midge and hence may be found in oilseed rape. *Platygaster munita*, although a closely-related species, has not been associated with the brassica pod midge. It is therefore unlikely to be found in rape crops and is not considered further here.

Key characters of *Platygaster subuliformis* ♀♀ (See also Fig. 2.9).

1. General appearance as in Fig. 3.45.
2. Length 1.7–1.9 mm.
3. Colour black except for tarsi and extremities of femora and tibiae, which are brown.
4. Wings transparent (hyaline) and colourless, not smoky grey-brown ('infuscated') as in *P. munita* and *P. tisias*. Surface of both wings covered with fine evenly-spaced hairs except for a bare patch near base of forewing (chalcid speculum). Forewing edge with short fringe of marginal hairs, a little longer distally and towards the trailing edge (posterior margin), but not markedly so (Fig. 3.49a, not b).
5. Scutellum domed and rounded (Fig. 3.48a).
6. Gaster with a forward-projecting protrusion of first sternite (Fig. 3.50a, not b). Gaster elongated (Fig. 3.51a), not like *P. iolas* (Fig. 3.51b). Gaster with third tergite wider than long (Fig. 3.51a) but less markedly so than in *P oebalus* (Fig. 3.51c) and *P. tisias* (Fig. 3.51d).
7. Antennae with 10 flagellar segments, only segments four and 10 longer than wide (Fig. 3.52a, not b, c, d).
8. Vertex (top of head) with fine reticulate (network-like) sculpture which looks more transverse (cross-ways to length of insect) behind the lateral ocelli. Vertex not with the strongly transverse and coarse sculpture that is present in *P. oebalus*.

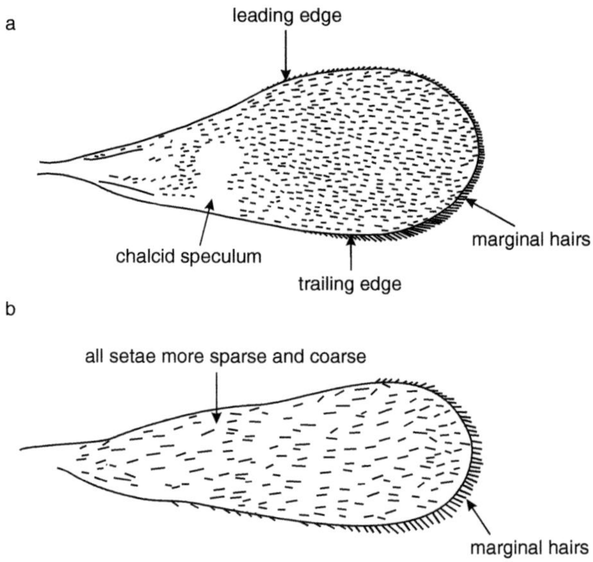

Fig. 3.49 Forewing of (**a**) *Platygaster subuliformis* (modified after Goulet and Huber 1993), and (**b**) another *Platygaster* sp

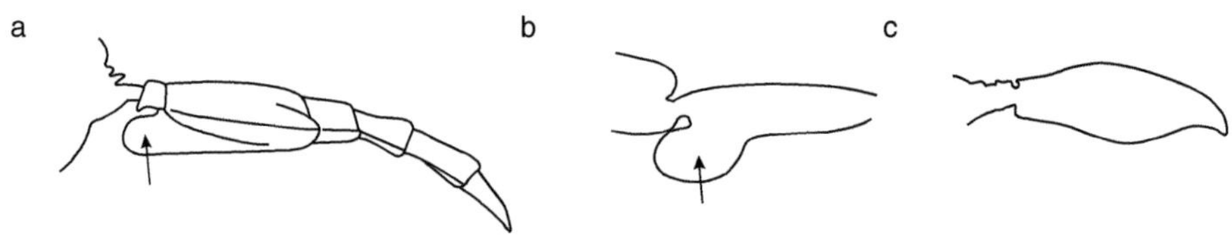

Fig. 3.50 Gasters (lateral views) of (**a**) *Platygaster subuliformis* ♀ (modified after Murchie et al. 1999), (**b**) another platygastrid species ♀ and (**c**) *Platygaster subuliformis* ♂. Protrusions of ♀ first sternite *arrowed*

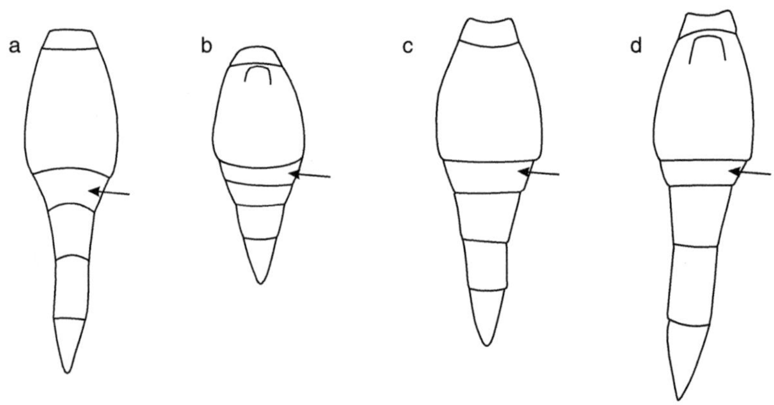

Fig. 3.51 Gasters (dorsal views) of ♀ (**a**) *Platygaster subuliformis*, (**b**) *Platygaster iolas*, (**c**) *Platygaster oebalus*, and (**d**) *Platygaster tisias*. Third tergites *arrowed* (**a** redrawn after Murchie et al. 1999; **b**, **c**, and **d** redrawn after Vlug 1985)

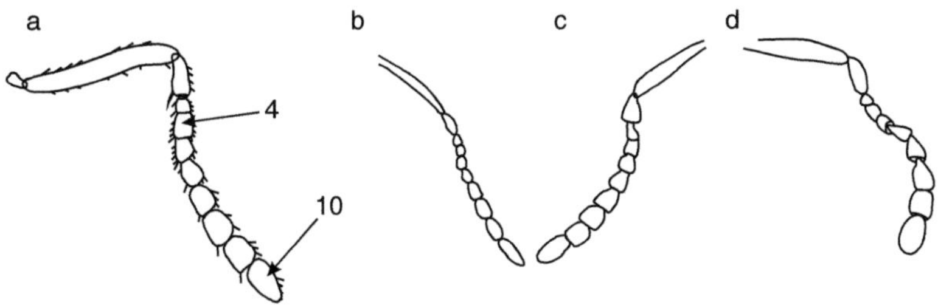

Fig. 3.52 Antennae of ♀ (**a**) *Platygaster subuliformis*, (**b**) *Platygaster iolas*, (**c**) *Platygaster oebalus*, and (**d**) *Platygaster tisias* (**a** redrawn after Murchie et al. 1999; **b**, **c**, and **d** redrawn after Vlug 1985)

Key characters of *Platygaster subuliformis* ♂♂

Males are more difficult to identify with confidence than females. To qualify as 'probable *P. subuliformis* males', specimens should, in addition to the key characters of the genus, also have the following characters:

1. Colour exactly as females. Specimens with brownish bodies or paler brown or red-brown legs are not *P. subuliformis*.
2. Wings as females. Specimens with brown-tinged wings, without a clear chalcid speculum, with coarser and less dense hairs, or with longer fringing hairs are not *P. subuliformis*.
3. Mesosoma as females.
4. Gaster not elongated in males (Fig. 3.50c).
5. Antennae as Fig. 3.53. In life, flagellum usually kinked at second segment. First segment of flagellum appears triangular in outline when viewed from a certain angle. Segments three to six of flagellum more globular than in female.
6. Head as females except reticulate sculpture on vertex may be more deeply embossed than in female and transverse component may be less marked.

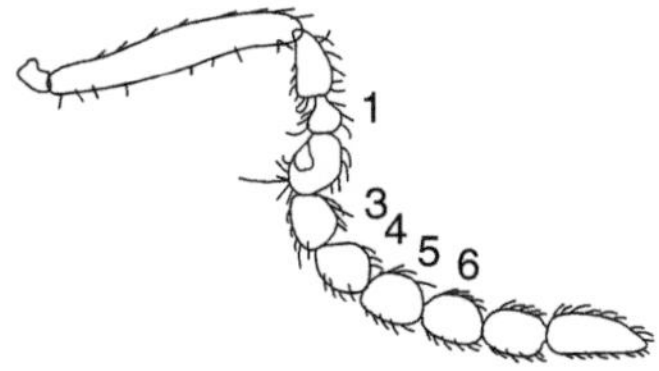

Fig. 3.53 Antenna of ♂ *Platygaster subuliformis*. Numbers indicate flagellar segments (redrawn after Murchie et al. 1999)

3.7 Parasitoids of the Family Proctotrupidae

Brachyserphus parvulus is the only member of the Proctotrupid family reported to be a parasitoid of a pest of oilseed rape; it can be a common parasitoid of the pollen beetle on spring rape in some years and sites.

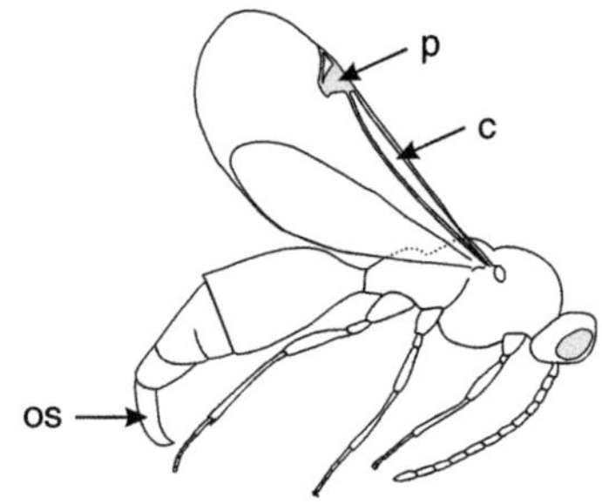

Fig. 3.54 *Brachyserphus parvulus* ♀. p = pterostigma, c = costal cell, os = ovipositor sheath

Key characters of *Brachyserphus parvulus*

1. General appearance as in Fig. 3.54.
2. Ovipositor usually tightly enclosed by its sheath which is short and broad, parallel-sided basally and curved in a ventral direction towards its tip (Fig. 3.54).
3. Pterostigma at least as wide as long (Fig. 3.54).

3.8 Glossary

Anellus (pl. anelli)	small segment(s) on antennae of chalcids, between pedicel and flagellum.
Apical	end of body or of appendage further from head.
Areolet	small cell in wing of ichneumonids, open apically in Tersilochinae.
Basal	end of body or of appendage nearer head.
Carina	ridge.
Cell	area of wing membrane enclosed partly or completely by veins.
Clypeus	lower part of face of insect, above mouthpart appendages.
Cornutus	elongate projection of petiole.
Costal cell	most anterior vein of wing, running along costal margin.
Coxa (pl. coxae)	first segment of leg.
Ectoparasitoid	parasitoid that feeds externally from its host.
Endoparasitoid	parasitoid that feeds within its host.
Flagellum	distal section of antenna, beyond pedicel.
Gaster	part of abdomen behind petiole in Parasitica.
Granulated	surface covered with small grain-like protruberances
Koinobiont	parasitoid that allows its host to continue to develop.
Mesepistenum	anterior part of mesopleuron.
Mesonotum	dorsal surface of second thoracic segment.
Mesopleuron	lateral and ventral part of mesothorax.
Mesoscutum	mesonotum without scutellum.
Mesothorax	second segment of thorax.

Metanotum	dorsal part of metathorax.
Metasoma	petiole plus gaster (second abdominal segment 2 onward)
Metathorax	third segment of thorax.
Notaulus (pl. notauli)	longitudinal groove on mesonotum.
Nucha	neck at apex of the propodeum
Ocellus (pl. ocelli)	light-sensitive, simple eyes. Three usually present in triangle between compound eyes at top of head.
Ovipositor	egg-laying structure in female.
Palp	segmented, sensory mouthpart arising from the maxilla or labium.
Pedicel	second segement of antenna, located between scape and flagellum.
Petiole	narrow waist or stalk in Parasitica between gaster and propodeum, comprised of second abdominal segment.
Pronotum	dorsal surface of prothorax.
Propodeum	first segment of abdomen fused with thorax.
Prothorax	first segment of thorax.
Pterostigma	pigmented area on margin of forewing towards apex of costal vein.
Reticulate	surface covered with net-like sculpture
Scape	basal segment of antenna.
Scutellum	middle part of mesonotum.
Seta (pl. setae)	bristle
Sternaulus	curved furrow or depression dividing lower part of mesopleruon.
Tarsus	distal part of leg.
Tegula	small lobe covering base of forewing.
Tergite	dorsal sclerite on abdomen.
Thyridiae	depressions of upper anterior corners of first tergite of gaster.
Trochanter	second segment of leg between coxa and femur.
Truncate	cut off squarely with straight edge.
Vertex	top of head, behind the frons.

Acknowledgements We are indebted to Prof. Dr. Klaus Horstmann, Universität Würzburg, Germany, for his generous sharing of expertise on the taxonomy of Tersilochinae, for examination of specimens and for discussions of characters for separating species. We are especially grateful for his advice on a draft version of the section of this chapter devoted to Tersilochinae, which is based to a great extent upon his keys (Horstmann 1971, 1981). We also thank Dr. Kees van Achterberg, Natuurhistorisch Museum, Leiden, Netherlands, for examination of specimens and the sharing of his expertise in identification of Braconid species, Dr. Gary Gibson, Agriculture and Agri-Food Canada, Ottawa, Canada, Dr. Hannes Baur, Natural History Museum, Bern, Switzerland, and Dr Stephan Vidal, Georg-August University, Göttingen, Germany, for similar assistance in separation of Pteromalid species. We thank Dr. Archie Murchie, Agri-Food and Biosciences Institute, Belfast, UK, for advice on the identification of *Platygaster subuliformis*, *Trichomalus perfectus* and *Omphale clypealis*. We acknowledge the support of the EU co-funded project: MASTER (QLK5-CT-2001-01447), for providing travel funds to attend project meetings for invaluable taxonomic discussions, particularly with Dr. Bernd Ulber, Georg-August Universität, Göttingen, Germany, Dr. Christer Nilsson and Britt Åhman, Swedish University of Agricultural Sciences, Alnarp, Sweden, Dr. Zdzisław Klukowski, Wroclaw University of Environmental and Life Sciences, Poland and Dr. Eve Veromann, Estonian University of Life Sciences, Tartu, Estonia. Preparation of the guide was funded by the UK Department for Food and Rural Affairs and the Estonian Targeting Financing Projects SF0172655s04 and SF0170057s09. Rothamsted Research is an institute of the Biotechnology and Biological Sciences Research Council of the United Kingdom.

References

Achterberg C van (1993) Illustrated key to the subfamilies of the Braconidae (Hymenoptera: Ichneumonidae). *Zoologische Verhandellingen* 283: 1–189.

Achterberg C van, Quicke DLJ (2000) In: Unpublished course notes from *Taxonomy and biology of parasitic Hymenoptera*, 12–20 April 2000, Natural History Museum and Imperial College, London.

Alford DV (ed.) (2003) *Biocontrol of oilseed rape pests*. Blackwell, Oxford, UK.

Askew RR (1968) Hymenoptera 2. Chalcidoidea. *Handbooks for the Identification of British Insects* 7(2b): 1–39. Royal Entomological Society, London.

Barari H, Ferguson AW, Piper RW, Smith E, Quicke DLJ, Williams IH (2005) The separation of two hymenopteran parasitoids, *Tersilochus obscurator* and *Tersilochus microgaster* (Ichneumonidae), of stem-mining pests of winter oilseed rape using DNA, morphometric and ecological data. *B Entomol Res* 95: 299–307.

Bouček Z, Rasplus J-Y (1991) Illustrated key to West-Palearctic genera of Pteromalidae (Hymenoptera: Chalcidoidea). INRA, Paris.

Delucci V, Graham MWR (1956) La revision du genre *Trichomalus* Thomson (1878) (Chalcidoidea: Pteromalidae). *Beitrage zur Entomol* 6: 543–576.

Fitton M, Notton D, Henderson S (2000) Key to subfamilies of Ichneumonidae In: Unpublished course notes from *Taxonomy and biology of parasitic Hymenoptera,* 12–20 April 2000, Natural History Museum and Imperial College, London.

Gibson GAP, Baur H, Ulmer B, Dosdall L, Muller F (2005) On the misidentification of chalcid (Hymenoptera: Chalcidoidea) parasitoids of the cabbage seedpod weevil (Coleoptera: Curculionidae) in North America. *Can Entomol* 135: 381–403.

Goulet H, Huber JT (eds.) (1993) *Hymenoptera of the world: An identification guide to families*. Agriculture Canada, Canada.

Graham MWRdeV (1959) Keys to the British genera and species of Elachertinae, Eulophinae, Entedontinae, and Euderinae (Hym., Chalcidoidea). *Trans SocBrit Entomol* 13: 169–204.

Graham MWRdeV (1963) Additions and corrections to the British list of Eulophidae (Hym., Chalcidoidea), with descriptions of some new species. *Trans Soc Brit Entomol* 15: 167–275.

Graham MWRdeV(1969) *The Pteromalidae of north-western Europe (Hymenoptera, Chalcidoidea)*. British Museum (Natural History) Entomology Supplement 16.

Horstmann K (1971) Revision der europäischen Tersilochinen I (Hymenoptera, Ichneumonidae). *Veröffenlichungen der Zoologischen Staatsamlung München* 15: 47–138.

Horstmann K (1981) Revision der europäischen Tersilochinen II (Hymenoptera, Ichneumonidae). *Spixiana Sup* 4: 3–76.

Medvedev GS (ed.) (1978) *Keys to the insects of the European part of the USSR, Volume III, Hymenoptera, Part II*. Nauka: Leningrad. [In Russian].

Murchie AK, Polaszek A, Williams IH (1999) *Platygaster subuliformis* (Kieffer) (Hym., Platygastridae) new to Britain, an egg-larval parasitoid of the brassica pod midge *Dasineura brassicae* Winnetrz (Dipt., Cecidomyiidae). *Entomologist's Monthly Magazine* 135: 217–222.

Noyes J (1985) Chalcidoids and biological control. *Chalcid Forum* 5: 5–10.

Noyes J (2000) Chalcids – Chalcidoidea and Mymarommatoidea. Chapter 7 In: Unpublished course notes from *Taxonomy and biology of parasitic Hymenoptera*, 12–20 April 2000. Natural History Museum and Imperial College, London.

Noyes J, Notton D, Polaszek A (2000) Chalcids – 'Proctos': Proctotrupoidea, Platygastroidea and Ceraphronoidea. Chapter 9 In: Unpublished course notes from *Taxonomy and biology of parasitic Hymenoptera*, 12–20 April 2000, Natural History Museum and Imperial College, London.

Rosen H von (1964) Untersuchungen über die Verbeitung und Biologie von zwei Pteromaliden in Rapsschoten (Hym., Chalcidoidea). *Statens Växtskyddsanstalt – Meddelanden* 12: 451–465.

Tobias WJ, Belokobylskii SA, Kotenko AG (eds.) (1986) [*Keys to the Fauna of the European Part of the USSR, Volume III, Hymenoptera, Part IV.*] Nauka: Leningrad. [In Russian].

Vidal S (2003) Identification of Hymenopterous parasitoids associated with oilseed rape pests. In: Alford DV (ed.) *Biocontrol of oilseed rape pests*. Blackwell, Oxford, UK.

Vlug HJ (1985) The types of Platygastridae (Hymenoptera, Scelionidae) described by Haliday and Walker and preserved in the National Museum of Ireland and in the British Museum (Natural History). 2. Keys to species, redescriptions, synonymy. *Tidschrift voor Entomologie* 127: 179–224.

Vlug HJ (1995) Catalogue of the Platygastridae (Platygastroidea) of the world. *Hymenopterorum Catalogus* 19: 1–168.

List of Other Keys Useful for Identification of Parasitoids

Achterberg C van (1990) Illustrated key to the subfamilies of the Holarctic Braconidae (Hymenoptera: Ichneumonoidea). *Zoologische Mededelingen Leiden* 64: 1–20.

Annecke DP, Doutt RL (1961) The genera of the Mymaridae Hymenoptera: Chalcidoidea. *Entomology Memoirs* 5. The Government Printer, Pretoria.

Aubert J-F, Jourdheuil P (1958) Nouvelle description et biologie de quelques Ichneumonides appartenant aux genres *Aneuclis* Först., *Isurgus* Först. et *Thersilochus* Holm. *Revue Path* 37: 175–193.

Buhl PN (1999) A synopsis of the Platygastridae of Fennoscandia and Denmark. *Entomofauna* 20: 17–52.

Dessart P, Cancemi P (1986) Tableau dichotomique des genres de *Ceraphronoidae (Hymenoptera)* avec commentaires et nouvelles espêces. *Frustula Entomologica, Nouvelles Séries* 7–8(20–21): 307–372.

Fitton MG, Graham MWRdeV, Bouček ZRL, Fergusson NDM, Huddleston T, Quinlan J, Richards OW (1978) A Check List of British Insects. Part 4: Hymenoptera. *Handbooks for the Identification of British Insects* 11(4): 1–159.

Gahan AB (1941) A revision of the parasitic wasps of the genus *Necremnus* Thomson (Eulophidae; Hymenoptera). *J Washington Acad Sci* 31: 196–203.

Gauld I, Bolton B (eds.) (1988) *The Hymenoptera*. British Museum of Natural History and Oxford University Press, London.

Gibson GAP (1989) Phylogeny and classification of Eupelmidae, with a revision of the world genera of Calostinae and Metapelmatinae (Hymenoptera: Chalcidoidea). *Mem Entomol Society of Canada* 149: 1–125.

Graham MWRdeV (1987) A reclassification of the European Tetrastichinae (Hymenoptera: Eulophidae), with a revision of certain genera. *B Brit Mus (Natural History) Entomology Series* 55(1): 1–392.
Haeselbarth E (1973) Die Blacus-Arten Europas und Zentral-Asiens (Hymenoptera, Braconidae). *Veröffentlichungen der Zoologischen Staatsammlung Munchen* 16: 60–170.
Hellén W (1958) Zur Kenntnis der Braconidae (Hym.) Finnlands. II. Subfamilie Helconinae (part). *Fauna Fennica* 4: 3–37.
Horstmann K (1967) Bemerkungen zur Taxonomie der Tersilochinen (Hym., Ichneumonidae). *Opuscula Entomol* 32: 123–130.
Horstmann K, Yu DS (1999) Bemerkungen zur Taxonomie und Nomenklatur westpaläarktischer Ichneumonidae (Hymenoptera). *Zeitschrift der Arbeitsgemeinschaft Österreichischer Entomolgen Wien* 50: 65–74.
Jourdheuil P (1960) Influence de quelques facteurs écologiques sur les fluctuations de population d'une biocénose parasitaire: étude relative á quelques Hyménoptères (Ophioninae, Diospilinae, Euphorinae) parasites de divers Coléoptères inféodés aux Crucifères. *Ann des Épiphyties* 11: 445–658.
Masner L, Huggert L (1989) World review and keys to genera of the subfamily Inostemmatinae with reassignment of the taxa to the Platygastrinae and Sceliotrachelinae (Hymenoptera: Platygastridae). *Memoirs Entomol Soc Can* 147: 1–214.
Medvedev GS (ed.) (1987) *Keys to the insects of the European part of the USSR, Volume III, Hymenoptera, Part II.* Oxonian Press, New Delhi. Translation of publication in Russian by Nauka Publishers, Leningrad 1978.
Medvedev GS (ed.) (1995) *Keys to the insects of the European Part of the USSR, Volume III, Hymenoptera, Part IV.* Lebanon, NH.
Nixon GEJ (1938) A preliminary revision of the British Proctotrupinae (Hym. Proctotrupoidea). *Trans R Entomol Soc* 87(19): 431–465.
Oelke J (1969) Beitrage zur Insektenfauna der DDR: Hymenoptera, Bestimmungstabellen bis zur Unterfamilie. *Beitrage zur Entomol* 19: 753–802.
Osborne P (1960) Observations of the natural enemies of *Meligethes aeneus* (F.) and *M. viridescens* (F.) [Coleoptera: Nitidulidae]. *Parasitology* 50: 91–110.
Perkins JF (1959) Hymenoptera, Ichneumonoidea, Ichneumonidae. *Handbooks for the Identification of British Insects* 7(2): 1–29. Royal Entomological Society, London.
Schmiedeknecht O (1930) Die Hymenopteren Nord- und Mitteleuropas. *Verlag Gustav Fischer 2. Aufl.* p. 630.
Stresemann E (2000) *Exkursionsfauna von Deutschland. Band 2: Wirbellose, Insekten.* Spektrum Akademischer Verlag Gustav Fischer, Heidelberg.
Yu DS, Horstmann K (1997) A catalogue of world ichneumonidae (Hymenoptera). *Mem Am Entomol Inst Gainesville* 58: 1–2.

Chapter 4
Ground Beetles as Predators of Oilseed Rape Pests: Incidence, Spatio-Temporal Distributions and Feeding

Ingrid H. Williams, Andrew W. Ferguson, Märt Kruus, Eve Veromann, and Douglas J. Warner

Abstract Ground beetles or carabids are amongst the most abundant invertebrate predators in fields of oilseed rape in Europe. The immature stages of the six major pests of oilseed rape i.e., cabbage stem flea beetle, pollen beetle, cabbage seed weevil, cabbage stem weevil, rape stem weevil and brassica pod midge, are vulnerable to predation by carabids when they are in or on the soil from mid-September to mid-July. About 42 species of carabid are common in rape fields. The community composition varies between countries, between spring and winter crops and with crop management, as species differ in distribution and habitat requirements. The ten most widespread and dominant species are *A. similata*, *Anchomenus dorsalis*, *Bembidion lampros*, *Harpalus affinis*, *Harpalus rufipes*, *Loricera pilicornis*, *Nebria brevicollis*, *Poecilus cupreus*, *Pterostichus melanarius* and *Trechus quadristriatus*. A field-scale study found within-field spatio-temporal coincidence of *T. quadristriatus* and *Pterostichus madidus* with cabbage stem flea beetle eggs, of *A. similata*, *N. brevicollis* and *Asaphidion* spp. with pollen beetle larvae, and of *A. dorsalis* with larvae of cabbage stem weevil, cabbage seed weevil and brassica pod midge. Carabid within-field distributions are influenced by their life cycles, habitat preferences, mobility and food availability. Evidence of predation from analysis of gut contents of field-collected individuals and laboratory feeding trials is reviewed in relation to carabid functional morphology, prey size and diel periodicity of activity. Feeding trials suggest that *Clivina fossor* also has potential as a predator of oilseed rape pests. The implications for biocontrol-based integrated pest management are discussed.

4.1 Introduction

Ground beetles or carabids are amongst the most abundant and important invertebrate predators in arable ecosystems. Their biocontrol potential against aphids in cereal crops has long been recognised (Vickerman and Sunderland 1975, Chiverton

I.H. Williams (✉)
Institute of Agricultural and Environmental Sciences, Estonian University of Life Sciences, Tartu 51014, Estonia
e-mail: ingrid.williams@bbsrc.ac.uk

I.H. Williams (ed.), *Biocontrol-Based Integrated Management of Oilseed Rape Pests*, DOI 10.1007/978-90-481-3983-5_4, © Springer Science+Business Media B.V. 2010

1986, Winder 1990, Sunderland 2002) but the study of their potential as natural enemies of the pests of oilseed rape was initiated only recently (Büchs and Nuss 2000, Warner et al. 2000).

Reviews of the predator taxa, including carabids, found in fields of oilseed rape in Europe (Büchs and Alford 2003), their taxonomy and identification (Alford et al. 2003a), methodology for sampling, trapping and rearing them (Büchs 2003a), the impact of on-farm landscape structures (Büchs 2003b) and their role in biocontrol of rape pests (Büchs 2003c) have recently been published. Here we present further information on the incidence and activity-densities of carabids in European rape fields, their spatio-temporal relationships with the major coleopteran and dipteran pests of oilseed rape in Europe and evidence of predation on these pests. We focus on the role of adult carabids only as little is yet known about predation by carabid larvae on the pests of oilseed rape.

Adult carabids are active terrestrial beetles. Most species feed on the surface of the soil, although a few venture below the surface (including *Harpalus rufipes* (De Geer) (Luff 1978) and *Clivina fossor* (L.) (Forsythe 2000)) or climb onto vegetation (including *Bembidion lampros* (Herbst)), *Trechus quadristriatus* (Schrank) and *H. rufipes* (Vickerman and Sunderland 1975), *Amara apricaria* (Paykull) (Thiele 1977, Forsythe 2000), *Poecilus cupreus* (L.) (Chiverton 1988, Ekbom et al. 1992, Mundy et al. 2000), *Amara similata* (Gyllenhal) (Luka et al. 1998), *Amara familiaris* (Dufschmid) and *Curtonotus aulicus* (Panzer) (Forsythe 2000). The pests of oilseed rape are therefore most vulnerable to predation by adult carabids when their eggs or larvae are on or near the soil surface although there is some evidence that some species of carabid e.g., *A. similata*, do seek prey in the oilseed rape crop canopy (Luka et al. 1998) and that others, e.g., *C. fossor* kill larvae or pupae below the soil surface (Schernéy 1959).

The six major pests of winter oilseed rape in Europe, namely the cabbage stem flea beetle (*Psylliodes chrysocephala* (Linnaeus), Coleoptera: Chrysomelidae), the pollen beetle (*Meligethes aeneus* (Fabricius), Coleoptera: Nitidulidae), the cabbage seed weevil (*Ceutorhynchus obstrictus* (Marsham) syn. *C. assimilis* (Paykull) Coleoptera: Curculionidae), the cabbage stem weevil (*Ceutorhynchus pallidactylus* (Marsham), syn. *C. quadridens* (Panzer), Coleoptera: Curculionidae), the rape stem weevil (*Ceutorhynchus napi* Gyllenhal, Coleoptera: Curculionidae) and the brassica pod midge (*Dasineura brassicae* Winnertz, Diptera: Cecidomyiidae) migrate to the crop in succession, attacking it at various growth stages and damaging different parts of the plant (Alford et al. 2003, Williams Chapter 1 this volume). All are univoltine except for the brassica pod midge which has two generations on winter rape and one on spring rape.

The immature stages of these pests, are found in or on the soil at various times during the growing season of winter rape and can provide a continuous and often plentiful supply of potential prey items for carabids to feed on from mid-September through to mid-July (Fig. 4.1).

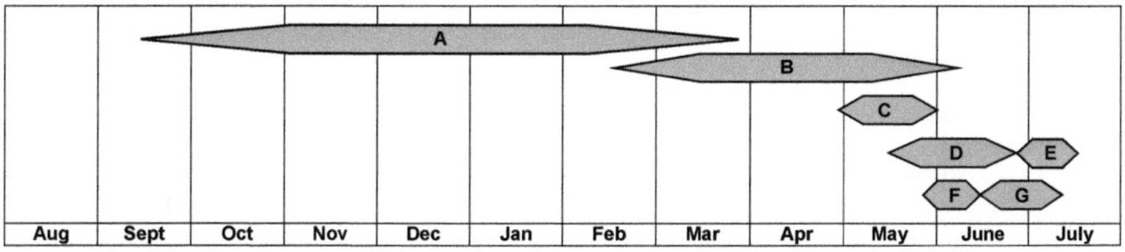

Fig. 4.1 Phenology of the immature stages of the major coleopteran and dipteran pests of winter oilseed rape vulnerable to epigeal predators. (*A*) eggs and young larvae, and (*B*) mature larvae of cabbage stem flea beetle. Mature larvae of (*C*) pollen beetle, (*D*) first generation brassica pod midge, (*E*) second generation brassica pod midge, (*F*) cabbage stem weevil and (*G*) cabbage seed weevil (1998/1999 data from Rothamsted Research, UK)

The cabbage stem flea beetle is the only one of these major pests to lay its eggs in the soil, around or on the lower parts of the rape plants. Most eggs are laid in the autumn, but oviposition can continue during warmer periods in winter and spring. Larvae enter the stems and lower leaf petioles where they feed, from September onwards. Mature larvae leave the stems from February onwards, and burrow into the soil to pupate. The other major pests lay their eggs in or on the rape plant: in the buds (pollen beetle), in or on the leaf petioles or stems (the stem weevils) or in the pods (the seed weevil and the pod midge) during the spring/summer. When mature, the larvae of these pests drop from the plant to the ground, where they burrow into the soil to pupate. The exact timing of larval drop varies with country and season but, in the UK, occurs in temporal succession from mid February to early July (Fig. 4.1). The rape stem weevil is not found in the UK, but, in Germany, its mature larvae drop to the ground during May and June. Densities of mature larvae dropping from the canopy of the winter oilseed rape crop can be high e.g., of cabbage stem flea beetle 134/m^2 (Ferguson et al. 2006), of pollen beetle 2,000/m^2 (Nuss and Büchs 2000) and 1,597/m^2 (Ferguson et al. 2003a), of cabbage seed weevil 820/m^2 (Nuss and Büchs 2000), and of brassica pod midge, 7,342 larvae/m^2 (Ferguson et al. 2004).

4.2 Incidence

There are about 2,700 species of carabid in Europe (Wachmann et al. 1995) but only *ca.* 42 species occur commonly in fields of oilseed rape (Table 4.1) and have potential to contribute significantly to biocontrol of pests of the rape crop. Thirty-seven of these species are known to be predators of pests, including coleopteran and dipteran pests, on other crops (Sunderland 2002). The density of carabids in oilseed rape fields can be high; 20–80 individuals/m^2 have been reported from Germany (Basedow 1973, Büchs and Nuss 2000).

Table 4.1 Carabid species most frequently found in oilseed rape crops in Europe[a](x) and their dominance in winter (W) and spring (S) crops in different countries from 2003 to 2005

	EUR W	DE W	SE W	PL W	UK W	UK W	UK W	EE W	EE W	EE S	EE S	EE S
Species[b]	–	2003	2003	2003	2003	2004	2005	2004	2005	2003	2004	2005
Agonum muelleri (Herbst)	x	–	–	–	–	–	–	–	+++	+	++	+
Amara aenea (De Geer)	x	–	–	–	–	–	–	–	–	–	–	–
Amara bifrons (Gyllenhal)	–	–	–	–	–	–	–	–	–	–	+++	–
Amara eurynota (Panzer)	–	–	–	–	–	–	–	–	++	–	+++	+
Amara familiaris (Dufschmid)	x	–	–	–	–	–	–	–	–	–	–	–
Amara fulva (Müller)	–	–	–	–	–	–	–	–	–	–	–	+
Amara ovata (Fabricius)	x	–	–	–	–	–	++	–	+	–	–	+
Amara plebja (Gyllenhal)	–	–	–	–	+	+	–	–	–	–	–	–
***Amara similata* (Gyllenhal)**	x	+++	+++	–	+++	+++	+++	–	+	–	–	–
***Anchomenus dorsalis* (Pontoppidan)**	x	+	+++	–	+	++	–	+	+	+	–	–
Asaphidion flavipes (L.)	–	–	–	–	+	+	–	–	++	–	–	+
***Bembidion lampros* (Herbst)**	x	–	+++	+	+	+	–	–	–	–	–	+
Bembidion properans (Stephens)	x	–	–	–	–	–	–	–	–	+	++	++
Bembidion tetracolum (Say)	x	–	–	–	–	–	–	–	–	–	–	–
Bembidion quadrimaculatum (L.)	–	–	–	–	–	–	–	–	–	–	–	+
Calathus ambiquus (Paykull)	–	–	–	–	–	–	–	–	–	–	+	+
Calathus cinctus Motschulsky	x	–	–	–	–	–	–	–	–	–	–	–
Calathus erratus (Sahlberg)	x	–	–	–	–	–	–	–	–	–	–	+
Calathus fuscipes (Goeze)	x	–	–	–	–	–	–	–	–	–	–	–
Calathus melanocephalus (L.)	–	–	–	–	–	–	–	–	–	+	–	++
Carabus auratus L.	x	–	–	–	–	–	–	–	–	–	–	–
Carabus cancellatus Illiger	–	–	–	–	–	–	–	+	+	–	–	–
Carabus granulatus L.	–	–	–	–	–	–	–	–	+	–	–	–
Clivina fossor (L.)	x	–	–	–	+	–	–	–	++	–	–	+
Curtonotus aulicus (Panzer)	–	–	–	–	–	–	–	–	–	–	+	–
***Harpalus affinis* (Schrank)**	x	+	+++	+	+	++	–	+	+	–	+	++

Table 4.1 (continued)

	EUR W	DE W	SE W	PL W	UK W	UK W	UK W	EE W	EE W	EE S	EE S	EE S
Harpalus brevicollis Seville	–	–	–	++	–	–	–	–	–	–	–	–
***Harpalus rufipes* (De Geer)**	x	+	+++	+	+	++	+	+	+	+++	+++	+
Harpalus tardus (Panzer)	x	–	–	–	–	–	–	–	–	–	–	–
Leistus spinibarbus (Fabricius)	–	–	–	–	+	–	+	–	–	–	–	–
***Loricera pilicornis* (Fabricius)**	x	+++	+	–	+	–	+	–	+	–	+	–
***Nebria brevicollis* (Fabricius)**	x	+	–	–	+++	–	+++	–	–	–	–	–
Nebria salina Fairmaire and Laboulbène	–	–	–	–	–	–	++	–	–	–	–	–
Notophilus biguttatus (Fabricius)	x	+	–	–	+	+	–	–	–	–	–	–
***Poecilus cupreus* (L.)**	x	++	+	+++	++	+	++	+++	+	+++	++	+
Poecilus versicolor (Sturm)	x	–	–	–	–	–	–	–	+	–	+	–
Pterostichus madidus (Fabricius)	x	–	–	–	+	+	–	–	–	–	–	–
***Pterostichus melanarius* (Illiger)**	x	+	++	++	+++	+	+++	+	++	+++	+++	+++
Pterostichus niger (Schaller)	–	–	–	–	–	–	–	–	–	+	+	+
Stomis pumicatus (Panzer)	–	++	–	–	–	–	–	–	–	–	–	–
Synuchus vivalis (Illiger)	–	++	–	–	–	–	–	–	–	–	+	+
***Trechus quadristriatus* (Schrank)**	x	–	–	–	–	–	–	–	–	–	–	–

EUR = Europe, DE = Germany, SE = Sweden, PL = Poland, UK = United Kingdom, EE = Estonia. + denotes species comprised 1–5% of catch, ++ denotes 5–10%, +++ denotes > 10%. *T. quadristriatus* occurs mainly in autumn. Key species in bold.

[a]Data for EUR from Büchs and Alford (2003), for DE, SE and PL from Büchs et al. (2006), for UK (Williams, unpublished), for EE from Tarang et al. (2004), Luik et al. (2006), Veromann et al. (2006).

[b]Nomenclature follows Luff (2007): *Anchomenus dorsalis* (Pontoppidan) (= *Agonum dorsale* (Pontoppidan)); *Curtonotus* (= *Amara*) *aulicus* (Panzer); *Harpalus* (= *Pseudoophonus*) *rufipes* (DeGeer) (= *H. pubescens* (Müller)); *Harpalus affinis* (Schrank) (= *H. aeneus* (Fabricius)); *Poecilus* (= *Pterostichus*) *cupreus* (L.); *Synuchus vivalis* (Illiger) (= *S. nivalis* (Panzer)).

The species composition of the carabid assemblage in a particular oilseed rape field is largely determined by its geographical position, local soil and crop conditions and by time of year (Luff 2002).

The carabid community in oilseed rape fields is not uniform across Europe and dominant species vary with country. Pitfall trapping in winter rape crops in five countries: Estonia, Germany, Poland, Sweden and the UK, during May–July 2003–2005, as part of the EU-funded MASTER project (Williams et al. 2005), identified 25 species that were dominant (>1% of the catch) (Tarang et al. 2004, Büchs et al. 2006, see Table 4.1). Only four species: *H. affinis*, *H. rufipes*, *P. cupreus* and *P. melanarius*, were dominant in all five countries. Other species were dominant in some countries but not others, e.g., *A. similata* and *A. dorsalis* were dominant species in Estonia, Germany, Sweden and the UK but not in Poland. Some were dominant in only one country but not elsewhere, e.g., *P. madidus* in the UK, *Stomis pumicatus* in Germany and *H. brevicollis* in Poland. Five species (*A. muelleri*, *A. eurynota*, *C. cancellatus*, *C. granulatus* and *P. versicolor*) were dominant only in Estonia. A study by Houpert (1983) found that the most active/abundant carabid species in winter oilseed rape in France were *A. dorsalis, P. cupreus* and *P. melanarius*. In both Estonia and the UK, where data is available from more than 1 year, relative activity/abundance of many species varied with year. In the UK, whereas *A. similata* represented >10% catch in all 3 years 2003–2005, *H. affinis* represented 5–10% in 2004, 1–5% in 2003 and <1% in 2005 (Table 4.1).

Comparison of carabid communities on winter and spring oilseed rape has also revealed differences. Thus, in Estonia, the assemblage of dominant carabid species was more species-rich in the spring than in the winter crop and its composition varied between winter and spring crops; of the 15 species dominant on winter rape, four were not dominant on spring rape and of 26 species dominant on spring rape, 12 were not dominant on winter rape. A study of carabids in spring turnip rape *Brassica rapa* (L.) fields in southern Finland during 1983–1985, 1993 and 1999–2000 identified over 40 species of which only five were consistently found: *H. rufipes, P. melanarius, C. melanocephalus*, *C. fossor* and *A. eurynota* (Hokkanen 2004).

The habitat preferences of different species of carabid determine the composition of the carabid communities of arable crops, like oilseed rape. These are so specific that carabids are often used as indicator species to characterise different habitats (Lövei and Sunderland 1996, Holland et al. 2002). The persistence in a habitat of a species depends on the habitat being suitable for the most vulnerable and least mobile stage of the life cycle, the larva. Thiele (1977) distinguished two main groups of carabids: 'field' and 'woodland', according to their temperature, light and moisture preferences; 'field' species, on the whole, prefer warmer and drier sites than 'woodland' species, although species within each group have different microclimate preferences. Forsythe (2000) classified the commoner carabid species into 11 ecological groups, largely according to their soil moisture preferences. Thirty-two of the 42 species found frequently in oilseed rape fields (Table 4.1) were classified by Forsythe. Twenty-four of these species prefer either well-drained dry soils or soils of intermediate moisture content, four (*A. flavipes*, *C. granulatus*, *L. pilicornis* and *P. cupreus*) prefer moist or very wet conditions, one (*Pterostichus niger*) prefers woodland and one (*C. fossor*) lives below the soil surface.

The carabid community of the oilseed rape crop can differ from that in cereal crops despite the fact that most oilseed rape crops are grown as break crops within a cereal rotation (Büchs 2003b). There are several reasons for this. The microclimate provided beneath the crop canopy varies with crop, its structure and its management, particularly tillage, crop density and weed control, and harvest. Further, the adults of most carabid species leave the field to overwinter in field margins and re-colonise fields afresh each spring; species vary as to their phenology, mobility, habitat and food preferences and consequently in the extent to which they disperse from field margins into different crops.

The management of any arable crop has potential to affect the composition of the carabid community within it, but there is relatively little published information on how carabids in oilseed rape crops are affected by different husbandry practices (Büchs et al. 1997, Büchs 2003b, Luik et al. 2006). Table 4.2 summarises results from some limited field trials in UK and in Estonia (Veromann et al. 2006) in which oilseed rape crops were grown under farming systems with different management intensity. Standard practice with tillage by ploughing was compared to an integrated crop management system with non-inversion tillage; insecticide was either not applied, applied when pest thresholds had been exceeded or applied prophylactically. Although carabid numbers were very variable between fields in the same trial, between years and between crops, in five of the six trials, carabid numbers were greatest in a field with an integrated than a standard system of management. This general trend concurs with results of other studies both on oilseed rape (Büchs et al. 1997, Büchs 2003b, Luik et al. 2006) and in cereals (Holland and Luff 2000, Hance 2002, Holland et al. 2002) that show that species richness and activity-densities increase with more extensive management systems, reduced tillage and lower pesticide inputs. However, it is also generally recognised that the

Table 4.2 Total numbers of carabids caught in pitfall traps in oilseed rape crops under different management systems in UK and Estonia (EE) during spring/summer 2003–2005. Carabid numbers in bold indicate the management system with the most abundant carabids in each crop/country/year combination

Crop management system			Standard practice		Integrated	
Tillage			Plough		Non-inversion	
Insecticide use			Prophylactic	According to pest threshold	According to pest threshold	None
Winter rape	UK	2003	170	–	–	**632**
		2005	763	523	**1,199**	345
	EE	2004	3,559	3,400	3,027	**3,654**
		2005	432	**736**	650	583
Spring rape	EE	2003	876	–	–	**2,169**
		2004	1,743	–	–	**2,329**

drawbacks of pitfall trapping as activity-density traps (Adis 1979) make the effects of husbandry on carabid abundance difficult to interpret.

In conclusion, on the basis of their incidence in fields of winter oilseed rape in Europe, ten species of carabid emerge as key species in carabid assemblages in the crop. These are *A. similata*, *A. dorsalis*, *B. lampros*, *H. affinis*, *H. rufipes*, *L. pilicornis*, *N. brevicollis*, *P. cupreus* and *P. melanarius* during the summer months and *T. quadristriatus* during the autumn. In spring rape, *A. muelleri* and *A. eurynota* may also be important, but to date, available information on spring rape comes largely from Estonia and Finland. With their wide distribution and dominance (usually more than 1% but more than 10% of the carabid assemblage in some years) these species have greatest potential as predators of oilseed rape pests.

4.3 Spatio-Temporal Relationships Between Carabids and Pests

Carabids have potential to contribute to biocontrol of a pest only if they coincide temporally and spatially with the pest stage that is vulnerable to predation. Information on the spatio-temporal relationships between carabids and pests and how these are influenced by environmental factors and crop husbandry practices is therefore key to the development of strategies aiming to enhance conservation biocontrol in a crop.

Despite the apparent homogeneity of an arable crop, such as oilseed rape, the habitat within it is far from stable and is not homogeneous, and consequently the temporal and spatial distributions of carabids, like that of the pests, are neither constant nor uniform within it. The cropped area of a field undergoes considerable change between crop establishment and harvest; consequently it provides a suitable habitat for most species of carabid only temporarily. Most species move between the more stable environment of the field margin into the cropped area of the field at different times of the year to make use of suitable habitats and the resources within them as they become available.

Most studies of within-field carabid-pest associations in agro-ecosystems have been in cereal crops, with particular emphasis on their impact on cereal aphid populations (e.g., Vickerman and Sunderland 1975, Winder 1990, Bohan et al. 2000, Winder et al. 2001, Sunderland 2002, Symondson et al. 2002a,b Thomas et al. 2002). The first major study of the within-field distributions of carabids in relation to those of the pests of oilseed rape is reviewed below.

4.3.1 Within-Field Distributions of Pests and Carabids: A Case Study

The first intensive field-scale study in Europe of the spatio-temporal distributions of carabids and the major coleopteran and dipteran pests of oilseed rape (Warner 2001, Warner et al. 2000, 2003, 2008) was conducted in a field of winter rape (2.4 ha) on

Rothamsted Farm, Hertfordshire, UK, in 1998/1999. The study focussed on identifying the carabid species spatially and temporally coincident with five pests, each at the stages of their life cycles most vulnerable to predation by adult carabids, viz., cabbage stem flea beetle eggs, which are laid in the soil around the rape plants, and their newly-hatched larvae before plant entry in the autumn, and the mature larvae of the pollen beetle, brassica pod midge, cabbage stem weevil and cabbage seed weevil as they dropped from the canopy of the crop to the soil to pupate during spring and summer.

Plant and insect samples were taken from each of 36 or 40 spatially-referenced sampling locations arranged as a grid across the crop to provide two-dimensional data on distributions. Adult cabbage stem flea beetles were caught during autumn in water trays placed on the soil. Rape plants were examined for flea beetle larval infestation in early December and again in mid March. Mature pollen beetle, weevil and pod midge larvae were collected, from May to July, as they dropped from the plants into water trays placed on the soil. Carabids were sampled using pitfall traps. The spatial distributions were analysed and compared using SADIE (Spatial Analysis by Distance IndicEs, Perry 1995, 1998a,b) to determine the degree of spatial association between carabids and pests.

4.3.1.1 Cabbage Stem Flea Beetle and Carabids

Cabbage stem flea beetle adults were caught from mid-September to late October (when the first ground frosts occurred) with a peak in mid-October; most eggs were probably laid during this period (Warner et al. 2003). Adults gradually colonised most of the crop area but were clustered in the south-western and central areas. Spatial association analyses found a positive relationship between the distributions of adult females (Fig. 4.2a) and flea beetle larvae (Fig. 4.2b); these distributions probably also reflect those of eggs and young larvae in the soil vulnerable to predation by carabids.

Three species of carabid dominated the pitfall trap catches in the autumn: in order of abundance, *T. quadristriatus*, *N. brevicollis* and *P. madidus*. The first and last of these species were most active/abundant in mid-October while *N. brevicollis* declined from late September to mid-October, and then increased gradually during the second half of October. Of these species, *T. quadristriatus* (Fig. 4.2c) and *P. madidus* (Fig. 4.2d) were spatially associated with flea beetle larvae during October.

4.3.1.2 Pollen Beetle and Carabids

Mature pollen beetle larvae dropped from the crop canopy during May with a peak between 8 and 20 May (Warner et al. 2008). They were clustered in the north-eastern and eastern part of the crop, extending towards its centre (Fig. 4.3a).

The five most active/abundant carabid taxa in pitfall traps (>5% of catch) during peak pollen beetle larval drop, were, in order of abundance, *A. similata*, *A. dorsalis*, *N. brevicollis*, *Asaphidion* spp. and *L. pilicornis*. Of these, *A. similata* (Fig. 4.3b),

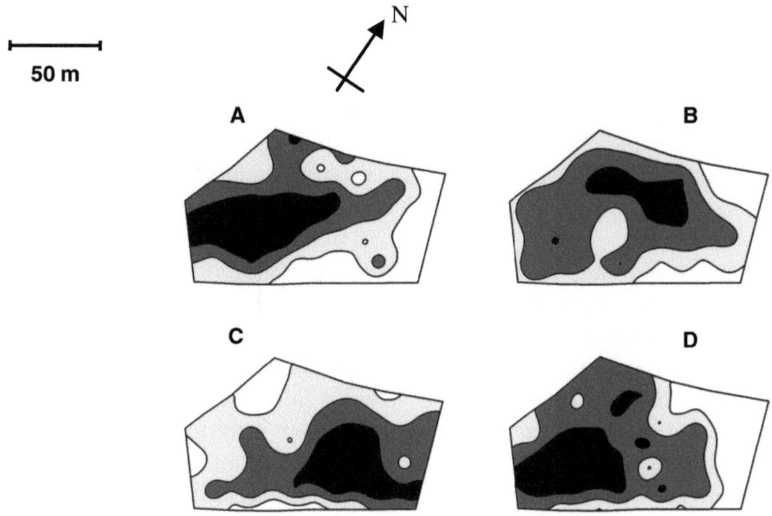

Fig. 4.2 SADIE plots of spatial distributions of (**a**) total female cabbage stem flea beetle caught in water traps on the ground, (**b**) total cabbage stem flea beetle larvae in plants, (**c**) total *T. quadristriatus* and (**d**) total *P. madidus* in a winter oilseed rape crop at Rothamsted, UK in 1998 (after Warner et al. 2003 Figs. 3a, b, 4a and b). All insects were sampled at 36 locations across the field. Black and white areas are identified by SADIE as clusters and gaps in insect distributions, respectively. In *dark grey* and *pale grey areas* the distributions show tendencies towards clustering and gappiness, respectively

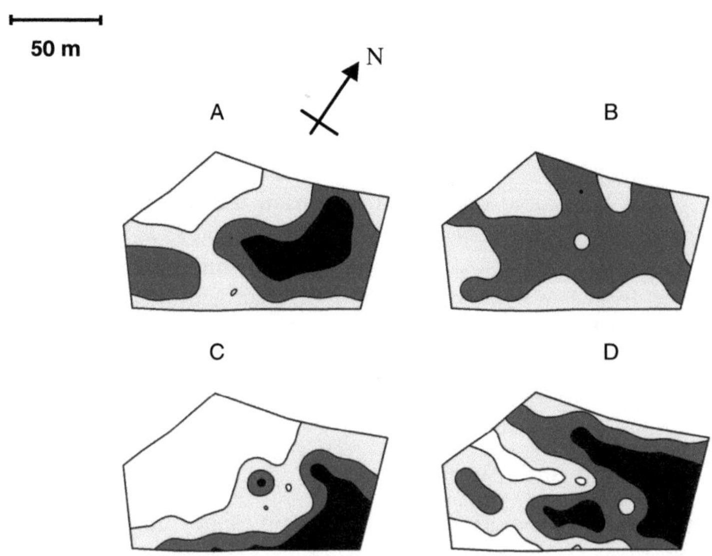

Fig. 4.3 SADIE plots of spatial distributions of (**a**) pollen beetle larvae and of (**b**) *A. similata*, (**c**) *N. brevicollis* and (**d**) *Asaphidion* spp. adults during peak larval drop in May 1999 in a winter oilseed rape crop at Rothamsted, UK (after Warner et al. 2008, Fig. 2 a-l with the three dates combined to give 8–20 May). All insects were sampled at 40 locations across the field. Other details as Fig. 4.2

N. brevicollis (Fig. 4.3c), and *Asaphidion* spp. (Fig. 4.3d) were spatially associated with pollen beetle larvae for at least part of the peak larval drop period. In a similar study in Germany, Felsman and Büchs (2006) reported finding spatial associations between pollen beetle larvae and *A. dorsalis*, *H. affinis*, *H. rufipes*, *P. cupreus* and *Pterostichus macer* (Marsham).

4.3.1.3 Cabbage Stem Weevil and Carabids

Mature weevil larvae dropped from the plants in June with two peaks in their abundance (Warner et al. 2008). The first peak (1–5 June), was assumed to be mostly stem weevil larvae as they mature before those of the seed weevil. At this time, the larvae were clustered at the eastern edge of the crop (Fig. 4.4a).

Carabid species active/abundant (>5% trap catch) during the stem weevil larval peak were *A. similata*, *N. brevicollis* and *A. dorsalis*, of which *A. dorsalis* (Fig. 4.4b) was spatially associated with the larvae.

4.3.1.4 Cabbage Seed Weevil and Carabids

The second peak of weevil larvae to drop from the plants (25–29 June) was assumed to be mostly seed weevil as they mature later than those of stem weevil (Warner et al. 2008). The larvae were fairly evenly distributed across the crop although more clustered around the edge, particularly at the eastern edge, than at the centre (Fig. 4.5a). Carabid species active/abundant (>5% trap catch) during the seed weevil larval peak were, in order of abundance, *A. similata*, *A. dorsalis*, *P. madidus* and *P. melanarius* of which only *A. dorsalis* was spatially associated with seed weevil larvae (Fig. 4.5b).

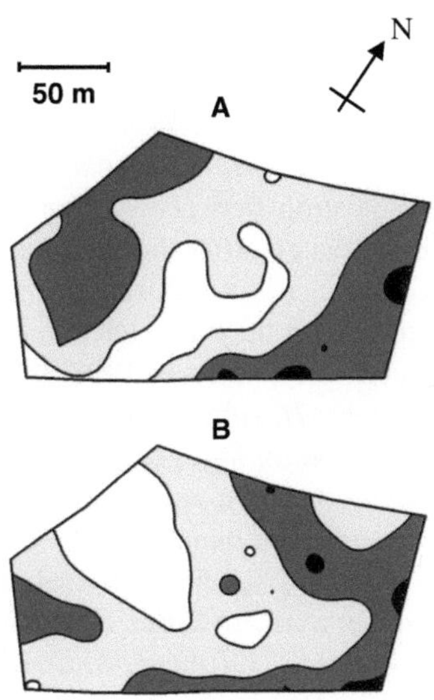

Fig. 4.4 SADIE plots of spatial distributions of (**a**) cabbage stem weevil larvae and of (**b**) *A. dorsalis* during peak larval drop in early June 1999 in a winter oilseed rape crop at Rothamsted, UK (after Warner et al. 2008, Fig. 3a, c). Other details as Fig. 4.3

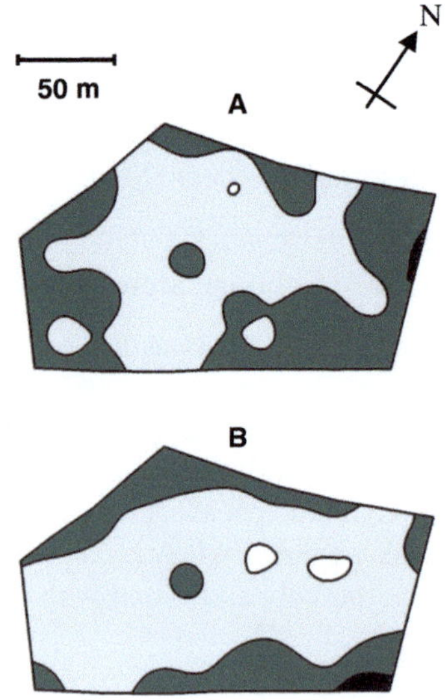

Fig. 4.5 SADIE plots of spatial distributions of (**a**) cabbage seed weevil larvae and (**b**) *A. dorsalis* during peak larval drop in late June 1999 in a winter oilseed rape crop at Rothamsted, UK (after Warner et al. 2008, Fig. 3b, d). Other details as Fig. 4.3

4.3.1.5 Brassica Pod Midge and Carabids

Mature pod midge larvae dropped from the pods into trays beneath the canopy from 16 May until the end of sampling on 19 July, just before harvest (Warner et al. 2000, Warner 2001). First generation larvae were most abundant from 1 to 5 June and second generation larvae from 3 to 7 July. Both generations of larvae had marked edge distributions, being more clustered within the first 20 m of the crop (Fig. 4.6a). Other studies (e.g., Free and Williams 1979) have also found the pod midge to be largely edge-distributed.

Carabid species most active/abundant during peak drop of first generation pod midge larvae (1–5 June) were, in order of abundance, *A. similata*, *A. dorsalis*, *N. brevicollis*, *L. pilicornis*, *B. lampros* and *A. flavipes*. Of these, only *A. dorsalis* (Fig. 4.6b) was spatially associated with the midge larvae. Although spatial association between *N. brevicollis* and midge larvae was not significant overall, immediately after the midge larval peak (5–9 June), *N. brevicollis*, which had been centre-distributed during the peak, became clustered in the eastern corner of the field suggesting they may have moved to where midge larvae were most abundant. In an earlier pilot study in 1998, temporal and overall spatial association was also found between *H. rufipes* and first generation midge larvae (Warner et al. 2000). Carabid predation on first generation midge larvae is probably more important than that on the second as it would reduce the number surviving to emerge as adult midges and cause further damage to the crop later that year.

Carabid species active/abundant during the drop of second generation midge larvae were, in order of abundance, *P. madidus*, *A. similata*, *P. melanarius* and *A. dorsalis*. Of these, *A. dorsalis* was spatially associated overall with the larvae in late June/early July, just before peak larval drop and it remained most active

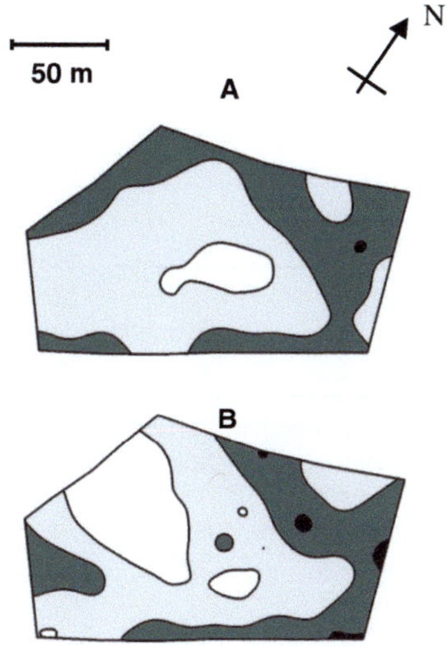

Fig. 4.6 SADIE plots of spatial distributions of (**a**) first generation brassica pod midge larvae and (**b**) *A. dorsalis* during peak midge larval drop in a winter oilseed rape crop at Rothamsted, UK, 1–5 June 1999 (after Warner et al. 2000 Fig. 2a, b). Other details as Fig. 4.3

along the north-eastern edge and in the northern and eastern corners where midge larvae were abundant. In Germany, Felsman and Büchs (2006) reported finding spatial associations between brassica pod midge larvae and *A. similata*, *H. rufipes* (= *Pseudoophanus rufipes*), *S. pumicatus* and *P. cupreus*. Spatial associations between pod midge larvae and those carabid species that can be spermophagous (such as *A. similata* and *H. rufipes,* Luff 1980, Jørgensen and Toft 1997) must be treated with some caution as they may also reflect an association between the carabid and oilseed rape seeds which fall to the ground from pods that split to release midge larvae.

4.3.2 Factors Affecting Within-Field Carabid Distributions

Many of the biotic and abiotic factors affecting the distribution of the carabid assemblage in oilseed rape fields are the same as those in cereal fields (most recently reviewed by Thomas et al. 2002), but relatively little is known about the factors affecting the distributions of individual species within the assemblage. Here, we examine how their life cycles, habitat preferences, mobility and food availability affect the distributions of carabid species active in fields of oilseed rape.

4.3.2.1 Life Cycles

Most of the carabid species associated with oilseed rape crops have an annual life cycle, reproducing either in the spring after adult overwintering, or in the

Table 4.3 Carabid species that were abundant in pitfall traps and temporally (T) or spatio-temporally (**TS**) coincident with the eggs/larvae of oilseed rape pests in the soil of a winter oilseed rape crop in UK (after Warner 2001, Warner et al. 2000, 2003, 2008). Breeding season after Thiele (1977)

Carabid		Pest					
		Cabbage stem flea beetle		Cabbage	Cabbage	Brassica pod midge larvae	
Taxa	Breeding season	eggs/ larvae	Pollen beetle larvae	stem weevil larvae	seed weevil larvae	1st generation	2nd generation
Amara similata	Spring	–	**TS**	T	T	T	T
Anchomenus dorsalis	Spring	–	T	**TS**	**TS**	**TS**	**TS**
Asaphidion spp.	Spring	–	**TS**	–	–	T	–
Bembidion lampros	Spring	–	–	–	–	T	–
Harpalus rufipes	Autumn	–	–	–	–	**TS**	–
Loricera pilicornis	Spring	–	T	–	–	T	–
Nebria brevicollis	Autumn	T	**TS**	T	–	**TS**	–
Pterostichus madidus	Autumn	**TS**	–	–	T	–	T
Pterostichus melanarius	Autumn	–	–	–	–	–	T
Trechus quadristriatus	Autumn	**TS**	–	–	–	–	–

autumn before larval overwintering (Table 4.3). A few, however, are biennial, or live and breed for two or more seasons and overwinter as both adults and larvae (Thiele 1977).

The spring-breeders typically overwinter as inactive adults within hibernacula in the sheltered semi-natural habitat of field margins. They recolonise fields by migration from the margins in the spring, generally between March and May (Sotherton 1984, 1985) and tend consequently to be more abundant on field headlands than crop centres. They are most active feeding and reproducing in the field during spring/early summer (March to June), after which the adults usually die (Luff 1986, Wallin 1985, 1989, Lys 1994, Zangger et al. 1994, Fadl and Purvis 1998). New generation adults appear in the autumn and most species migrate out of the field to field margins to overwinter, but some species, such as *A. flavipes* and *B. lampros*, may remain active in oilseed rape crops throughout the winter (Büchs 2003b). Within cereal fields, the distributions of some species (e.g., *B. lampros*, *P. melanarius* and *P. cupreus*) have been shown to be relatively stable through time (Holland et al. 1999, Thomas et al. 2001), suggesting strong attraction to suitable environments and/or resources within certain parts of the field; mobile species are also capable of moving rapidly to resource-rich patches.

Spring-breeding carabid species are probably more important than autumn-breeding carabid species as predators of oilseed rape pests as they are active in the crop early in the season when mature larvae of all the major pests are dropping from the plants to the soil to pupate. However, they probably have more impact at the edges than at the centres of crops. Thus, the spring-breeding *A. dorsalis*, which was spatially and temporally associated with cabbage stem weevil, cabbage seed weevil (Warner et al. 2008) and pod midge larvae (Warner et al. 2003), was largely edge-distributed. However, the spring-breeding *A. similata* and *Asaphidion* spp., which were both spatially and temporally associated with pollen beetle larvae (Warner et al. 2008), were more centre-distributed. Both of these carabid species have strong dispersal abilities; *A. similata* can migrate to rape by flight (Williams unpublished) and *A. flavipes* is capable of rapid migration into crops from overwintering sites in field margins (Thomas et al. 2002, Holland 2002). In winter wheat, *A. flavipes* was similarly found in small patches both within the field and its margins whereas *B. lampros*, another spring breeder, was mostly within 60 m of the field edge, close to a hedgerow (Holland et al. 1999, Thomas et al. 2001).

Autumn-breeders, typically overwinter as larvae within the field (Lyngby and Nielson 1980, Coombes and Sotherton 1986) and emerge as adults the following summer, e.g., adult *P. madidus* and *P. melanarius* emerge during July and August (Thiele 1977). These new generation adults tend to be more evenly distributed within the field than those of spring-breeders, as they do not need to migrate in from field margins. Reproducing in the late summer/autumn, they feed most actively at this time and are therefore potentially useful predators of eggs/larvae present in the soil later in the season. For example, Warner et al. (2003) found *T. quadristriatus* and *P. madidus* to be spatially and temporally associated with cabbage stem flea beetle eggs/larvae in the autumn.

All five of the autumn-breeding species most common in oilseed rape fields (*N. brevicollis*, *H. rufipes*, *T. quadristriatus*, *P. madidus* and *P. melanarius*) may survive as adults for a second year, often overwintering in field margins (Greenslade 1965, Luff 1980) as well as in autumn-sown cereal fields (Sotherton 1984). Like spring-breeding species, some adults may recolonise the crop from the margins in the spring and, if they survive the winter in large enough numbers, they can be potential predators of early-maturing pest larvae. Adult *N. brevicollis* and *T. quadristriatus* have also been reported to be active during the winter within oilseed rape fields (Büchs 2003b). In winter wheat, *P. madidus*, *P. melanarius* have been found in large patches within the cropped area of the fields (Holland et al. 1999, Thomas et al. 2001).

Nebria brevicollis adults emerge early (April and May), giving this species the temporal characteristics of a spring-breeder, i.e., it is active early in the season, but, as the larvae overwinter within the crop, its spatial distribution is governed by its autumn-breeding habit and it is not restricted to the field edge early in the season (Greenslade 1964, Penney 1966, 1969, Fernàndez Garcìa et al. 2000). Thus, although *N. brevicollis* was spatially associated with an edge-distributed patch of first generation pod midge larvae in early June (Warner et al. 2000), the carabid was as abundant in the centre of the crop as at its edge when centre-distributed pollen beetle larvae were dropping from plants in May (Warner et al. 2008). In winter wheat, *N. brevicollis* has similarly been found in large patches within the cropped area of the field (Holland et al. 1999). However, teneral adults have only a few weeks of activity on the crop during June because they have an obligatory aestivation of 5–6 weeks during July and August before reproducing (Penney 1966, 1969). Although *N. brevicollis* was trapped in large numbers in early June, by late June it was scarce in traps, probably because of the onset of summer aestivation (Warner et al. 2008). This limits its potential as a predator of the later-maturing larvae of the cabbage seed weevil and the second generation pod midge (Warner et al. 2000). It is also dependent on a woodland type habitat, e.g., a hedgerow, in which to shelter during aestivation (Fernàndez García et al. 2000).

Harpalus rufipes is biennial in the UK (Briggs 1965, Luff 1980). Larvae overwinter in the field but do not emerge as adults until late July/early August. They do not breed in their year of emergence but overwinter in autumn-sown crops or in field margins, migrating back into fields once more in May and June to breed (Sotherton 1984). Adults are most active during June–August (Briggs 1965). In June, Warner (2001) trapped large numbers of adult *H. rufipes*, probably second-year adults, in a rape field adjacent to a grass bank in which they had probably overwintered, and found they were temporally and spatially associated with edge-distributed first generation pod midge larvae (Warner et al. 2000). Similarly, in winter wheat, *H. rufipes* has been found predominantly within 60 m of the field edge, close to a hedgerow (Holland et al. 1999, Thomas et al. 2001).

Trechus quadristriatus adults may also survive more than one season (Mitchell 1963, Thiele 1977, Paul 1986). In the autumn they need to feed to build up fat reserves for the winter and remain active at low temperatures (3°C). Although generally autumn-breeding, adults that overwinter may also breed in the spring (Mitchell 1963). In the autumn, they were both temporally and spatially associated with cabbage stem flea beetle eggs/larvae (Warner et al. 2003) but they also have potential to be predators of this pest in the spring when mature larvae drop to the ground to pupate.

4.3.2.2 Habitat Preferences

Habitat requirements include physical factors such as microclimate (temperature, moisture and light; see also Section 4.2) as well as the need for field margins for overwintering (see Section 4.3.2.1 above), and the need for sufficient food (see Section 4.4).Within-field variability of environmental factors has been shown to affect carabid distributions (Thomas et al. 2002). In cereal fields carabid activity-densities and/or predation levels are related particularly to soil moisture content (Hengeveld 1979) and weed cover (Speight and Lawton 1976, Holland et al. 1999). Plant cover, from the crop stand or from weeds affects the physical environment providing shade, lowering temperature and increasing humidity as well as potentially providing habitat for prey or seed for spermatophagous species (see

Section 4.3.2.4 below). Other factors that influence distributions are soil organic content and its chemical and physical properties, e.g., pH, calcium content and particle size (Thomas et al. 2002). The impact of on-farm landscape structures on carabids has recently been reviewed by Büchs (2003b).

4.3.2.3 Mobility

Carabid species vary in their ability to move within the landscape, between overwintering sites and fields as well as between and within fields. Most carabids disperse by walking or running along the soil surface and can do so at speed; *P. niger* can move at up to 20 m/h in a cereal field (Wallin and Ekbom 1988). Hind wing development in carabids is variable, many are reluctant fliers or have vestigial wings and are incapable of flight (Lindroth 1974, Thiele 1977). However, some, among them *Bembidion* spp. *Amara* spp., *Harpalus* spp. and *L. pilicornis*, can fly. By flight, carabids are capable of moving rapidly over greater distances (Thiele 1977, van Huizen 1990, Luff 2007) and are thus adapted to disturbed habitats such as cultivated fields (Kromp 1999). Flight aids dispersal to new habitats and the search for food and mates. *Amara similata* overwinters in field margins but flies to crops in the spring (Thomas et al. 2002). The species appears to have an attraction for the rape crops over cereal crops, as they are more active/abundant in the former (Luka et al. 1998). They have been caught in window traps downwind of a winter rape crop during April (Williams unpublished) suggestive of upwind anemotactic flight towards it.

At landscape, farm and field scales, carabid movement can be impeded by physical barriers e.g., hedges, ditches, banks, fences, tracks, roads, railways and rivers, particularly for those species which walk or run (Thomas et al. 2002).

4.3.2.4 Food Availability

The availability of suitable food for adult and larval carabids and its spatial distribution within a field influences the spatial distribution of carabids.

Most species of carabid are polyphagous to some extent, eating both animal and plant matter (Lindroth 1992, Toft and Bilde 2002). They seek their food actively by random search. Visual cues are important for diurnal species (Forsythe 2000) but tactile, gustatory and olfactory cues are also used (Bryan and Wratten 1984, Wheater 1989, Vet and Dicke 1992, Kielty et al. 1996, Lövei and Sunderland 1996, Monsrud and Toft 1999, Mundy et al. 2000). They are opportunistic feeders, many eating whatever they can find (live or dead), handle and consume (Lövei and Sunderland 1996), although some species specialise in a particular diet (Toft and Bilde 2002). Dissection of the guts of 24 European carabid species revealed the remains of aphids, spiders, mites, harvestmen, Collembola, Heteroptera, and adults and larvae of Lepidoptera, Diptera and Coleoptera (Hengeveld 1980a, b). Other studies have shown that some also feed on worms, nematodes, Hymenoptera, centipedes, millipedes, molluscs, fungi, seeds and pollen (Pollet and Desender 1987, Sunderland et al. 1995). Sunderland (2002) has reviewed the pest species consumed by carabids. Diet can vary with season (Toft and Bilde 2002). Carabid larvae tend to have similar diets to the adults although the larvae of some species e.g., *H. rufipes* and *A. similata*, will develop normally when fed on seeds alone (Briggs 1965, Thiele 1977, Luff 1980, Toft and Bilde 2002).

Adult carabids are voracious feeders, consuming close to their own body weight of food daily (Thiele 1977). The food is used to build up fat reserves, especially for reproduction and hibernation (Thiele 1977) and a mixed diet of high-quality food has been shown to maximise reproductive potential in many species (Wallin et al. 1992, Bilde and Toft 1994, Zangger et al. 1994, Bilde et al. 2000, Toft and Bilde 2002).

Continuity of food supply is important if the carabid community in an arable field is to be retained. Brust (1990) suggested that carabids remain in the same foraging area within a crop for as long as the food supply is adequate and mark-recapture studies tend to confirm this, e.g., Thomas et al. (1998) recaptured the majority of *P. melanarius* within 55 m of their release site after 30 days. In the oilseed rape field the continuous supply of pest eggs/larvae potentially available as food from mid-September to mid-July (Fig. 4.1) should encourage carabids to remain in the field, enhancing their biocontrol potential. In cereal fields, some species have stable distributions (Thomas et al. 2002) but they can also respond to changes in the location of prey (Symondson et al. 2002a); *P. melanarius* has been shown to aggregate in areas of high slug (Symondson et al. 1996, Bohan et al. 2000) and aphid densities (Bryan and Wratten 1984, Winder et al. 2001). In oilseed rape, most pests are irregularly distributed (Ferguson et al. 2003b, Williams and Ferguson Chapter 8 this volume) and the carabids with greatest pest control potential are likely to be those predisposed by their habitat requirements to coincide with patches of prey or to move to them. *Nebria brevicollis* apparently moved from the centre of a rape field to its edge where pod midge larvae were more abundant (Warner et al. 2000).

The abundance and distribution within the field of alternative sources of food, such as Collembola and seed, will also affect the spatial distribution of carabids. Collembola are amongst the most numerous of soil arthropods (Wallwork 1976) and are abundant on the soil surface of cultivated fields (Joosse 1981). They form part of the diet of many species of carabid (Mitchell 1963, Sunderland and Vickerman 1980, Hance et al. 1990, Bilde et al. 2000, Toft and Bilde 2002) and some among them, e.g., *N. brevicollis* (Greenslade 1965, Sunderland 1975, Hengeveld 1980a, b) and *L. pilicornis* (Sunderland 1975), are specialist feeders on Collembola. *Nebria brevicollis* has been shown to respond to odour from Collembola in an olfactometer (Kielty et al. 1996). The remains of Collembola have been found in the guts of *A. flavipes*, *L. pilicornis*, *N. brevicollis* and *N. biguttatus* (Davies 1953) and *B. lampros* (Davies 1953, Mitchell 1963). Some carabid species may find some species of Collembola difficult to capture because of their ability to jump (Bilde et al. 2000, Mundy et al. 2000), but, in general, Collembola are probably of great importance in supporting carabid populations, particularly when other prey is scarce (Hengeveld 1980a,b, Pollet and Desender 1987, Bilde et al. 2000, Toft and Bilde 2002). Their distribution in the field is thought to be important in determining the movement of carabids into and within arable fields (Desender et al. 1984) and carabids feeding on Collembola may be more active/abundant both temporally and spatially when and where Collembola occur (Pollet and Desender 1986, Niemelä et al. 1986). In winter rape fields, both *T. quadristriatus* (Warner et al. 2003) and *N. brevicollis* (Warner et al. 2008) were spatially associated with Collembola. Although Collembola support populations of carabids, when present at the same time as pest larvae, they may compete with them as a food for carabids. As a result, pests may survive and reproduce more rapidly in patches where predators have access to alternative prey (Symondson et al. 2002a, Koss and Snyder 2005, Symondson et al. 2006). Carabid species which depend largely on Collembola as prey may not aggregate in response to patches of high prey density, unlike other carabids (Bryan and Wratten 1984). See also Section 4.4.

Seed, both from weeds and from the crop, is probably also an important alternative food source affecting the within-field distribution of many polyphagous predators, particularly seed-predatory carabids, such as *H. rufipes*, that may contribute to weed control (Tooley and Brust 2002). *Amara* spp. (as well as Collembola) have been found to be associated with weed cover in winter wheat (Holland et al. 1999, Fernàndez García et al. 2000) but it is not known whether these associations occur in oilseed rape fields.

4.4 Feeding

The carabids associated with oilseed rape crops are all generalist polyphagous predators, feeding on a wide range of invertebrate prey as well as plant matter. However, species differ in the composition of their diet. Their potential for biocontrol depends on the extent to which the pests of oilseed rape form part of that diet.

4.4.1 The Carabid Diet

Information about the composition of the diet of carabids comes largely from the analysis of the gut contents of field-collected individuals, feeding trials in the laboratory and the study of their functional morphology.

4.4.1.1 Gut Contents

Gut dissection and analysis of the contents is a useful technique for determining the diet of field-collected carabids but only in species that are fragment feeders, e.g., *A. dorsalis, A. similata, H. rufipes, H. affinis, L. pilicornis, N. brevicollis, P. cupreus* and *P. madidus*, and not in fluid feeders with extra-oral digestion, e.g., *T. quadristriatus* (Davies 1953, Thiele 1977, Forsythe 1982, 1983, Evans and Forsythe 1985, Ingerson-Mahar 2002, Toft and Bilde 2002). However, even with fragment feeders the technique has limitations: some food leaves no identifiable traces (e.g., nematodes); different body parts remain in the gut for different lengths of time; dissection cannot determine whether prey was alive or dead when eaten; some contents may be secondary, having been eaten by the prey before being consumed by the carabid (Davies 1953, Sunderland et al. 1987). Direct observation of feeding in the field is impractical, but a range of biochemical, immunological and molecular techniques are now becoming available for use in determining the diet of field-collected carabids (Symondson 2002).

4.4.1.2 Feeding Trials

Feeding trials conducted in the laboratory provide some clues about the foods that carabids will accept and which prey they are physically and behaviourally able to handle and kill. However results from feeding trials must be interpreted with caution. Adult carabids will not necessarily consume food under laboratory conditions that they feed on readily in the field and, conversely, food accepted in the laboratory may not be consumed in the field where a larger choice of food may be available (Luff 1974). Food preferences, as indicated by choice tests, reflect many interacting factors, such as ease of discovery, capture and handling, palatability and nutritional value (Toft and Bilde 2002).

4.4.1.3 Functional Morphology

Study of the functional morphology of carabid body size and form, locomotory adaptations, mouthparts, digestive tract and chemoreceptors can provide useful clues as to diet (Forsythe 2000, Ingerson-Mahar 2002). Carabid species differ greatly in size as do the invertebrates on which they prey. As the size of predator is closely related to the size of its preferred prey (Wheater 1988), the distribution of body sizes within an assemblage of carabids may have a direct impact on the role of that assemblage in biocontrol. Prey size and mobility affects the ability of the carabid to capture, handle and kill it. Penney (1966) examined the guts of *N. brevicollis* and found that they contained remains of diptera (38%), collembola (32%), mites (23%), spiders (4%) and small earthworms (4%) but there was strict selection on the basis of size with no evidence of prey over 4 mm in length. Bulbous compound eyes and narrowed pronota are associated with highly predaceous species, e.g., *A. flavipes*. Long legs and slender build allow movement at speed, probably useful for capturing active prey, such as species of Collembola which can jump; *Nebria, Anchomenus* and *Bembidion* spp. are rapid runners. Short strong legs with wide femora e.g., in *Harpalus* and *Pterostichus* give strength for horizontal pushing. Burrowing ground beetles e.g., *Clivina* spp. have a waisted cylindrical body shape with a muscular thorax, short legs with narrow femora and spines adapted for digging. There is a positive correlation between gape distance of carabid species and median prey size (Wheater 1988). Species with narrow mouths, long mandibles, little or no molar area and a groove with setae on the inner margin of the ventral side of the mandibles tend to be predaceous and fluid feeders. Those with short mandibles and increased molar areas tend toward herbivory. Fragment feeders e.g., *L. pilicornis*, tend to have shorter mandibles; this species has developed a unique hunting technique in which large setae on the underside of the head and antennae trap Collembola (Forsythe 1982).

4.4.1.4 Feeding Guilds

Toft and Bilde (2002) used evidence from gut dissection, laboratory feeding trials and functional morphology to group carabids into seven feeding guilds: generalist carnivores, generalist insectivores, granivores and specialist feeders on molluscs, microarthropods, caterpillars, ants and termites. The species most active in oilseed rape fields are classified by Toft and Bilde (2002) as generalist carnivores, generalist insectivores, microarthropod specialists or granivores (Table 4.4). The generalist carnivores are large and feed on a wide range of invertebrate prey (including earthworms, molluscs and arthropods) as well as plant matter and may also scavenge for food. The smaller generalist insectivores feed mainly on a wide variety of insect prey but may also take seed and other vegetable matter and may scavenge on dead insects. Collembola and mites constitute the majority of the diet of microarthopod specialists and plant matter is not consumed. By contrast, granivores predominantly feed on seeds but may take some insects, as indicated for *H. affinis*, *H. rufipes* and *A. similata* by feeding trials (Table 4.4).

4.4.2 Predation on Pests

Semi-field experiments have shown that emergence of new generation pollen beetle, cabbage stem and seed weevils and brassica pod midge can be substantially reduced (by 45–80%) in rape plots to which generalist polyphagous predators, including carabids, staphylinids and spiders, were added compared to those from which they were excluded (Büchs and Nuss 2000). Pest species differ in their vulnerability to predation by different carabid species and carabid species differ in the extent to which they consume different pest species.

4.4.2.1 Pest Vulnerability

Pest vulnerability to predation by carabids is influenced by their relative diel periodicity, burrowing behaviour and size.

Diel periodicity in the time of the drop of mature larvae from rape plants limits the time they are available to predators. Most cabbage seed weevil and brassica pod midge larvae drop from rape plants at night; Warner (2001) found that 54% of seed weevil larvae and 50% of pod midge larvae dropped between 22.00 and 04.00 h.

Larvae that drop at night are most vulnerable to predation by carabids that feed actively during the night (Table 4.4). Luff (1978) used a 24 h time-sorting pitfall trap to collect carabids in a field in UK. Of the carabid species common in oilseed rape fields (Table 4.1), he found that *H. rufipes*, *Pterostichus* spp., *N. brevicollis*, *A. dorsalis*, *A. apricaria* and *T. quadristriatus* were nocturnal whereas *A. flavipes* and *B. lampros* were diurnal; *L. pilicornis* was mainly nocturnal with a second peak of activity during the day. Autumn-breeding carabids are nocturnal whereas spring-breeders may be nocturnal or diurnal (Thiele and Weber 1968, Table 4.4).

Table 4.4 Carabid species that feed on eggs/larvae of major coleopteran and dipteran pests of oilseed rape. Results from laboratory feeding tests and gut dissection of field-collected carabids. √ denotes that the carabid has eaten a pest in feeding tests or been found to have pest remains in its gut, **x** denotes that a pest was not eaten or not found in the gut. Feeding guild according to Toft and Bilde (2002) and Sunderland (2002), body lengths of carabids according to Luff (2007) and diel periodicity according to Luff (1978)

Carabid				Pest						
				Cabbage stem flea beetle eggs	Pollen beetle larvae		Weevil larvae		Brassica pod midge larvae	
Species	Feeding guild	Length (mm)	Diel periodicity	Eaten	Gut	Eaten	Gut	Eaten	Gut	Eaten
Pterostichus madidus	Generalist carnivore	14–18	Nocturnal/Diurnal	x	√	√	–	√	–	√
Pterostichus melanarius	Generalist carnivore	13–17	Nocturnal/Diurnal	–	–	–	–	–	–	√
Harpalus rufipes	Granivore	11–16	Nocturnal	–	x	√	√	–	x	√
Nebria brevicollis	Microarthropod specialist	11–14	Nocturnal	x	√	√	–	–	–	√
Poecilus cupreus	Generalist insectivore	11–13	–	–	√	√	x	–	x	√
Harpalus affinis	Granivore	9–12	–	–	√	√	√	–	x	√
Amara similata	Granivore	8–9.5	Diurnal	–	x	√	x	√	√	√
Loricera pilicornis	Microarthropod specialist	6–8	Nocturnal/Diurnal	–	–	x	–	–	–	√
Anchomenus dorsalis	Generalist insectivore	6–8	Nocturnal	–	√	√	x	√	x	√
Clivina fossor	Generalist insectivore	6–6.8	–	–	–	√	–	–	–	–
Asaphidion flavipes	Generalist insectivore	3.9–4.7	Diurnal	–	–	x	–	x	–	√
Trechus quadristriatus	Generalist insectivore	3.6–4.1	Nocturnal	√	–	–	–	–	–	–
Bembidion lampros	Generalist insectivore	3–4	Diurnal	–	–	x	–	x	–	√

The mature larvae of rape pests remain on the soil surface for a short time only before they burrow beneath. Warner (2001) found that 60% of pollen beetle larvae, 75% of seed weevil larvae and 10% of brassica pod midge larvae that dropped from a winter rape crop canopy were concealed beneath the soil after 1 min; all pollen beetle, weevil and pod midge larvae were concealed after 8.5, 6.0 and 20.5 min, respectively. Once below the soil surface they are only accessible to those species that venture below the surface, such as *H. rufipes* (Luff 1978), *Bembidion* spp. (Ulber 1997) and *C. fossor* (Forsythe 2000). Warner (2001) compared feeding over 96 h by *A. dorsalis*, *A. similata* and *N. brevicollis* on pollen beetle and pod midge larvae 24 h after they had burrowed below a Cocofibre® substrate and found no evidence of feeding by the carabids once the pest larvae were below the surface. However, Schernéy (1959, 1961) reported that the burrowing carabid, *C. fossor*, did feed on pollen beetle larvae buried 6–7 cm below the soil surface.

There is a six-fold difference in size between the largest species, *P. madidus*, and the smallest species, *B. lampros*, commonly found in oilseed rape fields (Table 4.4, Fig. 4.7). The larger *Pterostichus* spp. are generalist carnivores, feeding on a wide range of invertebrate prey (including earthworms, molluscs and arthropods) as well as plant matter and may scavenge for food, whereas the smaller species (*T. quadristriatus* and *B. lampros*) are generalist insectivores feeding mainly on a wide variety of insect prey (both dead and alive) but probably also taking seed and other vegetable matter (Toft and Bilde 2002, Table 4.4).

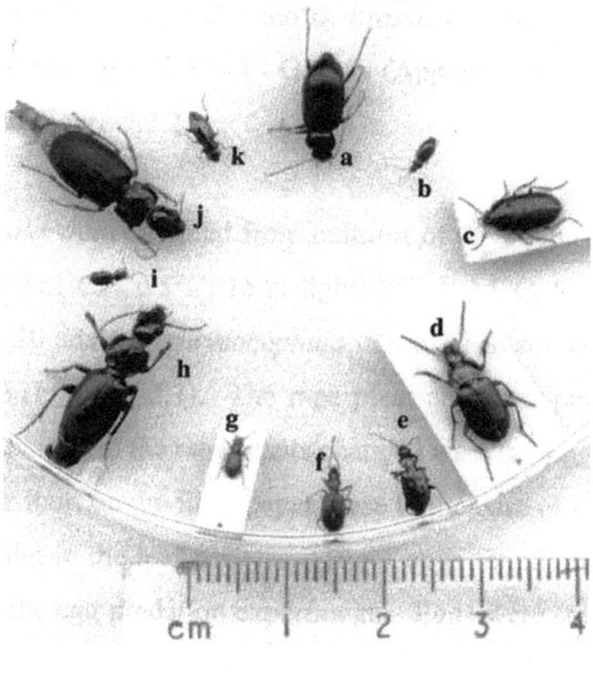

Fig. 4.7 Carabids frequently-occurring in oilseed rape fields: (*a*) *N. brevicollis*; (*b*) *T. quadristriatus*; (*c*) *A. similata*; (*d*) *P. cupreus*; (*e*) *L. pilicornis*; (*f*) *A. dorsalis*; (*g*) *A. flavipes*; (*h*) *P. madidus*; (*i*) *B. lampros*; (*j*) *H. rufipes*; (*k*) *N. biguttatus* (Photo: Douglas Warner)

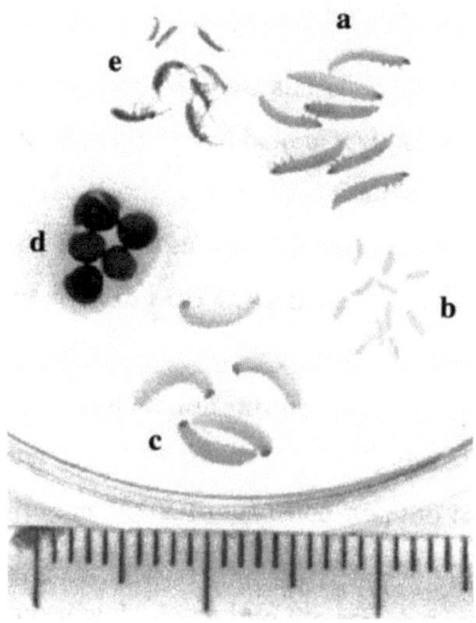

Fig. 4.8 Potential prey of carabids (*a*) pollen beetle larvae; (*b*) brassica pod midge larvae; (*c*) cabbage seed weevil larvae; (*d*) oilseed rape seeds; (*e*) Collembola, showing relative size against a cm scale (Photo: Douglas Warner)

The larvae of oilseed rape pests also vary considerably in size (Fig. 4.8). Warner (2001) weighed 50 mature field-collected larvae of pollen beetle, seed weevil and brassica pod midge larvae; mean live weights were 1.7, 2.2 and 0.2 mg, respectively. Feeding trials (Table 4.4) have shown that the larger and medium-sized carabid species (*P. madidus*, *A. similata* and *A. dorsalis*) will feed on the larger weevil larvae whereas the smaller species (*A. flavipes* and *B. lampros*) reject them but that all carabid species tested feed on the small brassica pod midge larvae.

4.4.2.2 Pest Consumption

Analyses of gut contents of field-collected individuals and feeding tests in the laboratory have provided evidence on the propensity of particular species within the carabid assemblage of oilseed rape fields to consume a particular pest species and have highlighted carabid food preferences.

Cabbage Stem Flea Beetle

In the UK, three species of carabid, *T. quadristriatus*, *N. brevicollis* and *P. madidus*, dominated pitfall trap catches in the autumn, the time when most cabbage stem flea beetle eggs are laid in the soil around rape plants (Warner et al. 2003). In laboratory no-choice feeding experiments, in which field-collected females of these three species were starved for 48 h, and then presented with flea beetle eggs, only *T. quadristriatus* ate the eggs, consuming a mean of six eggs in 24 h. *Nebria brevicollis* and *P. madidus* at most damaged only a few eggs (<1 per 24 h) (Warner et al. 2003). However, there is no evidence that *T. quadristriatus* ventures below the soil surface to feed and hence it is not clear whether cabbage stem flea beetle eggs are accessible to them in the field.

Mature cabbage stem flea beetle larvae leave rape plants to pupate in the soil from late February to early June. There is no information about which carabid species are active in rape fields during the early part of this period but during May and early June, in the UK, the most active/abundant species caught in pitfall traps, are *A. similata, A. dorsalis, N. brevicollis, A. flavipes* and *L. pilicornis* (Warner et al. 2008) and these species may feed on the mature flea beetle larvae at this time.

Pollen Beetle

Eight species of carabid have been recognised as potential predators of pollen beetle larvae (Table 4.4).

Examination and molecular analyses of the gut contents of field-collected carabids has identified five species that feed on pollen beetle larvae in the field (Table 4.4). Piper and Williams (2004) collected carabids from a field of winter oilseed rape and, on dissection, found the remains of pollen beetle larvae in the guts of *N. brevicollis* (28% of 165 dissected), *P. madidus* (15% of 27 dissected) and *P. cupreus* (6% of 35 dissected). Schlein et al. (2006) found the remains (mandibles and legs) of pollen beetle larvae in the guts of field-collected *P. cupreus, A. dorsalis* and *H. affinis* (22, 8, 6% respectively of 50 of each dissected) but not in those of *A. similata* or *H. rufipes* (50 of each dissected); they also confirmed feeding by *P. cupreus* on pollen beetle larvae (6% of 40 tested) using a molecular method (Polymerase Chain Reaction, PCR).

Feeding tests, where carabids are given one food only (no-choice) or more than one food (choice), have identified a further two species, *A. similata* (Warner 2001) and *H. rufipes* (Schlein and Büchs 2006a), that will feed on pollen beetle larvae, at least in the laboratory (Table 4.4). In addition, the burrowing carabid *C. fossor* has been reported to kill up to 65% of pollen beetle larvae introduced to 6–7 cm depth of soil in the laboratory (Schernéy 1959, 1961).

Carabid species vary in their voracity for pollen beetle larvae. Warner (2001) compared consumption of live pollen beetle larvae by seven species of field-collected carabids in no-choice tests following 48 h of starvation. Species varied in the mean time they took to make their first kill: *P. madidus* (26 s) < *N. brevicollis* (51 s) < *A. similata* (141 s) < *A. dorsalis* (422 s), and in the number of larvae they killed/consumed in 2 h: *P. madidus* (45) < *N. brevicollis* (24) < *A. similata* (7) < *A. dorsalis* (5); *A. flavipes*. *Bembidion lampros* and *L. pilicornis* did not kill any larvae.

Carabids exhibit different strengths of preference for particular prey. In choice tests in which carabids were offered both Collembola and pollen beetle larvae, *P. madidus* and *N. brevicollis* killed 38 and 51% fewer beetle larvae, respectively, than in tests where only pollen beetle larvae were offered. By contrast, the numbers of pollen beetle larvae killed by *A. dorsalis* and *A. similata* were not influenced by the presence of Collembola as alternative prey (Warner 2001).

The degree of carnivory/spermatophagy and preferences for food depend on the food choices available to a carabid. Schlein and Büchs (2006a) compared the feeding preferences of four species of carabids for pollen beetle larvae and oilseed rape seeds. Individual beetles, each starved for 48 h, were presented with 10 pollen beetle larvae and five oilseed rape seeds. All four species fed on both larvae and seeds, but they varied in the numbers of each that they ate (larvae: *H. rufipes* > *P. cupreus* > *A. similata* > *H. affinis*, mean range 7.5–3.2 per 24 h; seeds: *H. rufipes* > *P. cupreus* > *A. similata* > *H. affinis*, mean range 2.5–0.8 per 24 h). Comparing these results with similar trials using brassica pod midge larvae plus seeds instead of pollen beetle larvae plus seeds (see later), Schlein and Büchs (2006a) found that preferences for certain pest larvae became evident. Thus *H. rufipes* and *H. affinis* ate relatively more seeds when offered seeds with midge larvae than when offered seeds with pollen beetle larvae indicating an apparent preference for pollen beetle larvae over pod midge larvae. In contrast, *A. similata* preferred pod midge to pollen beetle larvae but *P. cupreus* showed no preference.

Cabbage Stem Weevil and Cabbage Seed Weevil

Five species of carabid have been identified that feed on weevil larvae (Table 4.4).

Gut dissection and molecular analyses of field-collected carabids has identified two species that feed on weevil larvae under field conditions (Table 4.4). Schlein et al. (2006) dissected 50 field-collected individuals of each of five carabid species; they found the remains (mandibles and head capsules) of weevil (*Ceutorhynchus* spp.) larvae in the guts of *H. rufipes* and *H. affinis* (10 and 6%, respectively), but not in those of *A. similata*, *A. dorsalis* and *P. cupreus*. They also found evidence of feeding by *H. affinis* on weevil (*Ceutorhynchus* spp.) larvae (one of 40 tested) using a molecular method (Polymerase Chain Reaction, PCR). Feeding tests in the laboratory have identified a further two species, *P. madidus* and *A. dorsalis* (Warner 2001), that will feed on weevil larvae when given no choice (Table 4.4).

Voracity for weevil larvae varies between carabid species and consumption rates can change when an alternative food is available. Warner (2001) compared consumption of live seed weevil larvae by four species of field-collected carabids after 48 h of starvation. Species varied in time to first kill: *P. madidus* (44 s) < *A. dorsalis* (491 s), and in the number of larvae killed/consumed in 2 h: *P. madidus* (31) > *A. dorsalis* (3). *Bembidion lampros* and *A. flavipes* did not feed. In comparable choice tests, in which the same species of carabid were offered seed weevil larvae together with Collembola and oilseed rape seeds, both *P. madidus* and *A. dorsalis* fed on all the foods offered and the numbers of seed weevil larvae consumed by *P. madidus* decreased by 41%. Schlein and Büchs (2004) presented each of 48 field-collected *A. similata* individually with a single oilseed rape seed and a cabbage stem weevil larva and reported that 36 beetles ate only the weevil, 12 beetles ate both and none ate only the seed.

Brassica Pod Midge

Eleven species of carabid have been found to feed on brassica pod midge larvae in the laboratory, but only one, *A. similata*, has also been shown to feed on them in the field (Table 4.4). Schlein et al. (2006) found the remains (spatulae) of pod midge larvae in the guts of field-collected *A. similata* (14% of 50 dissected). They found none in those of *A. dorsalis*, *H. affinis*, *P. cupreus* or *P. rufipes* (50 of each dissected), all of which will feed on the larvae in the laboratory.

Carabid species vary in their consumption capacities for midge larvae in laboratory feeding tests. Warner (2001) compared seven species of field-collected carabids in no-choice tests after 48 h of starvation. All seven species killed and consumed some midge larvae but they varied in mean time to first kill: *P. madidus* (26 s) < *H. rufipes* (31 s) < *N. brevicollis* (37 s) < *A. dorsalis* (63 s) < *A. similata* (77 s) < *A. flavipes* (288 s) < *L. pilicornis* (315 s) < *B. lampros* (431 s) and in the mean number of larvae killed in 2 h: *P. madidus* (72) > *P. rufipes* (34) > *N. brevicollis*, *A. dorsalis* and *A. similata* (22 each) > *L. pilicornis* (3) > *A. flavipes* and *B. lampros* (2 each). Schlein and Büchs (2004) conducted no-choice tests with five species of field-collected carabid given field-collected midge larvae killed by freezing prior to the experiments; all five species tested fed on the larvae, but mean consumption rates ranged between 11 and 4 per 24 h, differing between species: *P. melanarius* > *A. similata* > *P. cupreus* > *H. rufipes* > *A. dorsalis*.

Consumption of midge larvae may be reduced when carabids are presented with a choice of food in feeding tests. Warner (2001) found that when carabids were offered midge larvae together with Collembola and oilseed rape seeds, feeding on the midge larvae decreased in comparison with no-choice tests. Reductions in midge larva consumption differed between species: *A. similata* (89%) > *P. madidus* (79%) > *H. rufipes* (76%) > *N. brevicollis* (54%) > *A. dorsalis* (42%); *A. similata, P. madidus* and *H. rufipes* fed more on the rape seeds than on either of the insect prey. Schlein and Büchs (2004) found that, when presented with a choice of midge larvae killed by freezing and oilseed rape seeds, *H. rufipes* consumed significantly more rape seeds than larvae in 24 h while *A. similata* consumed more larvae than rape seeds.

Choice tests have also revealed feeding preferences of carabid species for different species of pest larvae. Schlein and Büchs (2006a) compared the feeding preferences of four carabid species using tests where pollen beetle or pod midge larvae were presented together with rape seeds. By comparing the relative numbers of larvae and of seeds eaten in experiments with pollen beetle and with pod midge larvae, preferences for particular prey larvae became evident: *A. similata* preferred pod midge larvae to pollen beetle larvae, *H. rufipes* and *H. affinis* preferred pollen beetle larvae to pod midge larvae, whereas *P. cupreus* showed no preference.

At least one carabid species, *A. similata*, has been identified that feeds not only on midge larvae and loose rape seeds but will also chew into rape pods to obtain these foods. In the field, feeding damage to the flowers, stem and young pods of rape plants by *A. similata* has been reported (Luka et al. 1998). In the laboratory, when individual, field-collected *A. similata* and *H. affinis* were given a choice of an uninfested oilseed rape pod and five loose oilseed rape seeds, *A. similata* fed on the pods as well as on loose seeds whereas *H. affinis* fed only on the loose seeds (Schlein and Büchs 2006b). Moreover, *A. similata* is able to detect pod midge larvae inside pods. Given a choice of uninfested oilseed rape pods and pods infested with pod midge larvae, most *A. similata* fed on the infested pods (infested 76%, uninfested 15%, both 9%) (Schlein and Büchs 2006b). The mechanism by which it detects the larvae is unknown but vibrotaxis or response to kairomones may be involved.

4.5 Implications for Biocontrol-Based Integrated Management of Rape Pests

Ground beetles or carabids are a widespread, abundant and diverse component of the generalist predator assemblages of oilseed rape fields in Europe. It is now widely recognised that naturally-occurring assemblages of generalist predators exert a considerable background level of pest control in crops (Sunderland 2002). The value of generalist predators, including carabids, for pest control in oilseed rape has been demonstrated in semi-field experiments, in which they have been either excluded from or augmented in plots of oilseed rape (Büchs and Nuss 2000). Biocontrol-based integrated management strategies for pests of oilseed rape seek to conserve and even enhance naturally-occurring populations of carabids by appropriate land and crop management (conservation biological control, Barbosa 1998).

Thirty-seven of the 42 species commonly found in oilseed rape fields are known to be predators of pests, including coleopteran and dipteran pests, on other crops (Sunderland 2002) and probably also contribute to predation of oilseed rape pests. On the basis of their widespread distribution and dominance in fields of winter oilseed rape in Europe, ten species of carabid emerge as key species in carabid assemblages in the crop. These are *A. similata*, *A. dorsalis*, *B. lampros*, *H. affinis*, *H. rufipes*, *L. pilicornis*, *N. brevicollis*, *P. cupreus* and *P. melanarius* and *T. quadristriatus.* Of these, seven species, *A. similata*, *A. dorsalis*, *B. lampros*, *H. rufipes*, *L. pilicornis*, *N. brevicollis*, *P. melanarius* and *T. quadristiatus*, as well as the less common and widespread *Asaphidion* spp., were both temporally and spatially associated with the pest eggs/larvae in study of a winter oilseed rape crop in UK. All of these species, and in addition, *P. madidus* and *C. fossor* have been shown to feed on the eggs/larvae of pests of oilseed rape, either in the field or in laboratory feeding tests. Carabids are mobile and can rapidly colonise a field and move to areas of high food density within it. Their opportunistic feeding habits allow them to make use of alternative non-pest food when pests are not available as well as exploit pest-prey as soon as they become available. However, more work is needed to determine whether these key carabid species make a significant contribution with other biocontrol agents to depress pest populations in oilseed rape crops sufficiently to be of economic importance to the farmer.

Many agricultural land management and crop husbandry practices are known to be detrimental to carabid populations generally, e.g., increase in field size, reduction in field boundary landscape elements, tillage and inputs of agrochemicals (Hance 2002, Lee and Landis 2002, Büchs 2003b). Managing agricultural landscapes and crops to provide habitats that favour carabids is critical to their conservation in agroecosytems and enhancement of their capacity for biocontrol of pests in the crop.

Carabid populations in fields can be conserved and enhanced by reducing field size to increase the proportion of field edge to field centre and by providing perennial habitats at field margins or as beetle banks within fields. Most species need suitable field margins or non-cropped habitats in which to overwinter or to provide a refuge from farming operations for at least a proportion of the population and thereby mitigate the negative consequences of disturbances to carabid populations by husbandry practices, such as pesticide applications and tillage. Beetle banks can be used to divide large fields into smaller areas. By ploughing two furrows together a raised bank is created with drier conditions and perennial vegetation favoured by carabids from which they can rapidly disperse in the spring covering the field more evenly.

Within the cropped area, organic matter provides shelter, alternative food for detritus-feeding insects, buffers microclimate extremes and improves soil texture; it can be boosted by returning crop residues to the soil after harvest and by applying manure. Weed cover increases heterogeneity of microclimatic conditions and niche availability, weed seed and alternative prey. Bare soil should be avoided and cover crops and conservation headlands favour carabids. Crop husbandry practices should aim to minimise disturbance to carabid populations. Minimal tillage/non-inversion tillage is preferable to ploughing; in addition to reducing mortality of carabids, it helps ensure that organic matter remains near the soil surface. Inputs of insecticides which kill carabids and their prey and inputs of herbicides which kill weeds that supply weed seed and prey items should be minimised.

Various agri-environment schemes in Europe offer options to create wildlife habitats on farmland and many are of potential benefit to carabid populations. In the UK, Environmental Stewardship has been introduced by the Department for Environment Food and Rural Affairs (Defra) whereby the land owner receives an area payment to compensate for loss of income incurred (e.g., due to reduced crop yield or increase in management costs) for land managed for environmental benefits (http://www.defra.gov.uk/erdp/). Environmental Stewardship options with potential to benefit carabid populations within cultivated land include hedgerow management, the creation of beetle banks, conservation headlands, grass margins and buffer zones, crop establishment by direct drilling and reduction of pesticide use and the leaving of stubble over winter.

Acknowledgements This review was supported by Estonian Targeting Financing Projects SF0172655s04 and SF0170057s09, Estonian Scientific Foundation grant nos. 2,655, 5,736, and 4,993, and the UK Department for Food and Rural Affairs. Rothamsted Research is an institute of the Biotechnology and Biological Sciences Research Council of the United Kingdom.

References

Adis J (1979) Problems of interpreting arthropod sampling with pitfall traps. *Zooloogischer Anzeiger Jena* 202: 177–184.

Alford DV, Büchs W, Prescher S, Kromp B, Ulber B (2003a) Taxonomy and identification of predators. In: Alford DV (ed.) *Biocontrol of oilseed rape pests*. Blackwell, Oxford, UK.

Alford DV, Nilsson C, Ulber B (2003b) Insect pests of oilseed rape crops. In: Alford DV (ed.) *Biocontrol of oilseed rape pests*. Blackwell, Oxford, UK.

Barbosa P (ed.) (1998) *Conservation biological control*. Academic Press, San Diego, USA.

Basedow T (1973) Der Einflussepigäischer Raubarthropoden auf die Abundanz phytophager Insekten in der Agrarlandschaft. *Pedobiologia* 13: 410–411.

Bilde T, Axelsen JA, Toft S (2000) The value of Collembola from agricultural soils as food for a generalist predator. *J Appl Ecol* 37: 672–683.

Bilde T, Toft S (1994) Prey preference and egg production of the carabid beetle *Agonum dorsale*. *Entomol Exp Appl* 73: 151–156.

Bohan DA, Bohan AC, Glen DM, Symondson WOC, Wiltshire CW, Hughes L (2000) Spatial dynamics of predation by carabid beetles on slugs. *J Anim Ecol* 69: 367–379.

Briggs JB (1965) Biology of some ground beetles (Coleoptera, Carabidae) injurious to strawberries. *B Entomol Res* 56: 79–93.

Brust JB (1990) Direct and indirect effects of four herbicides on the activity of carabid beetles (Coleoptera, Carabidae). *Pestic Sci* 30: 309–320.

Bryan KM, Wratten SD (1984) The responses of polyphagous predators to prey spatial heterogeneity: Aggregation by carabid and staphylinid beetles to their cereal aphid prey. *Ecol Entomol* 9: 251–259.

Büchs W (2003a) Sampling, trapping and rearing of predators. In: Alford DV (ed.) *Biocontrol of oilseed rape pests*. Blackwell, Oxford, UK.

Büchs W (2003b) Impact of on-farm landscape structure and farming systems on predators. In: Alford DV (ed.) *Biocontrol of oilseed rape pests*. Blackwell, Oxford, UK.

Büchs W (2003c) Predators as biocontrol agents of oilseed rape pests. In: Alford DV (ed.) *Biocontrol of oilseed rape pests*. Blackwell, Oxford, UK.

Büchs W, Alford DV (2003) Predators of oilseed rape pests. In: Alford DV(ed.) *Biocontrol of oilseed rape pests*. Blackwell, Oxford, UK.

Büchs W, Felsman DS, Klukowski Z, Luik A, Nilsson C, Ahman B, Ulber B, Williams IH (2006) Key predator species in oilseed rape crops – a review of literature and MASTER results. *CD-Rom Proc Int Symp Integrated Pest Management in Oilseed Rape, 3–5 April 2006, Göttingen, Germany*.

Büchs W, Harenberg A, Zimmerman J (1997) The invertebrate ecology of farmland as a mirror of the intensity of the impact of man? – an approach to interpreting results of field experiments carried out in different crop management intensities of a sugar beet and an oil seed rape rotation including set-aside. *Biol Agr Hort* 15: 83–108.

Büchs W, Nuss H (2000) First steps to assess the importance of epigaeic active polyphagous predators on oilseed rape insect pests with soil pupating larvae. *IOBC/wprs Bull* 23(6): 151–163.

Chiverton PA (1986) Predator density manipulation and its effects on populations of *Rhopalosiphum padi* (Hom:Aphididae) in spring barley. *Ann Appl Biol* 109: 49–60.

Chiverton PA (1988) Searching behaviour and cereal aphid consumption by *Bembidion lampros* and *Pterostichus cupreus*, in relation to temperature and prey density. *Entomol Exp Appl* 47: 173–182.

Coombes DS, Sotherton NW (1986) The dispersal and distribution of polyphagous Coleoptera in cereals. *Ann Appl Biol* 108: 461–474.

Davies MJ (1953) Crop contents of some British carabids. *Entomol Monthly Mag* 89: 18–23.

Desender K, Mertens J, D'Hustler M, Berbiers P (1984) Diel activity patterns of Carabidae, Coleoptera; Staphylinidae, Coleoptera; and Collembola in a heavily grazed pasture. *Rev Ecol Biol Sol* 21: 347–361.

Ekbom BS, Wiktelius S, Chiverton PA (1992) Can polyphagous predators control the bird cherry-oat aphid, *Rhopalosiphum padi,* in spring cereals. *Entomol Exp Appl* 65: 215–223.

Evans MEG, Forsythe TG (1985) Feeding mechanisms, and their variation in form, of some adult ground-beetles (Coleoptera: Caraboidea). *J Zool* 206: 113–143.

Fadl A, Purvis G (1998) Field observations on the lifecycles and season activity patterns of temperate carabid beetles (Coleoptera: Carabidae) inhabiting arable land. *Pedobiologia* 42: 171–183.

Felsman DS, Büchs W (2006) The spatio-temporal within-field distribution of pest larvae and key predators in Germany, and the predators' effect on emergence of new pest generations. *CD-Rom Proc Int Symp Integrated Pest Management in Oilseed Rape, 3–5 April 2006, Göttingen, Germany*.

Ferguson AW, Barari H, Warner DH, Campbell JM, Smith ET, Williams IH (2006) Distributions and interactions of the stem miners, *Psylliodes chrysocephala* and *Ceutorhynchus pallidactylus*, and their parasitoids in a crop of winter oilseed rape *(Brassica napus.)*. *Entomol Exp Appl* 119: 81–92.

Ferguson AW, Campbell JM, Warner DJ, Watts NP, Schmidt JEU, Williams IH (2003a) Spatio-temporal distributions of *Meligethes aeneus* and its parasitoids in an oilseed rape crop and their significance for crop protection. *Proc 11th Int Rapeseed Cong, 6–10 July 2003, Copenhagen, Denmark*, 1057–1059.

Ferguson AW, Campbell JM, Warner DJ, Watts NP, Schmidt JEU, Williams IH (2004) Phenology and spatial distributions of *Dasineura brassicae* and its parasitoids in a crop of winter oilseed rape: Implications for integrated pest management. *IOBC/wprs Bull* 27(10): 243–252.

Ferguson AW, Klukowski Z, Walczak B, Clark SJ, Mugglestone MA, Perry JN, Williams IH (2003b) Spatial distribution of pest insects in oilseed rape: Implications for integrated pest management. *Agr Ecosyst Environ* 95: 509–521.

Fernàndez Garcìa A, Griffiths GJK, Thomas CFG (2000) Density, distribution and dispersal of the carabid beetle *Nebria brevicollis* in two adjacent cereal fields. *Ann Appl Biol* 137: 89–97.

Forsythe TG (1982) Feeding mechanisms of certain ground beetles (Coleoptera: Carabidae). *Coleopts Bull* 36: 26–73.

Forsythe TG (1983) Mouthparts and feeding of certain ground beetles (Coleoptera: Carabidae. *Zool J Linn Soc-Lond* 79: 319–376.

Forsythe TG (2000) Ground beetles. *Naturalists' handbooks 8* 2nd ed. Richmond Publishing, Slough, UK.

Free JB, Williams IH (1979) The distribution of insect pests on crops of oilseed rape (*Brassica napus* L.) and the damage they cause. *J Agr Sci* 93: 139–149.

Greenslade PJM (1964) The distribution, dispersal and size of a population of *Nebria brevicollis* (F.) with comparative studies on three other Carabidae. *J Anim Ecol* 33: 311–333.

Greenslade PJM (1965) On the ecology of some British carabid beetles with special reference to life histories. *T Soc Brit Entomol* 16: 179.

Hance T (2002) Impact of cultivation and crop husbandry practices. In: Holland JM (ed.) *The agroecology of carabid beetles*. Intercept, Andover, UK.

Hance T, Gregoire-Wibo C, Lebrun P (1990) Agriculture and ground beetle populations. The consequence of crop types and surrounding habitats on activity and species composition. *Pedobiologia* 34: 337–346.

Hengeveld R (1979) The analysis of spatial patterns of some ground beetles (Col. Carabidae). In: Cormack RM, OrdJK (eds.) *Spatial and temporal analysis in ecology*. International Co-operative publishing house, Fairland.

Hengeveld R (1980a) Polyphagy, oligophagy and food specialization in ground beetles (Coleoptera; Carabidae). *Neth J Zool* 30: 564–584.

Hengeveld R (1980b) Food specialization in ground beetles (Coleoptera, Carabidae): An ecological or a phylogenetic process? *Neth J Zool* 30: 585–594.

Hokkanen HMT (2004) Impact of predators on pollen beetle *Meligethes aeneus* on rapeseed in Finland. *IOBC/wprs Bull* 27(10): 293–296.

Holland JM (2002) Carabid beetles: Their ecology, survival and use in agroecosystems. In: Holland JM (ed.) *The agroecology of carabid beetles*. Intercept, Andover, UK.

Holland JM, Frampton GK, Van der Brink P (2002) Carabids as indicators within temperate arable farming systems: Implications from scarab and link integrated farming systems projects. In: Holland JM (ed.) *The agroecology of carabid beetles*. Intercept, Andover, UK.

Holland JM, Luff ML (2000) The effects of agricultural practices on Carabidae in temperate ecosystems. *Integrated Pest Manage Rev* 5: 109–129.

Holland JM, Perry JN, Winder L (1999) The within-field spatial and temporal distribution of arthropods in winter wheat. *B Entomol Res* 89: 499–513.

Houpert G (1983) Les carabiques et la culture du colza en Lorraine. *6th Int Rapeseed Cong, Paris, 17–19 May*, 2: 1178–1183.

Ingerson-Mahar J (2002) Relating diet and morphology in adult carabid beetles. In: Holland JM (ed.) *The agroecology of carabid beetles*. Intercept, Andover, UK.

Joosse E (1981) Ecological strategies and population regulation of Collembola in heterogeneous environments. *Pedobiologia* 21: 346–356.

Jørgensen HB, Toft S (1997) Role of granivory and insectivory in the life cycle of the carabid beetle *Amara similata*. *Ecological Entomol* 22: 7–15.

Kielty JP, Allen-Williams LJ, Underwood N, Eastwood EA (1996) Behavioural responses of three species of ground beetle (Coleoptera: Carabidae) to olfactory cues associated with prey and habitat. *J Insect Behav* 9: 237–250.

Koss AM, Snyder WE (2005) Alternative prey disrupt biocontrol by a guild of generalist predators. *Bio Control* 32: 243–251.

Kromp B (1999) Carabid beetles in sustainable agriculture: A review on pest control efficacy, cultivation impacts and enhancement. *Agr Ecosyst Environ* 74: 187–228.

Lee JC, Landis DA (2002) Non-crop habitat management for carabid beetles. In: Holland JM (ed.) *The agroecology of carabid beetles*. Intercept, Andover, UK.

Lindroth CH (1974) Coleoptera: Carabidae, IV, part 2, Handbooks for the identification of British insects. Royal Entomological Society of London, UK.

Lindroth CH (1992) Ground beetles of Fennoscandia. Vols. 1–3. Intercept, Andover, Hampshire, UK.

Lövei GL, Sunderland KD (1996) Ecology and behaviour of ground beetles (Coleoptera: Carabidae). *Annu Rev Entomol* 41: 231–256.

Luff ML (1974) Adult and larval feeding habits of *Pterostichus madidus* (F.) (Col. Carabidae). *J Nat Hist* 8: 403–409.

Luff ML (1978) Diel activity patterns of some field Carabidae. *Ecol Entomol* 3: 53–62.

Luff ML (1980) The biology of the ground beetle *Harpalus rufipes* in a strawberry field in Northumberland. *Ann Appl Biol* 94: 153–164.

Luff ML (1986) Aggregation of some Carabidae in pitfall traps. In: den Boer PJ (ed.) *Carabid beetles*. Gustav Fisher, Stuttgart, Germany.

Luff ML (2002) Carabid assemblage organization and species composition. In: Holland JM (ed.) *The agroecology of carabid beetles*. Intercept, Andover, UK.

Luff ML (2007) *The Carabidae (ground beetles) of Britain and Ireland*. Handbooks for the Identification of British Insects 4 Part 2 (2nd ed.). The Field Studies Council, Shrewsbury, UK, for the Royal Entomological Society, UK.

Luik A, Veromann E, Kevväi R, Kruus M (2006) A comparison of the pests, parasitoids and predators on winter and spring oilseed rape crops in Estonia. *CD-Rom Proc Int Symp Integrated Pest Management in Oilseed Rape, 3–5 April 2006, Göttingen, Germany.*

Luka H, Pfiffner W, Wyss E (1998) *Amara ovata* und *A. similata* (Coleoptera, Carabidae), zwei phytophage Laufkäferarten in Rapsfeldern. *Mitt der Schweizer Entomol Gesellschaft* 71: 125–131.

Lyngby JE, Nielson HB (1980) The spatial distribution of carabids (Coleoptera: Carabidae) in relation to a shelterbelt. *Entomol Meddelelser* 48: 133–140.

Lys JA (1994) The positive influence of strip-management on ground beetles in a cereal field, increase, migration and overwintering. In: *Carabid beetles, ecology and evolution*. Kluwer Academic Publishers, Dordrecht, The Netherlands.

Mitchell B (1963) Ecology of two carabid beetles, *Bembidion lampros* (Herbst) and *Trechus quadristriatus* (Schrank). 1. Life cycles and feeding behaviour. *J Anim Ecol* 32: 289–299.

Monsrud C, Toft S (1999) The aggregative numerical response of polyphaagous predators to aphids in cereal fields: Attraction to what? *Ann Appl Biol* 134: 26–42.

Mundy CA, Allen-Williams LJ, Underwood N, Warrington S (2000) Prey selection and foraging behaviour by *Pterostichus cupreus* L. (Col., Carabidae) under laboratory conditions. *J Appl Entomol* 124: 349–358.

Niemelä J, Haila Y, Ranta E (1986) Spatial heterogeneity of carabid beetle dispersion in uniform forest of the Åland Islands, SW Finland. *Ann Zool Fenn* 23: 289–296.

Nuss H, Büchs W (2000) Zur Bedeutung epigäischer Raubarthropoden als Antagonisten von Blüten- und Schotenschädlingen des Winterrapses mit Verpuppung im Boden. *Mitt der Dtsch Ges Allg Angew Ent* 12: 141–147.

Paul WD (1986) Vergleich der epigäischen Bodenfauna bei wendender bzw. Nichtwendender Grundbodenbearbeitung. *Mitt Biol Bundesanst Land-forstwirtsch, Berlin-Dahlem* pp 232–290.

Penney MM (1966) Studies on certain aspects of the ecology of *Nebria brevicollis* (F.) (Coleoptera, Carabidae). *J Anim Ecol* 35: 505–512.

Penney MM (1969) Diapause and reproduction in *Nebria brevicollis* (F.) (Coleoptera, Carabidae. *J Anim Ecol* 38: 219–233.

Perry JN (1995) Spatial analysis by distance indices. *J Anim Ecol* 64: 303–314.

Perry JN (1998a) Measures of spatial pattern for counts. *Ecology* 79: 1008–1017.

Perry JN (1998b) Measures of spatial pattern and spatial association for counts of insects. In: Baumgartner J, Brandmayr P, Manly BFJ (eds.) *Population and community ecology for insect management and conservation. Proc 20th Int Cong Entomology, Florence, Italy, 25–31 August 1996*, 21–33. Balkema, Rotterdam.

Piper R, Williams IH (2004) Incidence and feeding activity of epigeic, predatory invertebrates within winter oilseed rape in the UK with comparisons between integrated and conventional crop management. *IOBC/wprs Bull* 27(10): 281–288.

Pollet M, Desender K (1986) Prey selection in carabid beetles, Col., Carabidae: Are diel activity patterns of predators and prey synchronised? *Med Fac Landbouww Rijksuniv Gent* 51: 957–971.

Pollet M, Desender K (1987) Feeding ecology of grassland inhabiting carabid beetles, Carabidae, Coleoptera: In relation to the availability to some prey groups. *Acta Phytopathol Hun* 22: 223–246.

Scherněy F (1959) *Unsere laufkäfer*. Ziemsen, Wittenberg, Germany.

Scherněy F (1961) Beiträge zur Biologie und ökonomischen Bedeutung räuberisch lebender Käferarten. *Z Angew Zool* 48: 163–175.

Schlein O, Büchs W (2004) Approaches to assess the importance of carnivorous beetles as predators of oilseed rape pests. *IOBC/wprs Bull* 27(10): 289–292.

Schlein O, Büchs W (2006a) Feeding capacity and food preferences of key species of carabid beetles in oilseed rape fields. *CD-Rom Proc Int Symp Integrated Pest Management in Oilseed Rape, 3–5 April 2006, Göttingen, Germany.*

Schlein O, Büchs W (2006b) The ground beetle *Amara similata* as a predator of pest larvae in oilseed rape fields; ignored but influential in biological control? *CD-Rom Proc Int Symp Integrated Pest Management in Oilseed Rape, 3–5 April 2006, Göttingen, Germany.*

Schlein O, Büchs W, Niepold F (2006) Proof of pest larval consumption by predators using the PCR-technique and microscopical gut dissection. *CD-Rom Proc Int Symp Integrated Pest Management in Oilseed Rape, 3–5 April 2006, Göttingen, Germany.*

Sotherton NW (1984) The distribution and abundance of predatory arthropods overwintering on farmland. *Ann Appl Biol* 105: 423–429.

Sotherton NW (1985) The distribution and abundance of predatory Coleoptera overwintering in field boundaries. *Ann Appl Biol* 106: 17–21.

Speight MR, Lawton JH (1976) The influence of weed-cover on the mortality imposed on artificial prey by predatory ground beetles in cereal fields. *Oecologia* 23: 211–223.

Sunderland KD (1975) The diet of some predatory arthropods in cereal crops. *J Appl Ecol* 12: 505–515.

Sunderland KD (2002) Invertebrate pest control by carabids. In: Holland JM (ed.) *The agroecology of carabid beetles*. Intercept, Andover, UK.

Sunderland KD, Crook NE, Stacey DL, Fuller BJ (1987) A study of aphid feeding by polyphagous predators on cereal aphids using ELISA and gut dissection. *J Appl Ecol* 24: 907–933.

Sunderland KD, Lövei GL, Feulon J (1995) Diets and reproductive phonologies of the introduced ground beetles *Harpalus affinis* and *Clivina australasiae* (Coleoptera: Carabidae) in New Zealand. *Aust J Zool* 43: 39–50.

Sunderland KD, Vickerman GP (1980) Aphid feeding by some polyphagous predators in relation to aphid density in cereal fields. *J Appl Ecol* 17: 389–396.

Symondson WOC (2002) Diagnostic techniques for determining carabid diets. In: Holland JM (ed.) *The agroecology of carabid beetles*. Intercept, Andover, UK.

Symondson WOC, Cesarini S, Dodd PW, Harper GL, Bruford MW, Glen DM, Wiltshire CW, Harwood JD (2006) Biodiversity vs. biocontrol: Positive and negative effects of alternative prey on control of slugs by carabid beetles. *B Entomol Res* 96: 637–645.

Symondson WOC, Glen DM, Ives AR, Langdon J, Wiltshire W (2002a) Dynamics of the relationship between a generalist predator and slugs over five years. *Ecology* 83: 137–147.

Symondson WOC, Glen DM, Wiltshire CW, Langdon CJ, Lidell JE (1996) Effects of cultivation techniques and methods of straw disposal on predation by *Pterostichus melanarius* (Coleoptera: Carabidae) upon slugs (Gastropoda: Pulmonata) in an arable field. *J Appl Ecol* 33: 741–753.

Symondson WOC, Sunderland KD, Greenstone MH (2002b) Can generalist predators be effective biocontrol agents? *Annu Rev Entomol* 47: 561–594.

Tarang T, Veromann E, Luik A, Williams I (2004) On the target entomofauna of an organic winter oilseed rape field in Estonia. *Latvijas Entomol* 41: 100–110.

Thiele H-U (1977) *Carabid beetles in their environments*. Springer-Verlag, Berlin, Germany.

Thiele H-U, Weber F (1968) Tagesrhythmen der Aktivität bei Carabiden. *Oecologia (Berlin)* 1: 315–355.

Thomas CFG, Holland JM, Brown NJ (2002) The spatial distribution of carabid beetles in agricultural landscapes. In: Holland JM (ed.) *The agroecology of carabid beetles*. Intercept Limited, Andover, Hampshire, UK.

Thomas CFG, Parkinson L, Griffiths GJK, Fernandez Garcia A, Marshall EJP (2001) Aggregation and temporal stability of carabid beetle distributions in field and hedgerow habitats. *J Appl Ecol* 38: 100–116.

Thomas CFG, Parkinson L, Marshall EJP (1998) Isolating the components of activity-density for the carabid beetle *Pterostichus melanarius* in farmland. *Oecologia* 116: 103–112.

Toft S, Bilde T (2002) Carabid diets and food value. In: Holland JM (ed.) *The agroecology of carabid beetles*. Intercept, Andover, UK.

Tooley J, Brust GE (2002) Weed seed predation by carabid beetles. In: Holland JM (ed.) *The agroecology of carabid beetles*. Intercept, Andover, UK.

Ulber B (1997) Some results from the investigations of Wolf-Schwerin on carabid-induced mortality of *Dasineura brassicae* and of Goltermann on carabid predation on *Meligethes aeneus* larvae in northern Germany. *Biocontrol of Oilseed Rape Insect Pests Newsletter 4*.

van Huizen THP (1990) 'Gone with the wind'. Flight activity in carabid beetles in relation to wind direction and to reproductive state of females in flight. In: Stork NE (ed.) *The role of ground beetles in ecological and environmental studies*. Intercept, Andover, UK.

Veromann E, Tarang T, Kevväi R, Luik A, Williams I (2006) Insect pests and their natural enemies on spring oilseed rape in Estonia: Impact of cropping systems. *Agr Food Sci* 15: 61–72.

Vet LEM, Dicke M (1992) Ecology of the infochemical use by natural enemies in tritrophic system. *Annu Rev Entomol* 37: 141–172.

Vickerman GP, Sunderland KD (1975) Arthropods in cereal crops: Nocturnal activity, vertical distribution and aphid predation. *J Appl Ecol* 12: 755–765.

Wachmann E, Platen R, Barndt D (1995) *Laufkäfer: Beobachtung, lebensweise*. Naturbuch Verlag, Augsburg.

Wallin H (1985) Spatial and temporal distribution of some abundant carabid beetles in cereal fields and adjacent habitats. *Pedobiologia* 28: 19–34.

Wallin H (1989) The influence of different age classes on the seasonal activity and reproduction of four medium-sized carabid species inhabiting cereal fields. *Holarctic Ecol* 12: 201–211.

Wallin H, Chiverton PA, Ekbom BS, Borg A (1992) The diet, fecundity and egg size in some polyphagous predatory carabid beetles. *Entomol Exp Appl* 65: 129–140.

Wallin H, Ekbom B (1988) Movements of carabid beetles (Coleoptera: Carabidae) inhabiting cereal fields: A field tracing study. *Oecologia* 77: 39–43.

Wallwork JA (1976) *The distribution and diversity of soil fauna*. Academic Press, London.
Warner DJ (2001) The potential of Carabidae in the control of insect pests of winter oilseed rape. PhD thesis, University of Hertfordshire.
Warner DJ, Allen-Williams LJ, Ferguson AW, Williams IH (2000) Pest-predator spatial relationships in winter rape: Implications for Integrated Crop Management. *Pest Manage Sci* 56: 977–982.
Warner DJ, Allen-Williams LJ, Warrington S, Ferguson AW, Williams IH (2003) Mapping, characterisation, and comparison of the spatio-temporal distributions of cabbage stem flea beetle (*Psylliodes chrysocephala*), carabids, and Collembola in a crop of winter oilseed rape (*Brassica napus*). *Entomol Exp Appl* 109: 225–234.
Warner DJ, Allen-Williams LJ, Warrington S, Ferguson AW, Williams IH (2008) Implications for conservation biocontrol of spatio-temporal relationships between carabid beetles and coleopterous pests in winter oilseed rape (*Brassica napus*). *Agr Forest Entomol* 10: 375–387.
Wheater CP (1988) Predator-size relationships in some Pterostichini (Coleoptera: Carabidae). *Coleopterists's Bull* 42: 237–240.
Wheater CP (1989) Prey detection by some predatory Coleoptera. *J Zool* 218: 171–185.
Williams IH, Büchs W, Hokkanen H, Menzler-Hokkanen I, Johnen A, Klukowski Z, Luik A, Nilsson C, Ulber B (2005) MASTER – Integrating biological control within IPM for winter oilseed rape across Europe. *Proc BCPC Int Cong Crop Science and Technology, Glasgow, 31 Oct–2 Nov 2005*, 1: 301–308.
Winder LH (1990) Predation of the cereal aphid *Sitobion avenae* by polyphagous predators on the ground. *Ecol Entomol* 15: 105–110.
Winder L, Alexander C, Holland JM, Woolley C, Perry JN (2001) Modelling the dynamic spatio-temporal response of predators to transient prey patches in the field. *Ecol Lett* 4: 568–576.
Zangger A, Lys JA, Nentwig W (1994) Increasing the availability of food and the reproduction of *Poecilus cupreus* in a cereal field by strip-management. *Entomol Exp Appl* 71: 111–120.

Chapter 5
Pests and Their Enemies in Spring Oilseed Rape in Europe and Challenges to Integrated Pest Management

Barbara Ekbom

Abstract Although oilseed rape cultivation in Europe is dominated by winter oilseed rape there are areas where spring oilseed rape is cultivated and interest in the crop is growing. Spring rape has a somewhat different set of insect pests. In addition, where spring rape and winter rape are cultivated in the same area, some pest problems may be exacerbated. A particular problem in spring rape cultivation is the establishment of the crop. Flea beetles of the genus *Phyllotreta* are a serious problem and generally have been controlled by seed treatments. The flea beetles are attacked by several enemies including parasitoids, nematodes and protozoa. Generalist predators such as carabids and spiders may play a role as predators, but there is presently little knowledge concerning this possible mortality factor. Agricultural practices such as seed bed preparation and seeding rates may also be important for controlling this pest. Certain pests such as the cabbage aphid, the pod midge and the pollen beetle may be affected by the close proximity of spring and winter oilseed rape. Phenologies of these pests and their enemies are examined in relation to spring rape cultivation. The implications for integrated pest management are discussed.

5.1 Introduction: Cultivation of Spring Oilseed Rape and Some Reasons for Using Spring Varieties

Although winter oilseed rape (WOSR) varieties are the dominant form of oilseed rape cultivated in many areas, spring-sown crops are important in northern areas such as Canada and Scandinavia. In addition, spring oilseed rape (SOSR) and turnip rape (*Brassica rapa*) are beginning to become viable alternatives in other European areas. The benefits can be seen in several ways. In areas at high latitudes, where growing seasons are short and winters often harsh, spring oilseed crops have long been the only option. Survival over the winter is a problem because low temperatures can damage or kill plants. With changing weather patterns the risk of not being able

B. Ekbom (✉)
Department of Ecology, Swedish University of Agricultural Sciences, Uppsala, Sweden
e-mail: barbara.ekbom@ekol.slu.se

I.H. Williams (ed.), *Biocontrol-Based Integrated Management of Oilseed Rape Pests*,
DOI 10.1007/978-90-481-3983-5_5,

to drill a crop at the proper time in the autumn may be increased whereas SOSR has a more flexible sowing window. Residues after harvest are generally lower in SOSR than in winter crops and this may facilitate drilling of winter wheat. Crop rotations which include OSR, increase yield in the first following wheat crop and help control grass weeds and reduces the level of cereal pathogens. Growing demands for OSR industrial and biofuel products also contribute to increasing interest in SOSR. Winter varieties have higher yields than spring crops, but lower inputs in SOSR, such as reduced nitrogen use, together with high oil yields mean that the return for SOSR is not far behind that for WOSR. When other advantages are weighed in, SOSR may be chosen by more growers.

The two forms of OSR crops in Europe share many insect pests such as pollen beetle (*Meligethes* spp.), stem weevils (*Ceutorhynchus* spp.), and brassica pod midge (*Dasineura brassicae*) (see also Williams Chapter 1 this volume). The cabbage stem flea beetle (*Psylliodes chrysocephala*) is, however, exclusive to autumn-sown forms as the larvae overwinter in this crop. Instead other species of flea beetle of the genus *Phyllotreta* are major pests in SOSR. These beetles are a key threat to crop establishment.

5.2 A Key Threat to Crop Establishment, Flea Beetles of the Genus *Phyllotreta*

5.2.1 Species and Damage Caused

The genus *Phyllotreta* is one of the largest alticine genera, but only a few species are found as pests in SOSR (Table 5.1). The larvae of most species live in the soil on the roots of Brassica host plants; the exception is *P. nemorum* that has larvae that mine leaves. *Phyllotreta nemorum* is more often found in vegetable Brassicas and seldom in SOSR where *P. undulata* or sometimes *P. atra* dominate (Sommer 1981,

Table 5.1 Species of *Phyllotreta* found in fields in Europe and North America (Ekbom and Kuusk 2005, Sommer 1981)

Flea beetle	Larval stage	Distribution	Colour
Phyllotreta undulata Kutschera	Root feeder	Europe	Yellow stripes
Phyllotreta striolata (Fabricius)	Root feeder	Europe, Canada	Yellow stripes
Phyllotreta vittula Redtenbacher	Mines stems and leaf petioles of cereals	Europe	Yellow stripes
Phyllotreta nemorum (L.)	Leaf miner	Not usually found in rape fields	Yellow stripes
Phyllotreta nigripes (Fabricius)	Root feeder	Europe	Black
Phyllotreta atra (Fabricius)	Root feeder	Europe	Black
Phyllotreta cruciferae (Goeze)	Root feeder	Canada, USA	Black

Ekbom 1990). A curiosity is *P. vittula*, the larvae of which mine stems and leaf petioles of cereals. Although this species does not reproduce on Brassicas, it can be found feeding as an adult on newly-emerged SOSR, sometimes in high numbers (Ekbom 1990).

Most of the *Phyllotreta* species commonly found in SOSR are between 2 and 2.5 mm long. The species all have one generation per year and their life cycles are similar (see also the life cycle description in Dosdall and Mason Chapter 6 this volume). Flea beetles overwinter as adults, often in hiding places away from the fields they came from. The beetles leave their overwintering sites between March and May depending on the temperature. As the weather gets warmer and newly-sown SOSR crops are emerging the beetles move to these fields where they are particularly active in warm, dry weather. High numbers of flea beetles can be devastating for crop establishment. Even if the compensation ability of OSR is high, early attacks will cause irreversible damage as the beetles may kill plants as they feed on the cotyledons and stems, sometimes even while plants are below ground (Jones and Jones 1984). Eggs are laid on the soil at the base of the plant, often in batches. Females have about 28 ovaries and this number of eggs (at most) can be laid at one time (Sommer 1981); new egg batches will then be matured. In laboratory experiments, up to 200 eggs/female could be laid. There are three larval instars. Young larvae are white; later the head becomes light brown. They live on the roots at a depth of 5–30 cm; younger larvae feed on fine roots while older larvae will attack larger roots (Sommer 1981). Fully developed larvae are about 5 mm long and development to pupation takes about 4 weeks (Jones and Jones 1984). Larvae build an earthen cell in which they pupate. Larvae do not appear to cause economically-important damage. After about 4 weeks as a pupa, adults of the new generation emerge in late summer; the timing will be temperature dependent.

5.2.2 Current Practices for Controlling Flea Beetles

Currently, seed dressing with systemic insecticides is the most common management practice used against flea beetles in areas with high populations. In areas in Sweden and Finland, it has recently been seen that seed dressing alone cannot protect against large and sustained attacks of flea beetles. Therefore one or more additional insecticide treatment may be applied by spraying. Even with this high insecticide pressure yield losses may occur. Risks for environmental damage and insecticide resistance with additional insecticide treatments are obvious. If insecticide sprays are to be used a control threshold is needed. In three countries, Sweden, Finland, and Canada, control thresholds have been developed; when 25–30% of the cotyledon area is eaten, application of an insecticide is recommended to avoid economic damage. Use of this threshold is somewhat difficult because attacks occur with surprising speed. It may be necessary to check fields every day during periods of warm, sunny weather when plants are beginning to emerge.

5.2.3 Importance of Agricultural Practices for Reducing Attack and Damage

There are cultural practices that can be used for reducing the risk and severity of flea beetle attack (see also Dosdall and Mason Chapter 6 this volume). Oilseed directly drilled into fields generally sustains less damage than oilseed on fields with minimum or normal tillage. Seeding rate and row spacing can also make a difference. Increases in seeding rate and row spacing tend to give the flea beetles extra food and the damage is more spread out. Although increased cost for seed would be the result of higher seeding rates, the resulting control may well be worth the extra cost. It is important to take temperature and soil moisture into consideration. Proper seed bed preparation and optimal depth of sowing to ensure good contact with moisture can contribute to rapid and even germination. If plants emerge at about the same time then damage to individual plants may be reduced. Rapid growth will aid plant development past the most sensitive cotyledon stage and thereby allow the plants to tolerate damage and prevent the flea beetles from killing plants.

5.2.4 Natural Enemies and Their Potential for Biocontrol

In a study by Sommer (1981), working from the European Station of the Commonwealth Institute of Biological Control in Switzerland, a number of parasites of *Phyllotreta* spp. were collected. One parasitoid, a braconid *Townesilitus bicolor* (Wesmael) (Wylie 1988) was chosen for release in Canada, but never established. There are only a few studies on the parasitism levels caused by *T. bicolor*, but apart from regularly high levels reported by Jourdheuil (1960) in France, levels of around 15% seem to be common (Sommer 1981, Ekbom 1990, 1991). The parasitoid lays its eggs in flea beetle adults in the late summer. The flea beetle adults then take the first-instar larva with them to overwintering sites. In the spring, when the flea beetles return to agricultural fields, the parasitoid resumes its development. The growing parasitoid larva completely inhibits maturation of ovaries and eggs in female hosts. Parasitoid larvae leave their host, killing it, and pupate in the soil. This indicates that wasps will probably emerge in oilseed fields 2–3 weeks later and begin to parasitize the new generation of flea beetles.

No studies have been done to ascertain whether parasitized flea beetles eat less than non-parasitized beetles, so it is not known if parasitism can lessen damage to plants. If parasitism levels of *T. bicolor* increased, a gradual reduction in *Phyllotreta* flea beetle populations might be achieved, but, at present, we know little about the reasons for low levels of parasitism. Some possible reasons for this may be that (i) parasitized individuals of *Phyllotreta* may be more susceptible to insecticide treatment and therefore more likely to die before the parasitoid completes its development, or (ii) insecticides used against pollen beetle may kill emerging parasitoids and limit parasitism of flea beetles.

Nothing is known about larval parasitoids of *Phyllotreta* spp. except for those parasitizing *P. nemorum* (Ulber and Williams 2003), a species not considered a problem in SOSR (Table 5.1). The most common parasitoid of *P. nemorum* larvae appears to be *Diospilus morosus* Reinhardt. This parasitoid also parasitizes larvae of other Coleoptera present in OSR, namely the cabbage stem flea beetle and the cabbage seed weevil (*Ceutorhynchus obstrictus* syn. *assimilis*). The larval stage of *P. nemorum* is a leaf miner and it is therefore more exposed to parasitoid attack and easier to study than the root-feeding larvae of the other *Phyllotreta* spp. found in SOSR. Because the root-feeding larvae are difficult to study there is a very large gap in our knowledge about mortality factors concerning juvenile stages of *Phyllotreta* spp. It is not unreasonable to suppose that egg or larval parasitoids may exist; however, finding them may require large and systematic inventories.

Predators, including carabids (Chiverton 1984), lacewings (Burgess 1980), nabids (Burgess 1982, Culliney 1986), pentatomids (Culliney 1986), wolf spiders (Sandra Öberg, *pers. obs*), and crickets (Burgess and Hinks 1987), have been observed feeding on adult flea beetles in the field. Although the beetles move quickly and jump readily when disturbed, it seems that they can be caught and consumed by a variety of generalist predators commonly found in agricultural fields. Eggs are laid in batches on the soil surface close to host plants. No-one has investigated possible predation of *Phyllotreta* spp. eggs, but we know that predation of cabbage root fly (*Delia radicum*) eggs (also laid on the soil surface near host plants) does take place and is performed by a wide range of predators (Finch 1996). There appears to be good reason to investigate the action of generalist predators against *Phyllotreta* flea beetles. Flea beetles arrive early in the season to SOSR and may be an important source of food for predators present in the field at a time when other prey may be scarce.

Intercellular parasites, such as protozoans, are a group of pathogens that do not usually cause direct mortality of the host. They may, however, weaken the host and lower fecundity and thereby have an impact on host population dynamics. Entomopathogenic nematodes (EPN) can also cause chronic infections in a host, but the effect of such pathogens on population numbers of the host is difficult to determine. A microsporidian, an eugregarine and a nematode, identified as *Nosema phyllotretae* Weiser, *Gregarina sp.*, and *Howardula phyllotretae* Oldham, respectively, have been recorded as parasites of *Phyllotreta* flea beetles (Sommer 1981). Multi-pathogen infections can be common and parasitoids were sometimes found together with pathogens (Lipa and Ekbom 2003).

Issi and Radishcheva (1979) found infection levels of *N. phyllotretae* in about 35% of collected *Phyllotreta* spp. adults. Prevalence levels of the microsporidian in insects collected over a period of more than 10 years in the Uppsala area of Sweden ranged from 25 to 76% (Lipa and Ekbom 2003). Although there is no information on the effect of the microsporidian on flea beetle survival or fecundity, it is possible that high levels of infection in a population could be detrimental to flea beetle population growth.

Eugregarine infections were quite common in *Phyllotreta* flea beetles collected both in central Europe (Sommer 1981), in Sweden (Lipa and Ekbom 2003) and in

Turkey (Yaman 2002) where infections in some collections of *P. undulata* reached over 90% and for *P. atra* over 70% in August. The eugregarine infections were often characterized by large numbers of gamonts and sporonts in the gut lumen and this may interfere with food digestion. It is, however, not known if the hosts are generally affected in a negative manner.

The nematode *H. phyllotretae* is a very common parasite of flea beetles in Europe. Its prevalence has been reported to be between 10 and 60% (Sommer 1981, Ekbom 1990, 1991, Lipa and Ekbom 2003). Dispersal of the nematode is facilitated by the fact that parasitic females live in the adult beetles and are carried to new sites. The life cycle is briefly described by Sommer (1981). After overwintering inside flea beetle adults, fertilized female nematodes deposit larvae into the host's haemoceol. Juveniles leave the host through the reproductive system, for instance during flea beetle oviposition. After leaving the host, the juveniles require about a week to develop into mature males and females. This means that fertilized females can attack flea beetle larvae that hatch from the eggs with which the nematodes were deposited. Female nematodes penetrate the host cuticle to enter the body cavity and remain in ovarial diapause as the flea beetle larvae grow, pupate and become adults. The female nematode does not release new larvae until the following spring, after overwintering inside the flea beetle.

These nematodes do not kill the host nor do they appear to reduce egg development in female flea beetles. The value of *H. phyllotretae* as a biological control agent would seem to be limited unless there are significant sublethal effects that influence population growth. These nematodes do not appear to be an important mortality factor for larvae of *Phyllotreta* flea beetles. Entomopathogenic nematodes have been shown to be useful against foliar pests and recently four species of commercially available EPNs were tested in the laboratory using adults of several species of *Phyllotreta* flea beetles (Trdan et al. 2008). Mortality varied between 44 and 77% depending on temperature, which nematode species was used and at what dose. A problem with using nematodes for direct control is that high mortality occurs only after 6–8 days. The beetles could continue to damage the crop in the interim. EPNs used as a method to reduce flea beetle populations might have some impact, especially if larvae are affected. The logistics and costs of such a strategy are probably not, however, feasible at present.

Entomopathogenic fungi are sometimes thought to have potential as substitutes for conventional insecticides. Antwi et al. (2007) tested a commercial product containing *Beauveria bassiana* against *Phyllotreta* flea beetles; damage and yield were not significantly different from untreated controls. Because flea beetles are most often a problem when weather is warm and dry, the likelihood that a fungus would be able to control adults is very low. A treatment with entomopathogenic fungi might, however, reduce larval populations or infect adults emerging in the late summer and increase overwintering mortality. Unfortunately, finding the right species and strain of fungi that will have high efficacy against a particular soil-living insect is not a trivial task (Vänninen et al. 1999). Neither are application techniques against larvae, or to infect emerging adults, available.

5.3 Pest Problems and Their Relation to Concurrent Cultivation of Spring and Winter Varieties of Oilseed Rape

5.3.1 Aphids

Three species of aphids, *Lipaphis erysimi* (Kaltenbach), *Brevicoryne brassicae* (L.) and *Myzus persicae* (Sulzer), can be of economic importance in oilseed crops. *Myzus persicae* is not restricted to crucifers as it is extremely polyphagous, but *B. brassicae* and *L. erysimi* are specialists on crucifers. Although *L. erysimi* is sometimes a devastating pest in India (Mandal et al. 1994), none of the aphid species are especially common on OSR in temperate zones. All three aphid species can cause direct damage when their numbers are high, and suction feeding on the plant can decrease plant vigour and cause deformation. In addition, *M. persicae* is an important vector of beet western yellows luteovirus (BWYV) and *B. brassicae,* which does not transmit the virus as efficiently as *M. persicae*, may act as a minor vector (Smith and Barker 1999). This virus has a substantial yield-limiting potential in OSR (Hardwick et al. 1994, Jones et al. 2007); BWYV reduces seed yield and oil content. Cauliflower mosaic caulimovirus (CaMV) and Turnip mosaic potyvirus (TuMV) have also been found in oilseed rape crops (Hardwick et al. 1994, Shahraeen et al. 2003). *Myzus persicae* and *B. brassicae* are known to be important vectors of these diseases (ICTVdB 2006a, b), and *L erysimi* is most likely an adequate vector also, at least of TuMV (Wang et al. 1998). If aphid attacks become more frequent, perhaps in conjunction with climate change, virus problems in OSR may become more serious.

In northern areas, WOSR can be an important overwintering site for *B. brassicae*. Aphids arrive at the crop and lay eggs in autumn. In the spring, when aphids hatch and begin reproducing, it is often cool and moist, which is not conducive to rapid aphid development. If there are large areas of WOSR in proximity to SOSR then populations of *B. brassicae* may build up on WOSR and when they are ready to migrate they can move to SOSR. The spring crops will often be at a relatively early developmental stage when *B. brassicae* arrive and at that time the crop will be very susceptible to aphid attack. At this time the crop would also be very susceptible to BWYV, which might be carried from WOSR or acquired from other sources during migration. Even if *B. brassicae* is not an efficient vector of the virus, in years with high populations of the aphid some virus spread may be possible. SOSR could, in a worst case scenario, provide a green bridge for the virus so that it could be carried back to WOSR. Turnip rape develops more quickly than SOSR, which will be more likely to avoid heavy damage. The canopy climate in SOSR early in the season is often warmer and drier than that in WOSR. This will promote aphid reproduction and high numbers may develop rapidly.

In a study of aphid parasitoids on WOSR in the autumn in France (Desneux et al. 2006), *Diaeretiella rapae* (McIntosh) and *Aphidius matricariae* (Haliday) were found, albeit in low numbers. For *B. brassicae* and *L. erysimi* the most important species was *D. rapae* while *A. matricariae* seemed to prefer *M. persicae*. Parasitism rates were low. There may be several reasons for this, not the least of which is that widespread use of insecticides may reduce parasitoid populations. Also when aphids migrate to new hosts they may leave their natural enemies behind (enemy-free space). It might be possible to use parasitoids in an augmentative biological control program; releases could be made as soon as aphids were detected. Both parasitoids are relatively easy to produce in culture. In cabbage, releases of *D. rapae* have been used successfully to control *B. brassicae* (Zhang and Hassan 2003).

Other known aphid enemies such as predaceous ladybird beetles and hoverfly larvae are probably important predators of aphids in OSR. How quickly these aphid specialist predators will respond to growing aphid densities is not known. Ladybird beetles and hoverflies will often lay eggs where aphids are available in order to provide resources for their larvae. Predatory bugs belonging to the Heteroptera, for example, Nabidae, Geocoridae and Anthocoridae, also consume *M. persicae* and *B. brassicae* (Snyder et al. 2008, Simonsen et al. 2009). Suppression of these two aphid species has been shown to be strongest when a diversity of natural enemies is present (Snyder et al. 2008). All management techniques to increase natural enemy diversity and abundance are therefore to be recommended for aphid biological control.

Entomopathogenic fungi have been used in some situations to combat aphids, but high humidity is a requirement for fungal infections. This is one of the reasons that it is not likely that aphids in OSR would be effectively controlled using fungal pathogens (Butt et al. 2001). Entomophthoralean fungi, which often have a much more restricted host range than hyphomycetous fungi, are know to cause natural epizootics in aphids and it has been suggested that these fungi have potential for conservation biological control (Nielsen et al. 2007). Lack of knowledge about fungal survival strategies and requirements for infecting aphids is, however, a major constraint to developing any biocontrol strategies to be used against aphids in the field.

5.3.2 Brassica Pod Midge

The brassica pod midge has several generations per year. The larvae overwinter in cocoons in the soil of fields where OSR was grown in the preceding year. Pupation takes place in the spring and when adults emerge they will move to WOSR fields and attack pods. The midges are small, fragile and short-lived, and therefore the new WOSR field should be close to the midge emergence site. The eggs laid will give rise to second generation adults about 4–6 weeks later, depending on the temperature. Second generation adults can lay eggs in pods in the field of their birth (WOSR) if suitable resources are available; they can also move to SOSR fields where they can give rise to a third generation (Williams et al. 1987). Thus, the concurrent cultivation of winter and spring crops allows the midge an extra generation and this may contribute to population growth such that more midges will be produced in an area. The pod midge is rarely found on SOSR in areas without winter crops as the first generation would emerge before pod resources are accessible. When winter and spring OSR fields are in close proximity the second generation of pod midge may attack the spring crop and this means that midge attack on the spring crop will be high (Axelsen 1992). It has also been found that larvae entering diapause early in the season (for example from the first generation) have a higher mortality than those entering diapause after the third generation (Axelsen 1992).

Parasitoids of the brassica pod midge are described elsewhere in this book (Ulber et al. Chapter 2 this volume). Unfortunately little is known about the phenology and biology of the various species. Parasitism rates vary widely in different studies (Williams 2003), which makes it impossible to say anything about the factors that might influence the efficacy of different parasitoids. Because midge larvae must fall to the ground to go into the soil to build their cocoons, they are exposed for a short while to predation by ground-living predators such as carabids and spiders (see also Williams et al. Chapter 4 this volume and Frank et al. Chapter 10 this volume). The importance of predation, however, as well as the possible impact of soil pathogens such as fungi is sparse. Without this information it is difficult to make any predictions about the potential of biological control to combat brassica pod midge.

5.3.3 Pollen Beetle

Close proximity of WOSR and SOSR crops could provide a prolonged resource base for pollen beetles (*Meligethes* spp.). Although the beetles are univoltine, they have a long oviposition period and eggs are matured continuously throughout the life of the adult female (Ekbom and Ferdinand 2003). Therefore, when the bud stages most susceptible to oviposition (bud size = 2–3 mm) are no longer available in WOSR, the beetles can continue to reproduce in SOSR. This extra resource supply could result in higher populations of pollen beetles.

Pollen beetles are also a greater threat to cultivation of SOSR than WOSR. The beetles can begin to arrive in the spring crop at a much earlier crop developmental stage than in the winter crop. As crops are particularly susceptible at early bud stages, SOSR will most often be at higher risk than WOSR (Nilsson 1987). This is illustrated by the fact that pest control thresholds for the pollen beetle are generally lower in SOSR than in WOSR (Alford et al. 2003, Williams Chapter 1 this volume).

Because of the phenology of spring crops, their tolerance to pollen beetle attack is lower and more insecticide sprays against pollen beetle may be used in SOSR. The result of higher insecticide pressure on the beetle can be the development of resistance (Kazachkova et al. 2007, Thieme et al. Chapter 12 this volume). In a questionnaire study by Richardson (2008), a significant correlation was found between the number of pollen beetle insecticide applications and the year of resistance development; more applications resulted in earlier resistance. Insecticide resistance reported for pollen beetles is by no means restricted to areas where SOSR and WOSR cultivation coincide (Richardson 2008), but a number of areas from where resistance has been reported do have both spring and winter crops. Insecticide treatment, first in WOSR crops and then in SOSR crops, increases the insecticide pressure on a population of pollen beetles as the insects may move from one crop to the other and thereby be exposed multiple times to insecticides.

The parasitoids and potential predators of pollen beetles are described elsewhere in this book (see Ulber et al. Chapter 2 this volume, and Williams et al. Chapter 4 this volume) so only mortality factors caused by entomopathogens will be discussed here. A few species of hyphomycetous fungi, some common in agricultural soils, are known to be pathogenic to pollen beetles: *Beauveria bassiana*, *Metarhizium anisopliae*, and *Paecilomyces fumosoroseus* (Hokkanen et al. 2003). Attempts to use fungi to considerably increase mortality when the new generation of beetles emerges from the soil have not, however, been successful. Entomopathogenic nematodes have been suggested as possible biological control agents of beetle larvae close to pupation time (Hokkanen et al. 2003), but, at this time, logistics (applications to the soil of crops at a late development stage) and costs would seem to be prohibitive. Only a few pathogens have been recorded from the pollen beetle. These include the microsporidian *Anncalia* (= *Nosema*) *meligethi* (Issi and Radishcheva 1979), the haplosporidian *Haplosporidium meligethi* (Lipa and Hokkanen 1991), and an eugregarine *Gregarina* sp. (Lipa and Hokkanen 1991, Lipa and Ekbom 2003). Hokkanen et al. (2003) suggested that the microsporidian, which is common in adult populations of pollen beetles found outside of OSR growing areas, could potentially play an important role in the population dynamics of the beetle because the pathogen has a major impact on overwintering mortality. The problem is that the pathogen is absent from most OSR cultivation areas, possibly due to the fact that infected individuals may be more susceptible to insecticides than uninfected individuals (Lipa and Hokkanen 1992). Introduction of the microsporidian could be an option, but only if the reason for its current absence in OSR growing areas is explained. In addition, transmission pathways and the life cycle of the pathogen in pollen beetles need to be studied in order to facilitate a successful introduction.

5.4 Prospects for Reducing Insecticide Use and Promoting Integrated Pest Management in Spring Oilseed Rape

The cultivation of SOSR can be advantageous from an agronomic and climatic point of view. It is, however, currently a difficult task to grow SOSR without the use of insecticides. Seed treatment with systemic insecticides is presently a necessity in many areas to avoid damage by *Phyllotreta* flea beetles. Under some conditions, complementary treatment by spraying insecticides is also called for. In Europe, where the pollen beetle is present, insecticide treatment is often used in SOSR at some time from early bud stage to just before flowering. Pest control thresholds are available for both flea beetles and pollen beetles, but, because they are often exceeded, insecticides may be used regularly. The potential insecticide load on SOSR is therefore very high. Few field crops in Europe would normally demand such high levels of insecticide use.

Promoting integrated pest management under these conditions is difficult. Until recently, seed treatments and a well-timed spray against pollen beetles would generally provide the protection needed to produce a good yield. The emergence of insecticide resistance, however, has changed the situation and the response has been to use different insecticides with disparate modes of action and often to spray more. Although we know that pest insects in OSR do have many natural enemies, we also know that insecticide use is often more detrimental to these enemies than to the pests themselves (see also Ulber et al. Chapter 13 this volume).

We lack basic knowledge about the impact of natural enemies. Although parasitism rates on pollen beetles can sometimes be high (Ulber et al. Chapter 2 this volume) and ways to enhance these rates are available (e.g., see Nilsson et al. Chapter 11 this volume), we still do not know how high these rates must be to drive beetle populations into decline or how to achieve them. Lowering populations of flea beetles and pollen beetles would undoubtedly improve conditions for SOSR, but presently no strategy for this is available. The need to immediately remove the attacking insects from the crop with insecticides has priority over the need to lower pest population levels over time. The potential for using insect pathogens in oilseed rape pest management seems very low. Among the reasons for this are a lack of information on pathogens attacking the insect pests in their natural environment and lack of economic incentives to develop pathogen-based programs. To date no highly virulent or especially prevalent pathogens have been found in OSR cropping systems.

For a crop such as SOSR, action by natural enemies early in the season is a necessity for biological control as flea beetles arrive at crop emergence and pollen beetles will migrate to the crop at an early and sensitive growth stage. The importance of early aphid predation by generalist predators in spring-sown cereals has been demonstrated (Östman et al. 2001). It is possible that generalist predators could also have such an impact on beetle pests in SOSR. The problem is that sustained and frequent use of insecticides in SOSR may be inhibiting this potential. We urgently need to explore the ability of generalist natural enemies to control *Phyllotreta* spp. and pollen beetles. Previous studies on the predation of *M. aeneus* by carabids (Nilsson and Andreasson 1987, Nuss and Büchs 2000, Hokkanen 2004) lack the direct evidence of predation on *M. aeneus*, which is only possible by direct observation, gut dissection, or molecular identification of prey in predatory diets. The number of spider species and their density in rape crops is relatively high compared to maize and wheat (Frank and Nentwig 1995), providing an additional source of potential natural enemies. Predation of *Phyllotreta* flea beetles in SOSR has hardly been studied at all. By using molecular methods to detect pollen beetle predation (Cassel-Lundhagen et al. 2009) and flea beetle predation, we might be able to identify key predators of these two insect pests. The body of knowledge concerning generalist predators in arable fields has been expanding over the past few decades and this information can help us devise strategies to enhance predator abundance once we know more about the key predators.

Despite the fact that there are numerous parasitoids and predators of OSR insect pests to be found in the arable landscape there are still serious problems with insect pests in SOSR. An outcome of theoretical population dynamics modeling is that, at low densities, insect pests may be kept in check by natural enemies, but if pest population growth increases (or natural enemy abundance decreases) then pest outbreak is imminent (Southwood and Comins 1976). Perhaps the populations of pest insects have become much too high to be brought under control by biological control. Hokkanen (2000) argues that insecticide treatment, at or above threshold levels, will not necessarily reduce the size of the next generation of pollen beetles (or reduce population growth rates), but most certainly will affect natural enemies and their potential impact. In a study by Veromann et al. (2008) it was shown that the application of insecticides promoted pollen beetle population increase. Considering these results, it appears that as long as insecticides are a constant in SOSR systems then successful biological control will not be possible. In a short-term perspective, stopping the use of insecticides would only result in almost complete destruction of the crop, which seems a waste of resources.

In order to try to return the advantage to the natural enemies a possible strategy could be to break off oilseed cultivation in a large area for 1 year. Flea beetles and pollen beetles are both univoltine. The most important resource for beetle reproduction is cultivated oilseed rape crops; other possible host plants are, by comparison, scarce and patchily distributed in the landscape. Thus a year without oilseed crops (both WOSR and SOSR) would reduce the number of new generation beetles radically. Because parasitoids would also be at a disadvantage when their host numbers decline one could consider sowing some 'parasitoid conservation' areas where pollen beetles would be allowed to lay eggs and parasitoids reproduce; a less preferred host plant, like white mustard, might be an option. Perhaps such 'parasitoid conservation' areas could be placed near to fields expected to have overwintering parasitoids. In addition to providing reservoirs for pollen beetle parasitoids, the flowering plants would also offer nectar and pollen for other beneficial insects. Generalist predators would not be negatively influenced by lower beetle populations as they are able to use alternative prey and the absence of insecticide treatment would be to their benefit. Such a plan would, naturally, require large scale coordination and possibly financial incentives. On the other hand, implementation might just turn the tables and give the natural enemies a chance to perform their important ecosystem service, biological control.

Acknowledgements This work was funded by the Swedish Research Council for Environment, Agricultural Sciences and Spatial Planning (FORMAS).

References

Alford DV, Nilsson C, Ulber B (2003) Insect pests of oilseed rape crops. In: Alford DV (ed.) *Biocontrol of oilseed rape pests*. Blackwell, Oxford, UK.

Antwi FB, Olson DL, Knodel JJ (2007) Comparative evaluation and economic potential of ecorational versus chemical insecticides for crucifer flea beetle (Coleoptera: Chrysomelidae) management in canola. *J Econ Entomol* 100: 710–716.

Axelsen J (1992) The population dynamics and mortalities of the pod gall midge in winter rape and spring rape in Denmark. *J Appl Entomol* 114: 463–471.
Burgess L (1980) Predation on adults of the flea beetle *Phyllotreta cruciferae* by lacewing larvae. *Can Entomol* 112: 745–746.
Burgess L (1982) Predation on adults of the flea beetle *Phyllotreta cruciferae* by the weater damsel bug, *Nabis alternatus*. *Can Entomol* 118: 731–732.
Burgess L, Hinks CF (1987) Predation on adults of the flea beetle *Phyllotreta cruciferae* by the northern fall field cricket, *Gryllus pennsylvanicus*. *Can Entomol* 119: 495–496.
Butt TM, Jackson CW, Magen N (2001) *Fungi as biological control agents; progress, problems and potential*. CABI Publishing, Wallingford, UK.
Cassel-Lundhagen A, Öberg S, Högfeldt C, Ekbom B (2009) Species-specific primers for predation studies of the pollen beetle, *Meligethes aeneus*. *Mol Ecol Res* 9: 1132–1134.
Chiverton PA (1984) Pitfall-trap catches of the carabid beetle *Pterostichus melanarius*, in relation to gut contents and prey densities, in insecticide treated and untreated spring barley. *Entomol Exp Appl* 36: 23–30.
Culliney TW (1986) Predation on adult *Phyllotreta* flea beetles by *Podisus maculiventris* and *Nabicula americolimbata*. *Can Entomol* 118: 731–732.
Desneux N, Rabasse J-M, Ballanger Y, Kaiser L (2006) Parasitism of canola aphids in France in autumn. *J Pest Sci* 79: 95–102.
Ekbom B (1990) Flea beetles (*Phyllotreta* spp.) in spring oilseed rape in Sweden. *IOBC/wprs Bull* 12(4): 57–61.
Ekbom B (1991) The effect of abiotic and biotic factors on flea beetle populations in spring rape in Sweden. *IOBC/wprs Bull* 14(6): 21–27.
Ekbom B, Ferdinand V (2003) Field oviposition rates of pollen beetles (*Meligethes aeneus* Fab.). *Agr Forest Entomol* 5: 247–252.
Ekbom B, Kuusk A-K (2005) Jordloppor i våroljeväxter. [Flea beetles in spring oilseed rape, in Swedish]. *Faktablad om växtskydd. Jordbruk* 45 J.
Finch S (1996) Effect of beetle size on predation of cabbage root fly eggs by ground beetles. *Entomol Exp Appl* 81: 199–206.
Frank T, Nentwig W (1995) Ground dwelling spiders (Araneae) in sown weed strips and adjacent fields. *Acta Oecologia* 16: 179–193.
Hardwick NV, Davies JML, Wright DM (1994) The incidence of 3 virus diseases of winter oilseed rape in England and Wales in the 1991/92 and 1992/93 growing seasons. *Plant Pathol* 43: 1045–1049.
Hokkanen HMT (2000) The making of a pest: Recruitment of *Meligethes aeneus* onto oilseed Brassicas. *Entomol Exp Appl* 95: 141–149.
Hokkanen H (2004) Impact of predators on pollen beetle *Meligethes aeneus* on rapeseed in Finland. *IOBC/wprs Bull* 27(10): 293–296.
Hokkanen HMT, Menzler-Hokkanen I, Butt TM (2003) Pathogens of oilseed rape pests. In: Alford DV (ed.) *Biocontrol of oilseed rape pests*. Blackwell, Oxford, UK.
ICTVdB Management (2006a) 00.015.0.01.001. Cauliflower mosaic virus. In: Büchen-Osmond, C (ed.) *ICTVdB – The universal virus database*, version 4. Columbia University, New York, USA. http://www.ncbi.nlm.nih.gov/ICTVdb/ICTVdB/ (accessed 3 April 2009).
ICTVdB Management (2006b). 00.057.0.01.072. Turnip mosaic virus. In: Büchen-Osmond, C (ed.), *ICTVdB – The Universal Virus Database*, version 4. Columbia University, New York, USA. http://www.ncbi.nlm.nih.gov/ICTVdb/ICTVdB/ (accessed 3 April 2009).
Issi IV, Radishcheva DF (1979) Microsporidioses of beetles noxious to cruciferous plants in Leningrad region. *Byul. Vses. Nauch. Issled. Instituta Zashchity Rastenii* 46: 19–23 (in Russian).
Jones RAC, Coutts BA, Hawkes J (2007) Yield-limiting potential of Beet western yellows virus in *Brassica napus*. *Aust J Agri Res* 58: 788–801.
Jones FGW, Jones MG (1984) *Pests of field crops*. Edward Arnold, London.
Jourdheuil P (1960) Influence de quelques facteurs écologiques sur les fluctuations de population d'une biocénose parasitaire: étude relative à quelques Hyménoptères (Ophioninae, Diospilinae,

Euphorinae) parasites de divers Coléoptères inféodés aux Crucifères. *Annales des Épiphyties* 11: 445–658.

Kazachkova N, Meijer J, Ekbom B (2007) Genetic diversity in pollen beetles (*Meligethes aeneus*) in Sweden: Population structure and insecticide resistance. *Agr Forest Entomol* 9: 259–269.

Lipa JJ, Ekbom B (2003) Microsporidian, Haplosporidian, and Eugregarine parasites present in populations of *Meligethes aeneus* in Sweden. *Acta Agr Scand B S P* 53: 87–89.

Lipa JJ, Hokkanen HMT (1991) A haplosporidian *Haplosporidium meligethi* sp.n. and a microsporidian *Nosema meligethi* I. et R., two protozoan parasites from *Meligethes aeneus*. *Acta Protozool* 30: 217–222.

Lipa JJ, Hokkanen HMT (1992) *Nosema meligethi* I. et R. in populations of *Meligethes* spp. in Europe. *Biocontrol Sci Techn* 2: 119–125.

Mandal SMA, Mishra BK, Patra AK (1994) Yield loss in rapeseed and mustard due to aphid infestation in respect to different varieties and dates of sowing. *Orissa J Agric Res* 7: 58–62.

Nielsen C, Jensen AB, Eilenberg J (2007) Survival of entomophthoralean fungi infecting aphids and higher flies during unfavorable conditions and implications for conservation biological control. In: Ekesi S, Maniania NK (eds.) *Use of entomopathogenic fungi in biological pest management*. Research Signpost, Trivandrum, India.

Nilsson C (1987) Yield losses in summer rape caused by pollen beetles (*Meligethes* spp.). *Swed J Agr Res* 17: 105–111.

Nilsson C, Andreasson B (1987) Parasitoids and predators attacking pollen beetles (*Meligethes aeneus* F.) in spring and winter rape in southern Sweden. *IOBC/wprs Bull* 10: 64–73.

Nuss H, Büchs W (2000) Importance of epigaeic predators as antagonists of soil dwelling insect pests of winter oilseed rape-inflorescences (buds, flowers and pods). *Mitteilungen der Deutschen Gesellschaft für allgemeine und angewandte Entomologie* 12: 141–147.

Östman Ö, Ekbom B, Bengtsson J (2001) Landscape heterogeneity and farming practice influence biological control. *Basic Appl Ecol* 2: 365–371.

Richardson DM (2008) Summary of findings from a participant country pollen beetle questionnaire. *EPPO Bulletin* 38: 68–72.

Shahraeen N, Farzadfar S, Lesemann DE (2003) Incidence of viruses infecting winter oilseed rape (*Brassica napus* ssp. *oleifera*) in Iran. *J Phytopathol* 151: 614–616.

Simonsen M-LR, Enkegaard A, Bang CN, Sigsgaard L (2009) Temperature effect on the predation rate of *Anthocoris nemorum* on cabbage aphids. *J Appl Entomol* 133: 198–200.

Smith HG, Barker H (eds.) (1999) *The luteoviridae*. CABI Publishing, Wallingford UK.

Snyder GB, Finke DL, Snyder WE (2008) Predator biodiversity strengthens aphid suppression across single- and multiple-species prey communities. *Biol Control* 44: 52–60.

Sommer G (1981) Biology and parasites of *Phyllotreta* spp. (Coleoptera, Halticinae). CIBC Report, Delemont.

Southwood TRE, Comins HN (1976) A synoptic population model. *J Anim Ecol* 45: 949–965.

Trdan S, Vidrih M, Valič N, Laznik Ž (2008) Impact of entomopathogenic nematodes on adults of *Phyllotreta* spp. under laboratory conditions. *Acta Agr Scand B S P* 58: 169–175.

Ulber B, Williams IH (2003) Parasitoids of flea beetles. In: Alford DV (ed.) *Biocontrol of oilseed rape pests*. Blackwell, Oxford, UK.

Veromann E, Kevväi R, Luik A, Williams IH (2008) Do cropping system and insecticide use in spring oilseed rape affect the abundance of pollen beetles (*Meligethes aeneus*) on the crop? *Int J Pest Manage* 54: 1–4.

Vänninen I, Hokkanen H, Tyni-Juslin J (1999) Screening of field performance of entomopathogenic fungi and nematodes against cabbage root flies. *Acta Agr Scand B S P* 49: 167–183.

Wang RY, Powell G, Hardie J, Pirone TP (1998) Role of the helper component in vector-specific transmission of potyviruses. *J Gen Virol* 79: 1519–1524.

Williams IH (2003) Parasitoids of the Brassica Pod Midge. In: Alford DV (ed.) *Biocontrol of oilseed rape pests*. Blackwell, Oxford, UK.

Williams IH, Martin AP, Kelm M (1987) The phenology of the emergence of brassica pod midge and its infestation of spring oil-seed rape. *J Agr Sci* 109: 309–314.
Wylie HG (1988) Release in Manitoba, Canada of *Townesilitus bicolor* an European parasite of *Phyllotreta* spp. *Entomophaga* 33: 25–32.
Yaman M (2002) *Gregarina phyllotretae* Hoshide 1953, a protozoan parasite of the flea beetles *Phyllotreta undulata* and *P. atra* in Turkey. *Appl Entomol Zool* 37: 649–653.
Zhang WQ, Hassan SA (2003) Use of the parasitoid *Diaeretiella rapae* to control the cabbage aphid *Brevicoryne brassicae*. *J Appl Entomol* 127: 522–526.

Chapter 6
Key Pests and Parasitoids of Oilseed Rape or Canola in North America and the Importance of Parasitoids in Integrated Management

Lloyd M. Dosdall and Peter G. Mason

Abstract The development of 'double low oilseed rape' or canola as a unique crop in the mid 1970s resulted in rapid expansion in the area devoted to its production in North America, especially in the western Canadian provinces of Alberta, Saskatchewan and Manitoba. The crop was soon subjected to attack by a number of insect herbivores, sometimes responsible for severe economic losses. The key pests of the crop comprise the flea beetles *Phyllotreta cruciferae* (Goeze) and *Phyllotreta striolata* (Fabricius), several root maggot species especially *Delia radicum* (L.), cabbage seedpod weevil, *Ceutorhynchus obstrictus* (Marsham), bertha armyworm, *Mamestra configurata* Walker, several species of lygus bugs, *Lygus* spp., and the diamondback moth, *Plutella xylostella* (L.). Most pest species are alien, introduced from Europe or Asia, and as a consequence they have been the subject of importation and release of non-indigenous parasitoids in attempts to achieve classical biological control. To date, such attempts have met with limited success, but nevertheless substantial progress has been made in certain situations. This chapter reviews the life history and crop damage caused by each of the key pest species, and the parasitoid fauna associated with each. Attempts at classical or augmentative biological control of the pests are documented, and a case history is provided for one invasive species, the cabbage seedpod weevil. Challenges to biological control with parasitoids are identified for North American cropping systems.

6.1 Introduction

The genetic transformation of rapeseed that occurred in the mid 1970s through development of varieties of *Brassica napus* L. and *Brassica rapa* L. with low erucic acid and glucosinolate levels in the seed (< 2% erucic acid and < 30 μmol/gm of glucosinolates in oil-free meal) prompted rapid expansion in the production of 'double low oilseed rape' or 'canola' in North America (Anonymous 2007a). Reductions

L.M. Dosdall (✉)
Department of Agricultural, Food and Nutritional Science, 4-10 Agriculture-Forestry Centre, University of Alberta, Edmonton, AB T6G 2P5, Canada
e-mail: lloyd.dosdall@ualberta.ca

I.H. Williams (ed.), *Biocontrol-Based Integrated Management of Oilseed Rape Pests*, DOI 10.1007/978-90-481-3983-5_6, © Springer Science+Business Media B.V. 2010

in erucic acid content in the oil and glucosinolate in the meal were keys to success of the crop because the oil could be used for human consumption and the meal was suitable for animal feed. A further reason for expansion of the crop in North America was climatic. Canola grows at relatively low temperatures and requires fewer heat units to mature than most other oilseed crops, making it well suited to the temperate climates of the western Canadian provinces of Alberta, Saskatchewan and Manitoba where most production occurs (Thomas 2002). Minor canola production also occurs in the provinces of British Columbia, Ontario and Québec. Production in Canada has gradually escalated, so that currently over 5.9 million ha of canola, 175,000 ha of mustard (*Brassica juncea* (L.) Czern. and *Sinapis alba* L.), and 12,000 ha of various cabbage relatives are planted in a wide range of ecoregions (Statistics Canada 2007a, 2007b). Canola production is much lower in the USA, accounting for approximately 472,000 ha principally in North Dakota (90% of all US production), Minnesota (3%), and Montana (1%) with minor production in Washington, Oregon, Oklahoma, Michigan, Kansas, Idaho, Colorado and Georgia (NASS 2007). The area devoted to canola production will likely increase further in both countries as demand increases to supply the biofuel industry (Goodwin 2006). No canola is currently grown in Mexico.

Climatic considerations require that the canola varieties grown in North America differ phenologically from those of Europe. In North America, the annual forms of *B. napus* and *B. rapa* are sown in spring and harvested approximately 100 d later, but most European production is from winter-dormant, biennial varieties sown in late summer and harvested in the following year (Downey 1983, Alford 2003a). Exceptions exist, however, because some spring canola is grown in northern and central Europe and some winter-dormant canola is grown in the Pacific Northwest and Great Lakes-St. Lawrence River regions of North America.

A complex of pests attacks all above- and below-ground portions of canola crops. On a broad scale, the most serious pests of the crop are members of the Coleoptera, but substantial damage is also inflicted by species of Hemiptera, Lepidoptera, and Diptera (Lamb 1989). In North America, crop losses from insect herbivory can be extremely damaging, and are usually most dramatic when outbreak densities develop. For example, an outbreak of diamondback moth, *Plutella xylostella* (L.) (Lepidoptera: Plutellidae), in Saskatchewan and Alberta in 1995 caused losses to producers estimated at $45–$52 million (CAD) (WCCP 1995). In 1998, an outbreak of *Lygus* spp. in southern Alberta resulted in insecticide treatment of 400,000 ha at a cost of approximately $9 million (CAD) (Braun et al. 2001). Such outbreaks inflict enormous crop losses at irregular intervals, but some North American canola pests are primarily chronic in their effects. For example, the flea beetles *Phyllotreta cruciferae* (Goeze) and *Phyllotreta striolata* (Fabricius) (Coleoptera: Chrysomelidae) occur regularly in many areas of canola production and are estimated to cause losses of more than $300 million (CAN) annually (Madder and Stemeroff 1988).

In North America, control of pest insects on brassicaceous oilseed crops is most commonly achieved through insecticidal applications. Depending on the pest species involved, this can occur as early as planting time in spring with insecticide coated onto seeds for short-term systemic activity at the seedling stage, or as late as pod-filling stages when foliar insecticide may be applied by aircraft or ground sprayers. Although insecticide applications have provided reliable and reasonably cost-effective control of insect pests in canola, there has recently been greater emphasis on minimizing pesticide applications in order to strengthen export markets and ensure greater environmental responsibility. To this end, cultural and biological management strategies have been pursued, and research has been directed toward developing canola varieties resistant to insect infestations. In recent years, there has also been considerable effort to embody integrated insect management within the concept of integrated crop management. This holistic approach is seen as a means of making the most efficient use of energy, various inputs like fertilizers and pesticides, and minimizing environmental impacts to enhance long-term agroecosystem sustainability for profitable production.

Classical biological control, or the introduction and management of self-regulating natural enemy populations, is an important approach for reducing infestations of key pests of this crop (Alford 2003b). In North America, brassicaceous oilseed crops are grown on a vast scale, usually in monocultures, with individual fields sometimes occupying more than 300 ha. These agroecosystems pose serious challenges to biological control, and strategies like inundation by natural enemies are neither economical nor practical. Several serious pests are introduced invasive species, so most biological control efforts have been directed toward importing and introducing exotic parasitoids in an effort to reconstruct a component of the natural enemy complexes found in their natural ranges. Recently efforts have also been directed toward enhancing the activity of native natural enemies.

In this chapter we summarize aspects of the biology of the six insect pest species or species complexes responsible for most economic damage to brassicaceous oilseed crops in North America. We identify current control strategies for each pest species, with emphasis on their natural enemies and efforts that have been expended to enhance biological control. We also present a case study of the invasion of the cabbage seedpod weevil in canola in western Canada, with a description of the effectiveness of natural enemies for reducing its impact, and the dilemma of pursuing classical biological control when native parasitoids have shifted from their usual hosts to attack the weevil. The chapter concludes with a summary of the challenges facing biological control with parasitoids in North American cropping systems, and needs for future research.

6.2 Key Pests and Their Parasitoids

6.2.1 *Flea Beetles,* **Phyllotreta cruciferae** *(Goeze) and* **Phyllotreta striolata** *(Fabricius) (Fig. 6.1) (Coleoptera: Chrysomelidae)*

A complex of eight flea beetle species can occur in brassicaceous oilseed crops in western Canada (Burgess 1980), but of these, only the crucifer flea beetle, *P. cruciferae* and the striped flea beetle, *P. striolata*, are significant pests. Both are

Fig. 6.1 Adult of *Phyllotreta striolata* (Photo: Lloyd Dosdall)

adventive Eurasian species. The striped flea beetle appears to have arrived as early as 1,700 (Bain and LeSage 1998). The crucifer flea beetle likely established in the 1920s and by the late 1930s and 1940s had become a serious pest of cultivated cruciferous crops in Canada (Westdal and Romanow 1972). Although both species occur sympatrically throughout much of their North American range, *P. striolata* populations are usually more abundant than *P. cruciferae* in the cooler subhumid conditions that dominate in the Aspen Parkland Ecoregion of central Canada, and *P. striolata* dominates the flea beetle fauna in the Peace Lowland Ecoregion of northwestern Alberta. *Phyllotreta cruciferae* is dominant throughout most of southern Canada and the USA, and in both countries *P. cruciferae* is considered the dominant pest of newly-germinated *B. napus* and *B. rapa* (Lamb and Turnock 1982, Weiss et al. 1991).

Both flea beetle species overwinter in leaf litter and turf beneath shelterbelts, native trees, fence rows, and grassy areas (Burgess 1977, 1981). Peak emergence of *P. cruciferae* occurs in late May as ground temperature reaches 15°C (Ulmer and Dosdall 2006a). Where sympatric populations occur, *P. striolata* adults emerge from overwintering sites several days before adults of *P. cruciferae* (Wylie 1982). Adults may disperse directly into early-seeded crops but often seek out brassicaceous weeds or volunteer crop plants before moving into canola crops as seedlings emerge. Eggs are laid during June and early July and larvae feed on root hairs and taproots of host plants (Westdal and Romanow 1972). Pupation occurs in small earthen cells, and emergence of new generation adults occurs from the mid August to September (Westdal and Romanow 1972, Ulmer and Dosdall 2006a). Adults feed on foliage and pods of maturing canola before migrating to overwintering sites (Feeny et al. 1970).

Greatest crop damage from *Phyllotreta* spp. occurs in spring when overwintered adults feed on the cotyledons and stems of seedlings. Early-season feeding deprives plants of photosynthate, and seedling stems can be completely severed by flea beetle feeding (Feeny et al. 1970, Westdal and Romanow 1972). Adult beetles can continue to cause some plant damage into the growing season; however, once plants are beyond the seedling stage they are much less vulnerable to attack (Gavloski and Lamb 2000). When infestations of new generation adults are significant, pod feeding late in the season can result in reduced seed production and quality (Knodel and Olson 2002). Crop losses from flea beetle attack include reduced crop stands and uneven plant growth, delayed maturity, lower seed yield, and insecticidal applications, with economic costs in excess of $300 million (CAD) annually (Westdal and Romanow 1972, Lamb and Turnock 1982, Madder and Stemeroff 1988).

The primary control strategy for flea beetles in North American canola crops is to apply insecticidal seed coatings with short-term systemic activity, followed by foliar sprays if beetle populations are still causing excessive damage (Lamb 1988). Alternatives to insecticide use have focused primarily on cultural practices and host plant resistance. Adopting a reduced or zero tillage regime has been recommended for reducing insecticide use because it provides canola seedlings with cooler and moister conditions than occur with conventional tillage (Dosdall et al. 1999). *Phyllotreta* spp. prefer warm, dry conditions (Tahvanainen 1972), so populations decline with reduced tillage. Increasing plant density and widening row spacing at seeding can also reduce flea beetle damage (Dosdall et al. 1999). Seeding into near-freezing soil in fall rather than on more routine spring seeding dates can enable seeds to remain dormant during winter and germinate early in spring. Early growth enables seedlings to progress beyond the vulnerable cotyledon stage before flea beetles invade (Dosdall and Stevenson 2005). Seeding *B. napus* and *B. rapa* with larger seeds produces seedlings of greater biomass that are more tolearant of flea beetle damage (Elliott et al. 2007, 2008). Species and cultivars of Brassicaceae differ in their susceptibilities to attack by flea beetles (Lamb 1980, 1984, 1988, Lamb and Palaniswamy 1990), and this has been used to drive research toward developing canola cultivars resistant, or partially resistant, to flea beetle attack (Gavloski et al. 2000). However, no flea beetle-resistant germplasm has yet been developed.

Several parasitoid species have been reared from flea beetles, and among these the most common was the braconid, *Microctonus vittatae* Muesebeck (Wylie and Loan 1984) (see Fig. 6.2). It may be native to North America or may have been introduced accidentally with its hosts. *Microctonus vittatae* parasitizes adults of both flea beetle species, although incidence of parasitism was lower in *P. striolata* (3–15%) than in *P. cruciferae* (15–53%) (Wylie 1982). Females can produce fertile eggs without mating, and males of *M. vittatae* are rare. First-instar parasitoid larvae overwinter within adult beetles, and resume their development in spring when their hosts leave their overwintering sites (Wylie 1982). Development is relatively rapid in spring, with parasitoid larvae emerging from beetle hosts in mid May. They then spin cocoons and parasitoid adults emerge in late May to early June. The parasitoid alters

Fig. 6.2 Adult of *Microctonus* sp. (Photo: Lloyd Dosdall)

normal behavior of its flea beetle hosts by causing the beetles to emerge earlier from overwintering sites than non-parasitized beetles, an adaptation that could enable *M. vittatae* to complete an extra generation during the season (Wylie 1982).

The incidence of parasitism of *Phyllotreta* spp. by *M. vittatae* was usually less than 5% in southern Manitoba, presumably due to a short adult life, a low oviposition rate, and poor temporal synchronization with host populations in mid summer (Wylie et al. 1984). Studies in Europe determined that *Townesilitus bicolor* (Wesmael) (Hymenoptera: Braconidae) parasitized approximately 50% of flea beetles in summer, and developed in both *P. striolata* and *P. cruciferae* (Sommer 1981). It was subsequently imported and released as a biological control agent in southern Manitoba. However, specimens of *T. bicolor* were never recovered at or near the release sites (Wylie 1988), and its establishment status is unknown.

6.2.2 Root Maggots, Delia *spp. (Diptera: Anthomyiidae)*

Root maggots or root flies are chronic and serious pests of some brassicaceous oilseed crops in temperate climates of North America (Griffiths 1986a, 1986b, Soroka et al. 2004) (Fig. 6.3). A complex of five root fly species occurs in oilseed crops, but most primary damage is inflicted by the cabbage maggot, *Delia radicum* (L.), and the turnip maggot, *Delia floralis* (Fallén) (Liu and Butts 1982, Griffiths 1986a, 1986b). *Delia planipalpis* (Stein) can also attack uninjured plants, but it occurs in low densities in oilseed crops (Griffiths 1991b). The seedcorn maggot,

Fig. 6.3 Adult of *Delia* sp. (Photo: Lloyd Dosdall)

Delia platura (Meigen), and the bean seed maggot, *Delia florilega* (Zetterstedt), are believed to feed on canola taproots only secondarily, after the plant is attacked by one of the species of primary feeders. Nevertheless, *D. platura* and *D. florilega* can sometimes occur in high densities in oilseed crops (Soroka et al. 2004, Broatch et al. 2006).

Adults emerge from overwintered puparia from mid-May to early July, and most oviposition occurs in mid to late June when canola plants are in the rosette stage (Dosdall et al. 1994, Griffiths 1986a). Root maggot larvae feed on the conductive and storage cells of the taproot phloem, periderm, and xylem, disrupting the transport of water and nutrients between the root and upper regions of the plant (McDonald and Sears 1992). This can lead to reductions in root weight and root sugar content, stunted growth, premature lodging, decreased raceme numbers, and reduced seed yields (McDonald and Sears 1991, Griffiths 1991a, Hopkins et al. 1999). Injured roots can be invaded by root rot fungi (*Fusarium* spp.), causing further yield reductions (Griffiths 1986a, 1986b). Root flies complete one or two generations per year in canola, depending on the species, geographic latitude, and year (Griffiths 1986a, Broatch et al. 2006). Root flies occur throughout oilseed cropping ecoregions of western North America, but are most damaging in western and northwestern Alberta and along the northern edge of the Parkland Ecoregion of Alberta, Saskatchewan, and Manitoba (Soroka et al. 2004). Consequently, root fly infestations have become major constraints in canola production, responsible for yield losses estimated as high as $100 million (CAD) annually (Soroka et al. 2004).

Currently there are no insecticides registered for control of root flies in oilseed crops. Studies investigating the effectiveness of insecticidal seed treatments, coated

Fig. 6.4 Larva of *Aleochara bilineata* on pupa of *Delia* sp. (Photo: Lloyd Dosdall)

with polymer for slow release in the soil, determined that the organophosphate and carbamate products only slowed the onset of feeding damage but did not prevent it (Griffiths 1991a). Cultural control strategies have therefore been an important consideration for canola producers. Crop damage from *Delia* spp. can be reduced by selecting the most resistant species and cultivars (Dosdall et al. 1994), applying soil tillage prior to seeding (Dosdall et al. 1996a), seeding to achieve plant stands of relatively high densities (Dosdall et al. 1996b), using wide rather than narrow row spacings (Dosdall et al. 1998), and delaying weed removal until crops are in the four- rather than the two-leaf stage of development (Dosdall et al. 2003). However, in view of continued crop losses from these pests, biological control holds considerable promise.

Several parasitoid species attack root flies in North America. A recent survey of commercial canola fields from several locations in western Canada by Hemachandra (2004) yielded no egg parasitoids, but seven species of larval and pupal parasitoids were recorded including the hymenopterans *Phygadeuon* sp. 1 and *Phygadeuon* sp. 2 (Ichneumonidae), *Aphaereta minuta* (Nees) (Braconidae), *Trichopria* sp. (Proctotrupidae), and *Trybliographa rapae* (Westwood) (Eucoilidae), and the coleopterans *Aleochara verna* Say and *Aleochara bilineata* Gyllenhal (Staphylinidae) (Fig. 6.4). Of these, all except *A. bilineata* and *T. rapae* are considered of minor importance due to their rare frequency of occurrence (Turnock et al. 1995a, Hemachandra 2004).

Adult *A. bilineata* are predators and parasitoids of several root maggot species including *D. radicum*, *D. platura*, *D. floralis*, *Delia antiqua* (Meigen), and *D. planipalpis* (Read 1962, Klimaszewski 1984, Tomlin et al. 1985, Maus et al.

1998). *Aleochara bilineata* is probably native to Europe, but was accidentally introduced to North America and is now widely distributed throughout the Nearctic region (Klimaszewski 1984). The beetle can consume large numbers of root maggot eggs and larvae, and a single adult can consume an average of 23.8 eggs or 2.6 larvae of *D. radicum* per day (Read 1962). Under optimal conditions a pair of *A. bilineata* adults can destroy approximately 1,210 eggs and 128 larvae in their lifetime. Larvae of *A. bilineata* are parasitoids of root maggot pupae. Each *A. bilineata* female can produce nine to 15 eggs per day or 700 in its lifetime (Colhoun 1953). Larvae hatch in 3–7 days, and the first instar chews an opening in the puparial wall, enters, and consumes the developing pupa within (Royer et al. 1998). Overwintering of *A. bilineata* occurs as a first instar within the puparium, and new generation adults emerge during the following spring (Colhoun 1953). Acting as both a predator and a parasitoid, *A. bilineata* is an important natural control agent of cabbage maggot populations (Mukerji 1971). In western Canada, parasitism of cabbage maggot puparia by *A. bilineata* in cole crops can be as high as 94% (Turnock et al. 1995a).

Trybliographa rapae follows a parasitism pattern directly related to the density of its root maggot hosts where females spend more time host-searching on patches with a high host density (Jones and Hassell 1988). Females of *T. rapae* oviposit into all instars of larvae of *D. radicum*, *D. floralis*, and *D. platura*, although first- and second-instar larvae may be attacked more frequently in some conditions (Wishart and Monteith 1954). Females parasitize hosts in soil to a depth of approximately 4 cm, but not 6 cm, enabling host larvae that occur in deeper soil to escape parasitization (Hemachandra et al. 2007b). *Trybliographa rapae* larvae develop endoparasitically in their hosts until pupariation occurs (Block et al. 1987). Third-instar parasitoid larvae then exit the host body cavity, and feed within the puparia as ectoparasitoids on the pupae (Wishart and Monteith 1954). Hemachandra et al. (2007a) cited *T. rapae* as the most common parasitoid of *D. radicum* in canola in Canada, but in central Alberta it is only a minor component of the parasitoid fauna and occurs much less frequently than *A. bilineata* (J. Broatch, *unpublished data*).

Initiatives directed toward enhancing biological control of *Delia* spp. in oilseed brassicaceous crops in North America have focused on the hymenopteran *Phygadeuon trichops* Thompson (Ichneumonidae) and the coleopteran *Aleochara bipustulata* (L.) (Staphylinidae). Species of *Phygadeuon* are pupal parasitoids and minor components of the parasitoid fauna of *D. radicum* in western Canada (Hemachandra 2004). In Europe, *P. trichops* parasitizes a comparatively small portion of the *D. radicum* pupal population (Hughes and Mitchell 1960, Ryan and Ryan 1980). Adults of *P. trichops* live approximately 50 days, and although eggs can be deposited over most of that period, most oviposition occurs within the first 20 days of emergence (Plattner 1974). As many as four eggs can be implanted per host pupa, but only a single larva develops. In spite of its low frequency of occurrence in European populations of *D. radicum*, individuals of *P. trichops* were released in western Canada from 1949 to 1954, but the species does not seem to have established (Soroka et al. 2002). In Europe, *A. bipustulata* usually occurs in a smaller proportion of *D. radicum* puparia than *A. bilineata* or *T. rapae* (Wishart et al. 1957), but it could complement biological control by the two dominant parasitoids currently found in North America. No North American releases of *A. bipustulata* have yet been made, but it is currently being evaluated for classical biological control of *D. radicum* (Riley et al. 2007, Andreassen et al. 2007).

6.2.3 Cabbage Seedpod Weevil, Ceutorhynchus obstrictus *(Marsham) [syn.* Ceutorhynchus assimilis *(Paykull)] (Coleoptera: Curculionidae)*

This insect has different common names in Europe and North America (Fig. 6.5). In Europe, it is called cabbage seed weevil, but it is termed cabbage seedpod weevil in approved lists of common names in Canada (Entomological Society of Canada 2007) and the USA (Entomological Society of America 2007). In this chapter it is termed cabbage seedpod weevil since the focus is on North American pests, parasitoids, and cropping systems. The species name also varies between Europe and North America. In Europe, it is *Ceutorhynchus assimilis* (Paykull) in most literature, but in North America it is most commonly referred to as *Ceutorhynchus obstrictus* (Marsham), as a result of a recent revision of the genus by Colonnelli (1993). A petition to conserve the name of *C. assimilis* was recently reviewed by the International Commission of Zoological Nomenclature (Alford 2006), and the Commission ruled that the appropriate name is *C. obstrictus* (ICZN 2007).

Fig. 6.5 Adult of *Ceutorhynchus obstrictus* (Photo: Lloyd Dosdall)

The cabbage seedpod weevil is native to Europe, and soon after its discovery near Vancouver, BC, Canada in the early 1930s (McLeod 1962), it dispersed south to the Pacific Northwest and California where it damaged seed crops of brassicaceous vegetables (Hanson et al. 1948, Crowell 1952). After canola production began in Idaho and Washington, the weevil was observed to cause crop damage (McCaffrey 1992). It is univoltine and completes its larval development only in some Brassicaceae (Bonnemaison 1957, Dmoch 1965). Adults overwinter primarily in shelterbelts and emerge when soil temperatures reach approximately 15°C (Ulmer and Dosdall 2006b). Early in the season (April to May) adults migrate to patches of early flowering brassicaceous weeds like wild mustard (*Sinapis arvensis* L.), pennycress (*Thlaspi arvense* L.), flixweed (*Descurainia sophia* [L.] Webb), and hoary cress (*Lepidium draba* L.) and feed primarily in their inflorescences (Fox and Dosdall 2003, Dosdall and Moisey 2004). Adults disperse to canola primarily when it reaches the bud and flowering stages. Most oviposition occurs when plants are still flowering but pods on lower racemes are elongating. Females excavate an opening in the wall of the developing pod with their mouthparts, turn, and deposit a single egg into the opening (Kozlowski et al. 1983). First instars are most abundant when seeds in lower pods begin to enlarge, and second instars are most abundant when seeds within the lower pods are fully enlarged (Dosdall and Moisey 2004). Third instars are most abundant when seeds in lower pods are green. When mature, the third-instar larva chews an opening in the pod wall ('exit hole'), drops to the soil surface, digs in, and pupates in an earthen cell. Adults emerge about 14 days later. Development from egg to adult requires approximately 31–60 days in spring canola in western Canada (Dosdall and Moisey 2004).

Cabbage seedpod weevil can inflict damage to canola crops as either adults or larvae. When adults invade crops in spring, they feed on developing flower buds causing them to desiccate and racemes to bear fewer pods (Dosdall et al. 2001). Larval feeding within pods destroys developing seeds, and pods with exit holes shatter before harvest more frequently than non-infested pods. Feeding by new generation adults through the pod walls can further reduce yield and crop quality (Buntin et al. 1995, Dosdall et al. 2001).

Control measures investigated for cabbage seedpod weevil include chemical insecticide applications, cultural strategies, and host plant resistance. Chemical control is achieved with applications of synthetic pyrethroid insecticides, applied when crops are in approximately 10% flower (Dosdall et al. 2001, Cárcamo et al. 2005). Surrounding the perimeter of commercial fields with trap crops seeded to flower earlier than the main crop within the perimeter, can be successful (Cárcamo et al. 2007), but extensive monitoring is required to ensure that weevils in the trap are destroyed before they can invade the main crop. Seeding later in spring, and at a relatively high plant density, can help reduce weevil infestations and damage (Dosdall et al. 2006a). Germplasm resistant to infestation by the weevil is not available to canola growers, but research is ongoing to develop such varieties for commercial production using *S. alba* as a source of resistance (McCaffrey et al. 1999, Dosdall and Kott 2006).

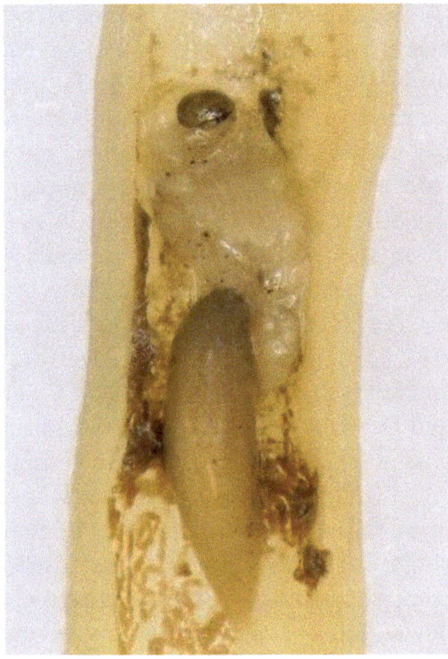

Fig. 6.6 Larva of *Necremnus tidius* on larva of *Ceutorhynchus obstrictus* (Photo: Lloyd Dosdall)

In the Pacific Northwest of Canada and the USA, surveys by Doucette (1948), Hanson et al. (1948), Carlson et al. (1951), McLeod (1952), and Walz (1957) found several parasitoid species associated with the cabbage seedpod weevil. Initially identifications of these species appeared to indicate that they included most of the dominant parasitoid fauna of the weevil in Europe: *Trichomalus perfectus* (Walker) and *Mesopolobus morys* L. (Hymenoptera: Pteromalidae), and *Necremnus duplicatus* Gahan (Hymenoptera: Eulophidae); however, subsequent examinations of voucher material and type specimens determined that the dominant North American fauna actually comprised the pteromalids *Trichomalus lucidus* (Walker) and *Mesopolobus moryoides* Gibson, and the eulophid *Necremnus tidius* (Walker) (Fig. 6.6) (Gibson et al. 2005). An additional species, the braconid *Microctonus melanopus* (Ruthe), was discovered attacking adult weevils in Idaho, and in some sites, parasitism levels by *M. melanopus* could be as high as 70% (Harmon and McCaffrey 1997).

In Alberta, no evidence of parasitism of adults or larvae of cabbage seedpod weevil was found in the initial 5-year period following its invasion (Kuhlmann et al. 2002), but in 2000, some adult weevils were found parasitized by *M. melanopus* (Fox et al. 2004). The parasitoid overwinters as a first instar within the adult weevil, and after emerging from its host in spring, it pupates in soil (Jourdheuil 1960). The new parasitoid generation attacks the same generation of overwintered weevils. There are therefore two generations per year of the parasitoid, but the weevil is univoltine (Harmon and McCaffrey 1997). Since 2001, levels of parasitism by *M. melanopus* have been low in western Canada (Fox et al. 2004, Dosdall unpublished data), and the parasitoid is not considered to be effective as a biological control agent for this pest.

In 2001, parasitoids of cabbage seedpod weevil larvae were discovered in southern Alberta, and during the next 6 years, a total of 14 parasitoid species representing four families were associated with the weevil (Gibson et al. 2005, Dosdall et al. 2006b, 2009). Of these, *N. tidius*, *T. lucidus*, *Chlorocytus* sp., and *Pteromalus* sp. were the species most frequently reared (Dosdall et al. 2006b, 2009). Parasitoids were also found attacking larvae of cabbage seedpod weevil in Georgia, USA. Investigations by Gibson et al. (2006a) found 13 species of Chalcidoidea representing five families. *Lyrcus maculatus* (Gahan) was the dominant species attacking the weevil and constituted about 96% of all Pteromalidae and 86% of the total parasitoid fauna. It was noted that some of these species are likely hyperparasitoids or emerged from insect contaminants of the mass-reared seedpods. As in western North America, earlier determinations of *T. perfectus* from Georgia were found to be misidentifications (Gibson et al. 2006a).

Classical biological control of cabbage seedpod weevil was attempted in North America with the release of three larval parasitoid species in British Columbia in 1949 (Gillespie et al. 2006). Recent re-examination of voucher material from the releases determined that the species released included *T. perfectus*, *M. morys*, and *Stenomalina gracilis* (Walker) (Pteromalidae) (Gibson et al. 2006b). Surveys near the original release sites and in central British Columbia found no evidence for establishment of *T. perfectus* or *M. morys*; however, recovery of specimens of *S. gracilis* near the release sites and hundreds of kilometres away confirm that this species has established (Gibson et al. 2006b, Gillespie et al. 2006).

6.2.4 Bertha Armyworm, Mamestra configurata *Walker (Lepidoptera: Noctuidae) (Fig. 6.7)*

Unlike most other pests of oilseed brassicaceous crops in North America, the bertha armyworm is indigenous and not an invasive alien species. It has a western distribution on the continent with a geographical range extending south to Mexico City, Mexico and north to Keg River, Canada (King 1928), but it causes crop damage only in western Canada (Mason et al. 1998a). It is polyphagous and recorded from some 40 species of dicotyledonous plants; its original hosts were possibly native species of Chenopodiaceae because it showed feeding and reproductive preferences for *Chenopodium album* L. in laboratory tests (King 1928, Bailey 1976, Turnock 1985, Dosdall and Ulmer 2004). The first outbreaks of bertha armyworm occurred in crops of flax (*Linum usitatissimum* L.), sweet clover (*Melilotus* spp.), and alfalfa (*Medicago sativa* L.) in western Canada in the late 1920s (Mason et al. 1998a). Outbreaks were rare and usually restricted to flax until 1944 when the first bertha armyworm outbreak was recorded in Saskatchewan on rapeseed. At this time, the crop was grown for industrial uses (Canola Council of Canada 2007), and the area devoted to its production was relatively small, but nevertheless a subsequent outbreak of 3-year duration occurred from 1947 to 1949. The development of new 'double low' oilseed varieties in the early 1970s and subsequent increases in crop production coincided with increasing regularity of bertha armyworm outbreaks (Mason et al. 1998a). Outbreaks usually last 3 years, and the economic costs can be considerable. The first year of the 1971–1972 outbreak resulted in economic losses of approximately $17 million (CAD) to producers (Reigert 1984), and the outbreak of 1995 caused losses of $30–$40 million with some 650,000 ha sprayed with insecticide for control of bertha armyworm (WCCP 1995, Mason et al. 1998a).

The bertha armyworm is univoltine and crop damage results from larval feeding. Overwintering occurs as pupae in the soil and adults emerge from mid June to early August (Mason et al. 1998a). Females are attracted to flowering canola for oviposition, and prefer to oviposit on *S. alba* relative to other commonly grown canola and mustard species (Turnock 1984, Ulmer et al. 2002). Eggs are laid in masses of 20–200 eggs on the undersides of host plant leaves, and larvae feed on foliage during their first four instars, usually causing only minor damage. However, by the time that larvae reach their fifth and sixth instars, leaves have usually senesced, and although larvae prefer to feed on leaf material (Bracken 1984), extensive crop damage can occur when these late instars feed on maturing pods (Mason et al. 1998a).

Few cultural options are currently available to help producers minimize the impact of bertha armyworm infestations. Adults are strong fliers so crop rotation can only be effective if practiced over a very broad geographical area (Mason et al. 1998a). Fall tillage can mechanically damage pupae or expose them to severe winter temperatures (Turnock and Bilodeau 1984), but this is incompatible with the

Fig. 6.7 Larva of *Mamestra configurata* (Photo: Lloyd Dosdall)

current movement toward zero or minimum tillage practices. Consequently, chemical insecticide applications have been the primary means of population control. The relationship between larval density and yield loss developed by Bracken and Bucher (1977) and Bracken (1987) is used to assess whether economic threshold levels have been reached, and if so, control can be implemented by application of synthetic pyrethroid or organophosphate insecticide (Wise and Dosdall 2002).

Parasitoids have been found associated with all pre-imaginal stages of the bertha armyworm. *Trichogramma inyoense* Pinto and Oatman (Hymenoptera: Trichogrammatidae) has been found to attack bertha armyworm eggs in Saskatchewan (Mason et al. 1998b), but its importance is not well understood because this life stage has only rarely been examined for parasitism. Approximately 15 species of Hymenoptera and Diptera, principally from the families Ichneumonidae and Tachinidae, have been reared from larvae and pupae (Wylie and Bucher 1977, Wylie 1979), but pupal parasitoids appear to have negligible importance for regulating natural populations of the armyworm. Its principal parasitoids are the ichneumonid *Banchus flavescens* Cresson (Fig. 6.8) and the tachinids *Athrycia cinerea* (Coquillette) and *Panzeria ampelus* (Walker) (Mason et al. 1998a).

Banchus flavescens is usually the most abundant parasitoid of the bertha armyworm. In the outbreak of 1971–1973, it was found in up to 95% of larvae in field populations in Saskatchewan (Arthur and Ewen 1975), and in 94% of larvae in Manitoba in 1981 (Turnock and Bilodeau 1984), although such high parasitism levels occur only near the end of the outbreak cycle. The parasitoid is univoltine, and overwinters as a prepupa in a cocoon in the soil, in obligatory diapause (Wylie and

Fig. 6.8 *Banchus flavescens* parasitizing larva of *Mamestra configurata* (Photo: Agriculture and Agri-Food Canada)

Bucher 1977, Arthur and Mason 1985). Adults emerge from mid June to the end of July and parasitize first, second, and third larval instars of the bertha armyworm. *Banchus flavescens* completes four larval instars in its host in approximately 22 days, and the fifth instar kills and egresses from its host through the fifth or sixth abdominal segment (Arthur and Mason 1985). Parasitism results in significantly decreased food consumption and lower biomass production of the host, although it does not reduce the time that the armyworm larvae occur in the crop (Mason et al. 2001).

Athrycia cinerea is univoltine and overwinters in the soil as a pupa in facultative diapause (Wylie and Bucher 1977, Wylie 1977c). Adults emerge in June and July, and females oviposit on third- to sixth-instar larvae of the bertha armyworm (Wylie 1977a). Parasitoid larvae hatch a few minutes later, burrow into their hosts and develop in the hemocoel within a respiratory funnel, either singly or gregariously (Wylie and Bucher 1977, O'Hara 1999). They kill their host larvae when hosts are in the fifth or sixth instar, and pupariate within the host remains or in soil (Wylie and Bucher 1977).

Panzeria ampelus is of minor importance as a parasitoid of the bertha armyworm, although it is widely distributed (Wylie and Bucher 1977, Arthur and Powell 1990). It is bivoltine in western Canada, and the second generation can parasitize larvae of *M. configurata* (Wylie 1977b). Females of *P. ampelus* attach their eggs to foliage in the vicinity of hosts, and eggs hatch in a few minutes (Tothill 1922, O'Hara 1999). Larvae are sensitive to vibrations caused by host larvae, and wave their bodies vigorously as hosts crawl nearby (Wylie 1977b). If contacted by a host larva, the parasitoid attaches to the host integument, pierces the exoskeleton, and forms a respiratory funnel at its point of entry. Generally only a single parasitoid develops in each host. Larval parasitoid development requires 2–3 weeks; final instars form puparia and overwintering occurs as pupae in the soil (Wylie 1977b).

Biological control attempts for the bertha armyworm in North America were initiated by studies of the parasitoid fauna associated with its Eurasian congener, *Mamestra brassicae* L., to determine whether niche gaps in the existing parasitoid guild could be filled by introductions from the Palaearctic region (Mason et al. 2002a). Eggs of *M. brassicae* in Europe are attacked by at least three parasitoid species, but all appeared inappropriate for release in North America because they were opportunists or had low affinity for *M. brassicae* eggs (Mason et al. 2002a). In European populations of *M. brassicae*, the ichneumonid, *Exetastes atrator* (Förster), appears to fill a niche similar to that of *B. flavescens*, so was not considered for introduction to North America (Turnock 1984). However, the tachinid, *Ernestia consobrina* (Meigen), occupies a niche similar to that of *P. ampelus*, and was considered appropriate for release because *P. ampelus* is not common in North American populations of bertha armyworm, and *E. consobrina* occurs in cooler parts of the European range of *M. brassicae* (Turnock 1984). Several thousand specimens of *E. consobrina* were released in southern Manitoba in 1986 and 1987 (Turnock and Carl 1995), but to date no evidence has been found of establishment, possibly because low host populations in the area have not permitted extensive evaluations of parasitism in the bertha armyworm (Mason et al. 2002a). The braconid, *Microplitis mediator* (Haliday), is a larval endoparasitoid of *M. brassicae* that can occur quite abundantly in Europe (Turnock 1984), and was considered for biological control in North America because of its potential to fill a vacancy in the parasitoid guild of the bertha armyworm (Mason et al. 2002a). Females attack the first three larval instars of *M. brassicae*, although parasitization success is considerably greater in the first two larval instars (Lauro et al. 2005). *Microplitis mediator* females also attack early instars of the bertha armyworm (Arthur and Mason 1986), and the species appears capable of surviving overwintering conditions in western Canada (Pivnick 1993). Several releases in Alberta and Saskatchewan of various life stages of *M. mediator* were made in the 1990s, but there is no evidence of successful establishment (Mason and Youngs 1994, Mason et al. 2002a).

6.2.5 Lygus Bugs, Lygus *spp. (Hemiptera: Miridae)*

In North American brassicaceous oilseed crops, species of *Lygus* (Hemiptera: Miridae) are chronic pests that can occasionally cause massive crop damage when their populations reach outbreak densities (Fig. 6.9). *Lygus* comprises 31 native species (Schwartz and Foottit 1998) of which 14 are recorded as crop pests (Maw et al. 2000). The most significant pests of canola include the tarnished plant bug, *Lygus lineolaris* (Palisot), the lucerne (pale legume) bug, *Lygus elisus* van Duzee, and *Lygus borealis* (Kelton) (Butts and Lamb 1990a, 1990b, Timlick et al. 1993). Occasionally populations of the western tarnished plant bug, *Lygus hesperus*

Fig. 6.9 Adult of *Lygus* sp. (Photo: Lloyd Dosdall)

Knight, and *Lygus keltoni* Schwartz also damage canola (Schwartz and Foottit 1992). Species assemblages in canola vary with location and year of collection (Leferink and Gerber 1997).

Adult lygus bugs overwinter beneath plant litter near the soil surface in field margins or shelterbelts (Craig and Loan 1987, Cermak and Walker 1992). In spring, adults disperse from overwintering sites to feed on host plants, often brassicaceous weeds in the late rosette to early flowering stages. In northern regions of canola production, the crop is colonized by overwintered adults, but in more southerly regions canola is colonized by adults of the first generation. In both situations, however, lygus adult population maxima occur when crops are in flower (Butts and Lamb 1991, Leferink and Gerber 1997). Females oviposit into plant stems (Broadbent et al. 2002), and the first nymphs appear from the end of flowering to the beginning of pod development, reaching maxima during mid-pod developmental stages (Leferink and Gerber 1997). There are five nymphal instars, and their development can occur rapidly – the first adults of the new generation produced in canola can appear during pod development. The number of generations of lygus bugs per year is affected by environmental conditions and host plant species, and varies between one and five (Schwartz and Foottit 1998). However, in canola only a single generation can complete development in the crop regardless of whether it is grown in northern Alberta, Canada (Butts and Lamb 1991) or in Georgia, USA (Buntin et al. 2007).

Damage by lygus bugs can occur to stems, buds, and pods. Adult feeding early in the season before flowering can cause lesions on stems, and when inflorescences develop, feeding can cause buds and flowers to abscise (Butts and Lamb 1990a). Feeding by final-instar nymphs and adults on seeds through the pod wall can cause seeds to collapse (Butts and Lamb 1990a, Turnock et al. 1995b). Plants can compensate for some damage by lygus bugs, especially damage to buds and flowers, but plants cannot compensate for collapsed seeds (Butts and Lamb 1990a). During the most severe outbreaks in western Canada, approximately 200,000 and 400,000 ha of canola cropland were treated with insecticide in 1997 and 1998, respectively, to control lygus bugs in Alberta and Saskatchewan (Braun et al. 2001).

Control of plant bugs in canola is currently dependent upon insecticidal applications. Cultural strategies are impractical given the migratory behavior and multivoltine life history of lygus bugs. Biological control therefore provides a potential alternative management approach to insecticide use.

Several native parasitoids attack the eggs, nymphs, and adults of lygus bugs in North America. Broadbent et al. (2002) listed 13 species reared from this pest complex in North America representing four families of Hymenoptera and comprising four egg parasitoids, five nymphal parasitoids, and four adult parasitoids in four families of Hymenoptera. However, the extant Nearctic parasitoid fauna was not considered to provide adequate biological control of lygus bugs in agroecosystems, and consequently attention turned to introductions of European parasitoid species to augment control by the native fauna. In Europe, nymphal parasitoids appear to have the greatest importance for providing biological control. The parasitoid, *Peristenus digoneutis* Loan (Hymenoptera: Braconidae) (Fig. 6.10), attacks primarily *Lygus*

Fig. 6.10 *Peristenus digoneutis* Loan attacking a nymph of *Lygus* sp. (Photo T. Haye)

rugulipennis Poppius and *Lygus pratensis* (L.) in Europe, but was found to successfully parasitize *L. lineolaris* (Lachance et al. 2001) The parasitoid was released in New Jersey in the early 1980s for biological control of *L. lineolaris* (Broadbent et al. 2002). It has established successfully in this species, and has dispersed northward so its range now encompasses northeastern USA and southern Québec, Canada (Day et al. 1990, Broadbent et al. 2002).

Peristenus digoneutis is most active in oviposition during the first 14 d of its adult life span, and although it can parasitize all nymphal instars of lygus bugs, maximum parasitization occurs in third and fourth instars (Haye 2004, Day 2007). Parasitoid development requires at least 14 days (Day 2007), and individuals overwinter as adults within cocoons at a soil depth of 1–5 cm (Haye 2004). *Peristenus rubricollis* (Thomson), a univoltine species introduced to eastern North America, is associated with *L. lineolaris* and *Adelphocoris lineolatus* (Goeze) (Hemiptera: Miridae) (Goulet and Mason 2006). A third species, *Peristenus relictus* (Ruthe) has established in California and is associated with *Lygus shulli* Knight (Pickett et al. 2007). *Peristenus relictus* has several desirable characteristics as a biological control agent, including facultative diapause, rapid developmental time, high parasitism levels, and ease of mass rearing (Broadbent et al. 2002). In Europe, *P. relictus* has been reared from a wide variety of Miridae (Haye et al. 2006, Goulet and Mason 2006), so it may not be a suitable biological control agent.

Releases of *P. digoneutis* and *P. relictus* were made in canola growing areas of Saskatchewan and Alberta in 1978 and 1981 (Craig and Loan 1984), and in 1985 and 1986 adults of *P. digoneutis* were released in central Saskatchewan (Soroka and Carl 2002). Neither species was recovered in subsequent collections, so additional releases were made from 1981 to 2000 (Broadbent et al. 2002). In eastern North America, *P. digoneutis* has extended its range from northeastern USA into canola growing areas in Ontario and Québec (Goulet and Mason 2006).

6.2.6 Diamondback Moth, **Plutella xylostella** *(L.) (Lepidoptera: Plutellidae) (Fig. 6.11)*

The diamondback moth is a serious pest of brassicaceous crops worldwide, and oilseed crops in North America have often been infested with damaging populations of this insect (Philip and Mengersen 1989, Buntin 1990, Brown et al. 1999). In addition to canola, diamondback moth larvae also feed on plants of white mustard, *S. alba* (Sarfraz et al. 2006), an oilseed crop grown in southern Alberta and Saskatchewan that is naturally resistant to several other crucifer specialists including *D. radicum* (Dosdall et al. 1994), *P. cruciferae* (Bodnaryk and Lamb 1991), and *C. obstrictus* (Kalischuk and Dosdall 2004). Diamondback moth adults prefer to oviposit on *S. alba* rather than on canola plants, and several fitness parameters like developmental times and pupal weights are higher on *S. alba* (Sarfraz et al. 2007). Crop damage from the moth occurs by larval feeding. Pale yellow eggs are laid singly or in small clusters, usually on the upper leaf surfaces (Justus et al. 2000), and hatch to first-instar larvae that mine spongy mesophyll of leaves (Harcourt 1957). Second-, third-, and fourth-instar larvae are surface feeders and consume leaf tissue

Fig. 6.11 Adult of *Plutella xylostella* (Photo: Lloyd Dosdall)

early in the season, usually causing only minor damage (Philip and Mengersen 1989). However, there can be three to five generations per year in North America (Harcourt 1957), and later generations can consume leaves, buds, flowers, pod walls, and green stem tissue (Anonymous 2007b).

Although some controversy exists regarding whether diamondback moth originated in Europe (Hardy 1938), Africa (Kfir 1998), or Asia (Liu et al. 2000), it has remarkable dispersal capabilities, and is now considered the most widely distributed of all Lepidoptera (Shelton 2004). The moth appears capable of overwintering in western Canada under favorable conditions (Dosdall 1994), but economically damaging populations in brassicaceous oilseed crops arise primarily via immigration from the south (Burgess et al. 1977, Philip and Mengersen 1989). For example, evidence was found linking a severe outbreak of the moth in oilseed crops in western Canada in 2001 to strong northerly airflow in early spring from southern USA and northern Mexico (Dosdall et al. 2004). The association between diamondback moth infestations in brassicaceous oilseed crops in temperate North America and air flow from southern regions of the continent is responsible for considerable variability in the magnitude of its infestations and subsequent economic losses. It usually causes minor economic damage each year, but in some years populations reach outbreak densities and substantial crop losses occur. In 1995 more than 1.25 million ha were sprayed with insecticide to control diamondback moth populations at an estimated cost to producers of $45–$52 million (Can.) (WCCP 1995). An outbreak on an even greater geographic scale occurred in 2001, with approximately 1.8 million ha treated with insecticide in western Canada (WCCP 2001). In southern and western USA, it is a principal cause of economic damage to the crop (Buntin 1990, Brown et al. 1999).

Current control strategies for the diamondback moth in North American oilseed crops involve applications of chemical insecticides, and both organophosphate and synthetic pyrethroid compounds are registered in Canada and the USA for control of the moth (Wise and Dosdall 2002, Kegley et al. 2007). Canola transgenic for expression of the *Bacillus thuringiensis* (*Bt*) cry1Ac gene was protected from diamondback moth attack (Ramachandran et al. 2000), but no *Bt*-canola crops are currently registered in North America. Consequently, biological control of the moth offers important opportunities for enhancing the sustainability of oilseed crop production.

Approximately 135 parasitoid species have been recorded from the diamondback moth worldwide (Delvare 2004), but the parasitoid complex in North America is dominated by three species (Braun et al. 2004). *Diadegma insulare* (Cresson) (Hymenoptera: Ichneumonidae) (Fig. 6.12) and *Microplitis plutellae* (Muesbeck) (Hymenoptera: Braconidae) attack the larvae, and *Diadromous subtilicornis* (Gravenhorst) (Hymenoptera: Ichneumonidae) attacks the prepupal and pupal stages (Harcourt 1986, Sarfraz et al. 2005). In western Canada, *D. insulare* is the principal parasitoid of the moth in canola, accounting for 30–45% of total parasitism; *M. plutellae* and *D. subtilicornis* are less common parasitoids, and were each found responsible for approximately 8–14% of parasitism (Braun et al. 2004, Dosdall et al. 2004).

Fig. 6.12 Adult of *Didegma insulare* (Photo: Lloyd Dosdall)

Diadegma insulare is a solitary larval endoparasitoid of the diamondback moth (Sarfraz et al. 2005). It can parasitize all four larval instars, but it kills and emerges from the pre-pupal stage of its host (Harcourt 1960). One host larva supports only one parasitoid larva (Harcourt 1960), and parasitized larvae consume less food than non-parasitized larvae (Okine et al. 1996). Field populations of *D. insulare* can be aggregated, in patterns correlated with distributions of hosts (Ulmer et al. 2005). The parasitoid spends more time in habitats with abundant food sources (Idris and Grafius 2001), and some flowering plants like alyssum, *Lobularia maritime* (L.) Desv. (Brassicaceae), provide good food sources for adult wasps (Johanowicz and Mitchell 2000). The parasitoid is attracted to brassicaceous crops damaged by feeding of diamondback moth larvae (Mitchell et al. 1999). *Diadegma insulare* cannot tolerate cold environmental conditions, and at low temperatures (4°C) cocoons soon lose viability (Okine et al. 1996), a factor that prevents overwintering in most North American regions of canola production.

Microplitis plutellae occurs transcontinentally in North America, and is also found in south-east Asia (Kirk et al. 2004, Braun et al. 2004, Sarfraz et al. 2005). Females can parasitize all four larval instars of the diamondback moth, but they kill and emerge from fourth instars (Sarfraz et al. 2005). *Microplitis plutellae* can be multivoltine, depending on the number of generations of its host. Unlike *D. insulare*, *M. plutellae* can enter pupal diapause (Putnam 1978). Its ability to diapause and overwinter in western Canada enables it to attack its hosts early in the season (Putnam 1978).

Diadromus subtilicornis occurs in North America and Asia, and is a solitary pupal endoparasitoid of the diamondback moth. Females oviposit in prepupae or in pupae, but the frequency of oviposition declines dramatically from 2 to 3 and 4 day old pupae (Tran and Takasu 2000a). Successful production of new generation parasitioids also decreases with increasing age of pupae at the time of oviposition. Females of *D. subtilicornis* engage in host-feeding: females can use their mouthparts to enlarge wounds made by the ovipositor during egg-laying, and may then feed on fluid oozing from the wound (Tran and Takasu 2000a). Developmental time from egg to adult emergence and longevity of adults is dependent on temperature, with slower development and shorter life spans occurring at higher (30°C) than moderate (25°C) temperatures (Tran and Takasu 2000b).

No introductions have been made of foreign parasitoids in attempts to improve biological control of the diamondback moth in oilseed brassicaceous crops. There are two reasons for this. First, members of the genus *Diadegma* are the most successful biological control agents of the moth worldwide (Sarfraz et al. 2005), and *D. insulare* already occurs widely in North America. Secondly, overwintering of the moth is not a common phenomenon in western Canada where most canola production occurs in North America, so releases would need to be made of a very cold hardy parasitoid. Most diamondback moth parasitoids evolved in tropical or subtropical habitats (Kfir 1998) and therefore lack the cold hardiness necessary for surviving the severe winters of the Northern Great Plains of North America.

6.3 The Cabbage Seedpod Weevil Invasion of North America: A Case Study of Invasion and Biological Control with Native and Introduced Parasitoids

Brassicaceous oilseed crops in North America harbor populations of several invasive insect species; in fact, of the dominant pest species described in the preceding section, only *M. configurata*, *D. floralis*, *D. platura*, and the *Lygus* species are indigenous. Prior to introduction of the cabbage seedpod weevil, the invasion dynamics of alien species was poorly known because invasions occurred early in the agricultural history of North America, and resources did not exist to study such events. The recent invasion by the cabbage seedpod weevil, however, offered an unprecedented opportunity to investigate temporal and geographical changes in distribution patterns, the ability of an invasive herbivore to complete development in a crop with a phenology different from its usual host in Europe, and a fauna of opportunistic parasitoids with the potential to shift from their native hosts to exploit a new and abundant resource of weevil larvae in canola pods.

The cabbage seedpod weevil was first discovered in North America in the southwestern coastal region of British Columbia in 1931 (McLeod 1962). During the next 25 years it was recorded from several additional localities in western North America including the states of Washington (Baker 1936), California (Hagen 1946),

Idaho (Walz 1957) and Oregon (Crowell 1952). Nearly 50 years after its discovery in British Columbia, it was found on the Canadian prairies and in eastern North America, in Maryland (Anonymous 1977), Tennessee (Boyd and Lentz 1994), Georgia (Buntin et al. 1995), and most recently in Québec (Brodeur et al. 2001) and Ontario (Mason et al. 2004). Based on analysis of variations in a 475-base pair fragment of the mitochondrial DNA *COI* gene among geographically distinct populations, Laffin et al. (2005) concluded that North American populations of the weevil likely resulted from at least two introductions. An introduction to western North America appears to have occurred from source populations in western or northern Europe, and these populations spread throughout most of North America. A second population established in Québec, apparently originating from source populations in northern Europe.

The cabbage seedpod weevil was discovered in Alberta in 1995 (Dosdall et al. 2001). When the first survey was conducted in 1997, the species' range already encompassed several hundred square kilometers (Dosdall et al. 2002), and no vegetational or geographical barriers existed to prevent it from expanding throughout the entire region of canola production of western Canada. A model incorporating ecological requirements of the species and meteorological data throughout western Canada indicated that the species would likely establish throughout the entire region (Dosdall et al. 2002). Rapid dispersal of the weevil was observed from 1997 to 2000, to the north and east from the region of southern Alberta where it was initially found, and in 2000 it was found in Saskatchewan for the first time.

The first major outbreak of cabbage seedpod weevil in North America occurred in 1999 over a vast region of southern Alberta encompassing approximately 100,000 ha (Dosdall et al. 2002). Weevil densities of 10–15 adults per 180° sweep net sample were common, an infestation level clearly responsible for major crop losses. In 2000, another outbreak occurred on an even greater geographical scale. Compared with the preceding year it encompassed at least a three-fold increase in canola cropland under severe attack, and was commonly associated with densities of 50–80 adults per sweep sample (Dosdall et al. 2002). The population expansion of the weevil occurred so quickly that no insecticides were yet registered in Canada for its control. Emergency registrations were therefore obtained from the Pest Management Regulatory Agency of Health Canada to enable producers to protect their crops from unacceptable yield losses. In the USA, by contrast, some organophosphate insecticides were registered and available for use against the cabbage seedpod weevil (McCaffrey 1992), but in Canada these products were considered to pose unacceptable environmental risks. Field studies subsequently undertaken to compare several potential insecticidal agents for efficacy against the weevil determined that synthetic pyrethroids performed better than a number of organophosphate agents tested (Cárcamo et al. 2005). This research led to registration of two synthetic pyrethroids for weevil control, deltamethrin and cyhalothrin-lambda (Saskatchewan Agriculture and Food 2007). A nominal economic threshold was recommended, using a mean of three to four weevil adults per 180° sweep net sample when crops were in early flower (Dosdall et al. 2001).

The economic costs of the invasion of western Canada by the cabbage seedpod weevil have been substantial. A central problem in deriving a comprehensive assessment of damage caused by an invasive species is the scarcity of available data to systematically track direct and indirect, market and non-market costs for pest species (Colautti et al. 2006). However, a conservative estimate based on costs incurred to apply insecticide on cropland indicates that the weevil caused estimated crop losses of at least $5 million in the first 10-year period following its discovery in Alberta and Saskatchewan (Colautti et al. 2006). To date, the weevil has not yet extended its range to encompass regions of central Alberta and Saskatchewan and southern Manitoba where most canola is produced. Crop losses from cabbage seedpod weevil will surely escalate in future years as its range expands to include these areas.

The cabbage seedpod weevil has continued to disperse and expand its range, but at a much slower pace than observed early in its invasion of southern Alberta and Saskatchewan (Dosdall et al. 2009). Drought throughout much of western Canada during 2001–2003 appears to have caused dramatic reductions in both population densities and the rate of weevil expansion, so that the magnitude of crop loss and quantities of insecticide applied to control weevil infestations never subsequently equaled levels observed during the outbreaks in 1999 and 2000. Nevertheless, research was continued in an effort to develop sustainable management strategies.

Cultural control tactics directed against the cabbage seedpod weevil proceeded on two fronts: research to use trap crops to reduce insecticide use, and studies to vary agronomic practices to favor development of the crop rather than enhance populations of the pest. Perimeter trap crops, seeded to flower before a main crop within, and sprayed with insecticide when weevils invaded the perimeter, could successfully maintain weevil populations below threshold levels in main crops in large, square fields, but were less effective in smaller or narrow fields (Cárcamo et al. 2007). Seeding in early May, rather than in mid to late April, and at recommended seeding rates, enabled development of plant stands best able to withstand crop damage from weevil attack (Dosdall et al. 2006a).

Host plant resistance research has been directed toward developing canola varieties resistant to infestation by the cabbage seedpod weevil. *Sinapis alba* was shown to be resistant (Doucette 1947, Kalischuk and Dosdall 2004), and served as the resistant parent in crosses with *B. napus* to develop intergeneric hybrid genotypes. Difficulties in crossing the two plant genera were overcome using 'embryo rescue' (Dosdall and Kott 2006), and resulted in production of a number of genotypes that appear to contain one or more genes for resistance from the *S. alba* parent. The resistance has been stable over a number of plant generations, and the most promising genotypes are being advanced in this breeding program to combine weevil resistance with agronomic quality (Dosdall and Kott 2006).

The discovery of the parasitoid, *M. melanopus*, in some Alberta populations of the cabbage seedpod weevil initially raised hope that population control could be achieved with this agent because *M. melanopus* attains relatively high levels of parasitism in some weevil populations in Idaho (Harmon and McCaffrey 1997), and the parasitoid is bivoltine in a univoltine host. In 2001, *M. melanopus* was found at 15 of 25 sites surveyed in southern Alberta (Fox et al. 2004), but since then extensive surveys have recovered the species only rarely (Dosdall unpublished data).

From 2001 to 2006 a total of 14 parasitoid species representing four families of Chalcidoidea were found associated with cabbage seedpod weevil in southern Alberta and Saskatchewan, including *N. tidius* and *Euderus albitarsus* (Zetterstedt) (Eulophidae), *T. lucidus*, *Chlorocytus* sp., *Pteromalus* sp., *M. moryoides*, *Mesopolobus bruchophagi* Gahan, *L. maculatus*, *Lyrcus incertus* (Ashmead), *Lyrcus perdubius* (Girault) and *Catolaccus aeneoviridis* (Girault) (Pteromalidae), *Conura torvina* (Cresson) and *Conura albifrons* (Walsh) (Chalcididae), and *Eurytoma tylodermatis* Ashmead (Eurytomidae) (Gibson et al. 2005, Dosdall et al. 2006b; Dosdall et al. 2009). Mass rearings of canola pods from Georgia, USA also recovered a diverse parasitoid fauna comprising 13 Chalcidoidea species from five families, including *C. torvina* (Chalcididae), *Euderus glaucus* Yoshimoto and *N. tidius* (Eulophidae), *Brasema allynii* (French) and *Eupelmus cyaniceps* Ashmead (Eupelmidae), *E. tylodermatis* (Eurytomidae), and *L. incertus*, *L. maculatus*, *L. perdubius*, *M. moryoides*, *Neocatolaccus tylodermae* (Ashmead), *Pteromalus cerealellae* (Ashmead) and *Pteromalus* sp. (Pteromalidae) (Gibson et al. 2006a).

The Chalcidoidea fauna appears to have distinct components in the western and eastern portions of the continent. In western North America, the dominant cabbage seedpod weevil parasitoids include *N. tidius*, *T. lucidus*, *Chlorocytus* sp., and *Pteromalus* sp. (Dosdall et al. 2006b, Dosdall et al. 2009), but *L. maculatus* is dominant in eastern North America (Gibson et al. 2006a).

In western Canada, parasitism of cabbage seedpod weevil larvae by Chalcidoidea increased from only 0.1% in 2002 to 5.0% in 2004 (Dosdall et al. 2006b). Parasitism increased further from 2004 to 2005, but was usually less than 15% for all species combined (Dosdall et al. 2009). These parasitism levels are much lower than those common in Europe where parasitism rates typically reach about 50% (Alford et al. 1996, Ulber and Vidal 1998), but can reach 90% and can be high even when larval densities are low (Murchie and Williams 1998). The situation in western Canada appears to reflect parasitoid populations that are building over time, by a parasitoid assemblage comprised primarily of indigenous species that have expanded their host ranges to exploit an abundant resource of cabbage seedpod weevil larvae (Dosdall et al. 2006b). In the recently invaded regions of Ontario and Québec, parasitoid faunal associations with the weevil are somewhat unstable and changing (Mason unpublished data).

Implementing a classical biological control program for the cabbage seedpod weevil could enable partial reconstruction of the natural enemy complex of this insect, and so prevent or reduce its continued dispersal across Canada. *Mesopolobus morys* and *T. perfectus* are considered the most effective parasitoids for controlling the pest in Europe (Williams 2003), and therefore have the greatest potential for achieving biological control in North America. Field host range studies of the principal parasitoids of the weevil in their native range determined that *T. perfectus* was the most specific and therefore most appropriate candidate for consideration (Muller et al. 2005). *Mesopolobus morys* was relatively host-specific on the cabbage seedpod weevil, although it also parasitized *Ceutorhynchus turbatus* Schultze, a species not known from North America, but under consideration as a biological control agent for the introduced weed, *L. draba*. Compatibility among biological control agents released for weeds and insect pests is essential (Kuhlmann et al. 2006a), so release of *M. morys* may not be appropriate.

Releases of European parasitoids may not be warranted when several species of native natural enemies already attack the weevil. However, current levels of parasitism in western Canada are variable: in some sites and years parasitism levels as high as 45% have been recorded, but usually levels have been much lower (Dosdall et al. 2007, Dosdall et al. 2009). Substantial year-to-year variation in parasitism levels has been found (Dosdall et al. 2007), suggesting that the system is in flux, perhaps as competition occurs among parasitoid species. It is not yet clear whether parasitism levels are still likely to increase over time, but parasitism should continue to be monitored, along with the dispersal biology of the weevil host, to better assess whether a classical biological control program should be implemented. If so, it appears that *T. perfectus* could be considered for re-release in Canada.

Sakai et al. (2001) outlined generalized steps for invasive species where initial transport and establishment of the invader in a new area is followed by a period of population build-up before further dispersal to new habitats. This process applies well to the cabbage seedpod weevil, which possesses several traits characteristic of successful invaders like small body size, rapid developmental rate, and a high intrinsic growth rate (Crawley 1986, Lawton and Brown 1986, Dosdall and Moisey 2004). Invasion of the weevil was also facilitated by the presence of an unoccupied niche of canola pods not extensively exploited by other herbivores, and able to provide protection and food for weevil larvae. Now well established in North America, it is evident that the cabbage seedpod weevil is destined to remain an important component of the insect fauna of canola agroecosystems. Moreover, models of climate change scenarios indicate that its range will only expand further as temperature and moisture conditions are altered (Olfert and Weiss 2006).

6.4 Challenges to Biological Control with Parasitoids in North American Cropping Systems

The ability of natural enemies to persist in the environment, to reproduce, and to disperse to new localities are important advantages of biological control (Greathead 1986), and when the process is successful, dramatic improvements can be achieved in economic and environmental sustainability. However, using parasitoids for biological control of insect pests in North American canola crops faces important challenges, such as pesticide use, new invaders and non-resident species, habitat availability and community interactions, compatibility with contemporary agronomic practices, regulatory oversight, and taxonomy.

6.4.1 Pesticide Use

Pesticides remain the primary management tool to control pest insects that infest canola crops. For example, 16 insecticides are registered for use in this crop (Manitoba Agriculture 2007). Most are registered for more than one species and a few are registered for use against as many as eight pests. It has been well documented that pesticides negatively impact beneficial species including parasitoids (Van Driesche and Bellows 1996), having both acute and sublethal effects (Mason et al. 2002b). These impacts can result in rebound of the pest species and/or increases of secondary pests to damaging levels (Van Driesche and Bellows 1996). Pesticide application strategies that conserve parasitoids such as those developed for long-term management of the wheat midge, *Sitodiplosis mosellana* (Géhin) (Diptera: Cecidomyiidae), in wheat (Elliott 1988), and the spruce budworm, *Choristoneura fumiferana* (Clemens) (Lepidoptera: Tortricidae), in forest systems (Nealis et al. 1992) are possible, but to date no similar techniques have been developed for canola in North America. Although there is currently little pressure to reduce reliance on chemical pesticides in field crop situations, the de-registration of products considered harmful will likely increase demand for new biological control agents (Floate et al. 2002). Developing innovative and comprehensive biological control programs for the complexes of pest species inhabiting canola may appear daunting, but the greenhouse industry provides a good model of how biological control can be developed successfully for multiple pest species (van Lenteren 2007, Shipp et al. 2007).

6.4.2 New Invaders and Non-resident Species

Canola agroecosystems are continually under assault by species recently introduced to the continent, or by species capable of expanding their geographic ranges to encompass areas of cropland. Some insect pest species of particular concern as recent arrivals, potential invaders, or periodic invaders include cabbage seedpod weevil, blossom beetle, swede midge, and diamondback moth. Implementing biological control for these species poses unique challenges.

The cabbage seedpod weevil, described in detail in Sections 6.2.6 and 6.3, is one example of a relatively new arrival of considerable economic importance. Although biological control with parasitoids is in a dynamic phase in Alberta, Saskatchewan, Ontario, and Québec (see Section 6.3), the weevil has been established for many more years in southwestern British Columbia with more stabilized impacts from parasitoids, and here these effects are insufficient to cause substantial reductions of the pest (Gillespie et al. 2006). Initiating a new classical biological control program for this insect may therefore be appropriate.

Meligethes viridescens (Fabricius) (Coleoptera: Nitidulidae), the bronzed or rape blossom beetle, is native to the western Palaearctic subregion and has established in eastern Canada (Mason et al. 2003). It has the potential to spread throughout North

American canola growing ecozones (Mason et al. 2003), even as boundaries of these areas shift due to climate change (Olfert and Weiss 2006). That *M. viridescens* favors spring-seeded crops puts the North American canola industry at great risk because most canola is spring-seeded. Parasitoids associated with *M. viridescens* in North America are unknown. Although knowledge of the parasitoid fauna associated with species of *Meligethes* in Europe is based primarily on studies of *Meligethes aeneus* (Fabricius) (Nilsson 2003), these studies suggest that the ichneumonid *Phradis morionellus* (Holmgren) and the braconids *Brachyserphus parvulus* (Nees), *Blacus nigricornis* Haeselbarth, and *Diospilus capito* (Nees) may be suitable candidates for introduction in the event that *M. viridescens* spreads to western regions of canola production.

The swede midge, *Contarinia nasturtii* (Kieffer) (Diptera: Cecidomyiidae), is native to Europe and Asia and was first reported in eastern Canada in 2000 (Hallett and Heal 2001). It has the potential to establish in all canola-growing regions (Olfert et al. 2006) and was recently reported from Saskatchewan (Canadian Food Inspection Agency 2007). Because swede midge feeds on a wide range of Brassicaceae, including weed and cultivated species, the potential for significant impact in canola is high. Parasitoids are unknown in the areas of North America where *C. nasturtii* has established, and in its area of origin few parasitoids have been associated with this species (Herrara 2006). Although prospects for biological control of *C. nasturtii* with parasitoids are uncertain, the success of establishing parasitoids in western Canada for biological control of a related species, the wheat midge, *S. mosellana* (Olfert et al. 2003), suggests that there is some potential for similar results with swede midge.

The diamondback moth is non-resident in most areas of North American canola production, and annually disperses northward on low-level winds from southern United States and northern and central Mexico (Dosdall et al. 2004). Seasonal infestations are highly unpredictable and may occur early or late in any region. Population regulation by parasitoids is also therefore quite unpredictable because the dominant parasitoid, *D. insulare*, evidently does not overwinter (Okine et al. 1996), so it must disperse northward along with its hosts. Models of climate change scenarios suggest that the northern extent of continuous reproduction of diamondback moth will shift northward in future years, the species will complete more generations per year throughout its range, and overwintering capability of the moth should also shift northward (Dosdall et al. 2008). The effects of climate change on environmental factors like air flow and on natural enemies cannot yet be predicted, but these could affect the economic damage incurred by this pest.

6.4.3 Habitat Availability and Community Interactions

Production systems in North America tend to be large scale in western Canada and USA (e.g., 80–300 ha), but are usually smaller in the east (e.g., 50 ha). Manipulation of habitats adjacent to production areas has been shown to enhance the establishment and sustainability of parasitoid populations in agricultural systems (Landis et al. 2000, Gurr et al. 2004). Although there is some evidence that the characteristics of field margins can influence parasitism levels in the middle of oilseed rape crops (Nilsson 2003), the challenge will be to determine how habitat manipulation can be developed for large-scale canola operations and to demonstrate what strategies are best suited for smaller scale operations. Some progress has been made; for example, habitat management strategies have been identified for *D. radicum* (Riley et al. 2007).

Intra-guild interactions are considered to be important constraints in many natural enemy systems (Brodeur and Boivin 2006). Direct interactions are comparatively easy to evaluate but indirect interactions such as competition and displacement are more difficult to detect and predict (Messing et al. 2006). Adult parasitoids may also feed on their hosts resulting in an increased longevity and fecundity of the female parasitoid but at the cost of progeny with reduced fitness (Heimpel and Collier 1996, Ferracini et al. 2006). Recent advances in methodology for evaluation of community-level interactions at the third trophic level (e.g., Snyder et al. 2006) may permit detection of these interactions prior to introduction of natural enemies. These methods may allow evaluation of the value of introducing more than one natural enemy species, and could be used to detect apparent competition between target and non-target hosts mediated between shared natural enemies. These interactions are extremely complex and are affected by plant diversity and mutualisms, nutritional ecology, hyperparasitism, reservoirs and refuges (Polis and Winemiller 1996, Strong and Pemberton 2001, Tscharntke and Hawkins 2002) and a great number of positive and negative interactions can occur within a food web where a new natural enemy invades. Progress has been made to document some of the native fauna and potential trophic interactions for *Lygus* spp. (Braun et al. 2001, Broadbent et al. 2006), *D. radicum* (Hemachandra et al. 2007a, 2007b), and the cabbage seedpod weevil (Gillespie et al. 2006, Dosdall et al. 2007), but given the diverse fauna of canola herbivores, it is clear that much more research is required in this area.

6.4.4 Compatibility with Contemporary Agronomic Practices

To be accepted by canola producers, biological control with parasitoids must be compatible with modern crop production practices. Some current crop management strategies are of particular concern because they could hamper establishment and survival of natural enemies. Some important practices that can affect parasitoid populations include rotational schemes, tillage regimes, and weed management techniques.

It is currently recommended that producers rotate canola annually, so that the crop is grown on a given parcel of land only once every 3–4 years (Thomas 2002). The benefits of rotation are substantial and are associated with improved yields, reductions in disease and insect pest pressure, and improved fertility and water management (Thomas 2002). However, rotations may not be beneficial to parasitoids.

For example, if large populations of *B. flavescens* and *A. cinerea* parasitize larvae of *M. configurata* in a canola field, the parasitoids would overwinter as pupae/puparia in soil. Rotation to a non-brassicaceous crop in the following year could harm parasitoid survival, or would at least require dispersal of new generation adults to new sites where their hosts are located.

Tillage regime is known to affect infestation levels of several insect pest species of canola like *P. cruciferae* (Dosdall et al. 1999), *Delia* spp. (Dosdall et al. 1996a, 1998), and *M. configurata* (Turnock and Bilodeau 1984), and although data are lacking on parasitoid effects, tillage is almost certain to affect overwintering success of some species. Cultivation before seeding could mechanically damage overwintering parasitoids, or move them nearer the soil surface where they may be attacked by other species or exposed to unfavourable environmental conditions. In the past 10 years, there has been a dramatic shift in tillage practices, resulting in a four-fold increase in land under zero tillage (Gamache 2007). In addition to the benefits of this practice in terms of reducing energy costs, preserving soil moisture, and minimizing erosion, less soil disturbance should also enhance survival of overwintering parasitoids.

In recent years, weed control practices in North American canola crops have undergone major changes with the introduction of genetically modified herbicide-tolerant varieties. By 2005, herbicide-tolerant transgenic cultivars accounted for 95% of canola production (Canola Council of Canada 2005). Such rapid adoption of this technology is explained by the convenience of using a single post-emergence herbicide to control a broad spectrum of weed species, resulting in increased yields and economic returns (Harker et al. 2003). Weeds are major constraints on canola yields; however, weeds also enhance biodiversity within agricultural systems and provide more complex and fractured microhabitats that can be beneficial to parasitoids and other natural enemies (Altieri and Letourneau 1982, Marshall et al. 2003). Small weedy backgrounds can also influence pest infestations. For example, Dosdall et al. (2003) found that root maggot egg deposition and subsequent root damage were reduced when weeds were removed with herbicide later rather than earlier in crop development, presumably because weeds disrupted the behavioural sequence performed by *D. radicum* females before oviposition. Reductions in weed biodiversity within canola agroecosystems may therefore restrict the ability of parasitoids to establish or sustain themselves in brassicaceous oilseed crops, and could also increase infestations of some pests like *D. radicum*.

6.4.5 Regulatory Oversight

Risk of non-target impacts in biological control is of increasing concern (Follett and Duan 1999, Wajnberg et al. 2001, Bigler et al. 2006), resulting in more stringent review of release proposals (Mason et al. 2005, De Clerck-Floate et al. 2006). Negative non-target impacts are rare; about 1.4% of introductions in biological control programs against arthropod pests carry negative ecological side-effects (van Lenteren et al. 2006). Problems in biological control of arthropod pests have arisen from failures to understand host-specificity, poor understanding of intra-guild interactions and poor understanding of food web and multi-trophic multi-species interactions (Rosenheim et al. 1995, Babendrier et al. 2006, Messing et al. 2006). The North American Plant Protection Organization has, as a consequence, generated standards for evaluating potential risks posed by exotic natural enemies of arthropod pests that are candidates for introduction (Anonymous 2000, De Clerck-Floate et al. 2006). However, unlike those promulgated for biological control agents of weeds, these standards do not yet prescribe specific methods and approaches for evaluating risks. Bigler et al. (2006) proposed methods for assessing risks in a broad sense, but even for simple host-range assessments, the validity of the methods and approaches have yet to be critically evaluated.

Generalist natural enemies are broadly considered undesirable candidates for introduction as most negative non-target impacts are considered to arise from such species (Simberloff and Stiling 1996). However, defining and recognizing host range in candidate species is difficult for arthropod natural enemies of arthropod pests. Kuhlmann et al. (2005, 2006b) proposed an approach for evaluating host-range using a combined phylogenetic and ecological centrifugal approach. Host range in parasitoids is presumed to be under strong selection pressure, and females should normally choose the optimum host species for their offspring. To date, work on the cabbage seedpod weevil has been successful in identifying key host-range testing issues (Kuhlmann et al. 2006a) and research on *Lygus* spp. showed the importance of field studies to validate laboratory findings on host range (Haye et al. 2005, 2006). Such work is time-consuming and labour-intensive, but nonetheless, it is important that this research be expanded to encompass a wider range of insect pests of canola agroecosystems.

6.4.6 Taxonomy of Parasitoids

Sound taxonomy is fundamental to the success of biological control using parasitoids (Huber et al. 2002). Apart from nomenclatural issues associated with pest species (see the previous discussion of *C. obstrictus* versus *C. assimilis*), identification of parasitoids associated with pests of canola has been a key challenge. Poor taxonomy not only leads to missed opportunities for biological control, but it also undermines its credibility.

One illustration of the importance of taxonomy in biological control programs for North American canola pests relates to *D. radicum*. The staphylinid predator-parasitoid *A. bipustulata* was supposedly introduced to western Canada from Europe for biological control in the 1950s (Andreassen et al. 2007), but no voucher specimens were held of release material. Prior to 1986, it was not possible to distinguish between *A. bipustulata* and several other closely related species, and without voucher specimens the species released remains unknown. Although some specimens of *Aleochara* held in collections from North American localities had been identified as *A. bipustulata*, it was found that they were actually *A. verna*

(Hemachandra et al. 2005). Consequently there is no evidence that *A. bipustulata* is present in North America, and it remains a candidate for introduction if assessments determine that it is an effective and safe biological control agent.

In another situation, examination of parasitoid voucher material deposited during releases for biological control of cabbage seedpod weevil revealed that specimens reported as *T. perfectus* were in fact different species, many of which are native to North America (Gibson et al. 2005, 2006a, 2006b). Further investigation determined that, although *T. perfectus* and another key parasitoid, *M. morys*, were released, they failed to establish (Gillespie et al. 2006). These examples point not only to the importance of retaining voucher material from every biological control project but also the need for routine post-release monitoring to ascertain whether or not a biological control agent establishes and has an impact on the target pest population. Of critical importance is the integral role of taxonomic expertise to validate species determinations made in association with biological control programs.

6.5 Conclusion

In North America, brassicaceous oilseed crops are associated with several herbivorous insect pests that can be responsible for enormous economic damage. Of the six most significant pest species or pest species complexes, most are oligophagous on Brassicaceae (e.g., *P. cruciferae*, *P. striolata*, *D. radicum*, *D. floralis*, *C. obstrictus*, *P. xylostella*), but a few species are polyphagous (e.g., *D. platura*, *M. configurata*, *Lygus* spp.). Most are alien, and were introduced from Europe or Asia, and relatively few of the major insect pests are indigenous to North America. In most cases, the parasitoid faunas of the pests are known, but much greater understanding is needed of parasitoid life histories, their host preferences, and overwintering strategies. Tritrophic interactions among the parasitoids, their herbivore hosts, and canola host plants are even more poorly understood. Several attempts have been made to introduce parasitoids to North America for classical biological control of canola pest insects, but to date none of these human interventions can be considered successful. Currently, minimal effort is being directed toward enhancing natural enemy populations in canola agroecosystems, in spite of the benefits that can be derived from such an approach. Serious challenges exist in using parasitoids for biological control of oilseed pests; however, new approaches and tools are being developed that will facilitate better understanding of host-parasitoid systems in North America. This progress is encouraging, because in spite of past failures and the daunting challenges that lie ahead, a primary goal is for biologically-based management strategies to ultimately become the preferred means for controlling pests of brassicaceous oilseed crops.

References

Alford DV (2003a) The oilseed rape crop. In: Alford DV (ed.) *Biocontrol of oilseed rape pests*. Blackwell, Oxford, UK.

Alford DV (ed) (2003b) *Biocontrol of oilseed rape pests*. Blackwell, Oxford, UK.

Alford DV (2006) *Curculio assimilis* Paykull, 1792 (currently *Ceutorhynchus assimilis*; Insecta, Coleoptera): Proposed conservation of the specific name. *B Zool Nomen* 63: 31–32.

Alford DV, Walters KFA, Williams IH, Murchie AK (1996) A commercially viable low-cost strategy for the management of seed weevil populations on winter oilseed rape in the UK. In: *Brighton Crop Protection Conf – Pests and Diseases, Vol. 2, 18–21 November 1996, Brighton, UK. British Crop Protection Council, Farnham*, pp 609–614.

Altieri MA, Letourneau DK (1982) Vegetation management and biological control in agroecosystems. *Crop Prot* 1: 405–430.

Andreassen L, Kuhlmann U, Mason PG, Holliday NJ (2007) Classical biological control of the cabbage maggot, *Delia radicum*. In: Canadian canola: An analysis of research needs. *CAB Reviews: Perspectives in Agriculture, Veterinary Science, Nutrition and Natural Resources* 2, no. 86, 13 pp.

Anonymous (1977) Cabbage seedpod weevil (*Ceutorhynchus assimilis*) – Maryland – new state record. *US Dep Agriculture, Cooperative Plant Pest Report* 2: 812.

Anonymous (2000) *Regional standards for phytosanitary measures (RSPM) RSPM # 12. Guidelines for petition for release of exotic entomophagous agents for the biological control of pests*. Secretariat of the North American Plant Protection Organization, Ottawa, Canada. http://www.nappo.org/Standards/OLDSTDS/RSPM12-e.pdf Cited 12 Nov 2007.

Anonymous (2007a) *Canola*. http://www.canola-council.org/PDF/canola/english/originhistory.pdf#zoom=100 Cited 23 Nov 2007.

Anonymous (2007b) *Diamondback moth*. http://www.agriculture.gov.sk.ca/Default.aspx?DN=688b2f99-ad99-423d-900c-c01a1c45d8a1. Cited 23 Nov 2007.

Arthur AP, Ewen AB (1975) Cuticular encystment: A unique and effective defense reaction by cabbage looper larvae against parasitism by *Banchus flavescens* (Hymenoptera: Ichneumonidae). *Ann Entomol Soc Am* 68: 1091–1094.

Arthur AP, Mason PG (1985) The life history and immature stages of *Banchus flavescens* (Hymenoptera: Ichneumonidae), a parasitoid of the bertha armyworm, *Mamestra configurata* (Lepidoptera: Noctuidae) in western Canada. *Can Entomol* 117: 1249–1255.

Arthur AP, Mason PG (1986) Life history and immature stages of the parasitoid *Microplitis mediator* (Hymenoptera: Braconidae), reared on the bertha armyworm *Mamestra configurata* (Lepidoptera: Noctuidae). *Can Entomol* 118: 487–491.

Arthur AP, Powell YM (1990) Description of the immature stages of *Panzeria ampelus* (Walker) (Diptera: Tachinidae), an occasional parasitoid of the bertha armyworm, *Mamestra configurata* (Walker), in western Canada. *Can Entomol* 122: 381–385.

Babendrier D, Bigler F, Kuhlmann U (2006) Current status and constraints in the assessment of non-target effects. In: Bigler F, Babendreier D, Kuhlmann U (eds.) *Environmental impact of invertebrates for biological control of arthropods: Methods and risk assessment*. CABI Publishing, Wallingford, UK.

Bailey CG (1976) A quantitative study of consumption and utilization of various diets in the bertha armyworm, *Mamestra configurata* (Lepidoptera: Noctuidae). *Can Entomol* 108: 1319–1326.

Bain A, LeSage L (1998) A late seventeenth century occurrence of *Phyllotreta striolata* (Coleoptera: Chrysomelidae) in North America. *Can Entomol* 130: 715–719.

Baker WW (1936) Notes on the European weevil, *Ceutorhynchus assimilis* Payk., recently found in the state of Washington. *Can Entomol* 8: 191–193.

Bigler R, Babendreier D, Kuhlmann U (eds.) (2006) *Environmental impact of invertebrates for biological control of arthropods: Methods and risk assessment*. CABI Publishing, Wallingford, UK.

Block W, Turnock WJ, Jones TH (1987) Cold resistance and overwintering survival of the cabbage root fly, *Delia radicum* (Anthomyiidae), and its parasitoid, *Trybliographa rapae* (Cynipidae), in England. *Oecologia* 71: 332–338.

Bodnaryk RP, Lamb RJ (1991) Mechanisms of resistance to the flea beetle, *Phyllotreta cruciferae* (Goeze), in mustard seedlings, *Sinapis alba* L. *Can J Plant Sci* 71: 13–20.

Bonnemaison L (1957) Le charançon des siliques (*Ceutorhynchus assimilis* Payk.) biologie et méthode de lutte. *Ann Épiphyties* 4: 387–543.

Boyd ML, Lentz GL (1994) Seasonal incidence of the cabbage seedpod weevil (Coleoptera: Curculionidae) on rapeseed in west Tennessee. *Environ Entomol* 23: 900–905.

Bracken GK (1984) Within plant preferences of larvae of *Mamestra configurata* (Lepidoptera: Noctuidae) feeding on oilseed rape. *Can Entomol* 116: 45–49.

Bracken GK (1987) Relation between pod damage caused by larvae of bertha armyworm, *Mamestra configurata* Walker (Lepidoptera: Noctuidae), and yield loss, shelling, and seed quality in canola. *Can Entomol* 119: 365–369.

Bracken GK, Bucher GE (1977) An estimate of the relation between density of bertha armyworm and yield loss on rapeseed, based on artificial infestations. *J Econ Entomol* 70: 701–705.

Braun L, Erlandson M, Baldwin D, Soroka J, Mason P, Foottit R, Hegedus D (2001) Seasonal occurrence, species composition, and parasitism of *Lygus* spp. in alfalfa, canola, and mustard. *Can Entomol* 133: 565–577.

Braun L, Olfert O, Soroka J, Mason P, Dosdall LM (2004) Diamondback moth biocontrol activities in Canada. In: Kirk AA, Bordat D (eds.) Improving biocontrol of *Plutella xylostella. Proc Int Symp, 21–24 October 2002, Montpellier, France*. pp 144–146.

Broadbent AB, Lachance S, Sears MK, Goulet H (2006) Native braconid parasitism of the tarnished plant bug (Hemiptera: Miridae) in southern Ontario. *Biocontrol Sci Technol* 16: 687–698.

Broadbent AB, Mason PG, Lachance S, Whistlecraft JW, Soroka JJ, Kuhlmann U (2002) *Lygus* spp., plant bugs (Hemiptera: Miridae). In: Mason PG, Huber JT (eds.) *Biological control programmes in Canada 1981–2000*. CABI Publishing, Wallingford, UK.

Broatch JS, Dosdall LM, Clayton GW, Harker KN, Yang RC (2006) Using degree-day and logistic models to predict emergence patterns and seasonal flights of the cabbage maggot and seed corn maggot (Diptera: Anthomyiidae) in canola. *Environ Entomol* 35: 1166–1177.

Brodeur J, Boivin G (eds.) (2006) *Trophic and guild interactions in biological control*. Springer Verlag, Amsterdam, The Netherlands.

Brodeur J, Leclerc LA, Fournier M, Roy M (2001) Cabbage seedpod weevil (Coleoptera: Curculionidae): A new pest of canola in northeastern North America. *Can Entomol* 133: 709–711.

Brown J, McCaffrey JP, Harmon BL, Davis JB, Brown AP, Erickson DA (1999) Effect of late season insect infestation on yield, yield components and oil quality of *Brassica napus*, *B. rapa*, *B. juncea* and *Sinapis alba* in the Pacific Northwest region of the United States. *J Agr Sci* 132: 281–288.

Buntin GD (1990) Canola insect issues in the southeast. *Proc Canola Crop Protection Symp, University of Tennessee, Memphis, TN*. pp 28–33.

Buntin D, Grey T, Harris GH, Phillips D, Prostko E, Raymer P, Smith N, Sumner P, Woodruff J (2007) *Canola production in Georgia*. http://pubs.caes.uga.edu/ caespubs/pubcd/B1331/B1331.htm. Cited 23 Nov 2007.

Buntin GD, McCaffrey JP, Raymer PL, Romero J (1995) Quality and germination of rapeseed and canola seed damaged by adult cabbage seedpod weevil, *Ceutorhynchus assimilis* (Paykull) [Coleoptera: Curculionidae]. *Can J Plant Sci* 75: 539–541.

Burgess L (1977) Flea beetles (Coleoptera: Chrysomelidae) attacking rape crops in the Canadian prairie provinces. *Can Entomol* 109: 21–32.

Burgess L (1980) Insect Pests. In: *Canola Canada's Rapeseed Crop*, Canola Council of Canada. Publication No. 56.

Burgess L (1981) Winter sampling to determine overwintering sites and estimate density of adult flea beetle pests of rape (Coleoptera: Chrysomelidae). *Can Entomol* 113: 441–447.

Burgess L, Dueck J, Petrie GA, Putnam LG (1977) Insect pests and diseases of rape and mustard. *Rapeseed Association Publ No. 48*, 32 pp.

Butts RA, Lamb RJ (1990a) Comparison of oilseed *Brassica* crops with high or low levels of glucosinolate and alfalfa as hosts for three species of *Lygus* (Hemiptera: Heteroptera: Miridae). *J Econ Entomol* 83: 2258–2262.

Butts RA, Lamb RJ (1990b) Injury to oilseed rape by mirid bugs (*Lygus*) (Heteroptera: Miridae) and its effect on seed production. *Ann Appl Biol* 117: 253–266.

Butts RA, Lamb RJ (1991) Pest status of *Lygus* bugs (Hemiptera: Miridae) in oilseed *Brassica* crops. *J Econ Entomol* 84: 1591–1596.

Canadian Food Inspection Agency (2007) *Swede Midge – Contarinia nasturtii.* http://www.inspection.gc.ca/english/plaveg/pestrava/connas/connase.shtml. Cited 26 Nov 2007.

Canola Council of Canada (2005) http://www.canola-council.org/ind_overview.html. Cited 21 Nov 2007.

Canola Council of Canada (2007) *Canola.* http://www.canola-council.org/pub_canola.html#eng. Cited 23 Nov 2007.

Cárcamo HA, Dosdall LM, Johnson D, Olfert O (2005) Evaluation of foliar and seed treatments for control of the cabbage seedpod weevil (Coleoptera: Curculionidae) in canola. *Can Entomol* 137: 476–487.

Cárcamo HA, Dunn R, Dosdall LM, Olfert O (2007) Managing cabbage seedpod weevil in canola using a trap crop – a commercial field-scale study in western Canada. *Crop Prot* 26: 1325–1334.

Carlson EC, Lange WH, Sciaroni RH (1951) Distribution and control of the cabbage seedpod weevil in California. *J Econ Entomol* 44: 958–966.

Cermak P, Walker GM (1992) *Tarnished plant bug: A major pest of strawberry. Fact Sheet No. 92–108*, Ontario Ministry of Agriculture, Food and Rural Affairs, Canada.

Colautti RI, Bailey SA, van Overdijk CDA, Amundsen K, MacIsaac HJ (2006) Characterised and projected costs of nonindigenous species in Canada. *Biol Invasions* 8: 45–59.

Colhoun EH (1953) Notes on the stages and the biology of *Baryodma ontarionsis* Casey (Coleoptera: Staphylinidae), a parasite of the cabbage maggot *Hylemya brassicae* Bouché. (Diptera: Anthomyiidae). *Can Entomol* 85: 1–8.

Colonnelli E (1993) The Ceutorhynchinae types of I.C. Fabricius and G. von Paykull (Coleoptera: Curculionidae). *Koleopterologische Rundschau* 63: 299–310.

Craig CH, Loan CC (1984) *Lygus* spp., plant bugs (Heteroptera: Miridae). In: Kelleher JS, Hulme MA (eds.) *Biological control programmes against insects and weeds in Canada 1969–1980.* Commonwealth Agricultural Bureaux, Farnham Royal, UK, pp 45–47.

Craig CH, Loan CC (1987) Biological control efforts on Miridae in Canada. In: Hedlund RC, Graham HM (eds.) *Economic importance and biological control of* Lygus *and* Adelphocoris *in North America.* ARS-64, United States Department of Agriculture, Agriculture Research Service, USA, pp 48–53.

Crawley MJ (1986) The population biology of invaders. *Philos T Roy Soc B* 314: 711–729.

Crowell HH (1952) Cabbage seedpod weevil control with parathion. *J Econ Entomol* 45: 545–546.

Day WH (2007) Effect of host instar on measuring parasitism of *Lygus* spp. (Hemiptera: Miridae) nymphs by *Peristenus* spp. (Hymenoptera: Braconidae). *Environ Entomol* 36: 1154–1158.

Day WH, Hedlund RC, Saunders LB, Coutinot D (1990) Establishment of *Peristenus digoneutis* (Hymenoptera: Braconidae), a parasite of the tarnished plant bug (Hemiptera: Miridae), in the United States. *Environ Entomol* 19: 1528–1533.

De Clerck-Floate RA, Mason PG, Parker DJ, Gillespie DR, Broadbent AB, Boivin G (2006) *Guide for the Importation and Release of Arthropod Biological Control Agents in Canada.* Minister of Supply and Services, A42–105/2006E.

Delvare G (2004) The taxonomic status and role of Hymenoptera in biological control of diamondback moth, *Plutella xylostella* (L.) (Lepidoptera: Plutellidae). In: Kirk AA, Bordat D (eds.) *Improving biocontrol of* Plutella xylostella. *Proc Int Symp, 21–24 October 2002, Montpellier, France.*

Dmoch J (1965) The dynamics of a population of the cabbage seedpod weevil (*Ceutorhynchus assimilis* Payk.) and the development of winter rape. Part I. *Ekol Pol–Pol J Ecol A* 13: 249–287.

Dosdall LM (1994) Evidence for successful overwintering of diamondback moth, *Plutella xylostella* (L.) (Lepidoptera: Plutellidae), in Alberta. *Can Entomol* 126: 183–185.

Dosdall LM, Clayton GW, Harker KN, O'Donovan JT, Stevenson FC (2003) Weed control and root maggots: Making canola pest management strategies compatible. *Weed Sci* 51: 576–585.

Dosdall LM, Dolinski MG, Cowle NT, Conway PM (1999) The effect of tillage regime, row spacing, and seeding rate on feeding damage by flea beetles, *Phyllotreta* spp. (Coleoptera: Chrysomelidae) in canola in central Alberta, Canada. *Crop Prot* 18: 217–224.

Dosdall LM, Florence LZ, Conway PM, Cowle NT (1998) Tillage regime, row spacing, and seeding rate influence infestations of root maggots (*Delia* spp.) (Diptera: Anthomyiidae) in canola. *Can J Plant Sci* 78: 671–681.

Dosdall LM, Gibson GAP, Olfert O, Keddie BA, Ulmer BJ (2007) Contributions to the life history, host range, and distribution of *Necremnus tidius* (Walker) (Hymenoptera: Eulophidae). *Ann Entomol Soc Am* 100: 861–868.

Dosdall LM, Gibson GAP, Olfert O, Mason PG (2009) Responses of Chalcidoidea (Hymenoptera) parasitoids to invasion of the cabbage seedpod weevil (Coleoptera: Curculionidae) in western Canada. *Biol Invasions* 11: 109–125.

Dosdall LM, Herbut MJ, Cowle NT (1994) Susceptibilities of species and cultivars of canola and mustard to infestation by root maggots (*Delia* spp.) (Diptera: Anthomyiidae). *Can Entomol* 126: 251–260.

Dosdall LM, Herbut MJ, Cowle NT, Micklich TM (1996a) The effect of tillage regime on emergence of root maggots (*Delia* spp.) (Diptera: Anthomyiidae) from canola. *Can Entomol* 128: 1157–1165.

Dosdall LM, Herbut MJ, Cowle NT, Micklich TM (1996b) The effect of seeding date and plant density on infestations of root maggots, *Delia* spp. (Diptera: Anthomyiidae), in canola. *Can J Plant Sci* 76: 169–177.

Dosdall LM, Kott LS (2006) Introgression of resistance to cabbage seedpod weevil to canola from yellow mustard. *Crop Sci* 46: 2437–2445.

Dosdall LM, Mason PG, Olfert O, Kaminski L, Keddie BA (2004) The origins of infestations of diamondback moth, *Plutella xylostella* (L.), in canola in western Canada. In: Endersby NM, Ridland PM (eds.) *The management of diamondback moth and other crucifer pests: Proc IV Int workshop, Melbourne, Australia, 26–29 November 2001*, pp 95–100.

Dosdall LM, Moisey DWA (2004) Developmental biology of the cabbage seedpod weevil, *Ceutorhynchus obstrictus* (Coleoptera: Curculionidae), in spring canola, *Brassica napus*, in western Canada. *Ann Entomol Soc Am* 97: 458–465.

Dosdall LM, Moisey D, Cárcamo H, Dunn R (2001) *Cabbage seedpod weevil fact sheet*. Alberta Agriculture, Food and Rural Development Agdex 622–21, 4 pp.

Dosdall LM, Moisey D, Kott L, Keddie BA, Good A, Rahman H, Mason PG, Olfert O, Cárcamo H, Kuhlmann U, McKenzie R, Hartman M, Fry K, Philip H, McCaffrey J (2006a) *Integrated Management of the Cabbage Seedpod Weevil*. Univ Alberta Tech Rep, 535 pp.

Dosdall LM, Stevenson FC (2005) Managing flea beetles (*Phyllotreta* spp.) (Coleoptera: Chrysomelidae) in canola with seeding date, plant density, and seed treatment. *Agron J* 97: 1570–1578.

Dosdall LM, Ulmer BJ (2004) Feeding, development, and oviposition of bertha armyworm (Lepidoptera: Noctuidae) on different host plant species. *Environ Entomol* 33: 756–764.

Dosdall LM, Ulmer BJ, Gibson GAP, Cárcamo HA (2006b) The spatio-temporal distribution dynamics of the cabbage seedpod weevil, *Ceutorhynchus obstrictus* (Coleoptera: Curculionidae), and its larval parasitoids in canola in western Canada. *Biocontrol Sci Techn* 16: 987–1006.

Dosdall LM, Weiss RM, Olfert O, Cárcamo HA (2002) Temporal and geographical distribution patterns of cabbage seedpod weevil (Coleoptera: Curculionidae) in canola. *Can Entomol* 134: 403–418.

Dosdall LM, Weiss RM, Olfert OO, Mason PG, Soroka JJ (2008) Diamondback moth, *Plutella xylostella* (L.), as a pest of canola in Canada: Its historical impact on the crop and predicted effects of climate change on its pest status. In: Shelton AM, Liu S (eds.) *The management of diamondback moth and other crucifer pests. Proc V Int Workshop, 24–27 October 2006. Beijing, China*, pp 112–121.

Doucette CF (1947) Host plants of the cabbage seedpod weevil. *J Econ Entomol* 40: 838–840.

Doucette CF (1948) Field parasitization and larval mortality of the cabbage seedpod weevil. *J Econ Entomol* 41: 763–765.

Downey RK (1983) The origin and description of the Brassica oilseed crops. In: Kramer JKG, Pigden WJ (eds.) *High and low erucic acid rapeseed oils*. Academic Press, Burlington, MA, pp 1–20.

Elliott RH (1988) Factors influencing the efficacy and economic returns of aerial sprays against wheat midge, *Sitodiplosis mosellana* (Diptera: Cecidomyiidae). *Can Entomol* 120: 941–954.

Elliott RH, Franke C, Rakow GFW (2008) Effect of seed size and seed weight on seedling establishment, vigour and tolerance of Argentine canola (*Brassica napus*) to flea beetles, *Phyllotreta* spp. *Can J Plant Sci* 88: 207–217.

Elliott RH, Mann LW, Olfert OO (2007) Effects of seed size and seed weight on seedling establishment, seedling vigour and tolerance of summer turnip rape (*Brassica rapa*) to flea beetles, *Phyllotreta* spp. *Can J Plant Sci* 87: 385–393.

Entomological Society of America (2007) *Common Names of Insects and Related Organisms*. Committee on the Common Names of Insects, http://www.entsoc.org/pubs/ common_names/index.htm. Cited 23 Nov 2007.

Entomological Society of Canada (2007) *Insect Common Names*. Committee on the Common Names of Insects, http://www.esc-sec.org/common_names_2006.pdf. Cited 17 January 2008.

Feeny P, Paauwe KL, Demong NJ (1970) Flea beetles and mustard oils: Host plant specificity of *Phyllotreta cruciferae* and *P. striolata* adults (Coleoptera: Chrysomelidae). *Ann Entomol Soc Am* 63: 832–841.

Ferracini C, Boivin G, Alma A (2006) Costs and benefits of host feeding in the parasitoid wasp *Trichogramma turkestanica*. *Entomol Exp Appl* 121: 229–234.

Floate KD, Bérubé J, Boiteau G, Dosdall LM, van Frankensuyzen K, Gillespie DR, Moyer J, Philip HG, Shamoun S (2002) Pesticides and biological control. In: Mason PG, Huber JT (eds.) *Biological control programmes in Canada, 1981–2000*. CABI Publishing, New York, pp 4–14.

Follett PA, Duan JJ (eds.) (1999) *Non-target effects of biological control*. Kluwer Academic Publishers, Dordrecht, the Netherlands, 316 pp.

Fox AS, Dosdall LM (2003) Reproductive biology of *Ceutorhynchus obstrictus* (Coleoptera: Curculionidae) on wild and cultivated Brassicaceae in southern Alberta. *J Entomol Sci* 38: 365–376.

Fox AS, Shaw SR, Dosdall LM, Lee B (2004) *Microctonus melanopus* (Ruthe) (Hymenoptera: Braconidae), a parasitoid of adult cabbage seedpod weevil (Coleoptera: Curculionidae): Distribution in southern Alberta and female diagnosis. *J Entomol Sci* 39: 350–361.

Gamache P (2007) *Direct seeding statistics for Alberta*. http://www.reducedtillage.ca/ article.aspx?a=359. Cited 22 Nov 2007.

Gavloski J, Ekuere U, Keddie A, Dosdall L, Kott L, Good A (2000) Identification and evaluation of flea beetle (*Phyllotreta cruciferae*) resistance within Brassicaceae. *Can J Plant Sci* 80: 881–887.

Gavloski JE, Lamb RJ (2000) Compensation by cruciferous plants is specific to the type of simulated herbivory. *Environ Entomol* 29: 1273–1282.

Gibson GAP, Baur H, Ulmer B, Dosdall L, Muller FJ (2005) On the misidentification of the chalcid (Hymenoptera: Chalcidoidea) parasitoids of the cabbage seedpod weevil (Coleoptera: Curculionidae) in North America. *Can Entomol* 137: 381–403.

Gibson GAP, Gates MW, Buntin GD (2006a) Parasitoids (Hymenoptera: Chalcidoidea) of the cabbage seedpod weevil (Coleoptera: Curculionidae) in Georgia, USA. *J Hym Res* 15: 187–207.

Gibson GAP, Gillespie DR, Dosdall L (2006b) The species of Chalcidoidea (Hymenoptera) introduced to North America for biological control of the cabbage seedpod weevil, and the first recovery of *Stenomalina gracilus* (Chalcidoidea: Pteromalidae). *Can Entomol* 138: 285–291.

Gillespie DR, Mason PG, Dosdall LM, Bouchard P, Gibson GAP (2006) Importance of long-term research in classical biological control: An analytical review of a release against the cabbage seedpod weevil in North America. *J Appl Entomol* 130: 401–409.

Goodwin M (2006) *Three potential sources for increased canola oil production in the Canadian prairies to meet the needs of biodiesel demand.* http://www.canola-council.org/biodiesel/Final%20Report-Canola_Oil_Feedstock_for_Biodiesel.pdf Cited 23 Nov 2007.

Goulet H, Mason PG (2006) Review of the Nearctic species of *Leiophron* and *Peristenus* (Hymenoptera: Braconidae: Euphorinae) parasitizing *Lygus* (Hemiptera: Miridae: Mirini). *Zootaxa* 1323: 1–118.

Greathead DJ (1986) Parasitoids in classical biological control. In: Waage J, Greathead D (eds.) *Insect parasitoids*. Academic Press, London, UK.

Griffiths GCD (1986a) Phenology and dispersion of *Delia radicum* (L.) (Diptera: Anthomyiidae) in canola fields at Morinville, Alberta. *Quaest Entomol* 22: 29–50.

Griffiths GCD (1986b) Relative abundance of the root maggots *Delia radicum* (L.) and *D. floralis* (Fallén) (Diptera: Anthomyiidae) as pests of canola in Alberta. *Quaest Entomol* 22: 253–260.

Griffiths GCD (1991a) Economic assessment of cabbage maggot damage in canola in Alberta. In: McGregor DI (ed.) *Proc GCIRC VIII Int Rapeseed Congress. Saskatoon, SK, Canada*, pp 528–535.

Griffiths GCD (1991b) *Flies of the Nearctic region. Cyclorrapha II. (Schizophora: Calyptratae) Anthomyiidae Volume 8 Part 2 Number 7*. Stuttgart, Germany: Schweizerbart.

Gurr GM, Wratten SD, Altieri MA (eds.) (2004) *Ecological engineering for pest management: Advances in habitat manipulation for arthropods*. CSIRO, Collingwood, Victoria, Australia, 232 pp.

Hagen KS (1946) The occurrence of *Ceutorhynchus assimilis* (Paykull) in California. *Pan-Pac Entomol* 22: 73.

Hallett R, Heal JD (2001) First Nearctic record of the swede midge (Diptera: Cecidomyiidae), a pest of cruciferous crops from Europe. *Can Entomol* 133: 713–715.

Hanson AJ, Carlson EC, Breakley EP, Webster RL (1948) Biology of the cabbage seedpod weevil in northwestern Washington. *State Coll Washington Agric Exp Station Bull* 498, 1–15.

Harcourt DG (1957) Biology of the diamondback moth, *Plutella maculipennis* (Curt.) (Lepidoptera: Plutellidae), in Eastern Ontario. II. Life-history, behaviour, and host relationship. *Can Entomol* 12: 554–564.

Harcourt DG (1960) Biology of the diamondback moth, *Plutella maculipennis* (Curt.) (Lepidoptera: Plutellidae), in eastern Ontario. III. Natural enemies. *Can Entomol* 92: 419–428.

Harcourt DG (1986) Population dynamics of the diamondback moth in southern Ontario. In: Talekar NS (ed.) Diamondback moth and other crucifer pests. *Proc Ist Int. Workshop, 11–15 March 1985. Tainan, Taiwan: Asian Vegetable Research and Development Center*. pp 3–16.

Hardy JE (1938) *Plutella maculipennis* Curt., its natural and biological control in England. *B Entomol Res* 29: 343–372.

Harker KN, Clayton GW, Blackshaw RE, O'Donovan JT, Stevenson FC (2003) Seeding rate, herbicide timing and competitive hybrids contribute to integrated weed management in canola (*Brassica napus*). *Can J Plant Sci* 83: 433–440.

Harmon BL, McCaffrey JP (1997) Parasitism of adult *Ceutorhynchus assimilis* (Coleoptera: Curculionidae) by *Microctonus melanopus* (Hymenoptera: Braconidae) in northern Idaho and eastern Washington. *J Agr Entomol* 14: 55–59.

Haye T (2004) *Studies on the ecology of European* Peristenus *spp. (Hymenoptera: Braconidae) and their potential for the biological control of* Lygus *spp. (Hemiptera: Miridae) in Canada.* Ph.D. Thesis, Christian Albrechts University, Kiel, Germany.

Haye T, Goulet H, Mason PG, Kuhlmann U (2005) Does fundamental host range match ecological host range? A retrospective case study of a *Lygus* plant bug parasitoid. *Bio Control* 35: 55–67.
Haye T, Kuhlmann U, Goulet H, Mason PG (2006) Controlling Lygus plant bugs with European *Peristenus relictus* (Hymenoptera: Braconidae) in Canada – Risky or not? *B Entomol Res* 96: 187–196.
Heimpel GE, Collier TR (1996) The evolution of host-feeding behaviour in insect parasitoids. *Biol Rev* 71: 373–400.
Hemachandra KS (2004) *Parasitoids of* Delia radicum *(Diptera: Anthomyiidae) in canola: Assessment of potential agents for classical biological control.* Ph.D. thesis, University of Manitoba, Winnipeg, MB.
Hemachandra KS, Holliday NJ, Klimaszewski J, Mason PG, Kuhlmann U (2005) Erroneous records of *Aleochara bipustulata* from North America: An assessment of the evidence. *Can Entomol* 137: 182–187.
Hemachandra KS, Holliday NJ, Mason PG, Soroka JJ, Kuhlmann U (2007a) Comparative assessment of the parasitoid community of *Delia radicum* in the Canadian prairies and Europe: A search for classical biological control agents. *Bio Control* 43: 85–94.
Hemachandra KS, Kuhlmann U, Mason PG, Holliday NJ (2007b) Spatial patterns of *Trybliographa rapae* parasitism of *Delia radicum* larvae in oilseed rape and cauliflower. *J Appl Entomol* 131: 338–346.
Herrara FRC (2006) *Seasonal development and natural enemies of an invasive exotic species, the swede midge* Contarinia nasturtii *(Kieffer), in Quebec.* M.Sc. Thesis, McGill University, Montreal, PQ.
Hopkins RJ, Griffiths DW, McKinlay RG, Birch ANE (1999) The relationship between cabbage root fly (*Delia radicum*) larval feeding and the freeze-dried matter and sugar content of Brassica roots. *Entomol Exp Appl* 92: 109–117.
Huber JT, Darbyshire S, Bissett J, Foottit RG (2002) Taxonomy and biological control. In: Mason PG, Huber JT (eds.) *Biological control programmes in Canada, 1981–2000.* CABI Publishing, New York, USA.
Hughes RD, Mitchell B (1960) The natural mortality of *Erioischia brassicae* (Bouché) (Dipt., Anthomyiidae): Life tables and their interpretation. *J Anim Ecol* 29: 359–374.
ICZN (International Commission on Zoological Nomenclature) (2007) Closure of cases. *B Zool Nomen* 64: 269.
Idris AB, Grafius E (2001) Evidence of *Diadegma insulare* (Cresson), a parasitoid of diamondback moth, *Plutella xylostella* (L.), present in various habitats. *Pakistan J Biol Sci* 1: 742–743.
Johanowicz DL, Mitchell ER (2000) Effects of sweet alyssum flowers on the longevity of the parasitoid wasps *Cotesia marginiventris* (Hymenoptera: Braconidae) and *Diadegma insulare* (Hymenoptera: Ichneumonidae). *Fla Entomol* 83: 41–47.
Jones TH, Hassell MP (1988) Patterns of parasitism by *Trybliographa rapae*, a cynipid parasitoid of the cabbage root fly, under laboratory and field conditions. *Ecol Entomol* 13: 309–317.
Jourdheuil P (1960) Influence de quelques facteurs écologiques sur les fluctuations de population d'une biocénose parasitaire: Étude relative à quelques Hyménoptères (Ophioninae, Diospilinae, Euphorinae) parasites de divers Coléoptères inféodés aux Crucifères. *Ann Épiphyties* 11: 445–658.
Justus KA, Dosdall LM, Mitchell BK (2000) Oviposition by *Plutella xylostella* (Lepidoptera: Plutellidae) and effect of phylloplane waxiness. *J Econ Entomol* 93: 1152–1159.
Kalischuk AR, Dosdall LM (2004) Susceptibilities of seven Brassicaceae species to infestation by the cabbage seedpod weevil (Coleoptera: Curculionidae). *Can Entomol* 136: 265–276.
Kegley S, Hill B, Orme S (2007) *PAN Pesticide Database*, Pesticide Action Network, North America, http://www.pesticideinfo.org. Cited 23 Nov 2007.
Kfir R (1998) Origin of the diamondback moth (Lepidoptera: Plutellidae). *Ann Entomol Soc Am* 91: 164–167.
King KM (1928) *Barathra configurata* Wlk., an armyworm with important potentialities on the northern prairies. *J Econ Entomol* 21: 279–292.

Kirk AA, Mercadier G, Bordat D, Delvare G, Pichon A, Arvanitakis L, Goudégnon AE, Rincon C (2004) Variability in *Plutella* and its natural enemies: Implications for biological control. In: Endersby NM, Ridland PM (eds.) The management of diamondback moth and other crucifer pests. *Proc IV Int Workshop, 26–29 November 2001. Melbourne*. Melbourne, Australia: Department of Natural Resources and Environment. pp 71–77.

Klimaszewski J (1984) A revision of the genus *Aleochara* Gravenhorst of America north of Mexico. (Coleoptera: Staphylinidae, Aleocharinae). *Mem Entomol Soc Can* 129: 1–211.

Knodel JJ, Olson DL (2002) *Crucifer flea beetle: Biology and integrated pest management*. NDSU Extension Service. North Dakota State University, Fargo, ND.

Kozlowski MW, Lux S, Dmoch J (1983) Oviposition behaviour and pod marking in the cabbage seed weevil, *Ceutorhynchus assimilis*. *Entomol Exp Appl* 34: 277–282.

Kuhlmann U, Dosdall LM, Mason PG (2002) *Ceutorhynchus obstrictus* (Marsham), Cabbage Seedpod Weevil (Coleoptera: Curculionidae). In: Mason PG, Huber JT (eds.) *Biological control programmes in Canada 1981–2000*. CABI Publishing, Wallingford, UK.

Kuhlmann U, Mason PG, Hinz HL, Blossey B, De Clerck-Floate RA, Dosdall L, McCaffrey JP, Schwarzlaender M, Olfert O, Brodeur J, McClay AS, Wiedenmann RN (2006a) Avoiding conflicts between insect and weed biological control: Selection of non-target species for test list to assess host specificity of cabbage seedpod weevil parasitoids. *J Appl Entomol* 130: 129–141.

Kuhlmann U, Schaffner U, Mason PG (2006b) Selection of non-target species for host specificity tests. In: Bigler F, Babendreier D, Kuhlmann U (eds.) *Biological control of arthropods using invertebrates: Methods for environmental risk assessment*. CABI Publishing, Wallingford, UK.

Kuhlmann U, Schafner U, Mason PG (2005) Selection of non-target species for host specificity testing of entomophagous biological control agents. In: Hoddle MS (Compiler) *Proc II Int Symp on Biological control of arthropods, Davos, Switzerland, 12–16 September 2005*, United States Department of Agriculture, Forest Service, Morgantown, WV, FHTET-2005–08, pp 566–583.

Lachance S, Broadbent AB, Sears MK (2001) In-host compatibility and in-host competition of exotic and native parasitoids of the tarnished plant bug (Heteroptera: Miridae). *Environ Entomol* 30: 1158–1163.

Laffin RD, Dosdall LM, Sperling FAH (2005) Population structure of the cabbage seedpod weevil, *Ceutorhynchus obstrictus* (Marsham) (Coleoptera: Curculionidae): Origins of North American introductions. *Environ Entomol* 34: 504–510.

Lamb RJ (1980) Hairs protect pods of mustard (*Brassica hirta* 'Giselba') from flea beetle feeding damage. *Can J Plant Sci* 60: 1439–1440.

Lamb RJ (1984) Effects of flea beetles, *Phyllotreta* spp. (Chrysomelidae: Coleoptera), on the survival, growth, seed yield and quality of canola, rape and yellow mustard. *Can Entomol* 116: 269–280.

Lamb RJ (1988) Assessing the susceptibility of crucifer seedlings to flea beetle (*Phyllotreta* spp.) damage. *Can J Plant Sci* 68: 85–93.

Lamb RJ (1989) Entomology of oilseed *Brassica* crops. *Ann Rev Entomol* 34: 211–229.

Lamb RJ, Palaniswamy P (1990) Host discrimination by a crucifer-feeding flea beetle, *Phyllotreta striolata* (F.) (Coleoptera: Chrysomelidae). *Can Entomol* 122: 817–824.

Lamb RJ, Turnock WJ (1982) Economics of insecticidal control of flea beetles (Coleoptera: Chrysomelidae) attacking rape in Canada. *Can Entomol* 114: 827–840.

Landis DA, Wratten SD, Gurr GM (2000) Habitat management to conserve natural enemies of arthropod pests in agriculture. *Annu Rev Entomol* 45: 175–201.

Lauro N, Kuhlmann U, Mason PG, Holliday NJ (2005) Interaction of a solitary larval endoparasitoid, *Microplitis mediator*, with its host, *Mamestra brassicae*: Host acceptance and host suitability. *J Appl Entomol* 129: 567–573.

Lawton JH, Brown KC (1986) The population and community ecology of invading insects. *Philos T Roy Soc B* 314: 607–616.

Leferink JHM, Gerber GH (1997) Development of adult and nymphal populations of *Lygus lineolaris* (Palisot de Beauvois), *L. elisus* Van Duzee, and *L. borealis* (Kelton) (Heteroptera: Miridae) in relation to seeding date and stage of plant development on canola (Brassicaceae) in southern Manitoba. *Can Entomol* 129: 777–787.

Liu HJ, Butts RA (1982) *Delia* spp. (Diptera: Anthomyiidae) infesting canola in Alberta. *Can Entomol* 114: 651–653.

Liu S, Wang X, Guo S, He J, Shi Z (2000) Seasonal abundance of the parasitoid complex associated with the diamondback moth, *Plutella xylostella* (Lepidoptera: Plutellidae) in Hangzhou, China. *B Entomol Res* 90: 221–231.

Madder DJ, Stemeroff M (1988) The economics of insect control on wheat, corn, and canola, 1980–1985. *B Entomol Soc Can* 20(suppl): 1–22.

Manitoba Agriculture (2007) *Guide to Field Crop Protection 2007*. http://www.gov.mb.ca/agriculture/crops/cropproduction/gaa01d01.html. Cited 23 Nov 2007.

Marshall EJP, Brown VK, Boatman ND, Lutman PJW, Squire GR, Ward LK (2003) The role of weeds in supporting biological diversity within crop fields. *Weed Res* 43: 77–89.

Mason PG, Arthur AP, Olfert OO, Erlandson MA (1998a) The bertha armyworm (*Mamestra configurata*) (Lepidoptera: Noctuidae) in western Canada. *Can Entomol* 130: 321–336.

Mason PG, Baute T, Olfert O, Roy M (2004) Cabbage seedpod weevil, *Ceutorhynchus obstrictus* (Marsham) (Coleoptera: Curculionidae) in Ontario and Québec. *J Entomol Soc Ontario* 134: 107–113.

Mason PG, Erlandson MA, Elliott RH, Harris BJ (2002b) Potential impact of Spinosad on parasitoids of *Mamestra configurata* (Lepidoptera: Noctuidae). *Can Entomol* 134: 59–68.

Mason PG, Erlandson MA, Youngs BJ (2001) Effects of parasitism by *Banchus flavescens* (Hymenoptera: Ichneumonidae) and *Microplitis mediator* (Hymenoptera: Braconidae) on bertha armyworm, *Mamestra configurata* (Lepidoptera: Noctuidae). *J Hym Res* 10: 81–90.

Mason PG, Flanders RG, Arrendondo-Bernal HA (2005) How can legislation facilitate the use of biological control of arthropods in North America? In: Hoddle MS (Compiler) *Proc II Int Symp on biological control of arthropods, Davos, Switzerland, 12–16 September 2005*, United States Department of Agriculture, Forest Service, Morgantown, WV, FHTET-2005–08, Vol II, pp 701–714.

Mason PG, Olfert O, Sluchinski L, Weiss RM, Boudreault C, Grossrieder M, Kuhlmann U (2003) Actual and potential distribution of an invasive canola pest, *Meligethes viridescens* (Coleoptera: Nitidulidae), in Canada. *Can Entomol* 135: 405–413.

Mason PG, Pinto JD, Long ZL, Harris JL (1998b) First record of *Trichogramma inyoense* Pinto & Oatman (Hymenoptera: Trichogrammatidae) attacking the bertha armyworm, *Mamestra configurata* Walker (Lepidoptera: Noctuidae). *Can Entomol* 130: 105–106.

Mason PG, Turnock WJ, Erlandson MA, Kuhlmann U, Braun L (2002a) *Mamestra configurata* Walker, bertha armyworm (Lepidoptera: Noctuidae). In: Mason PG, Huber JT (eds.) *Biological control programmes in Canada, 1981–2000*. CABI Publishing, New York, USA.

Mason PG, Youngs BJ (1994) Biological control of a Canadian canola pest, the bertha armyworm (*Mamestra configurata*), with the European parasitoid *Microplitis mediator*. *Norw J Agr Sci Suppl* 16: 405–406.

Maus C, Mittmann B, Peschke K (1998) Host records of parasitoid *Aleochara* Gravenhorst species (Coleoptera, Staphylinidae) attacking puparia of Cyclorrhapheous Diptera. Mittellungen Museum Naturkinde. *Deutsche Entomologische Zeitschrift* 45: 231–254.

Maw HEL, Foottit RG, Hamilton KGA, Scudder GGE (2000) *Checklist of the Hemiptera of Canada and Alaska. National Research Council of Canada*. NRC Research Press, Ottawa, Canada, 220 pp.

McCaffrey JP (1992) Review of U.S. canola pest complex: Cabbage seedpod weevil. In: *Proc 1992 U.S. Canola Conf, 5–6 March, Washington, D.C.*: 140–143.

McCaffrey JP, Harmon BL, Brown J, Brown AP, Davis JB (1999) Assessment of *Sinapis alba*, *Brassica napus* and *S. alba* x *B. napus* hybrids for resistance to cabbage seedpod weevil, *Ceutorhynchus assimilis* (Coleoptera: Curculionidae). *J Agr Sci* 132: 289–295.

McDonald RS, Sears MK (1991) Effects of root damage by cabbage maggot, *Delia radicum* (L.) (Diptera: Anthomyiidae), on yield of canola, *Brassica campestris* L., under laboratory conditions. *Can Entomol* 123: 861–867.

McDonald RS, Sears MK (1992) Assessment of larval feeding damage of the cabbage maggot (Diptera: Anthomyiidae) in relation to oviposition preference on canola. *J Econ Entomol* 85: 957–962.

McLeod JH (1952) Notes on the cabbage seedpod weevil, *Ceutorhynchus assimilis* (Payk.) (Coleoptera: Curculionidae), and its parasites. *Proc Entomol Soc BC* 49: 11–18.

McLeod JH (1962) Cabbage seedpod weevil – *Ceutorhynchus assimilis* (Payk.) Curculionidae. In: McLeod JH, McGugan BM, Coppel HC (eds.) *A review of the biological control attempts against insects and weeds in Canada*. Commonwealth Agricultural Bureau (CAB), Farnham Royal, Bucks, UK.

Messing R, Roitberg B, Brodeur J (2006) Measuring and predicting indirect impacts of biological control: Competiton, displacement and secondary interactions. In: Bigler FR, Babendreier D, Kuhlmann U (eds.) *Environmental impact of invertebrates for biological control of arthropods: Methods and risk assessment*. CABI Publishing, Wallingford, UK, pp 64–77.

Mitchell ER, Hu GY, Okine JS, Carpenter JE (1999) Parasitism of diamondback moth (Lepidoptera: Plutellidae) larvae by *Cotesia plutellae* (Hymenoptera: Braconidae) and *Diadegma insulare* (Hymenoptera: Ichneumonidae) in cabbage fields after inundative releases of *C. plutellae*. *J Entomol Sci* 34: 101–112.

Mukerji MK (1971) Major factors in survival of the immature stages of *Hylemya brassicae* (Diptera: Anthomyiidae) on cabbage. *Can Entomol* 103: 717–728.

Muller F, Cuenot E, Kuhlmann U (2005) *Biological control of cabbage seedpod weevil, Ceutorhynchus obstrictus*. Annual Report 2004/2005, Unpublished report, CABI Bioscience Switzerland Centre, Delémont, Switzerland.

Murchie AK, Williams IH (1998) A bibliography of the parasitoids of the cabbage seed weevil (*Ceutorhynchus assimilis* Payk.). *IOBC/wprs Bull* 21: 163–169.

NASS (National Agricultural Statistics Service) (2007) Acreage. http://usda.mannlib.cornell.edu/usda/current/Acre/Acre-06-29-2007.pdf Cited 17 Nov 2007.

Nealis VG, Frankenhuyzen K van, Cadogan BL (1992) Conservation of spruce budworm parasitoids following application of *Bacillus thuringiensis* var. *kurstaki*. *Can Entomol* 124: 1085–1092.

Nilsson C (2003) Parasitoids of pollen beetles. In: Alford DV (ed.) *Biocontrol of oilseed rape pests*. Blackwell, Oxford, UK.

O'Hara J (1999) Tachinidae (Diptera) parasitoids of bertha armyworm (Lepidoptera: Noctuidae). *Can Entomol* 131: 11–28.

Okine JS, Mitchell ER, Hu GY (1996) Low temperature effect on viability of *Diadegma insulare* (Hymenoptera: Ichneumonidae) and effect of this parasitoid on feeding rate of diamondback moth larvae (Lepidoptera: Plutellidae). *Fla Entomol* 7: 503–509.

Olfert O, Doane JF, Braun MP (2003) Establishment of *Platygaster tuberosula*, an introduced parasitoid of the wheat midge, *Sitodiplosis mosellana*. *Can Entomol* 135: 303–308.

Olfert O, Hallett R, Weiss RM, Soroka J, Goodfellow S (2006) Potential distribution and relative abundance of swede midge, *Contarinia nasturtii*, an invasive pest in Canada. *Entomol Exp Appl* 120: 221–228.

Olfert O, Weiss RM (2006) Impact of climate change on potential distributions and relative abundances of *Oulema melanopus*, *Meligethes viridescens* and *Ceutorhynchus obstrictus* in Canada. *Agr Ecosys Environ* 113: 295–301.

Philip H, Mengersen E (1989) *Insect pests of the prairies*. University of Alberta Press, Edmonton, AB.

Pickett CH, Rodriguez R, Brown J, Coutinot D, Hoelmer KA, Kuhlmann U, Goulet H, Schwartz MD, Goodell PB (2007) Establishment of *Peristenus digoneutis* and *P. relictus* (Hymenoptera: Braconidae) in California for the control of *Lygus* spp. (Heteroptera: Miridae). *Biocontrol Sci Techn* 17: 261–272.

Pivnick KA (1993) Diapause initiation and pupation site selection of the braconid parasitoid *Microplitis mediator* (Haliday). *Can Entomol* 125: 825–830.
Plattner HC (1974) Contributions to the biology of *Phygadeuon trichops* Thomson (Hym., Ichneumonidae). *Anzeiger für Schadlingskunde, Pflanzenschutz, Umweltschutz* 48: 56–60.
Polis GA, Winemiller KO (eds.) (1996) *Food webs: Integration of patterns and dynamics*. Chapman and Hall, New York.
Putnam LG (1978) Diapause and cold hardiness in *Microplitis plutellae*, a parasite of the larvae of the diamondback moth. *Can J Plant Sci* 58: 911–913.
Ramachandran S, Buntin DG, All JN, Raymer PL, Stewart CN (2000) Intraspecific competition of an insect-resistant transgenic canola in seed mixtures. *Agron J* 92: 368–374.
Read DC (1962) Notes on the life history of *Aleochara bilineata* (Gyll.) (Coleoptera: Staphylinidae), and on its potential value as a control agent for the cabbage maggot, *Hylemya brassicae* (Bouché) (Diptera: Anthomyiidae). *Can Entomol* 94: 417–424.
Reigert PW (1984) Insects on the Canadian plains. *Prairie Forum* 9: 327–344.
Riley KJ, Kuhlmann U, Mason PG, Whistlecraft J, Donald LJ, Holliday NJ (2007) Can mustard seed meal increase attacks by *Aleochara* spp. on *Delia radicum* in oilseed rape? *Biocontrol Sci Techn* 17: 273–284.
Rosenheim JA, Kaya HK, Ehler LE, Marois JJ, Jaffee BA (1995) Intraguild predation among biological control agents – Theory and evidence. *Bio Control* 5: 303–335.
Royer L, LeLannic J, Nenon JP, Boivin G (1998) Response of first-instar *Aleochara bilineata* larvae to the puparium morphology of its dipteran host. *Entomol Exp Appl* 87: 217–220.
Ryan J, Ryan MF (1980) Observations on the natural mortality of the overwintering pupa of the cabbage root fly, *Delia brassicae* (Wiedemann), in Ireland. *Pl Pathol* 29: 38–44.
Sakai AK, Allendorf FW, Holt JS, Lodge DM, Molofsky J, With KA, Baughman S, Cabin RJ, Cohen JE, Ellstrand NC, McCauley DE, O'Neil P, Parker IM, Thompson JN, Weller SG (2001) The population biology of invasive species. *Ann Rev Ecol Syst* 32: 305–332.
Sarfraz M, Dosdall LM, Keddie BA (2006) Diamondback moth–host plant interactions: Implications for pest management. *Crop Prot* 25: 625–639.
Sarfraz M, Dosdall LM, Keddie BA (2007) Resistance of some cultivated Brassicaceae to infestations by *Plutella xylostella* (L.) (Lepidoptera: Plutellidae). *J Econ Entomol* 100: 215–224.
Sarfraz M, Keddie BA, Dosdall LM (2005) Biological control of the diamondback moth, *Plutella xylostella*: A review. *Biocontrol Sci Techn* 15: 763–789.
Saskatchewan Agriculture and Food (2007) *Guide to crop protection*. http://www.agriculture.gov.sk.ca/Guide_to_Crop_Protection. Cited 21 Nov 2007.
Schwartz MD, Foottit RG (1992) *Lygus* species on oilseed rape, mustard, and weeds: A survey across the prairie provinces of Canada. *Can Entomol* 124: 151–158.
Schwartz MD, Foottit RG (1998) *Revision of the Nearctic species of the genus* Lygus *Hahn, with a review of the Palaearctic species (Heteroptera: Miridae)*. Associated Publishers, Gainesville, FL.
Shelton AM (2004) Management of the diamondback moth: Déjà vu all over again? In: Endersby NM, Ridland PM (eds.) *The management of diamondback moth and other crucifer pests. Proc IV Int. Workshop, 26–29 Nov 2001. Melbourne*. Melbourne, Australia: Department of Natural Resources and Environment. pp 3–8.
Shipp L, Elliott D, Gillespie D, Brodeur J (2007) From chemical to biological control in Canadian greenhouse crops. In: Vincent C, Goettel MS, Lazarovits G (eds.) *Biological control: A global perspective*. CABI Publishing, New York.
Simberloff D, Stiling P (1996) How risky is biological control? *Ecology* 77: 1965–1974.
Snyder WE, Snyder GB, Finke DL, Straub CS (2006) Predator biodiversity strengthens herbivore suppression. *Ecol Lett* 9: 789–796.
Sommer G (1981) *Biologie und Parasitenkomplex der Halticenen Gattung* Phyllotreta. Ph.D. Thesis, Alber-Ludwigs University of Freiburg.

Soroka JJ, Carl K (2002) *Adelphocoris lineolatus* (Goeze), alfalfa plant bug (Hemiptera: Miridae). In: Mason PG, Huber JT (eds.) *Biological control programmes in Canada, 1981–2000*. CABI Publishing, Wallingford, UK.

Soroka JJ, Dosdall LM, Olfert OO, Seidle E (2004) Root maggots (*Delia* spp., Diptera: Anthomyiidae) in prairie canola (*Brassica napus* L. and *B. rapa* L.): Spatial and temporal surveys of root damage and prediction of damage levels. *Can J Plant Sci* 84: 1171–1182.

Soroka JJ, Kuhlmann U, Floate KD, Whistlecraft J, Holliday NJ, Boivin G (2002) *Delia radicum* (L.), Cabbage Maggot (Diptera: Anthomyiidae). In: Mason PG, Huber JT (eds.) *Biological control programmes in Canada, 1981–2000*. CABI Publishing, New York, USA, pp 169–176.

Statistics Canada 2007a. *Field Crop Reporting Series Catalogue number 22-002-XIE.*

Statistics Canada 2007b. *Fruit and Vegetable Production. Catalogue number 22-003-XIB.*

Strong DR, Pemberton RW (2001) Food webs, risks of alien enemies and reform of biological control. In: Wajnberg E, Scott JK, Quimby PC (eds.) *Evaluating indirect ecological effects of biological control*. CABI Publishing, Wallingford, UK.

Tahvanainen JO (1972) Phenology and microhabitat selection of some flea beetles (Coleoptera: Chrysomelidae) on wild and cultivated crucifers in central New York. *Entomol Scand* 3: 120–138.

Thomas P (2002) *Canola Growers' Manual*. http://www.directfocus.com/canolamanual/ Cited 17 Nov 2007.

Timlick BH, Turnock WJ, Wise I (1993) Distribution and abundance of *Lygus* spp. (Heteroptera: Miridae) on alfalfa and canola in Manitoba. *Can Entomol* 125: 1033–1041.

Tomlin AD, Miller JJ, Harris CR, Tolman JH (1985) Arthropod parasites and predators of the onion maggot (Diptera: Anthomyiidae) in southwestern Ontario. *J Econ Entomol* 78: 975–981.

Tothill JD (1922) The natural control of the fall webworm (*Hyphantria cunea* Drury) in Canada together with an account of its several parasites. *B Can Dep Agric New Series* 3: 1–107.

Tran T, Takasu K (2000a) Host age selection by the host-feeding pupal parasitoid *Diadromus subtilicornis* (Gravenhorst) (Hymenoptera: Ichneumonidae). *Appl Entomol Zool* 35: 549–556.

Tran T, Takasu K (2000b) Life history of the pupal parasitoid *Diadromus subtilicornis* (Gravenhorst) (Hymenoptera: Ichneumonidae) as influenced by temperature, photoperiod, and availability of food and hosts. *Entomol Sci* 3: 255–264.

Tscharntke T, Hawkins BA (2002) *Multitrophic level interactions*. Cambridge University Press, UK.

Turnock WJ (1984) *Mamestra configurata* Walker, bertha armyworm (Leopidoptera: Noctuidae). In: Kelleher JS, Hulme MA (eds.) *Biological control programmes against insects and weeds in Canada 1969–1980*. Commonwealth Agriculture Bureaux, Slough, UK.

Turnock WJ (1985) Developmental, survival, and reproductive parameters of bertha armyworm, *Mamestra configurata* (Lepidoptera: Noctuidae) on four plant species. *Can Entomol* 117: 1267–1271.

Turnock WJ, Bilodeau RJ (1984) Survival of pupae of *Mamestra configurata* (Lepidoptera: Noctuidae) and two of its parasites in untilled and tilled soil. *Can Entomol* 116: 257–267.

Turnock WJ, Boivin G, Whistlecraft JW (1995a) Parasitism of overwintering puparia of the cabbage maggot *Delia radicum* (L.) (Diptera:Anthomyiidae), in relation to host density and weather factors. *Can Entomol* 127: 535–542.

Turnock WJ, Carl KP (1995) Evaluation of the Palaearctic *Eurithia consobrina* (Diptera: Tachinidae) as a potential biocontrol agent for *Mamestra configurata* (Leepidoptera: Noctuidae) in Canada. *Biocontrol Sci Techn* 5: 55–67.

Turnock WJ, Gerber GH, Timlick BH, Lamb RJ (1995b) Losses of canola seeds from feeding by *Lygus* species (Heteroptera: Miridae) in Manitoba. *Can J Plant Sci* 75: 731–736.

Ulber B, Vidal S (1998) Influence of host density and host distribution on parasitism of *Ceutorhynchus assimilis* by *Trichomalus perfectus*. *IOBC/wprs Bull* 21: 185–195.

Ulmer BJ, Dosdall LM (2006a) Emergence of overwintered and new generation adults of the crucifer flea beetle, *Phyllotreta cruciferae* (Goeze) (Coleoptera: Chrysomelidae). *Crop Prot* 25: 23–30.

Ulmer BJ, Dosdall LM (2006b) Spring emergence biology of the cabbage seedpod weevil (Coleoptera: Curculionidae). *Ann Entomol Soc Am* 99: 64–69.
Ulmer BJ, Dosdall LM, Gibson GAP (2005) Spatio-temporal distribution patterns of the diamondback moth and its principal parasitoid, *Diadegma insulare*, in canola crops and the occurrence of hyperparasitism. *Brassica* 6: 63–69.
Ulmer B, Gillott C, Erlandson M (2002) Oviposition preferences of bertha armyworm *Mamestra configurata* Walker (Lepidoptera: Noctuidae) on different crucifer cultivars and growth stages. *Environ Ent* 31: 1135–1141.
van Driesche RG, Bellows TS (1996) *Biological control*. Kluwer Academic Publishers, Dordrecht, The Netherlands.
van Lenteren JC (2007) Biological control for insect pests in greenhouses: An unexpected success. In: Vincent C, Goettel MS, Lazarovits G (eds.) *Biological control a global perspective*. CABI Publishing, New York, USA.
van Lenteren JC, Bale J, Bigler E, Hokkanen HMT, Loomans AM (2006) Assessing risks of releasing exotic biological control agents of arthropod pests. *Ann Rev Entomol* 51: 609–634.
WCCP (Western Committee on Crop Pests) (1995) *Minutes of the 34th Annual Meeting, October 19–21, Victoria, British Columbia*. 44 pp.
WCCP (Western Committee on Crop Pests) (2001) *Minutes of the 41st Annual Meeting, October 15–16, Banff, Alberta*. 71 pp.
Wajnberg E, Scott JK, Quimby PC (eds.) (2001) *Evaluating indirect ecological effects of biological control*. CABI Publishing, Wallingford, UK.
Walz AJ (1957) Observations on the biologies of some hymenopterous parasites of the cabbage seedpod weevil in northern Idaho. *Ann Entomol Soc Am* 50: 219–220.
Weiss MJ, McLeod P, Schatz BG, Hanson BK (1991) Potential for insecticidal management of flea beetles (Coleoptera: Chrysomelidae) on canola. *J Econ Entomol* 84: 1597–1603.
Westdal PH, Romanow W (1972) Observations on the biology of the flea beetle, *Phyllotreta cruciferae* (Coleoptera: Chrysomelidae). *Manitoba Entomol* 6: 35–45.
Williams IH (2003) Parasitoids of cabbage seed weevil. In: Alford DV (ed.) *Biocontrol of oilseed rape pests*. Blackwell, Oxford, UK.
Wise I, Dosdall L (2002) Oilseed crops. In: Johnson DL (ed.) *Guide to integrated control of plant pests. Technical Report, Western Committee on Crop Pests*. 199 pp.
Wishart G, Colhoun EH, Monteith AE (1957) Parasites of *Hylemya* spp. (Diptera: Anthomyiidae) that attack cruciferous crops in Europe. *Can Entomol* 89: 510–517.
Wishart G, Monteith E (1954) *Trybliographa rapae* (Westw.) (Hymenoptera: Cynipidae), a parasite of *Hylemya* spp. (Diptera: Anthomyiidae). *Can Entomol* 86: 145–154.
Wylie HG (1977a) Observations on *Athrycia cinerea* (Diptare: Tachinidae), a parasite of *Mamestra configurata* (Lepidoptera: Noctuidae). *Can Entomol* 109: 747–754.
Wylie HG (1977b) Observations on *Mericia ampelus* (Diptera: Tachinidae), an occasional parasite of bertha armyworm, *Mamestra configurata* (Lepidoptera: Noctuidae), in western Canada. *Can Entomol* 109: 1023–1024.
Wylie HG (1977c) Preventing and terminating pupal diapause in *Athrycia cinerea* (Diptera: Tachinidae). *Can Entomol* 109: 1083–1090.
Wylie HG (1979) Insect parasites reared from bertha armyworm, *Mamestra configurata* Walker, collected from artificial field populations near Winnipeg, Manitoba. *Manitoba Entomol* 11: 50–55.
Wylie HG (1982) An effect of parasitism by *Microctonus vitattae* (Hymenoptera: Braconidae) on emergence of *Phyllotreta cruciferae* and *Phyllotreta striolata* (Coleoptera: Chrysomelidae) from overwintering sites. *Can Entomol* 114: 727–732.
Wylie HG (1988) Release in Manitoba, Canada of *Townesilitus bicolor* (Hym.: Braconidae), an European parasite of *Phyllotreta* spp. (Col.: Chrysomelidae). *Entomophaga* 33: 25–32.

Wylie HG, Bucher GE (1977) The bertha armyworm, *Mamestra configurata* (Lepidoptera: Noctuidae). Mortality of immature stages on the rape crop 1972–1975. *Can Entomol* 109: 823–837.

Wylie HG, Loan C (1984) Five Nearctic and one introduced euphorine species (Hymenoptera: Braconidae) that parasitize adults of crucifer-infesting flea beetles (Coleoptera: Chrysomelidae). *Can Entomol* 116: 235–246.

Wylie HG, Turnock WJ, Burgess L (1984) *Phyllotreta* spp., flea beetles (Coleoptera: Chrysomelidae). In: Kelleher JS, Hulme MA (eds.) *Biological control programmes against insects and weeds in Canada 1969–1980*. Commonwealth Agricultural Bureaux, Slough, pp 73–76.

Chapter 7
Crop Location by Oilseed Rape Pests and Host Location by Their Parasitoids

Ingrid H. Williams and Samantha M. Cook

Abstract The behavioural ecology associated with location of the oilseed rape crop by its major coleopteran and dipteran pests and by their key hymenopterous parasitoids is reviewed. Results of studies investigating their responses to odour and colour cues from the crop, using olfactometers, wind tunnels, baited and/or coloured traps, and different plant lines, are presented. Host plant volatiles, particularly the isothiocyanates, carried downwind from the crop, and the colours yellow and green are important cues for orientation; odour cues can induce responses to colour cues. Species differ in the subset of cues they use and responses can vary with sex, age, generation, nutritional status, previous experience and time since last oviposition. Evidence suggests that odour-mediated upwind anemotaxis is used by the pests to locate the crop and their host plant, as well as by their parasitoids to locate the crop, the habitat of their host larvae, from a distance, while visual cues are important for orientation at closer range. Once the crop is located, pests use taste and tactile cues from the oilseed rape plant to accept or reject it while parasitoids use host-derived cues to locate their hosts. The implications for integrated pest management and conservation biocontrol are discussed.

7.1 Introduction

Oilseed rape (*Brassica napus* L.) is grown mostly as an annual break crop in a cereal rotation and is sown in a different field in each successive autumn (for winter rape) or spring (for spring rape). It is attacked by some generalist pests, such as slugs and pigeons, as well as by several crucifer-specialist pests which in turn are attacked by specific parasitioids. Most of the major coleopteran pests and their ectoparasitoids leave the crop to overwinter in non-cultivated areas of vegetation, whereas major dipteran pests and endoparasitoids overwinter in the soil of the field in which the

I.H. Williams (✉)
Institute of Agricultural and Environmental Sciences, Estonian University of Life Sciences, Tartu, 51014, Estonia
e-mail: ingrid.williams@bbsrc.ac.uk

I.H. Williams (ed.), *Biocontrol-Based Integrated Management of Oilseed Rape Pests*, DOI 10.1007/978-90-481-3983-5_7,

rape crop was grown and, in the spring, emerge in a cereal crop. So each year, the insect pests need to locate a new oilseed rape crop before they can make use of it as a resource for feeding and oviposition, and their parasitoids need to find the habitat of their hosts before they can utilise their hosts for oviposition.

Crop location by phytophagous insects, such as the pests of oilseed rape, involves a succession of yes/no decisions operating at the habitat, patch and host plant levels with the insect responding behaviourally to a different set of cues at each step (Miller and Strickler 1984). The relative importance of olfactory versus visual cues in host plant finding is still a matter of debate (Finch and Collier 2000, Couty et al. 2006). Olfactory cues from the host plant are probably most important from a distance but can also be involved at close range (Omura et al. 1999, Couty et al. 2006) while visual cues from the host plant often aid orientation at closer range (Prokopy and Owens 1983, Bernays and Chapman 1994, Finch and Collier 2000, Couty et al. 2006). The olfactory cues are termed kairomones, defined as behaviour-controlling chemicals that evoke a response that is adaptively favourable to the receiver (the insect) and not to the emitter (the plant) (Dicke and Sabelis 1988). Olfactory cues can induce responses to visual cues; appropriate odour-conditioned visual cues stimulate the insect to land on the plant (Prokopy 1986). Appropriate chemosensory (taste) and tactile cues from the plant then evoke either acceptance or rejection of the plant for feeding and oviposition (Finch and Collier 2000).

The key hymenopterous parasitoids of the major pests of oilseed rape are highly specific attacking only one or a few host species on the crop. Their search for hosts is a two-part process, first location of the habitat of the host (the crop) and then location of the host within that habitat (Vinson 1985). Cues derived directly from host larvae can be difficult to detect from a distance (Vet et al. 1995). Instead the parasitoids use more easily-detectable indirect olfactory and visual cues from the oilseed rape plant to find the habitat of their host first. Olfactory cues from the plant are termed synomones for the parasitoid: behaviour-controlling chemicals that evoke a response in the receiver (the parasitoid) that is adaptively favourable to both the receiver and the emitter (the plant). Interaction between the pest and its food plant, such as feeding damage, may also release synomones, which differ qualitatively and quantitatively from those emitted by the intact plant (Dicke and Sabelis 1988, Dicke and van Loon 2000). Moreover, the larvae of the major pests live on /in different parts of the oilseed rape plant: the stems, leaf petioles, buds, flowers or pods. To find them, the parasitoids probably use more specific cues from the different parts of the plant. Having located the habitat of their hosts, parasitoids search for their host larvae within it, probably responding to chemosensory cues of lower volatility, as well as to visual and tactile cues. Associative learning of host-finding cues, both olfactory and visual, is important too (Turlings et al. 1993). Cues that elicit no response in näive parasitoids can induce responses after they have been experienced in association with host contact. By associating highly reliable but less detectable host-derived cues with plant-derived cues parasitoids can increase their host encounter rate. As well as seeking host larvae on the crop for oviposition, adult parasitoids also exploit plants for food, often feeding on nectar from the flowers, including those of oilseed rape (Jervis et al. 1993). However, the range of plants exploited for food is poorly known as are the cues that the parasitoids use to locate them.

7.2 Crop Location by Pests and Parasitoids

7.2.1 Olfactory Cues

Cruciferous plants, such as oilseed rape, release a complex mixture of volatiles. The volatile profile is specific to plant species, cultivar, plant part and stage of development, and can vary with environmental conditions, time of day, and in response to damage caused by herbivory. The volatile plant or plant-host odours are carried downwind from the plant, as a kairomone plume in the air and disperse by turbulent diffusion in a spatially-complex manner (Murlis et al. 1992). Over flat terrain, puffs of ionised air, representing an odour plume, will travel at least 25 m from a source (Murlis and Jones 1981).

The green parts of the plant emit general 'green leaf volatiles' (acetates, aldehydes, aliphatic alcohols, esters, ketones and terpenes), as well as specific volatiles (isothiocyanates) from the breakdown of glucosinolates, defensive secondary compounds typical of plants of the family Brassicaceae (Kjaer 1976, Larsen 1981). The type of glucosinolates present, their distribution and concentration varies with plant species, cultivar and plant part (Clossais-Besnard and Larher 1991); this affects the volatile composition emitted.

Flowering oilseed rape emits fatty-acid derivatives, terpenoids, benzenoids, sulphides and nitrogen-containing compounds (including isothiocyanates) (Tollsten and Bergström 1988, Evans and Allen-Williams 1992, Robertson et al. 1993, Blight et al. 1995) and may do so rhythmically (Jakobsen et al. 1994). Jönsson et al. (2005) compared the volatile profiles from oilseed rape in bud and in flower and found that they differed both qualitatively and quantitatively. The dominant volatiles in both were the terpenes: sabinene, myrcene, limonene, and (*E,E*)-α-farnesene. However, of 20 identified compounds, only two ((*Z*)-3-hexenyl acetate and (*Z*)-3-hexenol), were produced in similar amounts from oilseed rape in bud and in flower. The remainder were all released in larger quantities from plants in flower than from those in bud, with some aromatic compounds (benzaldehyde, methylbenzoate, phenyl acetaldehyde and 2-phenyl ethanol) and some nitrogen-containing compounds (indole and benzyl cyanide) released mainly from rape in flower. Cook et al. (2007c) also found more (*E,E*)-α-farnesene, indole and phenylacetaldehyde in flower than in bud volatiles.

When an oilseed rape plant is damaged by an herbivorous insect, a systemic increase in indole glucosinolate and decrease in aliphatic glucosinolate concentration occurs (Koritsas et al. 1991). Further, as a result of tissue damage, cell myrosinases are released and hydrolyse glucosinolates to isothiocyanates so that a change occurs in the volatile profile emitted by the plant (Fahey et al. 2001). Jönsson and Anderson (2007) compared the volatile profiles of flowering racemes of oilseed rape plants, some infested and others not infested by the pollen beetle at the bud stage; infested racemes contained pollen beetle larvae. They obtained about 50% more volatiles from the infested than from the non-infested racemes with the former emitting increased amounts of two green leaf volatiles ((*Z*)-3-hexenyl acetate and (*Z*)-3-hexanol), 3-butenyl isothiocyanate, and various sesquiterpenes (β-elemene, α-farnesene, (*E,E*)-α-farnesene and two unidentified compounds). Infested racemes released less indole than uninfested racemes. Changes in the release of plant volatiles caused by herbivory can be important cues for parasitoids searching for their pest hosts (Dicke and van Loon 2000) and could be exploited in crop protection (Turlings and Ton 2006).

7.2.2 Visual Cues

Visual cues from the oilseed rape plant are less specific than the olfactory cues but are nevertheless important in crop location by both pests and their parasitoids. They include plant size, shape and colour, namely the blue-green colour of the vegetative parts in oilseed rape and the yellow colour of the petals during flowering. Plant parts, particularly flowers, may emit visual cues in the ultraviolet range (Horovitz and Cohen 1972), visible to insects but not man (see below). For example, the centres of oilseed rape flowers reflect in the near ultraviolet (350–400 nm) (Wäckers 1994, Omura et al. 1999) while the pollen is ultraviolet-absorbing (Lunau 1996) and therefore contrasts in colour with the petals.

7.2.3 Behavioural Responses to Olfactory and Visual Cues

The behavioural responses of insects to olfactory cues have been studied in the laboratory using olfactometers and wind tunnels, in semi-field arenas such as polytunnels and caged plots, as well as in the field using odour-baited traps and host plants. Studies of the physiological responses of both pests and parasitoids to olfactory cues using electroantennography and single cell recording techniques are not reviewed here; these have given valuable information about the chemicals that insects can detect with their sensilla but insects do not necessarily respond to them behaviourally during crop location.

Wind tunnel studies have shown that, on perceiving host plant odour, insects respond by oriented zigzag flight upwind towards the source (Kennedy 1977, Vet et al. 1995, Cardé 1996, Vinson 1998). This orientation to wind, stimulated by the presence of host-odour is termed host-plant-odour-mediated upwind anemotaxis (Kennedy 1977). The insect may be guided to the source from at least several metres away, or until it can respond visually to it (Drost and Cardé 1992, Bernays and Chapman 1994, Finch and Collier 2000). If it loses contact with the odour plume it makes wide lateral excursions cross-wind with counterturns and loops but without upwind progress until it renews contact with the odour (Kennedy 1983, Cardé 1996). Insects appear to use a subset of components from the volatile profile of the host plant rather than the whole profile as orientation cues. Some insects use volatiles that are species-specific to locate their hosts (e.g., the isothiocyanates) while the majority are thought to use certain ratios of ubiquitous compounds (Bruce et al. 2005).

The behavioural responses of insects to visual cues have been studied using different coloured traps, plant lines with different floral characteristics and olfactometer bioassays presenting colour stimuli. The visual spectrum of insects extends to shorter wavelengths of the daylight spectrum from ultraviolet to yellow-orange (300–650 nm) than that of man (400–700 nm); most insects studied to date have trichromatic vision with photoreceptors often peaking in sensitivity in the ultraviolet, blue and green wavelengths (Briscoe and Chittka 2001). Humans have blue, green and red receptors. As insect and human vision are so different, results from experiments using classification of colours as they appear to man must be treated with caution (Döring and Chittka 2007). Responses of insects to visual stimuli also depend on the colour contrast between the stimulus and its background and the physiological ability of the insect to discriminate between them, as well as on the visual angle of resolution (e.g., see Giurfa et al. 1996, Giurfa and Menzel 1997). This is relatively well-known for the honey bee. For example, the visual angle of resolution for bee-chromatic colours is between 5–15° and a corolla of a 5 cm flower would be detectable from about 45 cm, whereas a large corolla or an inflorescence 12 cm in diameter would be detectable from a distance of around 135 cm (Giurfa et al. 1996).

The behavioural response of an insect species to both olfactory and visual stimuli from its host plant can vary with sex, age, generation, nutritional status, previous experience and time since last oviposition (Bernays and Chapman 1994).

7.2.3.1 Pollen Beetle and Its Parasitoids

Pollen Beetle

The pollen beetle, *Meligethes aeneus* (Fabricius) (Coleoptera: Nitidulidae), is univoltine. Adults overwinter in field margins, woodland and hedgerows, emerging in the spring (March to June). They are polyphagous feeders at this stage and after emergence may feed on pollen from plants of different families for a week or two. However, at favourable temperatures, they seek cruciferous crops, like oilseed rape, for mating and oviposition, usually arriving at the green bud stage. They feed on pollen in the buds and flowers, and lay their eggs in the buds. Development from egg to adult takes about 1 month. The larvae also feed on pollen in the flowers, moving into younger flowers every few days. On maturity, larvae drop to the soil below and pupate. New generation beetles emerge in summer and feed on pollen from plants of many families. Their ovaries do not mature and they do not mate before seeking overwintering sites. For more information about this pest see Williams and Free (1978), Alford et al. (2003) and Williams (Chapter 1 this volume).

The pollen beetle may fly considerable distances between overwintering site and the oilseed rape crop. Marked beetles, released in the spring within an arable/forested landscape, were found to disperse in all directions and were recaptured on oilseed rape crops 200–300 m from the release point within 2 h of release, and on rape crops up to 13.5 km from the release point 10 days after release (Dlabola and Taimr 1965, Taimr et al. 1967).

Fig. 7.1 Experimental plot (20 m diameter) of winter oilseed rape used by Williams et al. 2007a,b to study upwind anemotaxis by pests and parasitoids. Plot is encircled, 5 m from its circumference, by eight window traps to catch pollen beetles alternating with eight Malaise traps to catch parasitoids. Masts with a wind vane and anemometers connected to a datalogger measured wind direction and wind speed (Photo: Ingrid Williams)

Pollen beetles fly upwind towards oilseed rape in the field, probably in response to olfactory cues from the crop, although responses to crop colour may also be involved. Evans and Allen-Williams (1994) released marked pollen beetles in the centre of a circular array (20 m radius) of yellow water traps. In the absence of odour in the traps, those crosswind or downwind from the release point caught most beetles but when traps were baited with rape odour, upwind traps caught most beetles. Williams et al. (2007b) sampled natural populations of beetles daily in spring and summer as they flew towards a plot of winter oilseed rape encircled by eight window traps positioned at different compass points (Fig. 7.1). They recorded the direction and strength of the wind during flight. Correlations between daily catch of beetles into traps and the wind volume into each trap were negative, indicating that flights by both overwintered and new generation beetles towards the plot were upwind (Fig. 7.2).

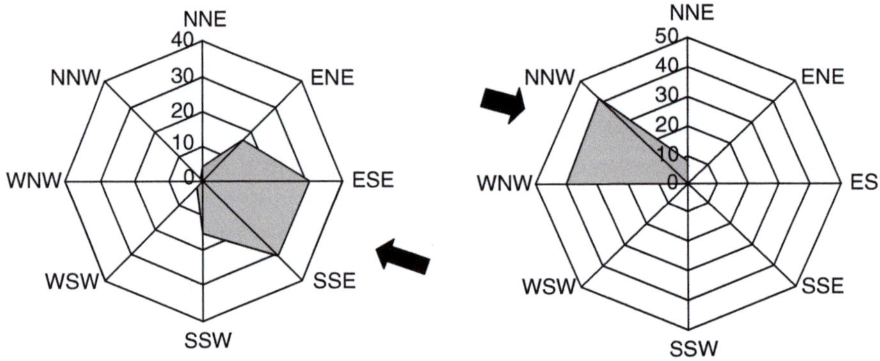

Fig. 7.2 Radar chart showing on the *left*, wind volume (1,000 m^3) into and, on the *right*, the numbers of overwintered-generation pollen beetles caught in the distal sides of window traps, placed at different compass points around a circular plot of winter oilseed rape, on 16 April. *Arrows* show mean wind direction (on *left*) and main direction of pollen beetle flight (on *right*) towards the crop (Based on Williams et al. 2007b)

Adult pollen beetles have a strong attraction to the colour yellow (Wasmann 1926, Nolte 1959, Laska et al. 1986, Buechi 1990) and yellow water and sticky traps are used to trap them in the field (Williams et al. 2003). Blight and Smart (1999) compared catches in different coloured water traps; they found yellow traps to be most attractive, yellow-green and white traps less attractive than yellow, and green, cream and black traps unattractive. Cook et al. (2006a) similarly found that the number of new generation beetles trapped on sticky cards varied with their colour. More beetles (33% of the total trapped) were caught on yellow cards than on white, green, red, blue or black cards (22, 15, 14, 10, and 6% of the total trapped, respectively). White cards caught more beetles than blue or black cards and green cards caught more than black cards.

The preference for the colour yellow attracts pollen beetles to flowering oilseed rape which normally has yellow petals. The influence of petal presence and colour has been investigated by comparing infestation of plant lines differing in petal characteristics (Fig. 7.3) and in plants with flowers dyed different colours. When compared, oilseed rape with yellow petals was more heavily infested than apetalous oilseed rape with no petals (Fig. 7.4, Frearson 2006, Frearson et al. 2006). However,

Fig. 7.3 Semi-field plot of potted oilseed rape plants used to compare the responses of pests and their parasitoids to plant lines with different floral characters as in Frearson et al. (2006) (Photo: Ingrid Williams)

Fig. 7.4 Oilseed rape plant lines with (**a**) yellow-petalled, (**b**) apetalous and (**c**) white-petalled flowers used to compare the responses of pests and their parasitoids to the presence and colour of petals as in Frearson et al. (2006) (Photo: Ingrid Williams)

Fig. 7.4 (Contd.)

this preference was evident even at the green bud stage before petals were visible, suggesting that some petal cues, perhaps odour, are detectable by the beetle at the bud stage. In one study, yellow-petalled lines of oilseed rape were more attractive to pollen beetles than lines of cream flower colour, which in turn were more attractive than white-flowered lines (Giamoustaris and Mithen 1996). However, in another study, they showed no preference for a plant line with yellow petals over an isogenic one with white petals (Fig. 7.4, Frearson 2006, Frearson et al. 2006). Cook et al. (2006a) compared pollen beetle infestation of white-flowered oilseed rape with petals dyed different colours through uptake of a solution of food colourants. In wind tunnel tests, more beetles were recorded on racemes with yellow- and white- and red-petalled flowers than on those with blue-petalled flowers (32, 28, 28, and 12% respectively, of total recorded). Similarly in the field, more beetles colonised whole plants with yellow and white than red and blue petals (44, 24, 18, and 15% respectively, of total recorded). Marking yellow petals with black dots, to simulate the presence of other pollen beetles on them, stimulates pollen beetles to land (Free and Williams 1978).

Pollen beetles, like many other insects, probably also respond to colour in the ultraviolet range. The centres of oilseed rape flowers reflect in the near ultraviolet (350–400 nm) (Wäckers 1994) while the pollen is ultraviolet-absorbing (Lunau 1996) providing a contrast in colour with the petals. However, the responses of pollen beetles to ultraviolet cues have not been studied.

Plant growth stage influences the spatial distribution of pollen beetles on the oilseed rape crop. Although the presence of flowers with yellow petals provides strong cues for the beetle during crop location, and they prefer plants in early-flower to those in bud (Cook et al. 2006b, 2007c), it is the abundance of buds, into which the females lay their eggs, that is an important determinant of residence time on plants and affects their spatial distribution on the crop (Nilsson 1988 a, b, Frearson et al. 2005).

Consistent with their polyphagy, adult pollen beetles are attracted to volatiles of plants from many different families. In the field, they are attracted to traps baited with extracts of plants from the Brassicaceae, such as oilseed rape, *Sinapis arvensis* and *Alliaria petiolata* (Free and Williams 1978, Bartlet 1996). In laboratory bioassays, they have been attracted to volatiles from plants from the Brassicaceae, Solanaceae, Gramineae and Asteraceae (Ruther and Thiemann 1997), Asteraceae and Fabaceae (Cook 2000) and to volatiles from the stamens of plants from Asteraceae, Rosaceae and Papaveraceae (Charpentier 1985). They are attracted to the odour of rape leaves, stems and buds (Görnitz 1953, Nolte 1959, Free and Williams 1978, Evans and Allen-Williams 1994, Ruther and Thiemann 1997, Cook et al. 2007c, Jönsson et al. 2007) as well as to the odour of flowers, floral parts and pollen (Charpentier 1985, Evans and Allen-Williams 1994, Cook et al. 2002). Pollen cues are important in host plant acceptance by the pollen beetle; adults were more numerous on male-fertile oilseed rape plants with pollen than male-sterile hybrids without pollen and more male-fertile than male-sterile buds were accepted for oviposition (Cook et al. 2004a).

Most studies have involved responses to rape odours in various types of olfactometers (Fig. 7.5, Evans and Allen-Williams 1994, Ruther and Thiemann 1997, Cook et al. 2002, Jönsson et al. 2005, Mauchline et al. 2005), to plants in wind tunnels (Fig. 7.7, Cook et al. 2006a, 2007c) or to traps baited with rape odour in the field (Free and Williams 1978, Evans and Allen-Williams 1994, Blight and Smart 1999, Smart and Blight 2000).

Many components of the odour of flowering oilseed rape attract the pollen beetle, including isoprenoids and derivatives of amino acids (such as the isothiocyanates and nitriles) and fatty acids (Table 7.1). Other components are inactive (3-butenenitrile, benzyl alcohol, hexanoic acid, 1-hexanol, (*Z*)-3-hexen-1-ol),

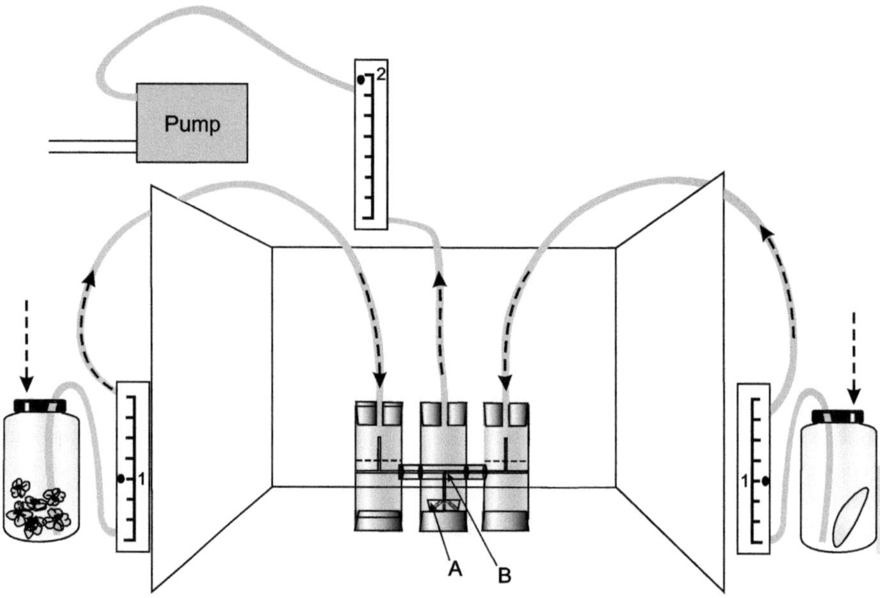

Fig. 7.5 Linear track olfactometer used to test pollen beetle responses to volatiles from oilseed rape and turnip rape. The pump draws air through two odour sources (in bottles behind the screens) and into the olfactometer. *Dotted arrows* show air flow through the apparatus. *A* = holding pot where beetles are introduced. Beetles climb up the vertical wire to the T-junction (*B*) where the two air streams meet and turn either left or right onto the *horizontal* wire into one air stream (Adapted from Cook et al. (2002) with permission from Blackwell Publishing)

(1-pentanol, 1–8-cineole, and (*E,E*)-α-farnesene) or even repellent (1-hexanol, (*Z*)-3-hexen-1-ol, cis-jasmone, 1-octen-3-ol, 1-pentanol) depending on release rate (Smart and Blight 2000). However, these compounds have mostly been tested singly; in the complex floral profile of oilseed rape they may act synergistically or antagonistically to provide host plant location cues for the beetle.

Plant odour can modify the behavioural response of the pollen beetle to colour. Blight and Smart (1999) found that baiting yellow water traps with a mixture of 2-propenyl, 3-butenyl, 4-pentenyl and 2-phenylethyl isothiocyanates enhanced trap catch up to 3.3 times.

The stage of development of the pollen beetle can affect its response to colour and odour cues. Jönsson et al. (2007) compared the responses of overwintered and new generation beetles to oilseed rape odour (bud and flower), colour (green and yellow) and combinations of odour and colour cues in an olfactometer. In the absence of colour, overwintered beetles moved towards bud odour but not towards flower odour, whereas new generation beetles moved towards both. Neither generation however, showed a preference when bud odour was tested against flower odour. In the absence of odour, neither generation showed a preference between green and yellow colour. When odour and colour cues were presented together, the two generations again differed in their responses. Both generations preferred flower odour combined with

Table 7.1 Volatile compounds attractive to the pollen beetle *Meligethes aeneus*

Compound	References
Nitrogenous amino acid derivatives	
2-Propenyl isothiocyanate	Free and Williams (1978), Lerin (1984), Evans (1991), Smart et al. (1995)
2-Phenylethyl isothiocyanate	Smart et al. (1995)
3-Butenyl isothiocyanate	Smart et al. (1995)
4-Pentenyl isothiocyanate	Smart et al. (1995)
Benzyl isothiocyanate	Smart and Blight (2000)
Indole	Smart and Blight (2000), Cook et al. (2007b)
Phenylacetonitrile	Smart and Blight (2000)
3-Phenylproprionitrile	Smart and Blight (2000)
4-Pentenenitrile	Smart and Blight (2000)
5-Hexenenitrile	Smart and Blight (2000)
Non-nitrogenous amino acid derivatives	
2-Phenylethanol	Smart and Blight (2000)
Benzaldehyde	Smart and Blight (2000)
p-Anisaldehyde	Smart and Blight (2000)
Phenylacetaldehyde	Smart and Blight (2000), Cook et al. (2007b)
Methyl salicylate	Smart et al. (1995)
Fatty acid derivatives	
Hexanal	Smart and Blight (2000)
(*E*)-2-hexenal	Evans (1991), Smart and Blight (2000)
(*Z*)-3-hexenyl acetate	Smart and Blight (2000)
Isoprenoids	
(*E*)-4,8-dimethyl-1,3,7-nonatriene	Smart and Blight (2000)
Linalool	Smart and Blight (2000)

yellow over green colour alone; overwintered generation beetles preferred bud odour combined with green over yellow alone, whereas new generation beetles did not. Further, only new generation beetles preferred flower odour combined with yellow over bud odour combined with green; thus adding colour stimuli changed their preference towards yellow and flower odour. Olfactory cues from the oilseed rape plant appeared to be relatively more important for overwintered generation beetles which seek buds for oviposition, while colour cues were relatively more important for new generation beetles which seek flowers of many plant families from which to feed.

Parasitoids of the Pollen Beetle

The key parasitoids of the pollen beetle are *Phradis interstitialis* (Thomson), *Phradis morionellus* (Holmgren), and *Tersilochus heterocerus* Thomson (Nilsson 2003, Ulber et al. Chapter 2 this volume) all (Hymenoptera: Ichneumonidae). They migrate to the oilseed rape crop in the spring during flowering. They are endoparasitoids of the eggs/larvae of the pollen beetle and seek their hosts in the buds and flowers of oilseed rape. There is niche segregation between the species; *P. interstitialis* oviposits primarily into beetle eggs within green buds, *P. morionellus* oviposits into larvae within green and yellow buds and open flowers and *T. heterocerus*

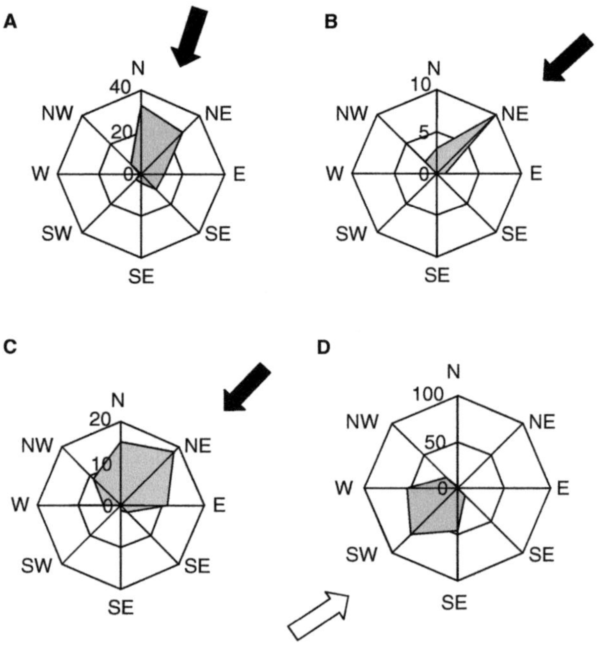

Fig. 7.6 Radar chart showing numbers of (**a**) *Phradis interstitialis*, (**b**) *Platygaster subuliformis*, and (**c**) *Tersilochus obscurator*, caught in the distal sides of Malaise traps, placed at different compass points around a circular plot of winter oilseed rape, and (**d**) the wind volume (1,000 m^3) into each trap, on 27 May. *Arrows* show predominant directions of parasitoid flight (**a**, **b**, **c**) and wind towards the crop (Based on Williams et al. 2007a)

oviposits into second instar larvae within flowers (Osborne 1960, Winfield 1963, Nilsson 2003, Jönsson 2005). All three are univoltine like their host and koinobiont, i.e., the host larva continues to develop with the parasitoid inside it; most parasitoid larval growth and pupation occurs inside the host's pupal chamber in the soil. About a month after pupation, it develops into an adult, which remains inside its cocoon until the following spring or early summer.

Migration to the oilseed rape crop in the spring by all three species of parasitoid is probably by odour-mediated upwind anemotaxis. Natural populations of *P. interstitialis* fly upwind towards oilseed rape probably in response to olfactory cues from the crop (Figs. 7.1 and 7.6a, d, Williams et al. 2007a) and it seems likely that the other key parasitoids respond similarly.

All three key parasitoid species are attracted to odour from oilseed rape in the bud stage when tested in an olfactometer (Jönsson et al. 2005) but species differ in their responses to odour from flowering rape; *T. heterocerus* was attracted towards it while the two *Phradis* species avoided it. When odour from rape in flower was tested against that from rape in the bud stage, *T. heterocerus* preferred flower odour while the *Phradis* spp. preferred bud odour.

The three key parasitoid species differ in their responses to insect-infested oilseed rape plants. Jönsson (2005) compared their responses to odours from buds, infested and non-infested with pollen beetle eggs, in an olfactometer. Both *Phradis* species preferred infested to non-infested buds while *T. heterocerus* showed no preference. He also compared the responses of *P. morionellus* and *T. heterocerus* to odours from flowering oilseed rape, infested and non-infested with pollen beetle larvae. Both species showed a preference for infested flowers. When non-infested flowering rape was compared with air, *T. heterocerus* preferred flowering rape but *P. morionellus* showed no preference. *Phradis morionellus* also showed no preference for infested flowers over air. Of the volatiles released in greater quantities from flowering racemes of oilseed rape infested by pollen beetle larvae than those not infested, (*Z*)-3-hexenylacetate, (*Z*)-3 hexenol, 3-butenyl isothiocyanate and (*E,E*)-α-farnesene were detected by *P. morionellus* antennae in gas chromatography-electroantennodetection analyses (Jönsson and Anderson 2007) and may have a role in host habitat location by this parasitoid.

The colour yellow is a strong visual cue for the parasitoids. In the field, all three key species are caught in yellow water traps (Williams et al. 2003, Johnen et al. 2006). In laboratory olfactometer bioassays using colour cues, all three are attracted more to flower yellow than to bud green (Jönsson et al. 2005).

Comparison of parasitoid incidence and parasitism of beetle larvae on oilseed rape plant lines with different petal characteristics has shown that the yellow petals of oilseed rape flowers are particularly attractive to those parasitoid species seeking host larvae in flowers (Frearson 2006, Frearson et al. 2006). Adult *T. heterocerus* were more numerous on petalous (with yellow petals) than on apetalous (without petals) plants, but neither *P. interstitialis* nor *P. morionellus* showed a preference. However, despite these differences, endoparasitism rates of beetle larvae by *T. heterocerus* and *P. interstitialis* did not differ between petalous and apetalous lines, so larval encounter rate in these species was unaffected by petal presence. Numbers of *P. interstitialis* and *P. morionellus* were similar on plant lines with yellow and white petals. Parasitism of pollen beetle larvae by *P. interstitialis* and *T. heterocerus* were assessed but only *T. heterocerus* showed higher percentage parasitism on the yellow-petalled line than on the white-petalled line.

Colour cues can enhance parasitoid attraction to olfactory cues but responses vary with species. In olfactometer bioassays, when yellow colour and flower odour were presented together, and tested against green colour and bud odour presented together, *P. interstitialis* was attracted equally to both stimuli, *T. heterocerus* showed an increased preference for flower odour, but the addition of green colour did not enhance the preference of *P. morionellus* for bud over flower odour (Jönsson et al. 2005).

These differences in response by the three key parasitoids of the pollen beetle to odour and colour cues from the buds and flowers of oilseed rape, are consistent with the different micro-habitats and host stage of development they seek and are therefore probably involved in microhabitat niche segregation between the species, minimising interspecific competition and facilitating coexistence. *Phradis interstitialis*, which searches and lays its eggs primarily into pollen beetle eggs within green buds, prefers bud to flower odour, avoids flower odour, prefers buds infested to those not infested with pollen beetle eggs and shows no preference for petalous over apetalous flowers or yellow over white petal colour. *Phradis morionellus*, which lays its eggs into larvae within green and yellow buds and open flowers, shows no preference for petalous over apetalous flowers or for yellow over white-petalled flowers. It prefers bud to flower odour and prefers plants infested with pollen beetle eggs and with beetle larvae to those without. *Tersilochus heterocerus*, which lays its eggs into second instar larvae within flowers, is attracted by the odour of flowering rape and prefers it to bud odour. Its preference for flower odour is enhanced by the colour yellow. It prefers petalous to apetalous flowers and has a higher percentage parasitism on yellow than white-petalled flowers. It prefers flowers infested with pollen beetle larvae to those without but does not show a preference for those infested with beetle eggs.

7.2.3.2 Cabbage Seed Weevil and Its Parasitoids

Cabbage Seed Weevil

The cabbage seed weevil, *Ceutorhynchus obstrictus* syn. *C. assimilis* Paykull (Coleoptera: Curculionidae), is univoltine and oligophagous on *Brassica* species. Adults emerge from overwintering sites in spring (May–June) and seek cruciferous plants on which to feed, mate and oviposit. They usually migrate to winter oilseed rape during flowering. The female lays her eggs, usually singly in young pods. The egg hatches 1–2 weeks later. The larva feeds on the developing seeds inside the pod, destroying about five. When mature, it chews a hole through the pod wall, exits through it and drops to the ground to pupate in the soil. New generation weevils emerge in late summer, feed for a week or two on cruciferous plants and then seek overwintering sites in perennial vegetation and leaf litter of field margins and woodlands. The ovaries do not mature and they do not mate before winter diapause. For more information about this pest see Alford et al. (2003) and Williams (Chapter 1 this volume).

Seed weevils may fly considerable distances between their overwintering sites and the oilseed rape crop. Marked weevils have been recaptured up to 5.9 km from a release point within 10 days of release in the spring (Taimr et al. 1967).

Seed weevils are attracted to the odour of oilseed rape and fly upwind towards its source. In olfactometers, they are attracted towards the odour of rape flowers and extracts of rape flowers (Evans and Allen-Williams 1989a, Bartlet et al. 1993, Cook et al. 2006b). In the field, attraction to extracts of oilseed rape was first reported by Görnitz (1953). In a field cage within a winter rape crop, weevils flew upwind in response to rape odour irrespective of wind speed; however, outside the host plant odour plume, weevils moved downwind at windspeeds of 1.5 m/s or more and towards light at lower speeds (Kjaer-Pedersen 1992). Evans and Allen-Williams (1993) released marked weevils in the centre of a circular array of yellow water traps in the field. When traps were baited with extracts of oilseed rape marked weevils were predominantly caught in upwind traps 20 m away demonstrating odour-mediated upwind anemotaxis. When traps were unbaited, female weevils dispersed randomly while male weevils dispersed downwind from the release point.

Response of the seed weevil to specific compounds varies during migration to, colonisation of and dispersal from the crop. Moreover, it seems likely that, in crop location, response is not to individual compounds but to the correct ratios of key components in the volatile profile from the host plant.

The seed weevil is attracted by a range of volatiles emitted by a wide spectrum of plant families, such as phenylacetonitrile, benzyl alcohol, terpenes (α-farnesene, α-pinene, sabinene and caryophyllene) which are major volatiles from flowers, the green leaf volatile, *cis*-3-hexenyl acetate and the aromatic, methyl salicylate (Evans and Allen-Williams 1989a, Evans 1991, Bartlet 1995, Smart and Blight 1997). In olfactometer studies, (*Z*)-3-hexen-1-ol and methyl salicylate attract walking weevils (Bartlet et al. 1997) while hexan-1-ol repels them (Evans and Allen-Williams 1989a). However, in the field, during crop colonisation, (*Z*)-3-hexen-1-ol lowered yellow water trap catch compared to the unbaited trap and 2-phenylethanol, methyl salicylate and hexan-1-ol were neither attractive nor repellent (Smart and Blight 1997).

As a crucifer-specialist, the weevil is also attracted to crucifer-specific glucosinolate metabolites, such as the isothiocyanates and nitriles. In olfactometer studies, weevils walk towards a mixture of 2-phenylethyl, 3-butenyl and 4-pentenyl isothiocyanates but not towards 2-phenylethyl isothiocyanate alone (Bartlet et al. 1993); they are also attracted to phenylacetonitrile, 4-pentenenitrile and 5-hexenenitrile (Bartlet et al. 1997). Combining an isothiocyanate mixture with phenylacetonitrile increased attraction but combining with methyl salicylate did not (Bartlet et al. 1997). In the field, during spring migration to crops, yellow water traps baited with phenylacetonitrile, 2-phenylethyl, 3-butenyl and 4-pentenyl isothiocyanate or with a mixture of these three isothiocyanates plus 2-propenyl isothiocyanate attracted more weevils than unbaited traps (Bartlet et al. 1993, Smart and Blight 1997, Smart et al. 1997). However, in other studies, baiting traps with either 2-propenyl or phenyl isothiocyanates has failed to enhance their attractiveness over unbaited traps (Free and Williams 1978, Lerin 1984, Smart and Blight 1997). The mixture of four isothiocyanates (2-phenylethyl, 3-butenyl, 4-pentenyl and 2-propenyl) and 2-phenylethyl isothiocyanate was not attractive to weevils once they had colonised rape plants (Smart and Blight 1997).

Seed weevils are attracted to the colour yellow, but responses to other colours are inconsistent between studies. Goos et al. (1976) and Laska et al. (1986) caught weevils in both yellow and white water traps. Yellow sticky traps have caught more weevils than light green or dark green traps (Buechi 1990) or blue traps (Ekbom and Borg 1993) whereas white, green and black traps are unattractive (Smart et al. 1997).

Petal absence and colour can affect infestation. In semi-field choice experiments, adult cabbage seed weevil consistently preferred yellow-petalled over white-petalled plants and infested petalous (with yellow petals) more than apetalous (without petals) plants, a preference later reflected by the distribution of weevil larvae in pods (Frearson 2006, Frearson et al. 2006).

Parasitoids of the Cabbage Seed Weevil

The key parasitoids of the cabbage seed weevil are *Trichomalus perfectus* (Walker), *Stenomalina gracilis* (Walker) and *Mesopolobus morys* (Walker), all Hymenoptera: Pteromalidae (Williams 2003a, Ulber et al. Chapter 2 this volume). They are all solitary idiobionts, ectoparasitic on the larva of the weevil and, like their host, are probably univoltine. Females migrate to oilseed rape crops in the spring (April/May) some weeks after their host, and seek pods containing weevil larvae. They pierce the pod wall with the ovipositor and lay a single egg on the host larva. The weevil larva is immobilised and discolours while the parasitoid larva feeds externally from it for seven to 10 days. The parasitoid larva then pupates alongside the host remains without forming a cocoon. On emergence from the pupa, the adult chews its way out through the pod wall. Mating occurs soon after emergence. Adults possibly overwinter in evergreen foliage and sheltered crevices.

The responses of seed weevil parasitoids to olfactory cues from the rape plant have not been studied, but something is known of their response to visual cues. All three parasitoid species are caught in yellow water traps (Williams et al. 2003, Johnen et al. 2006). Host larval encounter rate by seed weevil parasitoids appears unaffected by petal colour or presence. In semi-field choice experiments, ectoparasitism rates of cabbage seed weevil larvae by Pteromalids (probably mainly *T. perfectus*) did not differ between oilseed rape plant lines whose flowers had yellow petals and those with white petals or between those with yellow petals and those without (Frearson 2006, Frearson et al. 2006).

7.2.3.3 Brassica Pod Midge and Its Parasitoids

Brassica Pod Midge

The brassica pod midge, *Dasineura brassicae* (Winnertz) (Diptera: Cecidomyidae) is multivoltine, with two generations on winter rape and one on spring rape. Adults emerge in the spring after overwintering as larvae in cocoons in fields that grew oilseed rape in previous years. They mate at the emergence site and only the mated females migrate to oilseed rape crops. They lay their eggs in batches of 20–30 in pods, often in those previously damaged by the cabbage seed weevil. The eggs hatch after a few days and the larvae feed on the inner pod wall for up to a month. Midge-infested pods split open prematurely, releasing the larvae, which drop to the ground. A proportion pupate immediately, emerging as new generation adults a few weeks later and infest the crop anew, while others enter winter diapause to emerge the following year or even up to 5 years later. For more information about this pest see Alford et al. (2003) and Williams (Chapter 1 this volume).

The ability of mated female pod midge to locate an oilseed rape crop probably depends on the strength and direction of the wind during migration from the emergence site. In spring, female brassica pod midge have been caught in traps mainly downwind from the field in which they overwintered as pupae and in the absence of windbreaks infested crop headlands downwind of the emergence site (Evans and Allen-Williams 1989a, b). Radioactively-labelled midge released in the field dispersed mainly with the wind, but in sheltered areas, females approached and entered oilseed rape fields from adjacent cereal fields by flying upwind in the upper layers of vegetation, consistent with olfactory-mediated upwind anemotaxis (Sylvén 1970).

Brassica pod midge is attracted towards olfactory cues from the oilseed rape crop. The capture efficiency of water traps was increased by the addition of an extract of rape seed with high glucosinolate content (Erichsen and Daebeler 1987).

Response to olfactory stimuli varies with sex and physiological state of the pod midge. In a wind tunnel, mated females fly towards oilseed rape odour (Evans 1991). In olfactometer studies, mated female but not male pod midge were attracted to odour of macerated rape leaves and rape pods (Pettersson 1976), and female but not male midge were attracted to the odour of crushed rape leaves (Williams and Martin 1986). Mated but not virgin females responded positively to rape leaf odour, and more strongly when rape leaf odour was presented together with the yellow colour of rape petals than when either stimulus was presented alone (Evans and Allen-Williams 1989a). Mated female midges were attracted to the green leaf volatile, hexan-1-ol (Evans and Allen-Williams 1989a) and to the potassium salt of the glucosinolate sinigrin in water (Pettersson 1976).

In the field, yellow water traps baited with rape leaf or rape flower extracts caught more pod midge at both emergence and crop sites than unbaited traps, but few midge at the emergence site were mated females, suggesting that after mating, the females are briefly unresponsive to stimuli that later attract them (Evans and Allen-Williams 1989a). Field traps baited with 2-propenyl (allyl) isothiocyanate caught more midge than traps baited with 2-phenylethyl isothiocyanate or unbaited traps when placed in the canopy of a flowering oilseed rape crop (Murchie et al. 1997).

Brassica pod midge respond to visual cues from the oilseed rape plant. Males, virgin females and mated females respond positively to yellow rape flowers in an olfactometer (Evans and Allen-Williams 1989a, b). In semi-field choice experiments comparing different isogenic plant lines infestation by the midge was greater on petalous than on the apetalous plants (Frearson 2006, Frearson et al. 2006).

Parasitoids of the Brassica Pod Midge

The key parasitoids of the brassica pod midge are *Platygaster subuliformis* (Kieffer) (Hymenoptera: Platygastridae) and *Omphale clypealis* (Thomson) (Hymenoptera: Eulophidae) (Williams 2003b, Ulber et al. Chapter 2 this volume). Like their host, both are probably multivoltine. After overwintering in the field where rape was grown the previous year, the adults emerge over an extended period in the spring/early summer and migrate to oilseed rape crops. They seek their hosts in the pods. Both are endoparasitoids and oviposit into their hosts through the pod wall; *P. subuliformis* lays its eggs singly into midge eggs whereas *O. clypealis* lays its eggs singly into midge larvae. *Platygaster subuliformis* is a koinobiont, a parasitoid that allows its host to continue to develop after parasitisation; the parasitoid egg hatches and the larva feeds when its host has developed into a prepupa or pupa within a cocoon in the soil and on maturity, the parasitoid larva pupates within its host's larval skin. *Omphale clypealis* feeds within its host during its larval and pupal stages. In both species, some adults emerge the same year while others enter diapause to emerge the following spring. Mating occurs soon after emergence.

Natural populations of *P. subuliformis* fly upwind towards oilseed rape in the field, probably in response to olfactory cues from the crop. Malaise traps encircling a plot of winter oilseed rape were used to sample adult parasitoids as they flew towards the plot during spring and summer (Fig. 7.1, Williams et al. 2007a). Correlations between daily catch of *P. subuliformis* into traps and wind direction were negative, confirming that flights towards the plot were by upwind anemotaxis (Fig. 7.6b, d). Olfactory cues associated with oilseed rape attract both key species of parasitoid. Both *P. subuliformis* and *O. clypealis*, are, like their host, attracted to field traps baited with isothiocyanates placed in the canopy of a flowering oilseed rape crop (Murchie et al. 1997). However they differ in their responses: traps baited with allyl isothiocyanate caught more female *O. clypealis* than traps baited with 2-phenylethyl isothiocyanate or unbaited traps. Traps baited with 2-phenylethyl isothiocyanate caught more male and female *P. subuliformis* than traps baited with allyl isothiocyanate or unbaited traps.

The parasitoids may also use visual cues in crop location. Both species are caught in yellow water traps (Williams et al. 2003, Johnen et al. 2006). In semi-field choice experiments comparing different isogenic plant lines, adult *P. subuliformis* were more numerous on petalous than on apetalous plants and tended to be more numerous on yellow-petalled than on white-petalled plants, but not always so (Frearson 2006, Frearson et al. 2006).

7.2.3.4 Stem-Mining Pests and Their Parasitoids

Stem-Mining Pests

The three major stem-mining pests of oilseed rape: the cabbage stem flea beetle, *Psylliodes chrysocephala* (L.) (Coleoptera: Chrysomelidae), the rape stem weevil, *Ceutorhynchus napi* Gyllenhal (Coleoptera: Curculionidae) and the cabbage stem weevil, *Ceutorhynchus pallidactylus* (Marsham) (Coleoptera: Curculionidae) are all univoltine and oligophagous on cruciferous plants. They attack the oilseed rape crop in succession. For more information about these pests see Alford et al. (2003) and Williams (Chapter 1 this volume).

The cabbage stem flea beetle migrates to emerging winter oilseed rape crops in early autumn (late August/early September) to feed on the cotyledons and leaves. After mating, the females lay their eggs in soil close to rape plants. Egg-laying continues throughout autumn and winter if the weather is mild. The larvae mine the leaf petioles and later enter the stems and growing points. On maturity in the spring, they leave the plant and pupate in the soil. New generation adults emerge in late spring, feed on the leaves, stems and pods of oilseed rape and other brassicas, and in mid-summer, enter a period of aestivation before becoming active again in the autumn.

Rape stem weevil adults migrate to oilseed rape in early spring (February/March). Females lay their eggs singly in the stems, during stem elongation, close to a terminal bud. Eggs hatch in 1–2 weeks. The larvae feed within the stem for 3–5 weeks before leaving the plant to pupate in the soil. New generation adults remain in their earthen chambers overwinter and emerge the following spring.

The cabbage stem weevil migrates to oilseed rape in early spring (March/April). The females lay their eggs in small groups on the underside of leaf petioles and sometimes in young shoots. The larvae feed for 3–5 weeks, first in the petioles and later in the stems and lateral shoots. On maturity, they exit the plant and pupate in the soil. New generation adults emerge in the summer and feed on cruciferous plants, grazing on the undersides of leaf petioles and veins before entering hibernation.

The responses of the stem-mining pests to odour-baited traps suggest that olfactory cues from the oilseed rape plant are important orientation cues for them. More cabbage stem flea beetle have been caught in the autumn in water traps (grey) that were baited with rape seedlings or a mixture of three isothiocyanates (2-phenylethyl, 3-butenyl and 4-pentenyl) than in unbaited traps (Bartlet et al. 1992). Water traps caught more rape stem weevil when an extract of rape seed with high glucosinolate content was added to the trapping fluid (Erichsen and Daebeler 1987). Addition of 2-phenylethyl isothiocyanate to yellow water traps, yellow sticky traps or fluorescent green cone traps increased catch of the cabbage stem weevil and the rape stem weevil (Walczak et al. 1998). The importance of visual cues is less well understood as comparative studies with colour or other visual cues have not been made with stem weevils.

Parasitoids of the Stem-Mining Pests

The key larval endoparasitoids of cabbage stem flea beetle, cabbage stem weevil and rape stem weevil are *Tersilochus microgaster* (Szepligeti), *Tersilochus obscurator* Aubert and *Tersilochus fulvipes* (Gravenhorst) (all Hymenoptera: Ichneumonidae), respectively (Ulber 2003, Ulber and Williams 2003, Ulber et al. Chapter 2 this volume). Like their hosts, they are univoltine. Overwintered adults emerge in the spring (March/May) from fields in which oilseed rape was grown the previous year. They may feed from nectar-producing plants at field edges before migrating to oilseed rape, shortly before or at the start of flowering. Females antennate the stems and leaf petioles and probe with their ovipositors to find host larvae. They lay their eggs singly into their hosts. The tersilochines are koinobiont: the parasitoid larva hatches but allows its host to continue development. The parasitoid larva remains in its first instar until its host larva is mature and has left the plant to pupate, when it develops rapidly and kills the host prepupa. The adult parasitoid diapauses in its pupal cocoon which lies in the earthen cell formed by the host larva, and emerges the following spring.

Natural populations of *T. obscurator* fly upwind towards oilseed rape in the field, probably in response to olfactory cues from the crop. Malaise traps encircling a plot of winter oilseed rape were used to sample adult *T. obscurator* as they flew towards and away from the plot during spring and summer (Fig. 7.1, Williams et al. 2007a). Correlations between daily catch of *T. obscurator* into traps and wind direction were negative, confirming that flights towards the plot were by upwind anemotaxis (Fig. 7.6c, d). It seems likely that the other species of tersilochine behave in the same way.

The tersilochines are attracted by olfactory cues associated with the oilseed rape crop; both *T. obscurator* and *T. microgaster* are attracted to traps baited with isothiocyanate in the field. At a winter wheat emergence site, traps baited with 2-phenylethyl isothiocyanate have caught more post-diapause males and females than unbaited traps (Ulber and Wedemeyer 2006). However, in crops of oilseed rape, baited and unbaited yellow water traps caught fewer parasitoids than at the emergence site, and the tested concentration of 2-phenylethyl isothiocyanate did not attract either male or female parasitoids, perhaps because the odour was masked by volatiles emitted from the surrounding rape plants (Ulber and Wedemeyer 2006).

Visual cues are also attractive to the tersilochines. All three key *Tersilochus* spp. are caught in yellow water traps (Williams et al. 2003, Johnen et al. 2006) but the sexes appear to respond differently. Ulber and Wedemeyer (2006) found that, in the spring, at a post oilseed rape/winter wheat emergence site, yellow water traps caught 37–60 times more *T. microgaster* and *T. obscurator* males than females, probably indicating that, on emergence, males fly more than females and respond more strongly to the yellow colour of water traps.

7.3 Host Location by Parasitoids

Having located a potential host habitat, the parasitoid searches for its hosts within the habitat. Plant architecture can affect the distribution and accessibility of host larvae within the plant stand and thereby affect the searching efficiency of the parasitoid (Ulber and Fischer 2006).

Kairomones or contact chemicals of low volatility, such as host salivary gland or mandibular gland secretions, host frass, and cuticular secretions may be involved in host location, as may visual, acoustic and tactile cues. In other tritrophic systems, where the effects of plant-derived and host-derived cues have been studied together (Meiners and Hilker 1997, Hilker and Meiners 2002), the volatiles produced by herbivory/oviposition combine with cues from the oviduct secretion on eggs of the host to attract parasitoids (Hilker and Meiners 2002, Hilker et al. 2002, Colazza et al. 2004). Jönsson (2005) identified alkenes and fatty acids from air entrainments of pollen beetle larvae but none elicited responses from the antennae of its larval parasitoid, *P. morionellus*, indicating that they are probably not involved in host location. Some parasitoids perceive vibrations in the microhabitat caused by movement of the host larva and use these to locate the host (vibrotaxis) (Meyhöfer and Casas 1999).

Parasitoids may exhibit various changes in behaviour on finding a host patch: a decrease in walking speed with more time standing still, an increase in turning rate with sharper turns at the edge of the patch, antennation or drumming of the substrate (antennal search) or probing of the substrate with the ovipositor (ovipositor search) (Vet and van Alphen 1985). Recognition of the host by the parasitoid is usually dependent on further host-associated stimuli and is a prerequisite for the release of oviposition behaviour. Host size and shape may be important for acceptance.

The behaviour of *T. perfectus* searching for seed weevil larvae in the pods of oilseed rape followed by host acceptance and oviposition has been described (Dmoch and Rutkowska-Ostrowska 1978, Dmoch 1998) and is consistent with responses to chemosensory cues of low volatility detected by sensilla on the antennae and on the tip of the ovipositor. After cleaning of the antennae, head and thorax, legs, abdomen and wings, searching behaviour by the female consisted of 'radaring' (walking along the pod with antennae outstretched), 'touching', 'drumming', and 'stroking' the pod wall with the antennae, and 'tapping' the pod with the end of the abdomen. If a host larva is not found, these antennal movements are brief and the female moves to another pod. If the pod contains a host larva, the intensity, duration and frequency of the 'drumming' increases until the larva is located. Frass produced by the third instar larva promotes egg-laying; the female pierces the pod with her ovipositor and lays an egg on the larva.

Search modes and recognitions cues have not been studied in other key parasitoids of the pests of oilseed rape.

7.4 Implications for Biocontrol-Based Integrated Pest Management

Understanding the behavioural and chemical ecology underlying crop location by oilseed rape pests and host location by their parasitoids has important implications for integrated management of pests on oilseed rape, particularly those aiming to incorporate conservation biological control and parasitoids (Williams et al. 2005).

Understanding the effect of wind direction on the flight direction of pests (Williams et al. 2007b) and their parasitoids (Williams et al. 2007a) during migration to the oilseed rape crop and their responses to airborne host plant volatiles has potential to inform management of rotations on-farm to manipulate parasitoid populations, improve the precision of forecasting pest and parasitoid arrival and subsequent distributions on a crop and provide decision support for the use of pesticides on the crop. For example, growing oilseed rape upwind of the previous year's crop, where the parasitoids are overwintering, would increase their chances of contacting the odour plume from the crop on emergence and using upwind anemotaxis to locate the new crop and their hosts within it (Williams et al. 2007a). The airborne host plant volatiles used by the pests in host plant location and by the parasitoids in host habitat location can be used to bait traps for monitoring the arrival and abundance of pests and their parasitoids on the crop (Smart et al. 1993, Murchie et al. 1997). The response of parasitoids to plant-derived behaviour-controlling chemicals, may further have potential for attracting them to pest-infested crops. Both pests and parasitoids are usually patchily distributed on the crop and often more concentrated on the downwind side in relation to the direction of the prevailing wind during immigration (Ferguson et al. 2003, 2006) holding potential for the spatial targeting of insecticide without destroying all parasitoids.

'Push-pull' strategies in integrated pest management exploit behaviour-modifying stimuli, particularly those associated with host plant location, to manipulate the distribution of pests and their natural enemies on a crop (Cook et al. 2007b). Attractive stimuli (e.g., preferred host plants, aggregation or sex pheromones, visual, oviposition or gustatory stimulants) are used to 'pull' the pests to a trap crop while repellent and/or deterrent stimuli (e.g., less-preferred host plants, antifeedants, oviposition-deterring, epideictic or alarm pheromones) are used to 'push' pests from the protected crop while concentrating natural enemies on it for biocontrol.

The push-pull strategy being developed for oilseed rape utilises turnip rape (*Brassica rapa* L.) as the trap crop to 'pull' pests (Cook et al. 2006b, 2007b). Turnip rape is preferred to oilseed rape as a host plant for oviposition and/or feeding by the cabbage stem flea beetle, cabbage stem weevil, rape stem weevil, pollen beetle, and cabbage seed weevil (Hokkanen et al. 1986, Hokkanen 1989, 1991, Buechi 1990, Büchi 1995, Nilsson 2004, Barari et al. 2005, Cook et al. 2006b, Cárcamo et al. 2007, Cook et al. 2007c). Visual and olfactory cues related to growth-stage are partly responsible for the preference for turnip rape over oilseed rape; turnip rape grows faster, flowers earlier and has a more attractive colour and odour than

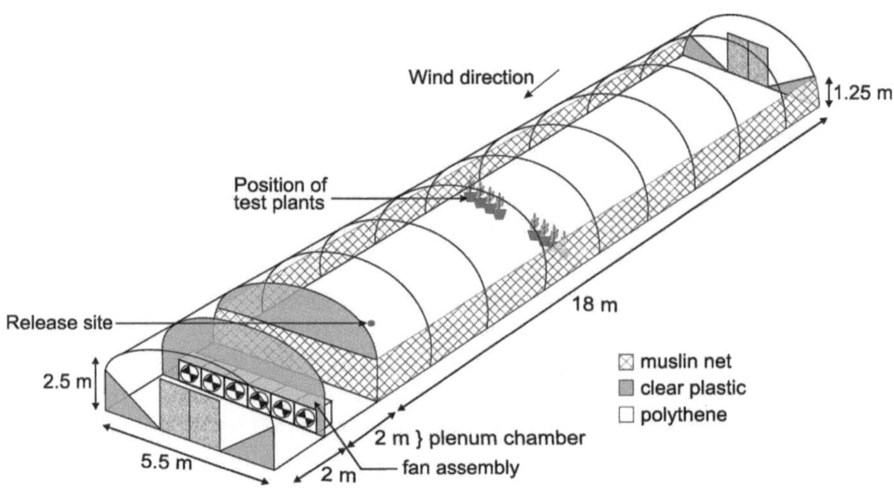

Fig. 7.7 Schematic diagram of polytunnel and layout of bioassay used to investigate the visual and olfactory cues involved in pest attraction to oilseed rape and turnip rape plants (Source: Cook et al. 2007c with kind permission of Springer Science and Business Media)

Fig. 7.8 Replicated field experiment testing the effect of surrounding oilseed rape plots with turnip rape trap crop borders on pest infestation. The oilseed rape is at the green bud stage while the turnip rape is in flower (Photo: Rothamsted Research)

oilseed rape (Fig. 7.7, Buechi 1990, Nilsson 2004, Cook et al. 2007c). Sown as a seed admixture with oilseed rape (Buechi 1990, Büchi 1995, Nilsson 2004) or as a border to the oilseed rape crop (e.g., Fig. 7.8, Cook et al. 2004b), field trials have shown that turnip rape can reduce the abundance of cabbage stem flea beetle (Barari et al. 2005) pollen beetle (Cook et al. 2004b) and seed weevil (Cook et al. 2004b, Cárcamo et al. 2007) in the oilseed rape crop and reduce the need for insecticide. Oil of lavender, *Lavendula angustifolia*, may have potential as a 'push' component on the main crop; it is repellent to the pollen beetle and masks host plant apparency (Mauchline et al. 2005).

Push-pull strategies aim to integrate biological control by natural enemies and components of such strategies should therefore not deter natural enemies. In field trials, there was no evidence that *T. obscurator*, a key parasitoid of the cabbage stem flea beetle, preferred turnip rape to oilseed rape (Barari et al. 2005). Lavender oil volatiles, including two compounds (linalool and linalyl acetate) known to be repellent to the pollen beetle, elicited electrophysiological, but not behavioural (olfactometry) responses from two key parasitoids of the pollen beetle, *P. interstitialis* and *P. morionellus* (Cook et al. 2007a). This suggests that turnip rape trap crops used to concentrate the cabbage stem flea beetle or oilseed rape plants treated with lavender oil to repel the pollen beetle would probably not adversely influence host habitat location by their key parasitoids or their effectiveness for biocontrol.

Advances in plant breeding techniques offer potential for the development of new cultivars of oilseed rape with altered olfactory and visual cues that are less attractive to pests and more attractive to their parasitoids. Increasing the proportion of indolyl glucosinolates, that do not catabolise to isothiocyanates, would make it more difficult for pests to locate the crop while maintaining defence against generalist herbivores (Bartlet et al. 1999b, Cook et al. 2006b). Male-sterile oilseed rape plants may reduce pest infestation (Cook et al. 2004a) and apetalous cultivars have potential for manipulating distributions of inflorescence pests on the crop without negative effects on parasitism (Frearson et al. 2006). Creation of blue- or red-flowered lines of oilseed rape would be less-attractive to pests (Cook et al. 2006a). Less attractive cultivars have potential for incorporation as a 'push' component into the 'push-pull' strategies which are currently under development (Cook et al. 2006b).

Before locating the oilseed rape crop, both pests and parasitoids may feed on other Brassicaceae or plants of other families. Removal of alternative host plants for pests and provision of food plants for parasitoids from the landscape surrounding the crop may be a useful strategy for crop protection (Baggen et al. 1999, Lavandero et al. 2006). In fields with wild flower strips, parasitism of pollen beetle larvae by *Phradis* spp. at 3 m from the crop edge increased over a 3-year period (Büchi 2002) and incidence of parasitisation of the pollen beetle is higher while that of crop damage is lower in oilseed rape crops growing in structurally-complex landscapes compared with those in simple landscapes (Thies and Tscarntke 1999). However, more information is needed about the without-crop foraging habits of both pests and parasitoids before this approach can be incorporated into integrated pest management.

Acknowledgements Writing of this reveiw was supported by the Estonian Targeting Financing Project SF0172655S04 and by the UK Department for Food and Rural Affairs. Rothamsted Research is an institute of the Biotechnology and Biological Sciences Research Council of the United Kingdom.

References

Alford DV, Nilsson C, Ulber B (2003) Insect pests of oilseed rape crops. In: Alford DV (ed.) *Biocontrol of oilseed rape pests*. Blackwell, Oxford.

Baggen LR, Gurr GM, Meats A (1999) Flowers in tri-trophic systems: Mechanisms allowing selective exploitation by insect natural enemies for conservation biological control. *Entomol Exp Appl* 91: 155–161.

Barari H, Cook SM, Clark SJ, Williams IH (2005) Effect of turnip rape (*Brassica rapa*) trap crop on stem-mining pests and their parasitoids in winter oilseed rape (*Brassica napus*). *BioControl* 50: 69–86.

Bartlet E (1995) *Semiochemicals and behaviour of the two oilseed rape pests* Ceutorhynchus assimilis *and* Psylliodes chrysocephala. PhD thesis, University of London, UK.

Bartlet E (1996) Chemical cues to host-plant selection by insect pests of oilseed rape. *Agr Zool Rev* 7: 89–116.

Bartlet E, Blight MM, Hick AJ, Williams IH (1993) The responses of the cabbage seed weevil (*Ceutorhynchus assimilis*) to the odour of oilseed rape (*Brassica napus*) and to some volatile isothiocyanates. *Entomol Exp Appl* 68: 295–302.

Bartlet E, Blight MM, Lane P, Williams IH (1997) The responses of the cabbage seed weevil *Ceutorhynchus assimilis* to volatile compounds from oilseed rape in a linear track olfactometer. *Entomol Exp Appl* 85: 257–262.

Bartlet E, Williams IH, Blight MM, Hick AJ (1992) Response of the oilseed rape pests, *Ceutorhynchus assimilis* and *Psylliodes chrysocephala*, to a mixture of isothiocyanates. In: Menken SBJ, Visser JH, Harrewijn P (eds.) *Proc 8th Int Symp Insect-Plant Relationships*. Kluwer Academic Publishers, Dordrecht, pp 103–104.

Bartlet E, Williams IH, Pickett JA (1999b) The ideal glucosinolate profile for pest resistance in oilseed rape. *IOBC/wprs Bull* 22(10): 13–17.

Bernays EA, Chapman RF (1994) *Host-plant selection by Phytophagous insects*. Chapman and Hall, London, UK.

Blight MM, Pickett JA, Wadhams LJ, Woodcock CM (1995) Antennal perception of oilseed rape, *Brassica napus* (Brassicaceae), volatiles by the cabbage seed weevil *Ceutorhynchus assimilis* (Coleoptera, Curculionidae). *J Chem Ecol* 21: 1649–1664.

Blight MM, Smart LE (1999) Influence of visual cues and isothiocyanate lures on capture of the pollen beetle, *Meligethes aeneus* in field traps. *J Chem Ecol* 25: 1501–1516.

Briscoe AD, Chittka L (2001) The evolution of color vision in insects. *Annu Rev Entomol* 46: 471–510.

Bruce TJA, Wadhams LJ, Woodcock CM (2005) Insect host location: A volatile situation. *Trends Plant Sci* 10: 269–274.

Buechi P (1990) Investigations on the use of turnip rape as trap plant to control oilseed rape pests. *IOBC/wprs Bull* 13(4): 32–39.

Büchi R (1995) Combination of trap plants (*Brassica rapa* var. *silvestris*) and insecticide use to control rape pests. *IOBC/wprs Bull* 18(4): 102–121.

Büchi R (2002) Mortality of pollen beetle (*Meligethes* spp.) larvae due to parasitoids and predators in rape fields and the effect of conservation strips. *Agr Ecosyst Environ* 90: 255–263.

Cárcamo HA, Dunn R, Dosdall LM, Olfert O (2007) Managing cabbage seedpod weevil in canola using a trap crop – A commercial field scale study in western Canada. *Crop Prot* 26: 1325–1334.

Cardé RT (1996) Odour plumes and odour-mediated flight in insects. *Olfaction in Mosquito-Host Interactions*, Ciba Foundation Symp 200, Wiley, UK, pp 54–70.

Charpentier R (1985) Host plant selection by the pollen beetle *Meligethes aeneus*. *Entomol Exp Appl* 38: 277–285.

Clossais-Besnard N, Larher F (1991) Physiological role of glucosinolates in *Brassica napus*. Concentration and distribution pattern of glucosinolates among plant organs during a complete life cycle. *J Agr Food Chem* 56: 25–38.

Colazza S, McElfresh JS, Millar JG (2004) Identification of volatile synomones, induced by *Nezara viridula* feeding and oviposition on bean species that attract the egg parasitoid *Trissolcus basalis*. *J Chem Ecol* 30: 945–964.

Cook S (2000) *The use of pollen cues in resource location by a pollinator and a pest* PhD thesis, The University of Nottingham, UK.

Cook SM, Bartlet E, Murray DA, Williams IH (2002) The role of pollen odour in the attraction of pollen beetles to oilseed rape flowers. *Entomol Exp Appl* 104: 43–50.

Cook SM, Jönsson M, Skellern P, Murray DA, Anderson P, Powell W (2007a) Responses of *Phradis* parasitoids to volatiles of lavender, *Lavendula angustifolia* – a possible repellent for their host, *Meligethes aeneus*. *BioControl* 52: 591–598.

Cook SM, Khan ZR, Pickett JA (2007b) The use of 'push-pull' strategies in integrated pest management. *Annu Rev Entomol* 52: 357–400.

Cook SM, Murray DA, Williams IH (2004a) Do pollen beetles need pollen? The effect of pollen on oviposition, survival, and development of a flower-feeding herbivore. *Ecol Entomol* 29: 164–173.

Cook SM, Rasmussen HB, Birkett MA, Murray DA, Pye BJ, Watts NP, Williams IH (2007c) Behavioural and chemical ecology underlying the success of turnip rape (*Brassica rapa*) trap crops in protecting oilseed rape (*Brassica napus*) from the pollen beetle (*Meligethes aeneus*). *Arthropod-Plant Interactions* 1: 57–67.

Cook SM, Skellern MP, Smith M, Williams IH (2006a) Responses of pollen beetles (*Meligethes aeneus*) to petal colour. *IOBC/wprs Bull* 29(7): 151–160.

Cook SM, Smart LE, Martin JL, Murray DA, Watts WP, Williams IH (2006b) Exploitation of host plant preferences in crop protection strategies for oilseed rape (*Brassica napus*). *Entomol Exp Appl* 119: 221–229.

Cook SM, Watts NP, Hunter F, Smart LE, Williams IH (2004b) Effects of a turnip rape trap crop on the spatial distribution of *Meligethes aeneus* and *Ceutorhynchus assimilis* in oilseed rape. *IOBC/wprs Bull* 27(10): 199–206.

Couty A, van Emden H, Perry JN, Hardie J, Pickett JA, Wadhams LJ (2006) The roles of olfaction and vision in host-plant finding by the diamondback moth, *Plutella xylostella*. *Physiol Entomol* 31: 134–145.

Dicke M, Sabelis MW (1988) Infochemical terminology:based on a cost benefit analysis rather than origin of compounds? *Funct Ecol* 2: 131–139.

Dicke M, van Loon JJA (2000) Multitrophic effects of herbivore-induced plant volatiles in an evolutionary context. *Entomol Exp Appl* 97: 237–249.

Dlabola J, Taimr L (1965) Some results obtained with the application of the tracer method in insect migration and dispersion studies. *Acta Entomol Bohemos* 62: 413–420.

Dmoch J (1998) Kairomones and searching behaviour of *Trichomalus perfectus* Walker. *IOBC/wprs Bull* 21(5): 171–176.

Dmoch J, Rutkowska-Ostrowska Z (1978) Host-finding and host-acceptance mechanism in *Trichomalus perfectus* Walker (Hymenoptera, Pteromalidae). *B Acad Pol Sci* 26: 317–323.

Drost YC, Cardé RT (1992) Use of learned visual cues during habitat location by *Brachymeria intermedia. Entomol Exp Appl* 64: 217–224.

Döring T, Chittka L (2007) Visual ecology of aphids – a critical review on the role of colours in host finding. *Arthropod Plant Interactions* 1: 3–16.

Ekbom B, Borg A (1993) Predators, *Meligethes* and *Phyllotreta* in unsprayed spring oilseed rape. *IOBC/wprs Bull* 16(9): 175–184.

Erichsen E, Daebeler F (1987) Zur Überwachung der Kohlschoten mücke (*Dasyneura brassicae* Winn.) im winterraps. *Nachr für den pflanzenschutzdienst der DDR* 41: 33–34.

Evans KA (1991) *The role of secondary plant metabolites in host-plant location by insect pests of oilseed rape*. Ph.D. thesis, Hatfield Polytechnic, UK.

Evans KA, Allen-Williams LJ (1989a) The response of the cabbage seed weevil (*Ceutorhynchus assimilis* Payk.) and the brassica pod midge (*Dasineura brassica*e Winn.) to flower colour and volatiles of oilseed rape. *Aspects of Applied Biol* 23: 347–353.

Evans KA, Allen-Williams LJ (1989b) Location and infestation of winter oilseed rape by the brassica pod midge, *Dasineura brassicae* Winn. (Diptera, Cecidomyiidae). *Mededelingen van de Faculteit Landbouwwetenschappen Rijkuniversiteit Gent* 54(3a): 717–725.

Evans KA, Allen-Williams LJ (1992) Electroantennogram responses of the cabbage seed weevil, *Ceutorhynchus assimilis*, to oilseed rape, *Brassica napus* spp. *oleifera,* volatiles. *J Chem Ecol* 18: 1641–1659.

Evans KA, Allen-Williams LJ (1993) Distant olfactory response of the cabbage seed weevil, *Ceutorhynchus assimilis*, to oilseed rape odour in the field. *Physiol Entomol* 18: 251–256.

Evans KA, Allen-Williams LJ (1994) Laboratory and field response of the pollen beetle, *Meligethes aeneus*, to the odour of oilseed rape. *Physiol Entomol* 19: 285–290.

Fahey JW, Zalcmann AT, Talalay P (2001) The chemical diversity and distribution of glucosinolates and isothiocyanates among plants. *Phytochemistry* 56: 5–51.

Ferguson AW, Barari H, Warner DH, Campbell JM, Smith ET, Williams IH (2006) Distributions and interactions of the stem miners, *Psylliodes chrysocephala* and *Ceutorhynchus pallidactylus*, and their parasitoids in a crop of winter oilseed rape *(Brassica napus). Entomol Exp Appl* 119: 81–92.

Ferguson AW, Klukowski Z, Walczak B, Clark SJ, Mugglestone MA, Perry JN, Williams IH (2003) Spatial distribution of pest insects in oilseed rape: Implications for integrated pest management. *Agr Ecosyst Environ* 95: 509–521.

Finch S, Collier RH (2000) Host-plant selection by insects – a theory based on 'appropriate/inappropriate landings' by pest insects on cruciferous plants. *Entomol Exp Appl* 96: 91–102.

Frearson DJT (2006) *The movement and host-selection of three inflorescence insect pests of oilseed rape and their parasitoids in relation to host-plant characteristics and environmental conditions*. PhD thesis, University of Reading, UK.

Frearson DJT, Ferguson AW, Campbell JM, Williams IH (2005) The spatial dynamics of pollen beetles in relation to inflorescence growth stage of oilseed rape: Implications for trap crop strategies. *Entomol Exp Appl* 116: 21–29.

Frearson DJT, Ferguson AW, Laugier N, Williams IH (2006) The influence of petals on numbers of pests and their parasitoids in oilseed rape. *Proc Symp Integrated Pest Management in Oilseed Rape, 3–5 April 2006, Göttingen, Germany.*

Free JB, Williams IH (1978) The responses of the pollen beetle, *Meligethes aeneus* and the seed weevil, *Ceuthorynchus assimilis* to oilseed rape, *Brassica napus*, and other plants. *J Appl Ecol* 15: 761–774.

Giamoustaris A, Mithen R (1996) The effect of flower colour and glucosinolates on the interaction between oilseed rape and pollen beetles. *Entomol Exp Appl* 80: 206–208.

Giurfa M, Menzel R (1997) Insect visual perception: Complex abilities of simple nervous systems. *Current Opinion in Neurobiology* 7: 505–513.

Giurfa M, Vorobyev M, Kevan P, Menzel R (1996) Detection of coloured stimuli by honeybees: Minimum visual angles and receptor specific contrasts. *J Comp Physiol A* 178: 699–709.

Goos M, Deptuch S, Raligowska K (1976) Introductory studies on collecting insects using colour traps in field experiments. *Polskie Pismo Entomol* 46: 829–834.

Görnitz K (1953) Untersuchungen über in cruciferen enthaltene insekten-attractivstoffe. *Nachr für den Deutschen Pflanzenschutsdienst N.F.* 7: 81–95.

Hilker M, Meiners T (2002) Induction of plant responses to oviposition and feeding by herbivorous arthropods: A comparison. *Entomol Exp Appl* 104: 181–192.

Hilker M, Rohfritsch O, Meiners T (2002) The plant's response towards insect egg deposition. In: Hilker M, Meiners T (eds.) *Chemoecology of insect egg and egg deposition*. Blackwell, Berlin, Germany.

Hokkanen HMT (1989) Biological and agrotechnical control of the rape blossom beetle *Meligethes aeneus* (Coleoptera, Nitidulidae). *Acta Entomol Fenn* 53: 25–29.

Hokkanen HMT (1991) Trap cropping in pest management. *Annu Rev Entomol* 36: 119–138.

Hokkanen HH, Granlund H, Husberg G-B, Markkula M (1986) Trap crops used successfully to control *Meligethes aeneus* (Col., Nitidulidae), the rape blossom beetle. *Ann Entomol Fenn* 52: 115–120.

Horovitz A, Cohen Y (1972) Ultraviolet reflectance characteristics in flowers of crucifers. *Am J Bot* 59: 706–713.

Jakobsen HB, Friis P, Nielsen JK, Olsen CE (1994) Emission of volatiles from flowers and leaves of *Brassica napus* in situ. *Phytochemistry* 37: 695–699.

Jervis MA, Kidd NAC, Fitton MG, Huddleston T, Dawah HA (1993) Flower-visiting by hymenopterous parasitoids. *J Nat Hist* 27: 67–105.

Johnen A, Williams IH, Ferguson AW, Büchs W, Klukowski Z, Luik A, Nilsson C, Ulber B (2006) MASTER: Construction of phenological models of key parasitoids in Europe and prospects for spray windows compatible with their conservation in winter oilseed rape. *Proc Symp Integrated Pest Management in Oilseed Rape, 3–5 April 2006, Göttingen, Germany.*

Jönsson M (2005) *Responses to oilseed rape and cotton volatiles in insect herbivores and parasitoids*. PhD thesis, University of Agricultural Sciences, Sweden.

Jönsson M, Anderson P (2007) Emission of oilseed rape volatiles after pollen beetle infestation; behavioural and electrophysiological responses in the parasitoid *Phradis morionellus*. *Chemoecology* 17: 201–207.

Jönsson M, Lindkvist A, Anderson P (2005) Behavioural responses in three ichneumonid pollen beetle parasitoids to volatiles emitted from different phenological stages of oilseed rape. *Entomol Exp Appl* 11: 363–369.

Jönsson M, Rosdahl K, Anderson P (2007) Responses to olfactory and visual cues by over-wintered and summer generations of the pollen beetle, *Meligethes aeneus*. *Physiol Entomol* 32: 188–193.

Kennedy JS (1977) Olfactory responses to distant plants and other odour sources. In: Shorey HH, McKelvey JJ (eds.) *Chemical control of insect behaviour*. Wiley, New York.

Kennedy JS (1983) Zigzagging and casting as a response to wind-borne odour: A review. *Physiol Entomol* 8: 109–120.

Kjaer A (1976) Glucosinolates in the cruciferae. In: Vaughan JG, Macleod AJ, Jones BMG (eds.) *The biology and chemistry of the cruciferae*. Academic Press, UK.

Kjaer-Pedersen C (1992) Flight behaviour of the cabbage seedpod weevil. *Entomol Exp Appl* 62: 61–66.

Koritsas VM, Lewis JA, Fenwick GR (1991) Glucosinolate responses of oilseed rape, mustard and kale to mechanical wounding and infestation by cabbage stem flea beetle. *Ann Appl Biol* 118: 209–221.

Larsen PO (1981) Glucosinolates. In: Stump PK, Conn EE (eds.) *The biochemistry of plants*. Academic Press, USA.

Lavandero B, Wratten SD, Didham RK, Gurr G (2006) Increasing floral diversity for selective enhancement of biological control agents: A double-edged sward? *Basic Appl Ecol* 7: 236–243.

Lerin J (1984) Effect de deux isothiocyanates sur les niveaux de capture en cuvettes jaunes d'insectes ravageurs du colza. *Acta Oecol-Oec Appl* 5: 61–70.

Lunau K (1996) Signalling functions of floral colour patterns for insect flower visitors. *Zool Anz* 235: 11–30.

Lāska P, Zelenkovā I, Bičik V (1986) Colour attraction in species of the genera: *Delia* (Diptera, Anthomyidae), *Ceutorhynchus*, *Meligethes*, and *Phyllotreta* (Coleoptera: Curculionidae, Nitidulidae, Chrysomelidae). *Acta Entomol Bohemos* 83: 418–424.

Mauchline AL, Osborne JL, Martin AP, Poppy GM, Powell W (2005) The effects of non-host plant essential oil volatiles on the behaviour of the pollen beetle *Meligethes aeneus*. *Entomol Exp Appl* 114: 181–188.

Meiners T, Hilker M (1997) Host location in *Oomyzus gallerucae* (Hymenoptera: Eulophidae), an egg parasitoid of the elm leaf beetle *Xanthogaleruca luteola* (Coleoptera: Chrysomelidae). *Oecologia* 112: 87–93.

Meyhöfer R, Casas J (1999) Vibratory stimuli in host location by parasitic wasps. *J Insect Physiol* 45: 967–971.

Miller JR, Strickler KL (1984) Finding and accepting host plants. In: Bell WJ, Cardé RT (eds.) *The chemical ecology of insects*. Chapman & Hall, London, UK.

Murchie AK, Smart LE, Williams IH (1997) Responses of *Dasineura brassicae* Winn. (Diptera:Cecidomyiidae) and its parasitoids *Platygaster subuliformis* Kieffer (Hymenoptera: Platygastridae) and *Omphale clypealis* Thomson (Hymenoptera:Eulophidae) to traps, baited with organic isothiocyanates, in the field. *J Chem Ecol* 23: 917–926.

Murlis J, Elkinton JS, Cardé RT (1992) Odour plumes and how insects use them. *Annu Rev Entomol* 37: 505–532.

Murlis J, Jones CD (1981) Fine scale structure of odour plumes in relation to insect orientation to distant pheromone and other sources. *Physiol Entomol* 6: 71–86.

Nilsson C (1988a) The pollen beetle (*Meligethes aeneus* F.) in winter and spring rape at Alnarp 1976–1978. II Oviposition. *Växtskyddsnotiser* 52: 139–144.

Nilsson C (1988b) The pollen beetle (*Meligethes aeneus* F.) in winter and spring rape at Alnarp 1976–1978. III Mortality factors. *Växtskyddsnotiser* 52: 139–144.

Nilsson C (2003) Parasitoids of pollen beetles. In: Alford DV (ed.) *Biocontrol of oilseed rape pests*. Blackwell, Oxford.

Nilsson C (2004) Trap plants to avoid insecticide application against pollen beetles in oilseed rape. *IOBC/wprs Bull* 27(10): 215–221.

Nolte HW (1959) Untersuchungen zum Farbsehen des Rapsglanzkäfers auf Farben und die ökologische Bedeurung des Farbsehens. *Biol Zbl* 78: 63–107.

Omura H, Honda K, Hayashi N (1999) Chemical and chromatic bases for preferential visiting by the cabbage butterfly, *Pieris rapae*, to rape flowers. *J Chem Ecol* 25: 1895–1906.

Osborne P (1960) Observations on the natural enemies of *Meligethes aeneus* (F.) and *M. viridescens* (F.) [Coleoptera: Nitidulidae]. *Parasitology* 50: 91–110.

Pettersson J (1976) Ethology of *Dasyneura brassicae* Winn. (Dipt., Cecidomyidae). 1. Laboratory studies of olfactory reactions to the host plant. *Symp BiolHungarica* 16: 203–208.

Prokopy RJ (1986) Visual and olfactory stimulus interaction in resource finding by insects. In: Payne TL, Birch MC, Kennedy CE (eds.) *Mechanisms in insect olfaction*. Oxford University Press, New York, USA.

Prokopy RJ, Owens ED (1983) Visual detection of plants by herbivorous insects. *Annu Rev Entomol* 28: 337–364.

Robertson GW, Griffiths DW, MacFarlane Smith W, Butcher RD (1993) The application of thermal-dissorption-gas chromatography-mass spectrometry to the analysis of five varieties of oilseed rape (*Brassica napus* spp. *oleifera*). *Phytochem Analysis* 4: 152–157.

Ruther J, Thiemann K (1997) Response of the pollen beetle *Meligethes aeneus* to volatiles emitted by the intact plants and conspecifics. *Entomol Exp Appl* 84: 183–188.

Smart LE, Blight MM (1997) Field discrimination of oilseed rape, *Brassica napus* volatiles by cabbage seed weevil, *Ceuthohynchus assimilis*. *J Chem Ecol* 23: 2555–2566.

Smart LE, Blight MM (2000) Response of the pollen beetle, *Meligethes aeneus*, to traps baited with volatiles from oilseed rape, *Brassica napus*. *J Chem Ecol* 26: 1051–1064.

Smart LE, Blight MM, Hick AJ (1993) Development of a monitoring system for the cabbage seed weevil and the pollen beetle. *IOBC/wprs Bull* 16(10): 351–354.

Smart LE, Blight MM, Hick AJ (1997) Effect of visual cues and a mixture of isothiocyanates on trap capture of cabbage seed weevil, *Ceuthorhynchus assimilis*. *J Chem Ecol* 23: 889–902.

Smart LE, Blight MM, Ryan J (1995) Response of pollen beetles, *Meligethes* spp., to volatiles from *Brassica napus*. *Proc 9th Int Rapeseed Congress, Cambridge, UK, 4–7 July 1995*, pp 1040–1042.

Sylvén E (1970) Field movement of radioactively labeled adults of *Dasyneura brassicae* Winn. (Dipt., Cecidomyiidae). *Entomol Scand* 1: 161–187.

Taimr L, Sedivy J, Bergmannova E, Hanker I (1967) Further experience obtained in studies on dispersal flights of *Meligethes aeneus* F., marked with P^{32} (Coleoptera). *Acta Entomol Bohemos* 64: 325–332.

Thies C, Tscarntke T (1999) Landscape structure and biological control in agroecosystems. *Science* 285: 893–895.

Tollsten L, Bergström G (1988) Headspace volatiles of whole plants and macerated plant parts of *Brassica* and *Sinapis*. *Phytochemistry* 27: 2073–2077.

Turlings TCJ, Ton J (2006) Exploiting scents of distress: The prospect of manipulating herbivore-induced plant odours to enhance the control of agricultural pests. *Curr Opin Plant Biol* 9: 421–427.

Turlings TCJ, Wäckers FL, Vet LEM, Lewis WJ, Tumlinson JH (1993) Learning of host-finding cues by hymenopterous parasitoids. In: Papaj DR, Lewis AC (ed.) *Insect learning*. Chapman & Hall, New York, USA.

Ulber B (2003) Parasitoids of ceutorhynchid stem weevils. In: Alford DV (ed.) *Biocontrol of oilseed rape pests*. Blackwell, Oxford.

Ulber B, Fischer K (2006) Effect of plant density and host plant architecture on the abundance and within-plant distribution of stem-mining pests and the level of parasitism. *Proc Int Symp Integrated Pest Management in Oilseed Rape, 3–5 April 2006, Göttingen, Germany*.

Ulber B, Wedemeyer R (2006) Responses of *Trichomalus microgaster* and *Trichomalus obscurator* (Hymenoptera: Ichneumonidae) to volatile 2-phenyl-isothiocyanate. *Proc Symp Integrated Pest Management in Oilseed Rape, 3–5 April 2006, Göttingen, Germany*.

Ulber B, Williams IH (2003) Parasitoids of flea beetles. In: Alford DV (ed.) *Biocontrol of oilseed rape pests*. Blackwell, Oxford.

Vet LEM, Lewis WJ, Cardé RT (1995) Parasitoids foraging and learning. In: Cardé RT, Bell WJ (eds.) *Chemical ecology of insects*. Chapman & Hall, London, UK.

Vet LEM, van Alphen JJM (1985) A comparative functional approach to the host detection behaviour of parasitic wasps. I. A quantitative study of 'Eucoilidae and Alysiinae'. *Oikos* 44: 478–486.

Vinson SB (1985) The behaviour of parasitoids. In: Kerkut GA, Gilbert LI (eds.) *Comprehensive insect physiology, biochemistry and pharmacology vol. 8*. Pergamon Press, Oxford, UK.

Vinson SB (1998) The general host selection behaviour of parasitoid Hymenoptera and a comparison of initial strategies utilized by larvaphagous and oophagous species. *Biol Control* 11: 79–96.

Wäckers FL (1994) The effect of food deprivation on the innate visual and olfactory preferences in the parasitoid *Cotesia rubecula. J Insect Physiol* 40: 641–649.

Walczak B, Kelm M, Klukowski Z, Smart LE, Ferguson AW, Williams IH (1998) The effect of trap design and 2-phenylethyl isothiocyanate on catches of stem weevils (*Ceuthorynchus pallidactylus* Marsh. and *C. napi* Gyll.) in winter oilseed rape. *IOBC/wprs Bull* 21(5): 141–146.

Wasmann E (1926) Versuche über den Farbensinn des Rapskäfers (*Meligethes aeneus* L.). *Zeit den Wissenschaftlichen Insektenbiologie* 21: 147.

Williams IH (2003a) Parasitoids of cabbage seed weevil. In: Alford DV (ed.) *Biocontrol of oilseed rape pests*. Blackwell, Oxford.

Williams IH (2003b) Parasitoids of brassica pod midge. In: Alford DV (ed.) *Biocontrol of oilseed rape pests*. Blackwell, Oxford.

Williams IH, Büchs W, Hokkanen H, Menzler-Hokkanen I, Johnen A, Klukowski Z, Luik A, Nilsson C, Ulber B (2005) MASTER – Integrating biological control within IPM for winter oilseed rape across Europe. *Proc BCPC Int Cong, Crop Science & Technology, Glasgow, 31 October–2 November 2005,* 1: 301–308.

Williams IH, Buechi R, Ulber B (2003) Sampling, trapping and rearing oilseed rape pests and their parasitoids. In: Alford DV (ed.) *Biocontrol of oilseed rape pests*. Blackwell, Oxford.

Williams IH, Frearson DJT, Barari H, McCartney A (2007a) First field evidence that parasitoids use upwind anemotaxis for host habitat location. *Entomol Exp Appl* 123: 299–307.

Williams IH, Frearson D, Barari H, McCartney A (2007b) Migration to and dispersal from oilseed rape by the pollen beetle, *Meligethes aeneus,* in relation to wind direction. *Agri Forest Entomol* 9: 279–286.

Williams IH, Free JB (1978) The feeding and mating behaviour of pollen beetles (*Meligethes aeneus* Fab.) and seed weevils (*Ceutorhynchus assimilis* Payk.) on oil-seed rape (*Brassica napus* L.). *J Agr Sci* 91: 453–459.

Williams IH, Martin AP (1986) Evidence for a female sex pheromone in the brassica pod midge, *Dasineura brassicae. Physiol Entomol* 11: 353–356.

Winfield AL (1963) A study on the effects of insecticides on parasites of larvae of blossom beetles (*Meligethes aeneus* F. Coleoptera: Nitidulidae). *Entomol Exp Appl* 6: 309–318.

Chapter 8
Spatio-Temporal Distributions of Pests and Their Parasitoids on the Oilseed Rape Crop

Ingrid H. Williams and Andrew W. Ferguson

Abstract Good co-incidence in time and space between parasitoids and the pests they attack is essential for effective biocontrol. The six major pests of oilseed rape in Europe migrate to the crop in succession, attacking the crop at various growth stages and damaging different parts of the plant. Research on the phenologies and within-field spatial distributions of the pests and their key parasitoids on oilseed rape crops is reviewed. Pest distributions are non-uniform, with differing irregular patterns of aggregation, often edge-distributed. Parasitoid distributions do not necessarily completely reflect those of the host stage they attack. Crop and environmental factors affecting these distributions are discussed. The spatio-temporal distributions have implications for biocontrol-based integrated management of the pests. These include accurate sampling for pest monitoring, the temporal and spatial targeting of pests with insecticide to avoid killing parasitoids and strategies for manipulating pest and parasitoid distributions through the siting, layout and surround of the crop to enhance parasitoid populations and increase their effectiveness.

8.1 Introduction

Information on the distributions of pests and their parasitoids on any crop, both within-field and within-plant, and the factors that influence them, are invaluable for the development of effective biocontrol-based integrated pest management on that crop. This is particularly so for a push-pull strategy using a trap crop, where the aim is to alter the distribution of pests and/or their parasitoids on the crop.

Initial colonisation of a rape crop by pests and their parasitoids is influenced largely by the location of overwintering sites and the direction of migratory flights from these to the crop (see Williams and Cook Chapter 7 this volume). Spread

I.H. Williams (✉)
Institute of Agricultural and Environmental Sciences, Estonian University of Life Sciences, Tartu 51014, Estonia
e-mail: ingrid.williams@bbsrc.ac.uk

I.H. Williams (ed.), *Biocontrol-Based Integrated Management of Oilseed Rape Pests*,
DOI 10.1007/978-90-481-3983-5_8,

within the crop is influenced by temperature, population size, crop size, the growth stage and structure of plants within the crop and their infestation.

Parasitism of pest larvae within the plants is influenced by the size, growth stage and architecture of the plants, the relative phenologies of both parasitoids and their host larvae and the spatio-temporal within-plant distributions of host larvae. Understanding these interactions informs crop husbandry practices, such as cultivar choice, crop layout, sowing density, row width and interplant distances which alter crop canopy structure and composition and have potential to decrease pest attack and to enhance parasitoid efficacy by increasing the vulnerability of pest larvae to attack by parasitoids.

8.2 Within-Field Spatio-Temporal Distributions

The six major pests of winter oilseed rape in Europe, namely the cabbage stem flea beetle (*Psylliodes chrysocephala* (Linnaeus), Coleoptera: Chrysomelidae), the pollen beetle (*Meligethes aeneus* (Fabricius), Coleoptera: Nitidulidae), the cabbage seed weevil (*Ceutorhynchus obstrictus* (Marsham) syn. *C. assimilis* (Paykull), Coleoptera: Curculionidae), the cabbage stem weevil (*Ceutorhynchus pallidactylus* (Marsham), syn. *C. quadridens* (Panzer), Coleoptera: Curculionidae), the rape stem weevil (*Ceutorhynchus napi* Gyllenhal, Coleoptera: Curculionidae) and the brassica pod midge (*Dasineura brassicae* Winnertz, Diptera: Cecidomyidae) migrate to the crop in succession, attacking it at various growth stages and damaging different parts of the plant (Alford et al. 2003, Williams Chapter 1 this volume).

The spatial distribution pattern for a single insect species has, in the past, often been measured using the relationship between variance and mean (Taylor 1984, Clark and Perry 1994) and associations between species have been measured by the correlation coefficient (Murchie 1996). These measures do not utilize information about the location of insect counts. In the past decade, the development of novel spatial statistics, notably Spatial Analysis by Distance Indices (SADIE, Perry 1995, 1998a and b) has enabled the spatial information in a two-dimensional array of sample counts to be used as part of the analysis, giving a much more detailed and informative picture of the pattern of crop colonization by an insect species, and further has enabled any associations between different species to be analysed.

8.2.1 Pollen Beetle and Its Parasitoids

The pollen beetle is univoltine. Adults emerge from overwintering sites in spring (March–June). They feed on flowers from different plant families for a few weeks during which the ovaries mature and mating takes place. When temperatures exceed 12–15°C, they migrate to oilseed rape. They lay their eggs in the buds. There are two larval instars, the first developing for 5–10 days within the bud and the second for about 14 days in open flowers, moving up the raceme to younger flowers. When mature, they drop to the soil to pupate. New generation adults emerge 1–5 weeks later. They feed in the flowers of a diversity of plant families for a few weeks and then seek overwintering sites in moist debris in woodland. For more detailed accounts of this pest see Alford et al. (2003) and Williams (Chapter 1 this volume).

The key parasitoids of the pollen beetle are *Phradis interstitialis* (Thomson), *Phradis morionellus* (Holmgren), and *Tersilochus heterocerus* Thomson; all are Hymenoptera: Ichneumonidae and, like their host, are univoltine (Nilsson 2003, Ulber et al. Chapter 2 this volume). They migrate to the oilseed rape crop in the spring. *Phradis interstitialis* is the first to appear on the crop, doing so while the pollen beetle is colonising, followed by *T. heterocerus* and *P. morionellus* which arrive when pollen beetle larvae are present (Fig. 8.1). There is niche segregation between the species; *P. interstitialis* oviposits primarily into beetle eggs within green buds, *P. morionellus* oviposits into larvae within green and yellow buds and open flowers and *T. heterocerus* oviposits into second instar larvae within flowers (Osborne 1960, Winfield 1963, Nilsson 2003, Jönsson 2005).

Pollen beetles first infest plants at the edges of a rape crop and later those nearer the centre. Sampling along line transects across many crops of both winter and spring rape, or from points at their edges and centres, Free and Williams (1979a) showed that the initial infestation of crop edges was irrespective of the presence of windbreaks. The edge effect was more pronounced on large than on small crops. Later the beetle populations spread towards crop centres and the proportion at the edges diminished. Pollen beetle larvae were more evenly distributed across the crop

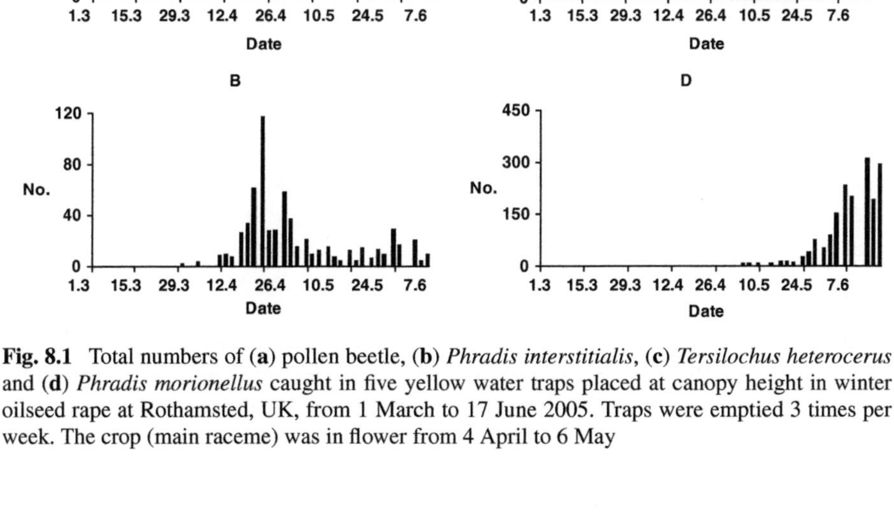

Fig. 8.1 Total numbers of (**a**) pollen beetle, (**b**) *Phradis interstitialis*, (**c**) *Tersilochus heterocerus* and (**d**) *Phradis morionellus* caught in five yellow water traps placed at canopy height in winter oilseed rape at Rothamsted, UK, from 1 March to 17 June 2005. Traps were emptied 3 times per week. The crop (main raceme) was in flower from 4 April to 6 May

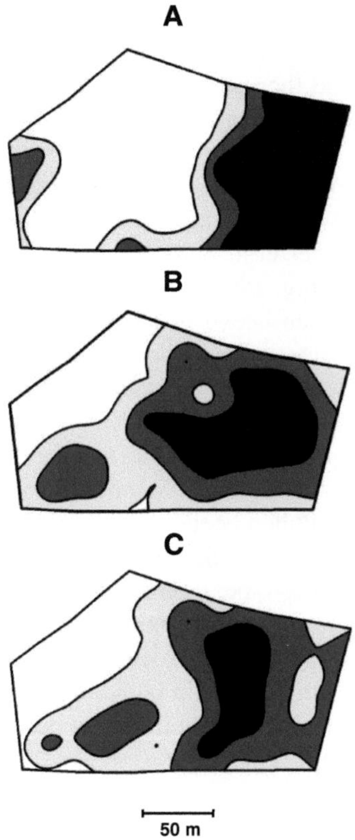

Fig. 8.2 SADIE plots of distributions of pollen beetle in a winter oilseed rape crop at Rothamsted, UK, in 1999. (**a**) overwintered adults beaten from plants, (**b**) larvae collected into water traps as they dropped from plants to pupate in the soil and (**c**) new generation adults caught on emergence from the soil in emergence traps. All insects were sampled at 40 locations across the field. *Black* and *white areas* are identified by SADIE as clusters and gaps in insect distributions, respectively. In *dark grey* and *pale grey areas* the distributions show tendencies towards clustering and gappiness, respectively

than the adults. Sampling at 40 locations across a winter rape crop, Ferguson et al. (2003a, b) found that overwintered pollen beetles mainly colonised the downwind half of a field and that their distribution, that of their larvae and of the resulting new generation to emerge in the summer remained clustered in the downwind part of the field and were associated with each other; the upwind half of the field remained relatively uninfested (Fig. 8.2).

Ferguson et al. (2003a) also compared the spatial distributions of pollen beetle larvae and larvae parasitized by *P. interstitialis* and *T. heterocerus* on the crop. They found that the distribution of pollen beetle larvae parasitized by the larvae of *P. interstitialis* showed irregular patterns of aggregation and was associated with that of their host larvae (Fig. 8.3); by contrast, pollen beetle larvae containing the eggs of *T. heterocerus* were spread evenly across the crop and showed little pattern.

The arrival points of the pollen beetle and its parasitoids during the crop colonisation phase are probably largely determined by wind direction during their migratory flights. Both the beetle and *P. interstitialis* use upwind anemotaxis to locate oilseed rape, flying upwind towards it probably in response to olfactory cues from the crop

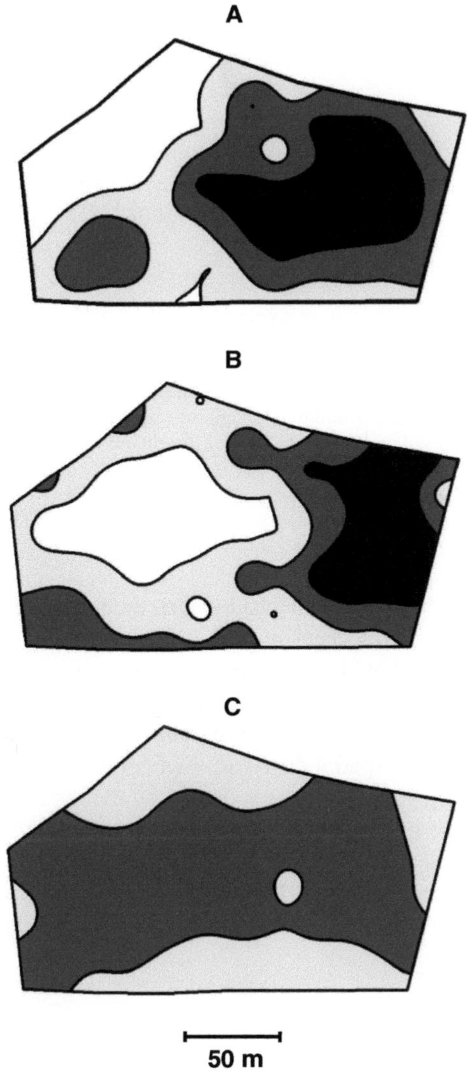

Fig. 8.3 SADIE plots of distributions of (**a**) all pollen beetle larvae, (**b**) larvae parasitized by *P. interstitialis* and (**c**) larvae parasitized by *T. heterocerus* in a winter oilseed rape crop at Rothamsted, UK, in 1999. Larvae were collected in water traps as they dropped from plants to the soil to pupate and parasitism was assessed by dissection. Other details as Fig. 8.2

(Williams et al. 2007a, b, Fig. 8.4), and it seems likely that other parasitoids of the pollen beetle also do so.

On arrival at or near the crop, visual cues probably aid orientation and stimulate the beetles to land. They show a strong visual landing response to the presence of plants with yellow flowers in the crop (Cook et al. 2006a, 2007b), but it is the abundance of buds, into which the females lay their eggs, that is an important determinant of residence time on plants (Frearson et al. 2005). Hence, during flowering, the search by female pollen beetles for buds of the right size (2–3 mm) in which

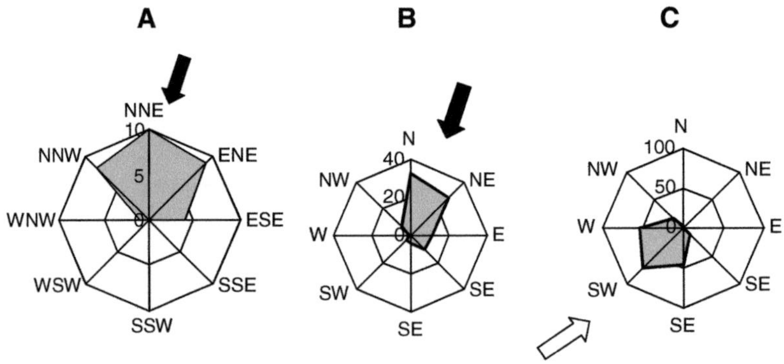

Fig. 8.4 Radar charts showing numbers of (**a**) pollen beetle caught in window traps (**b**) *Phradis interstitialis* caught in Malaise traps and (**c**) the wind volume (1,000 m^3) into each trap, on 27 May. Traps were placed at different compass points around a circular plot of winter oilseed rape at Rothamsted, UK. *Arrows* show predominant directions of insect flight (**a**, **b**) and wind (**c**) towards the crop

to oviposit probably induces them to spread from the edges towards the centres of the crop. Ferguson et al. (2003b) found a positive relationship between the numbers of pollen beetle larvae in winter rape plants and crop growth stage, suggesting that adults arrived early in plant development relative to their preferred growth stage and sought the most mature plants for oviposition.

The spatial distribution of parasitoids of the pollen beetle on the crop after arrival, is probably influenced most by the distribution of their hosts and how efficient they are in their search for them. The three key species, *P. interstitialis*, *P. morionellus* and *T. heterocerus*, search for their hosts in different micro-habitats within the flowering canopy and differ in their responses to olfactory and visual cues from the buds and flowers that contain their host eggs/larvae (see Williams and Cook Chapter 7 this volume).

8.2.2 Cabbage Seed Weevil and Its Parasitoids

The cabbage seed weevil is univoltine. Adults emerge from overwintering sites in the spring (April-June) and migrate to rape crops or early flowering cruciferous weeds when temperatures exceed 15 °C. The eggs are laid singly into young pods and under field conditions, hatch after 5–13 days. Each larva consumes about five seeds, reducing pod yield by about 18%. On maturity, the larva chews a hole in the pod wall, exits through it and drops to the soil to pupate. New generation adults emerge from pupation in late summer, often after winter rape has been harvested. They feed for a few weeks on cruciferous plants and then seek overwintering sites in vegetation and the leaf litter of field margins and woodland. For more detailed accounts of this pest see Alford et al. (2003) and Williams (Chapter 1 this volume).

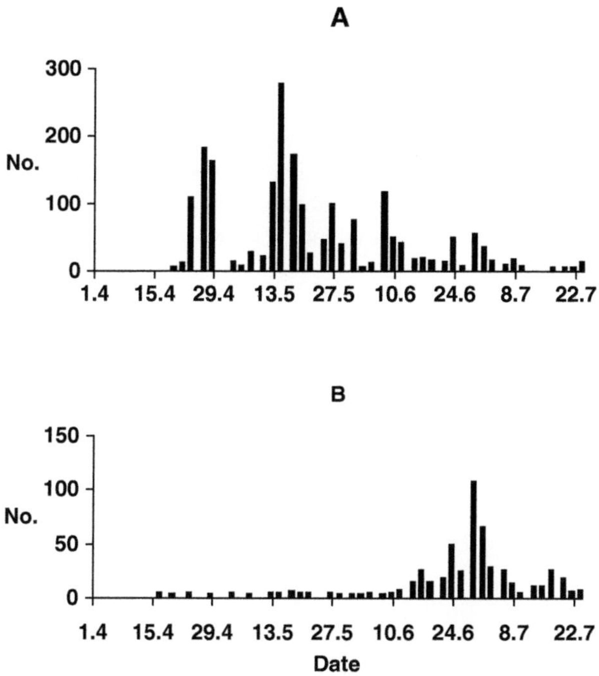

Fig. 8.5 Total numbers of (**a**) cabbage seed weevil and (**b**) *Trichomalus perfectus* caught in five yellow water traps placed at canopy height in winter oilseed rape at Rothamsted, UK, from 1 April to 23 July 2004. Traps were emptied three times per week. The crop (main raceme) was in flower from 12 April to 19 May

Trichomalus perfectus (Walker) (Hymenoptera: Pteromalidae) is the most widely distributed and important larval ectoparasitoid of the cabbage seed weevil in Europe (Williams 2003, Ulber et al. Chapter 2 this volume). Like its host it is univoltine. It migrates to the crop in the spring, some weeks later than the weevil, and when its targets for parasitisation, the second and third instar larvae of the weevil, are feeding within the pods (Fig. 8.5).

The spatial distribution of the cabbage seed weevil on oilseed rape crops is complex and dynamic, changing throughout its phases of immigration to, oviposition on and emigration from the crop. It is more abundant at the edges than at the centres of both winter and spring crops, irrespective of wind breaks; the edge distribution is particularly marked during immigration and relatively greater on large than on small crops (Risbec 1952, Thiem 1970, Kühne 1977, Free and Williams 1979a, Murchie et al. 1999b, Ferguson et al. 2000).

Free and Williams (1979a) found that, following the immigration phase (April/May), the proportion of cabbage seed weevil at the edge, gradually diminished as weevils moved further into the crop (June/July) and pods with seed weevil larvae were more evenly distributed over the crop than the adults. Free and Williams (1979b) found that when infestation was small, the edges of a crop tended to be relatively more heavily infested than crop centres, but with heavier infestations, centres could become more heavily infested than edges, presumably reflecting the overlapping of invading pest populations from opposite edges as they converged at the crop centre. Ferguson et al. (2000) sampled from the nodes of a rectangular grid across a winter rape crop and obtained a two-dimensional picture of crop colonisation. They confirmed that adults weevils first infested the edge of the crop (Fig. 8.6); however, the population was irregular and aggregated at all times and large areas of the crop, including some edges, remained relatively unpopulated.

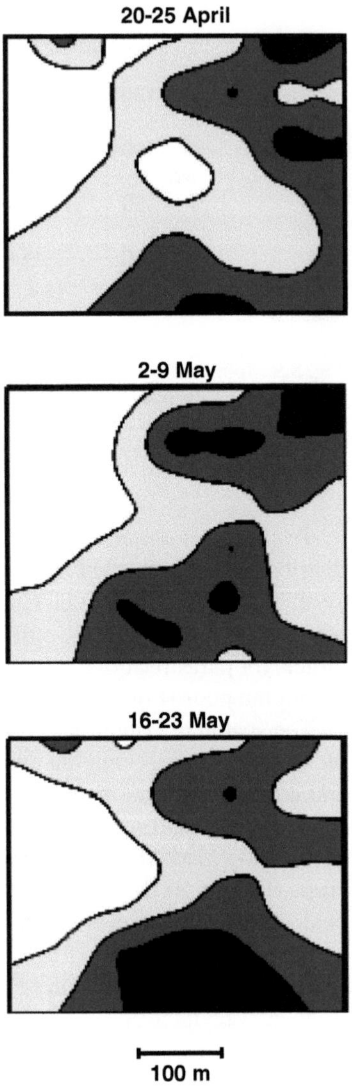

Fig. 8.6 SADIE plots of distributions of cabbage seed weevil in 36 yellow flight traps at crop canopy height on three dates during their immigration to a winter oilseed rape crop at Woburn, UK, in 1995. Other details as Fig. 8.2

The distribution patterns of male and female cabbage seed weevils have been found to be similar and strongly associated, particularly during immigration (early May to early June) when they are most abundant (Ferguson et al. 2000) although Murchie et al. (1999b) caught relatively more of the females than of the males at the edge of the crop and suggested that, once they start to oviposit, females may fly less than males. Ferguson et al. (2000) found that, although the distributions of females and of weevil larvae were spatially associated, they were not coincident in all parts of the crop. The availability of suitable pods for egg-laying probably influences the movement of fecund females on the crop; Ferguson et al. (2003b) found that the numbers of adults was inversely related to main raceme growth stage during flowering, when females would have been searching for young pods into which to oviposit. During the emigration phase, adult numbers declined simultaneously in all parts of the crop (Ferguson et al. 2000).

The spatial distributions of *T. perfectus* larvae and seed weevil larvae coincide but those of their adults caught in flight traps may not. Murchie et al. (1999b) found that the adult parasitoid had an edge distribution on a winter oilseed rape crop only during its early migration, whereas its adult host was strongly edge-distributed thoughout its immigration phase. By contrast, spatial distributions of *T. perfectus* larvae and seed weevil larvae were found to be associated (Ferguson et al. 2000, Fig. 8.7). The proportion of host larvae parasitized was uniform over the crop; there

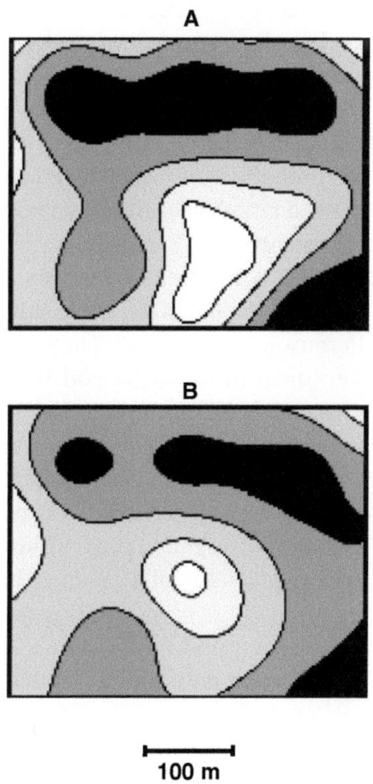

Fig. 8.7 Density distributions of (**a**) all and (**b**) parasitized cabbage seed weevil larvae in a winter oilseed rape crop at Woburn, UK, in 1995. Larval numbers were assessed in 400 pods at each of 19 locations. Contours delimit five density classes in each distribution and are equally spaced on log scales

was no evidence of density dependence of parasitoid with host suggesting that the parasitoid disperses throughout the range of its host and that its efficiency at finding hosts does not vary within that range.

8.2.3 Brassica Pod Midge and Its Parasitoids

The brassica pod midge is multivoltine with two generations on winter rape (Williams et al. 1987a) and one on spring rape (Williams et al. 1987b). Adults emerge from overwintering from mid-May to mid-July and mate at the emergence site. Males die soon after mating while the mated females migrate to rape crops. Each female starts to oviposit immediately it finds a suitable pod with walls its ovipositor can penetrate, frequently using pods already punctured by the cabbage seed weevil. In the laboratory, a female can lay 13–25 eggs; in the field, several females may lay into the same pod. The eggs hatch in 3–4 days and the larvae feed on the pod wall. There are three larval instars. Infested pods become yellowed and swollen and split prematurely, shedding larvae and seed. Mature larvae drop to the ground and burrow into the soil to pupate. A proportion emerges as adults the same year while the remainder enter diapause for up to 5 years. The second generation on winter rape has a greater proportion of larvae that enter diapause. For more detailed accounts of this pest see Alford et al. (2003) and Williams (Chapter 1 this volume).

Two key species of endoparasitoid: *Omphale clypealis* (Thomson) (Hymenoptera, Eulophidae) and *Platygaster subuliformis* (Kieffer) (Hymenoptera, Platygastridae) attack brassica pod midge larvae (Murchie et al. 1999a, Williams 2003, Ulber et al. Chapter 2 this volume). Like their host, adults of both species are probably multivoltine and emerge in the spring/early summer from soil in which oilseed rape has been grown previously and migrate to rape crops. In the UK, spring emergence of *P. subuliformis* in winter wheat following oilseed rape and its appearance on winter rape coincides closely with that of its host in May (Murchie 1996, Ferguson et al. 2004), while that of *O. clypealis* occurs about a month later (Fig. 8.8, Ferguson et al. 2004). They seek their hosts in the pods and lay their eggs singly into them through the pod wall; *P. subuliformis* lays into midge eggs whereas *O. clypealis* lays into midge larvae. *Platygaster subuliformis* is a koinobiont, its host larva continuing to develop after parasitisation; the parasitoid egg hatches and the larva feeds when its host has developed into a prepupa or pupa within a cocoon in the soil and on maturity, the parasitoid larva pupates within its host's larval skin. *Omphale clypealis* feeds within its host during its larval and pupal stages. In both species, some adults emerge the same year while others enter diapause to emerge the following spring. Mating occurs soon after emergence.

Brassica pod midge adults and larvae are usually strongly edge-distributed on both winter and spring oilseed rape crops (Ankersmit 1956, Free and Williams 1979a, Ferguson et al. 2003b, 2004). By contrast, on emergence in spring, after winter diapause in the soil of the previous year's oilseed rape crop, where they had been edge-distributed, Ferguson et al. (2004) found that the adults were no longer

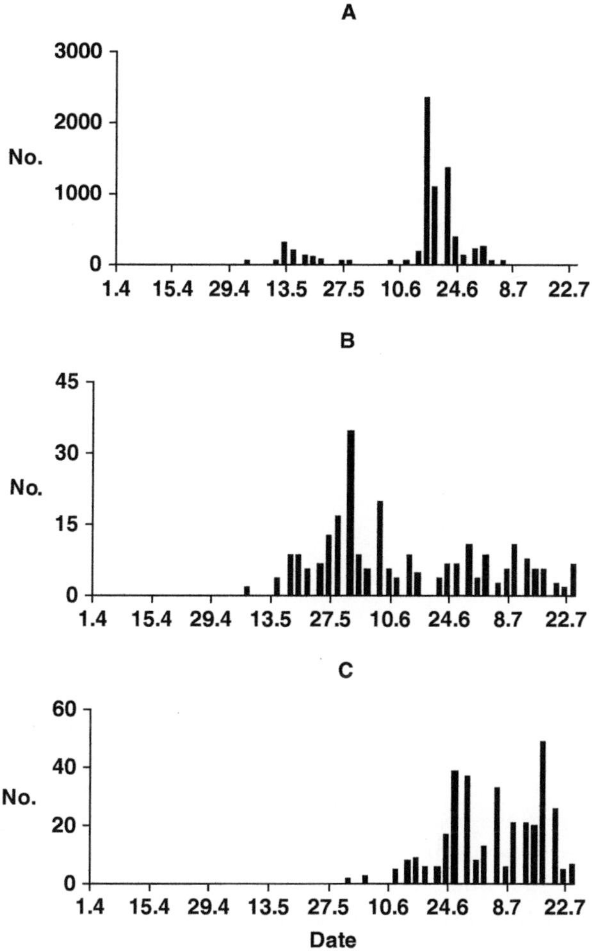

Fig. 8.8 Total numbers of (**a**) brassica pod midge, (**b**) *Platygaster subuliformis* and (**c**) *Omphale clypealis* caught in five yellow water traps placed at canopy height in winter oilseed rape at Rothamsted, UK, from 1 April to 22 July 2004. Traps were emptied three times per week. The crop (main raceme) was in flower from 12 April to 19 May

clustered at the edge of the field; they attributed this to soil-associated mortality factors which were themselves edge-distributed, for example, predators feeding on the pupating or overwintering larvae (Fig. 8.9).

Infestation of a crop by the brassica pod midge and its within-field spatial distribution is probably influenced by its distance from the emergence site of the midge, the strength and direction of the wind during migration, and the distribution on the crop of pods punctured by the cabbage seed weevil, through which the midge lays its own eggs. The midge mate at the emergence site and only mated females migrate to the rape crop in the spring. They are weak flyers. In mark-recapture experiments, they dispersed mainly with the wind (Sylvén 1970) and, in the absence of windbreaks, infested crop headlands downwind of the emergence site (Evans and Allen-Williams 1989a, b). However, in sheltered areas, they were seen to approach

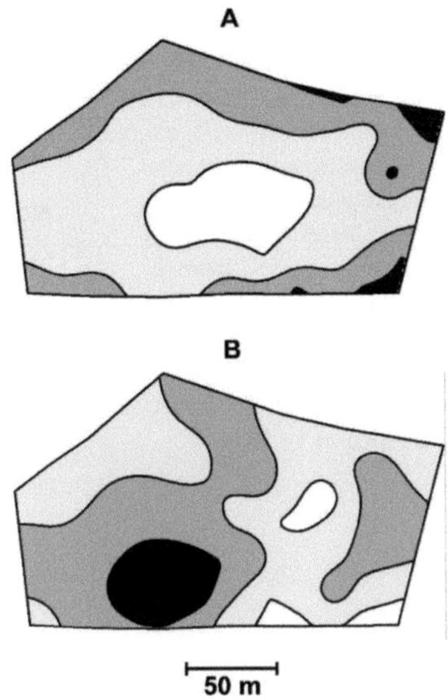

Fig. 8.9 SADIE plots of distributions of (**a**) pod midge larvae collected into water traps as they dropped from plants to pupate in the soil in a winter oilseed rape crop at Rothamsted, UK, in 1999 and (**b**) pod midge adults caught in emergence traps as they emerged in 2000 after winter diapause in the soil in the same field. Other details as Fig. 8.2

and enter oilseed rape crops by flying upwind in the upper layers of vegetation (Sylvén 1970), presumably using odour-mediated upwind anemotaxis. Visual cues may also be important close to the crop, as they are attracted by the yellow colour of the petals (Evans and Allen-Williams 1989a, b, Frearson 2006, Frearson et al. 2006). Ferguson et al. (2004) found no association between the distributions of seed weevil larvae and pod midge larvae on a winter oilseed rape crop they studied, suggesting an overabundance of punctured pods on the crop for the midge.

On a winter oilseed rape crop, the start of emergence of the new generations of both *O. clypealis* and *P. subuliformis* was coincident with the emergence of new generation adult pod midge but the emergence of parasitoids was more prolonged and continued until crop harvest (Ferguson et al. 2004). Both parasitoids were strongly edge-distributed, closely matching their host (Fig. 8.10). However, in the winter wheat following rape, emerging adult *P. subuliformis* had, like their hosts, lost their edge-distribution whereas adult *O. clypealis* remained edge-distributed.

8.2.4 Stem-Mining Pests and Their Parasitoids

The cabbage stem flea beetle is univoltine. Adults emerge from summer aestivation in mid- to late August and migrate to emerging winter rape crops during September and October. They feed on the cotyledons and leaves for about 2 weeks while their ovaries mature. Females lay their eggs in cracks in the soil around or on the lower parts of the rape plants. Most are laid in the autumn, but oviposition can continue during warmer periods in winter and spring. Larvae are found in plants

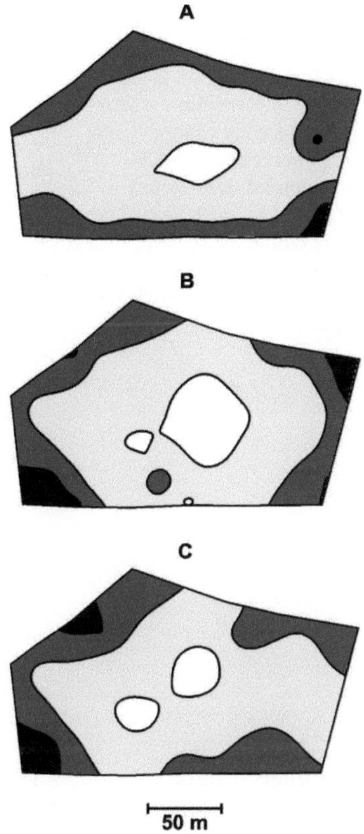

Fig. 8.10 SADIE plots of distributions of (**a**) first generation pod midge larvae collected in water traps as they dropped from plants to pupate in the soil, and (**b**, **c**) *O. clypealis* and *P. subuliformis* adults, respectively, collected in emergence traps on emergence from first generation midge pupae in the soil, in a winter oilseed rape crop at Rothamsted, UK, in 1999. Other details as Fig. 8.2

from September onwards. They feed while tunnelling the stems and lower leaf petioles, moving from older to younger tissue, causing damage. Mature larvae leave the stems from February onwards, and burrow into the soil to pupate. New adults emerge 8–12 weeks later from late May onwards. They feed for several weeks on the stems, leaves and pods of cruciferous plants. From mid-July onwards, they enter summer aestivation for 1–2 months in sheltered areas of vegetation in hedgerows and woodland.

The cabbage stem weevil is univoltine. Adults emerge from hibernation in the spring (March/April) and migrate to rape crops (Fig. 8.11). After a period of feeding and mating, females lay their eggs into the petioles or mid-ribs of the leaves (March to June). Eggs hatch after 6–11 days. There are three larval instars; first and second instars tunnel inside the leaf petioles and midribs, but later move into the stems. When mature (July and August), they leave the stems through exit holes and drop to the soil to pupate. New generation adults emerge from mid-July onwards and, after feeding on cruciferous plants, seek overwintering sites, where they remain until the spring.

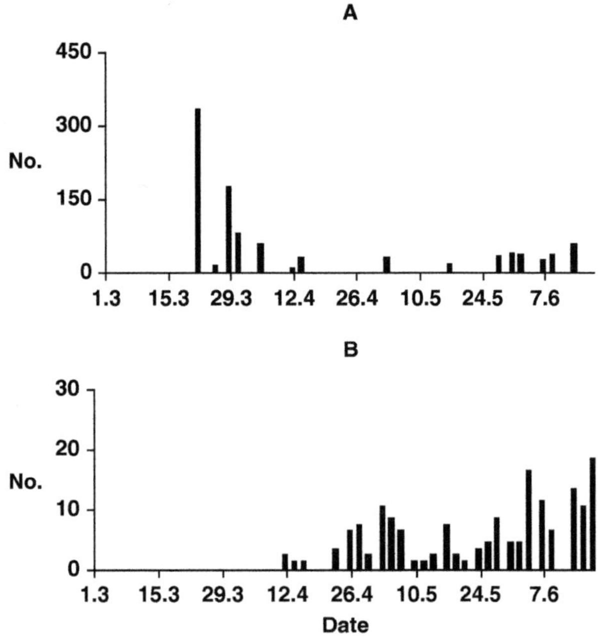

Fig. 8.11 Total numbers of (**a**) cabbage stem weevil and (**b**) *Tersilochus obscurator* caught in five yellow water traps placed at canopy height in winter oilseed rape at Rothamsted, UK, from 1 March to 17 June 2005. Traps were emptied three times per week. The crop (main raceme) was in flower from 4 April to 6 May

The rape stem weevil is univoltine. Adults migrate to winter rape in early spring (February/March). Females lay their eggs singly into the upper part of the main stem, close to the terminal bud, during stem elongation (March/April). This causes deformation which can lead to the stem splitting. Eggs hatch in 1–2 weeks. The larvae feed inside the stems for 3–5 weeks causing them to weaken, often bending and breaking. Last instar larvae leave the stems and pupate in the soil, remaining in their earthen chambers over winter to emerge the following spring. For more detailed accounts of the stem-mining pests see Alford et al. (2003) and Williams (Chapter 1 this volume).

The larvae of the cabbage stem flea beetle, the cabbage stem weevil and the rape stem weevil are attacked by the endoparasitoids *Tersilochus microgaster* (Szeplegeti), *Tersilochus obscurator* Aubert and *Tersilochus fulvipes* (Gravenhorst) (all Hymenoptera: Ichneumonidae), respectively (Ulber 2003, Ulber and Williams 2003, Ulber et al. Chapter 2 this volume). Like their hosts, they are univoltine. Overwintered adults emerge in the spring (March/May) from fields in which oilseed rape was grown the previous year. They may feed from nectar-producing plants at field edges before migrating to oilseed rape, shortly before or at the start of flowering (Fig. 8.11). Females lay their eggs singly into their hosts. The tersilochines are koinobiont: the parasitoid larva hatches but allows its host to continue development. The parasitoid larva remains in its first instar until its host larva is mature and has left the plant to pupate, when it develops rapidly and kills the host prepupa. The adult parasitoid diapauses in a pupal cocoon within the earthen cell formed by the host larva, and emerges the following spring.

The within-field spatio-temporal distributions of the stem-mining pests on winter oilseed rape are complex, with differing irregular patterns of aggregation rather than simple edge concentrations; those of adult pests are dynamic, changing throughout the immigration and egg-laying phases of crop colonisation (Free and Williams 1979a, Thioulouse 1987, Ferguson et al. 2003b, 2006, Klukowski 2006).

Thioulouse (1987) described the spatial and temporal structure of a population of the cabbage stem flea beetle on three scales: between plant, within-plot and landscape; he found a high degree of aggregation at all levels. He attributed spatial heterogeneity in infestation at the plant level to the egg-laying behaviour of the females which lay their eggs in the soil in batches of 2–16, the host plant location behaviour of the newly-emerged larvae towards the nearest plant, mortality of larvae in over-infested plants and dispersal of larvae during host-plant senescence into surrounding plants. At the plot level, he attributed spatial heterogeneity to proximity of aestivation sites, interactions with plant density and stem quality, differential survival rates and insect movements within the plot. At the landscape scale, he postulated that the location of aestivation sites relative to the plot and the direction of oriented invasion flights were the most important factors.

Klukowski (2006) investigated colonisation of winter rape crops by the cabbage stem weevil and the rape stem weevil. He found that males arrived before females, both were aggregated and more concentrated at the crop edge, and that the females tended to stay at the crop edge for longer than the males before spreading further into the crop during the period of egg-laying. Both pests however, remained essentially edge-distributed and crop centres never became as heavily infested as edges. Free and Williams (1979a) also reported more cabbage stem weevil at crop edges than at crop centres. Rate of spread of females into the crop from the edge is probably influenced by crop size and the availability of suitable plants at the edge for egg-laying.

Ferguson et al. (2006) found that, in a crop of winter rape, the distributions of the cabbage stem flea beetle and the cabbage stem weevil were polarised with respect to each other and suggested that this indicated an interaction between them (Fig. 8.12). In the autumn, adult cabbage stem flea beetle were clustered in the south-western and central parts of the crop away from the crop edge; the spatial distributions of males and females were closely associated. There was no evidence that the prevailing wind direction during immigration had influenced the pattern of colonisation. The distribution of larvae was closely associated with that of female flea beetles in the previous autumn and that of new generation beetles that emerged in late spring. By contrast, the cabbage stem weevil was clustered in the north-eastern end of the crop, where numbers of cabbage stem flea beetle were low. Upwind anemotaxis towards the crop during immigration in the spring could account for the distribution of cabbage stem weevil as winds were from the south-west during its immigration period. Ferguson et al. (2006) suggested that the polarisation of the two pests may have resulted from the stem weevil avoiding plant stems already infested by the flea beetle for it own oviposition, thus avoiding interspecific competition. The cabbage stem weevil exhibits within-plant spatial partitioning with the rape stem weevil; its

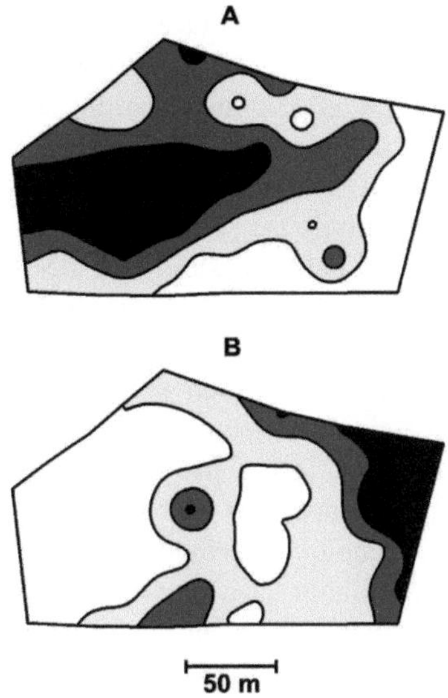

Fig. 8.12 SADIE plots of distributions of (**a**) cabbage stem flea beetle caught in 36 water traps on the ground in a winter oilseed rape crop at Rothamsted, UK, in autumn 1998 and (**b**) new generation cabbage stem weevil caught on emergence from the soil in emergence traps at 40 locations in the same crop in summer 1999. Other details as Fig. 8.2

larvae feed close to the base of the stem while those of the rape stem weevil feed in the mid-section of the stem (Dechert and Ulber 2004).

Variation in plant density in the crop influences the distribution of stem-mining pests through its effects on plant morphology. Length and diameter of stems, and the numbers of lateral racemes and leaves are inversely related to plant density (Ferguson et al. 2003b, Ulber and Fischer 2006). Thicker stems provide more food for stem-mining larvae than thin ones and may be more heavily infested by them (Dechert and Ulber 2004, Nuss and Ulber 2004, Ulber and Fischer 2006). Within-field distribution of cabbage stem weevil (Ferguson et al. 2003b) and emergence of new generation cabbage stem flea beetle (Ferguson et al. 2006) have both been found to be inversely related to area of high plant density.

The within-field spatial relationships of the larval endoparasitoids, *T. microgaster* and *T. obscurator*, with that of their hosts, the larvae of cabbage stem flea beetle and the cabbage stem weevil, respectively, were also investigated by Ferguson et al. (2006). The distribution of *T. microgaster* larvae was associated with that of *P. chrysocephala* larvae, indicating that the parasitoid was efficient at finding its hosts (Fig. 8.13). However, parasitisation rate declined with increase in host density, implying either that parasitoids are less efficient at finding hosts as their density increases or that there were too few parasitoids to exploit the greater abundance of larval hosts. *Tersilochus obscurator* was concentrated at one end of the crop (Fig. 8.14), a distribution probably partly due to the similarly polarised distribution of its host, the cabbage stem weevil, and partly accounted for by upwind anemotaxis towards the crop in spring when winds were from the south-west; *T. obscurator* has been shown to fly upwind to winter rape (Williams et al. 2007a). Emergence of new generation *T. obscurator* was greatest in the same area of the crop as emergence of its host, but was less edge-distributed. The lack of significant association

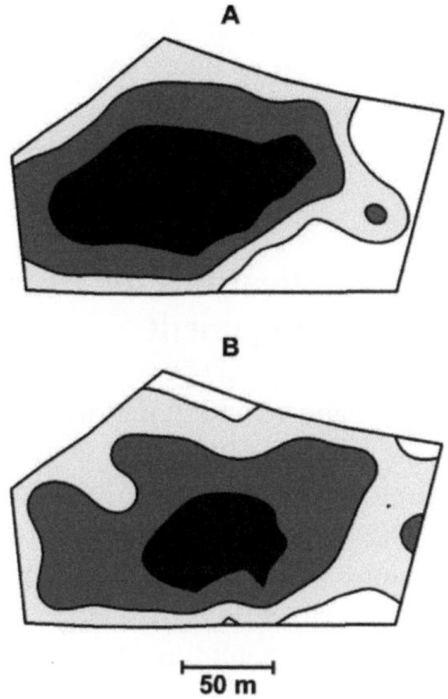

Fig. 8.13 SADIE plots of distributions of (**a**) all cabbage stem flea beetle larvae and (**b**) larvae parasitized by *T. microgaster* in a winter oilseed rape crop at Rothamsted, UK, in 1999. Larvae were collected in water traps as they dropped from the plants to the soil to pupate. Other details as Fig. 8.2

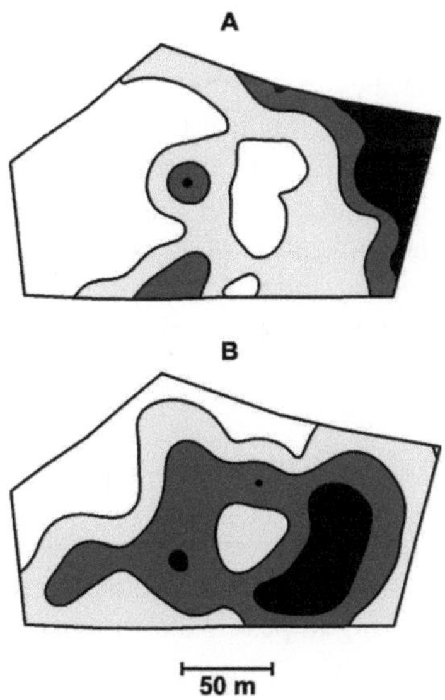

Fig. 8.14 SADIE plots of new generation adults of (**a**) cabbage stem weevil and (**b**) their parasitoid *T. obscurator* collected on emergence from the soil in emergence traps in a winter oilseed rape crop at Rothamsted, UK, in 1999. Other details as Fig. 8.2

between the distributions of emerging new generation adult host and parasitoid for either host-parasitoid pair, was probably an indication of the effectiveness of the parasitoids in suppressing their host population.

Variation in plant density in the crop influences not only the distribution of stem-mining pest larvae but also their parasitisation. Thicker stems provide pest larvae with structural refugia beyond the reach of parasitoid ovipositors. Ulber and Fischer

(2006) found that parasitism of the cabbage stem weevil by *T. obscurator* was greatest in lateral racemes and greater in the lower leaf petioles than in the main stem; parasitism of rape stem weevil by *Tersilochus fulvipes* increased from the base to the top of the main stem, and was also high in the lateral racemes.

8.3 Implications for Biocontrol-Based Integrated Pest Management

The spatio-temporal distributions of pest and parasitoid populations on the oilseed rape crop have important implications for the development of biocontrol-based IPM strategies for the crop. Parasitoids can provide an effective and environmentally-benign means of biocontrol of the pests but this is dependent on good coincidence in time and in space between adult parasitoids and the pest host stage that they attack. Any within-field heterogeneity in pest and parasitoid distributions has implications for sampling, for decision-making, and for crop yields. Further, it may allow temporal and/or spatial targeting of insecticides to kill pests while conserving parasitoids.

Precision farming aims to use within-field spatial information about the crop to target husbandry measures more precisely and specifically to the crop (Sylvester-Bradley et al. 1999). Technological advances, such as Global Positioning Systems and Geographical Information Systems enable spatial information about crop characteristics to be collected, utilized and modeled; they have already been applied to the application of fertilizers, herbicides and seeding densities, although not yet to the application of insecticides. Insect populations cannot yet be measured by remote sensing and the intensity of sampling required by other means is not feasible in commercial production.

8.3.1 Sampling

The non-uniform nature of the distributions of pests and parasitoids on the oilseed rape crop has implications for the accurate sampling of the crop to support decision-making in pest management. Where the distribution is aggregated, sampling along a line transect into the crop could lead to an inaccurate estimate of population size (Ferguson et al. 2000). This probably accounts for the recognised unreliability of some procedures that are currently in use. For example, in the UK, the recommended procedure for sampling populations of the cabbage seed weevil on a rape crop for monitoring and advisory purposes is to count the weevils obtained from beating the tops of each of 20 plants over a white tray, selecting the plants at random along a line transect into the crop (Lane and Walters 1993, Walters and Lane 1994). An assessment is made on three occasions during flowering and the largest mean number of weevils per plant from the three assessments is used to determine whether the economic threshold for control has been exceeded. The accuracy of population estimates for decision support would be improved if the non-uniform nature of pest spatial distributions were taken into account in sampling procedures. More accurate estimates would also enable more accurate modelling of potential yield loss.

8.3.2 Application of Insecticide

Biocontrol-based IPM strategies for the oilseed rape crop aim to minimise insecticide use and to conserve parasitoid populations. The six major pests of winter oilseed rape in Europe migrate to the crop in succession, attacking the crop at various growth stages and damaging different parts of the plant. The timing of insecticide applications must take account of the timing of their arrival and abundance on the crop relative to that of their parasitoids and the vulnerability of the crop if control is to be effective. Any insecticides applied to the crop should reduce pest populations to below the economic threshold for damage to the crop. The aggregated nature of most pest distributions on the crop means that pest damage and subsequent yield loss is also likely to be heterogenous on the crop. Areas of high pest density will have a greater impact on yield than predicted for an equivalent pest population randomly- or uniformly-distributed on the crop (Bardner and Fletcher 1974, Hughes and McKinley 1988). Ferguson et al. (2003b) found that the yield of plants showed no relationship with the numbers of pollen beetle larvae present; by contrast, the number of pods per plant that split and shed seed prematurely just before harvest were inversely related to the numbers of pollen beetle larvae, probably because early infestation and damage by the beetle delayed maturation of pods on infested plants. Any spatial pattern in the pre-flowering infestation of a crop by the beetle may therefore result in a spatial pattern in the rate of plant maturation within the crop, causing difficulties in determining the optimum harvest date. Application of insecticide to areas of low pest density, where the economic threshold is not reached, can be wasteful as damaged plants may compensate for any injury they sustain without yield loss. Further, broad-spectrum insecticides applied to the crop, particularly during or after flowering, kill many parasitoids directly or indirectly through killing their hosts on the crop. Temporal and spatial targeting of insecticide applications has potential for both minimising insecticide use and for conserving parasitoids.

Currently, insecticides may be applied to the winter oilseed rape crop several times during its life (Winfield 1992, Anonymous 1996, Lane and Gladders 2000, Alford et al. 2003, see also Williams Chapter 1 this volume and Thieme et al. Chapter 12 this volume). Seed is usually supplied treated with insecticide to control the cabbage stem flea beetle. Until recently the organochlorine insecticide, gamma-HCH was used. However, concerns about its safety to human operators and its deleterious effects on non-target organisms resulted in seed treatment with this chemical being withdrawn in some countries and replaced with carbosulfan and other insecticides. From autumn 2001, seed treatment with imidacloprid plus beta-cyfluthrin (Chinook, Bayer) has been registered in the UK for use on winter rape. The use of a pyrethroid insecticide spray, such as alpha-cypermethrin, during early crop emergence when adult or larval damage by the cabbage stem flea beetle is first seen and then again 1 month later is recommended but often crops are sprayed prophylactically. Such autumn treatments also aim to kill aphids, potential vectors of viruses on the crop.

Chemical control of spring and summer pests is now dominated by the use of synthetic pyrethroids, e.g., alpha-cypermethrin, deltamethrin and esfenvalerate. Cabbage stem weevil and rape stem weevil infestations are usually controlled in early spring, before flowering, pollen beetle infestations at the green-yellow bud stage. In the UK, insecticide usage against cabbage seed weevil and brassica pod midge changed during the 1990s from mostly the organophosphate triazophos, applied post-flowering, to synthetic pyrethroids, applied during flowering. This is often applied prophylactically in a tank-mix with fungicides. Insecticide application is usually targeted against populations of cabbage seed weevil on which infestation by brassica pod midge is largely dependent.

8.3.2.1 Temporal Targeting of Insecticides

The time of colonisation of the oilseed rape crop by the pollen beetle relative to the growth stage of the crop has implications for the need to control the pest by the application of insecticide. If the adults arrive before flowering, when the crop is in the green/yellow bud stage of growth, they chew into the buds to feed and lay their eggs causing bud abscission (Williams and Free 1978, Nilsson 1987); neither adults nor larvae cause economic injury to open flowers (Williams and Free 1979).

Timing the application of insecticides to avoid periods of parasitoid activity is a potentially valuable element of conservation biocontrol. Such 'spray windows' are present for the oilseed rape crop where it is attacked predominantly by only one or two of its six major pests (see also Johnen et al. Chapter 15 this volume). Thus application of insecticide in the autumn to control infestation by the cabbage stem flea beetle does not directly affect its parasitoid, *T. microgaster*, or other tersilochine parasitoids of the stem-mining pests, which do not emerge from overwintering before the spring. However, any spring or summer applications to the crop, when the parasitoids are present, are likely to be harmful to them (Fig. 8.11).

Insecticide application at green bud to control the pollen beetle may also kill one of its key parasitoids, *P. interstitialis*, which is the earliest to arrive on the crop and often temporally coincident with the beetle, but not those of *P. morionellus* or *T. heterocerus*, which arrive later during crop flowering (Fig. 8.1, Ferguson et al. 2003a).

The temporal dissociation between the immigration flights of the cabbage seed weevil and its key parasitoid *T. perfectus* provides an opportunity during flowering for the temporal targeting of insecticide to kill the weevil while conserving its parasitoid (Fig. 8.5, Murchie et al. 1997). This approach was proposed as part of an integrated strategy for the management of weevil populations on winter rape (Alford et al. 1996). In the UK, the decline during the 1990s, in the use of insecticide (triazophos) against the seed weevil post-flowering, when the parasitoid is most abundant, appeared to result in substantially increased rates of seed weevil parasitism (Alford et al. 1996).

Pyrethroid insecticides applied to winter rape to control the cabbage seed weevil during crop flowering are also aimed to control any first generation brassica pod midge present (Alford et al. 2003). At this time, *P. subuliformis* adults are also actively searching for their midge larval hosts in the pods, and are therefore vulnerable to insecticide, whereas, *O. clypealis* adults which are active later, are not at risk (Fig. 8.8, Ferguson et al. 2004).

There may be also potential to conserve parasitoids by applying insecticide at times of day when they are less active in the crop, for example in the early morning. Holdgate et al. (2006) found that the parasitoids *T. obscurator* and *P. subuliformis* showed marked diel periodicity on a crop of winter rape. Their activity was most strongly correlated with solar energy recorded 2 h previously and with temperature, few being active at mean temperatures below 15°C.

However, where several pests on the crop require control through insecticide application, the temporal succession of pests and parasitoids in the crop, makes it difficult to find spray windows to conserve all parasitoid species. All the key parasitoid species except *T. microgaster* and *O. clypealis*, are active during crop flowering and, as a whole, the parasitoid population is particularly vulnerable to insecticides at this time (Fig. 8.15).

8.3.2.2 Spatial Targeting of Insecticides

Insecticide (pyrethroid) applications are usually applied to the whole area of the oilseed rape crop (Alford et al. 2003) but the irregular and patchy distributions of

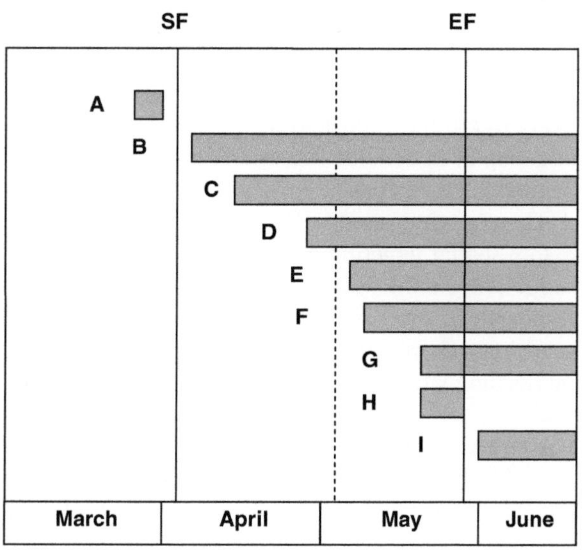

Fig. 8.15 Phenology of nine key parasitoids of oilseed rape pests caught in yellow water traps placed at canopy height in winter oilseed rape at Rothamsted, UK, from 1 March to 17 June 2005, in relation to crop flowering. (*A*) *Tersilochus microgaster*, (*B*) *Phradis interstitialis*, (*C*) *Tersilochus obscurator*, (*D*) *Tersilochus heterocerus*, (*E*) *Platygaster subuliformis*, (*F*) *Phradis morionellus*, (G) *Trichomalus perfectus*, (H) *Mesopolobus morys*, (I) *Omphale clypealis*. (SF) start of flowering, (EF) end of flowering, *dotted line* indicates end of flowering of the main raceme

the pests within the crop suggest scope for reducing insecticide inputs by spatially targeting it to areas of high pest density only.

The strong tendency for edge-distribution during the immigration phase indicated for most pests, particularly cabbage stem weevil (e.g., Free and Williams 1979a, Klukowski 2006), pollen beetle (Free and Williams 1979a), cabbage seed weevil (e.g., Free and Williams 1979a, Murchie et al. 1999b, Ferguson et al. 2000) and brassica pod midge (e.g., Free and Williams 1979a, Ferguson et al. 2003b), suggests that application of insecticide to crop edges only at this time would kill a large proportion of the adults, and be particularly effective on large crops where whole-field application would be wasteful as much of the crop may be uninfested. The later more complex aggregated distribution of adults in the crop suggests potential for even more precise spatial targeting of insecticide to areas where pests are most abundant. However, at present, it is not feasible for the grower to measure patterns of crop infestation by pests and consequently, it is difficult to know how to target crop protection measures to areas of high pest infestation, except by the use of trap crops or headland applications.

Spatial targeting of insecticide to areas of high pest density may also have potential for the conservation of some parasitoids, particularly those where pest and parasitoid distributions do not coincide completely. Thus, because *T. heterocerus* is more evenly spread across the crop than its host, the pollen beetle, any spatial targeting of insecticide to areas of high beetle density, such as to crop borders, would conserve those parasitoids in areas of low beetle density (Ferguson et al. 2003a); by contrast, the spatial association of *P. interstitialis* with its host probably makes it more vulnerable to any spatially-targeted insecticide treatments applied to control the beetle. Similarly, the close spatial associations between the cabbage stem flea beetle and its parasitoid, *T. microgaster* (Ferguson et al. 2006), between the cabbage stem weevil and its parasitoid, *T. obscurator* (Ferguson et al. 2006), between the cabbage seed weevil and *T. perfectus* (Murchie et al. 1999b, Ferguson et al. 2000), between the brassica pod midge and its parasitoids *P. subuliformis* and *O. clypealis* (Ferguson et al. 2004), offer little potential to conserve the parasitoids by spatial targeting of the pests; any spatially-targeted application of insecticide would be as likely to kill the parasitoid as its host. However, there may be potential for spatial targeting of insecticide in association with the use of a trap crop of turnip rape (see Section 8.3.3.2 below).

Greater understanding of the drivers influencing spatial distributions is needed before spatial targeting can become a practical proposition. The pattern of colonisation reflects the interactions between environmental factors and the behavioural responses of the pest, particularly those involved in crop location and host plant selection. These include landscape factors such as the location of overwintering/feeding sites and windbreaks, climatic variables, as well as within-crop factors such as plant density and interactions with other pests and plants. Effective spatial targeting is also dependent on the collection of fully spatially-referenced data. Models of pest spatial distributions could then be constructed to aid prediction of areas of the crop at most risk of pest infestation, pest injury and yield loss.

8.3.3 Manipulating Pest and Parasitoid Distributions

8.3.3.1 Crop Location

There is probably potential for manipulating the spatial distributions of pests and their parasitoids on a landscape scale through the use of on-farm crop rotations. Pests and their parasitoids use odour-mediated upwind anemotaxis to locate the oilseed rape crop and are often more abundant on the downwind side of the prevailing wind, particularly during the phase of immigration to the crop (see Williams and Cook Chapter 7 this volume). There is potential for manipulating their populations by making it less likely for pests or more likely for the parasitoids to find the crop. For example, locating the crop downwind of the overwintering sites of pests may mean they are less likely to contact the odour plume from the crop during their migration phase. Locating the crop upwind of the previous year's crop where most of the parasitoids overwinter may enhance their chances of locating the odour plume from the crop and increase their numbers on the crop.

8.3.3.2 Crop Layout

'Push-pull' strategies exploit behaviour-modifying stimuli, particularly those associated with host plant location, to manipulate the distribution of pests and their parasitoids on a crop (Cook et al. 2007a). Attractive stimuli (e.g., preferred host plants, aggregation or sex pheromones, visual, oviposition or gustatory stimulants) are used to 'pull' the pests to a trap crop and retain them on it while repellent and/or deterrent stimuli (e.g., less-preferred host plants, antifeedants, oviposition-deterring, epideictic or alarm pheromones) are used to 'push' pests from the protected crop while concentrating natural enemies on it for biocontrol.

The push-pull strategy being developed for oilseed rape utilises turnip rape (*Brassica rapa* L.) as the trap crop to 'pull' pests (Cook et al. 2007a, b). Turnip rape is preferred to oilseed rape as a host plant for oviposition and/or feeding by all the major coleopterous pests of oilseed rape (see also Williams and Cook Chapter 7 this volume). It can be sown as a seed admixture with oilseed rape (Buechi 1990, Büchi 1995, Nilsson 2004) or as a border to the oilseed rape crop (e.g., Cook et al. 2006b), the arrangement that simulation modelling suggests would be the most effective (Potting et al. 2005). Field trials have shown that a turnip rape trap crop border can reduce the abundance of cabbage stem flea beetle (Barari et al. 2005) and pollen beetle (Cook et al. 2006b) in the oilseed rape crop and reduce the need for insecticide. Push-pull strategies aim to integrate biological control by parasitoids; there is no evidence that parasitoids prefer turnip rape to oilseed rape (Barari et al. 2005), which suggests it would probably not adversely influence their effectiveness for biocontrol.

8.3.3.3 Crop Boundary

Populations of parasitic Hymenoptera can be enhanced by the sowing of weed strips with a diversity of flowering plants near a crop (e.g., Molthan and Ruppert 1988). In oilseed rape fields with wild flower strips, parasitism of pollen beetle larvae by

Phradis spp. at three metres from the crop edge increased over a 3 year period (Büchi 2002). Incidence of parasitisation of the pollen beetle has also been found to be greater while that of crop damage is lower in oilseed rape crops growing in structurally-complex landscapes compared with those in simple landscapes (Thies and Tscharntke 1999). Hausammann (1996) showed that the major pests (pollen beetle, cabbage seed weevil, winter stem weevil, cabbage stem weevil and brassica pod midge) which invade the rape crop from its edges, are not increased when such strips are sown within the crop. Although in his study parasitisation rates of pest larvae were low and not generally greater near the weed strips, he recommended within-field weed strip management as a useful component of integrated strategies aimed to conserve natural enemies while not enhancing pests.

Acknowledgements This review was supported by Estonian Targeting Financing Project SF0172655s04 and the UK Department for Food and Rural Affairs. Rothamsted Research is an institute of the Biotechnology and Biological Sciences Research Council of the United Kingdom.

References

Alford DV, Nilsson C, Ulber B (2003) Insect pests of oilseed rape crops. In: Alford DV (ed.) *Biocontrol of oilseed rape pests*. Blackwell, Oxford, UK.

Alford DV, Walters KFA, Williams IH, Murchie AK (1996) A commercially viable low-cost strategy for the management of seed weevil populations on winter oilseed rape in the UK. *Proc British Crop Protection Conf, Pests and Diseases, Brighton* pp 609–614.

Ankersmit GW (1956) De levenswijze en de bestrijding van de koolzaadsnuitkever (*Ceuthorrhynchus assimilis* Payk.) en de koolzaadgalmug (*Dasyneura brassicae* Winn). *Verslagen van Landbouwkundige Onderzoekingen* 62(9): 1–59.

Anonymous (1996) *Guideline on good plant protection practice. Rape. EPPO Bull* 26: 349–367.

Barari H, Cook SM, Williams IH (2005) Effect of a turnip rape (*Brassica rapa*) trap crop on stem-mining pests and their parasitoids in winter oilseed rape (*Brassica rapa*). *BioControl* 50: 69–86.

Bardner R, Fletcher KE (1974) Insect infestations and their effects on the growth and yield of field crops: A review. *B Entomol Res* 64: 141–160.

Buechi P (1990) Investigations on the use of turnip rape as trap plant to control oilseed rape pests. *IOBC/wprs Bull* 13(4): 32–39.

Büchi R (1995) Combination of trap plants (*Brassica rapa* var. *silvestris*) and insecticide use to control rape pests. *IOBC/wprs Bull* 18(4): 102–121.

Büchi R (2002) Mortality of pollen beetle (*Meligethes* spp.) larvae due to parasitoids and predators in rape fields and the effect of conservation strips. *Agr Ecosyst Environ* 90: 255–263.

Clark SJ, Perry JN (1994) Small sample estimation for Taylor's power law. *Environ Ecol Stat* 1: 287–302.

Cook SM, Khan ZR, Pickett JA (2007a) The use of 'push-pull' strategies in integrated pest management. *Annu Rev Entomol* 52: 357–400.

Cook SM, Rasmussen HB, Birkett MA, Murray DA, Pye BJ, Watts NP, Williams IH (2007b) Behavioural and chemical ecology underlying the success of turnip rape (*Brassica rapa*) trap crops in protecting oilseed rape (*Brassica napus*) from the pollen beetle (*Meligethes aeneus*). *Arthropod-Plant Inte* 1: 57–67.

Cook SM, Skellern MP, Smith M, Williams IH (2006a) Responses of pollen beetles (*Meligethes aeneus*) to petal colour. *IOBC/wprs Bull* 29(7): 151–160.

Cook SM, Smart LE, Martin JL, Murray DA, Watts WP, Williams IH (2006b) Exploitation of host plant preferences in crop protection strategies for oilseed rape (*Brassica napus*). *Entomol Exp Appl* 119: 221–229.

Dechert G, Ulber B (2004) Interactions between the stem-mining weevils *Ceutorhynchus napi* Gyll. and *Ceutorhynchus pallidactylus* (Mrsh.) in oilseed rape. *Agr Forest Entomol* 6: 193–198.

Evans KA, Allen-Williams LJ (1989a) The response of the cabbage seed weevil (*Ceutorhynchus assimilis* Payk.) and the brassica pod midge (*Dasineura brassicae* Winn.) to flower colour and volatiles of oilseed rape. *Aspects of Appl Biol* 23: 347–353.

Evans KA, Allen-Williams LJ (1989b) Location and infestation of winter oilseed rape by the brassica pod midge, *Dasineura brassicae* Winn. (Diptera, Cecidomyiidae). *Mededelingen van de Faculteit Landbouwwetenschappen Rijkuniversiteit Gent* 54(3a): 717–725.

Ferguson AW, Barari H, Warner DH, Campbell JM, Smith ET, Williams IH (2006) Distributions and interactions of the stem miners, *Psylliodes chrysocephala* and *Ceutorhynchus pallidactylus*, and their parasitoids in a crop of winter oilseed rape *(Brassica napus.)*. *Entomol Exp Appl* 119: 81–92.

Ferguson AW, Campbell JM, Warner DJ, Watts NP, Schmidt JEU, Williams IH (2003a) Spatio-temporal distributions of *Meligethes aeneus* and its parasitoids in an oilseed rape crop and their significance for crop protection. *Proc 11th Int Rapeseed Cong, 6–10 July 2003, Copenhagen, Denmark*, 1057–1059.

Ferguson AW, Campbell JM, Warner DJ, Watts NP, Schmidt JEU, Williams IH (2004) Phenology and spatial distributions of *Dasineura brassicae* and its parasitoids in a crop of winter oilseed rape: Implications for integrated pest management. *IOBC/wprs Bull* 27(10): 243–252.

Ferguson AW, Klukowski Z, Walczak B, Clark SJ, Mugglestone MA, Perry JN, Williams IH (2003b) Spatial distribution of pest insects in oilseed rape: Implications for integrated pest management. *Agr Ecosyst Environ* 95: 509–521.

Ferguson AW, Klukowski Z, Walczak B, Perry JN, Mugglestone MA, Clark SJ, Williams IH (2000) The spatio-temporal distribution of adult *Ceutorhynchus assimilis* in a crop of winter oilseed rape in relation to the distribution of their larvae and that of the parasitoid *Trichomalus perfectus*. *Entomol Exp Appl* 95: 161–171.

Frearson DJT (2006) *The movement and host-selection of three inflorescence insect pests of oilseed rape and their parasitoids in relation to host-plant characteristics and environmental conditions*. PhD thesis, University of Reading, UK.

Frearson DJT, Ferguson AW, Campbell JM, Williams IH (2005) The spatial dynamics of pollen beetles in relation to inflorescence growth stage of oilseed rape: Implications for trap crop strategies. *Entomol Exp Appl* 116: 21–29.

Frearson DJT, Ferguson AW, Laugier N, Williams IH (2006) The influence of petals on numbers of pests and their parasitoids in oilseed rape. *Proc Symp Integrated Pest Management in Oilseed Rape, 3–5 April 2006, Göttingen, Germany.*

Free JB, Williams IH (1979a) The distribution of insect pests on crops of oil-seed rape (*Brassica napus* L.) and the damage they cause. *J Agr Sci* 92: 139–149.

Free JB, Williams IH (1979b) The infestation of crops of oil-seed rape (*Brassica napus* L.) by insect pests. *J Agr Sci* 92: 203–218.

Hausammann A (1996) Strip-management in rape crop: Is winter rape endangered by negative impacts of sown weed strips? *J Appl Entomol* 120: 505–512.

Holdgate R, Ferguson AW, Mason NS, Clark SJ, Williams IH (2006) The diel periodicity of within-crop flight of key parasitoids of oilseed rape pests. *CD Proc SympIntegrated Pest Management in Oilseed Rape, 3–5 April 2006, Göttingen, Germany.*

Hughes G, McKinley RG (1988) Spatial heterogeneity in yield-pest relationships for crop loss assessment. *Ecol Model* 41: 67–73.

Jönsson M (2005) *Responses to oilseed rape and cotton volatiles in insect herbivores and parasitoids*. PhD thesis, University of Agricultural Sciences, Sweden.

Klukowski Z (2006) Practical aspects of migration of stem weevils on winter oilseed rape. *CD Proc Int Symp Integrated Pest Management in Oilseed Rape, 3–5 April 2006, Goettingen, Germany.*

Kühne W (1977) Untersuchungen zur Befallsverteilung der Rüsselkäfer *Ceutorhynchus napi* Gyll., *Ceuthorhynchus quadridens* Panz. und *Ceutorhynhcus assimilis* Payk. innerhalb grossflachinger Winterrapsbestände. *Arch Phytopathol Pfl* 13: 109–115.

Lane A, Gladders P (2000) Pests and diseases of oilseeds, brassica seed crops and field beans. In: Alford DV (ed.) *Pest and disease management handbook*. Blackwell, Oxford, UK.

Lane A, Walters KFA (1993) Recent indices and cost effective control of pests of oilseed rape in England and Wales. *IOBC/wprs Bull* 16(9): 185–192.

Molthan J, Ruppert V (1988) Significance of flowering wild herbs in boundary strips and fields for flower-visiting beneficial insects. In: Parey P (ed.) *Effects of crop edges*. Biol. Bundesanstalt, Land- und Forstwirt, Berlin-Dahlem, Berlin and Hamburg.

Murchie AK (1996) *Parasitoids of Cabbage Seed Weevil and Brassica Pod Midge in Oilseed Rape*. PhD thesis, Keele University, UK.

Murchie AK, Polaszek A, Williams IH (1999a) *Platygaster subuliformis* (Kieffer) (Hym., Platygastridae) new to Britain, an egg-larval parasitoid of the Brassica pod midge *Dasineura brassicae* Winnertz (Dipt., Cecidomyiidae). *Entomologist's Monthly Magazine* 135: 1624–1627.

Murchie AK, Williams IH, Alford DV (1997) Effects of commercial insecticide treatments to winter oilseed rape on parasitism of *Ceutorhynchus assimilis* Paykull (Coleoptera: Curculionidae) by *Trichomalus perfectus* (Walker) (Hymenoptera Pteromalidae). *Crop Prot* 16: 199–202.

Murchie AK, Williams IH, Perry J (1999b) Edge distributions of *Ceutorhynchus assimilis* (Paykull) and its parasitoid *Trichomalus perfectus* (Walker) in a crop of winter oilseed rape (*Brassica napus* L.). *BioControl* 44(4): 379–390.

Nilsson C (1987) Yield losses in summer rape caused by pollen beetles (*Meligethes* spp.). *Swedish J Agric Res* 18: 113–118.

Nilsson C (2003) Parasitoids of pollen beetles. In: Alford DV (ed.) *Biocontrol of oilseed rape pests*. Blackwell, Oxford, UK.

Nilsson C (2004) Trap plants to avoid insecticide application against pollen beetles in oilseed rape. *IOBC/wprs Bull* 27(10): 215–221.

Nuss H, Ulber B (2004) Effect of sowing density of oilseed rape on the abundance and within-plant distribution of *Psylliodes chrysocephala* L. *IOBC/wprs Bull* 27(10): 225–226.

Osborne P (1960) Observations on the natural enemies of *Meligethes aeneus* (F.) and *M. viridescens* (F.) [Coleoptera: Nitidulidae]. *Parasitology* 50: 91–110.

Perry JN (1995) Spatial analysis by distance indices. *J Anim Ecol* 64: 303–314.

Perry JN (1998a) Measures of spatial pattern for counts. *Ecology* 79: 1008–1017.

Perry JN (1998b) Measures of spatial pattern and spatial association for counts of insects. In: Baumgartner J, Brandmayr P, Manly BFJ (eds.) *Population and community ecology for insect management and conservation. Proc 20th Int Cong Entomology, Florence, Italy, 25–31 August 1996*. Balkema, Rotterdam, pp 21–33.

Potting RPJ, Perry JN, Powell W (2005) Insect behavioural ecology and other factors affecting the control efficacy of agro-ecosystem diversification strategies. *Ecol Model* 182: 199–216.

Risbec J (1952) Contribution à l'étude de *Ceuthorhynchus assimilis* Payk. Charançon des siliques de colza. *Rev Path Veg Ent Agr* 31: 179–180.

Sylvén E (1970) Field movement of radioactively labeled adults of *Dasyneura brassicae* Winn. (Dipt., Cecidomyiidae). *Entomol Scand* 1: 161–187.

Sylvester-Bradley R, Lord E, Sparkes DL, Scott RK, Wiltshire JJJ, Orson J (1999) An analysis of the potential of precision farming in northern Europe. *Soil Use Manage* 15: 1–8.

Taylor LR (1984) Assessing and interpreting the spatial distributions of insect populations. *Annu Rev Entomol* 29: 321–357.

Thiem E (1970) Die Bedeutung der Feldrandbenhandlung für de Bekämpfung von *Dasineura brassicae* Winn.(Dipt., Cecidomyiidae) und *Ceuthorrhynchus assimilis* Payk. (Col., Curculionidae) im blühenden Raps und der Einfluss dieser Insektizidmassnahme auf das Vorkommen von *Trichomalus perfectus* Walker (Hym., Chalcidoidea) an den Larven des Kohlschotenrüsslers. *Arch Phytopath Pfl* 6: 83–98.

Thies C, Tscharntke T (1999) Landscape structure and biological control in agroecosystems. *Science* 285: 893–895.

Thioulouse J (1987) Space-time structures in a winter rape pest population, *Psylliodes chrysocephala* (Col. Chrysomelidae): Methodological proposals and biological interpretations. *J Appl Ecol* 24: 435–450.

Ulber (2003) Parasitoids of ceutorhynchid stem weevils. In: Alford DV (ed.) *Biocontrol of oilseed rape pests*. Blackwell, Oxford, UK.

Ulber B, Fischer K (2006) Effect of plant density and host plant architecture on the abundance and within-plant distribution of stem-mining pests and the level of parasitism. *CD Proc Int Symp Integrated Pest Management in Oilseed Rape, 3–5 April 2006, Goettingen, Germany.*

Ulber B, Williams IH (2003) Parasitoids of flea beetles. In: Alford DV (ed.) *Biocontrol of oilseed rape pests*. Blackwell, Oxford, UK, pp 125–138.

Walters KFA, Lane A (1994) The development of an improved management strategy for the cabbage seed weevil on oilseed rape. In: Leather SR, Mills N, Watt AD, Walters KFA (eds.) *Individuals, populations and patterns in ecology*. Intercept Press, Andover, pp 187–197.

Williams I (2003) Parasitoids of brassica pod midge. In: Alford DV (ed.) *Biocontrol of oilseed rape pests*. Blackwell, Oxford, UK.

Williams IH, Frearson DJT, Barari H, McCartney A (2007a) First field evidence that parasitoids use upwind anemotaxis for host habitat location. *Entomol Exp Appl* 123: 299–307.

Williams IH, Frearson D, Barari H, McCartney A (2007b) Migration to and dispersal from oilseed rape by the pollen beetle, *Meligethes aeneus,* in relation to wind direction. *Agr Forest Entomol* 9: 279–286.

Williams IH, Free JB (1978) The feeding and mating behaviour of pollen beetles (*Meligethes aeneus* Fab.) and seed weevils (*Ceuthorhynchus assimilis* Payk.) on oil-seed rape (*Brassica napus* L.). *J Agr Sci* 91: 453–459.

Williams IH, Free JB (1979) Compensation of oil-seed rape (*Brassica napus* L.) plants after damage to their buds and pods. *J Agr Sci* 92: 53–59.

Williams IH, Martin AP, Kelm M (1987a) The phenology of the emergence of brassica pod midge (*Dasineura brassicae* Winn.) and its infestation of winter oil-seed rape (*Brassica napus* L.). *J Agr Sci* 108: 579–589.

Williams IH, Martin AP, Kelm M (1987b) The phenology of the emergence of brassica pod midge (*Dasineura brassicae* Winn.) and its infestation of spring oil-seed rape (*Brassica napus* L.). *J Agr Sci* 109: 309–314.

Winfield AL (1963) A study on the effects of insecticides on parasites of larvae of blossom beetles (*Meligethes aeneus* F. Coleoptera: Nitidulidae). *Entomol Exp Appl* 6: 309–318.

Winfield AL (1992) Management of oilseed rape pests in Europe. *Agr Zool Review* 5: 51–95.

Chapter 9
Biological Rape Pest Control in Spatio-Temporally Changing Landscapes

Carsten Thies and Teja Tscharntke

Abstract Community structure and trophic interactions depend on landscape context. We analysed trophic interactions of the pollen beetle (*Meligethes aeneus*) and its parasitoids on oilseed rape (*Brassica napus*) in 15 agricultural landscapes differing in structural complexity (~50–100% arable land) and interannual changes of rape crop area (with ~7% maximum expansion of rape crop area and ~8% maximum contraction of rape crop area from year to year). A patch of potted rape plants was placed in the centre of each landscape for standardized measurement. Parasitism decreased and herbivory increased as the percentage of arable land in the surrounding landscape increased. Thus, semi-natural habitats appeared to support parasitoid populations contributing to the reduction of populations of the pollen beetle. In addition, parasitism decreased following rape crop expansion, and increased following rape crop contraction, indicating interannual dilution and concentration effects of the higher trophic level populations. When semi-natural habitat area dropped below a value of ~20% of the landscape, or when the expansion of rape crop area between years exceeded ~5% of the landscape, respectively, parasitism dropped below a threshold value of about 32–36%, below which success in classical biological control has never been reported. In a geographic scale analysis using five spatial scales ranging from 0.5 to 3 km diameter, parasitism and herbivory showed the best correlations with both the percentage of arable land and the percentage of interannually changing rape crops at the same spatial scales, i.e., landscape sectors of 1–2 km diameter, thereby suggesting that this 'functional spatial scale' indicates their dispersal abilities.

9.1 Introduction

The importance of landscape context for understanding local population dynamics and trophic interactions has been increasingly recognized in recent decades

C. Thies (✉)
Department of Crop Science, Agroecology, Georg-August-University, Göttingen 37073, Germany
e-mail: carsten.thies@agr.uni-goettingen.de

I.H. Williams (ed.), *Biocontrol-Based Integrated Management of Oilseed Rape Pests*,
DOI 10.1007/978-90-481-3983-5_9, © Springer Science+Business Media B.V. 2010

(Kareiva 1990, Pimm 1991, McCauley et al. 1993, Wiens et al. 1993, Kareiva and Wennergren 1995, Pickett and Cadenasso 1995, Rosenzweig 1995, Holt 1996, Wiens et al. 1997, Ricketts 2001, Cronin and Reeves 2005, Tscharntke et al. 2005). Landscape approaches explicitly address the composition and configuration of spatial elements, and spatial scales that are much larger than those of a single habitat. A major task in spatial ecology is to identify the spatial scale experienced by populations and communities. Such knowledge is crucial for the decision which scales in space and time have to be managed.

Agricultural landscapes are characterized by high levels of disturbance owing to the annual nature of many crop species. Arable fields are characterized by annual harvesting, soil cultivation and crop rotation, thereby erasing many communities of herbivores and natural enemies. Hence, annual recolonization of crop fields from semi-natural habitats in the surrounding landscape is a fundamental feature of agricultural landscapes. Many natural enemy species depend on or profit from resources in non-crop habitats providing shelter from agricultural practices, overwintering refuges, alternative host plants and host, and/or food sources that are not available at a substantial level in crop fields (Landis et al. 2000, Bianchi et al. 2006). Landscapes dominated by arable fields have lost many species and associated trophic interactions (Tscharntke and Brandl 2004). Higher trophic level organisms such as natural enemies of pest insects with a low dispersal ability experience particularly high losses, thereby releasing high-dispersal herbivores from natural biocontrol. There are several examples published that pest populations of annual crops can be successfully regulated by natural enemies (Halaj and Wise 2001), with landscape management playing a major role (Altieri et al. 1993, Burel and Baudry 1995, Van Driesche and Bellows 1996, Matson et al. 1997, Menalled et al. 1999, Thies and Tscharntke 1999, Östman et al. 2001, Tscharntke et al. 2002, 2005). In contrast, crops can also increase populations of natural enemies (Thies et al. 2005, Vollhardt et al. 2008) which may spill over across cropland-natural habitats (Rand et al. 2006).

Here, we present two experiments on plant-herbivore-parasitoid interactions on oilseed rape (*Brassica napus* L.) and the 'functional spatial scales' at which the organisms respond to the spatio-temporally landscape context using the pollen beetle, *Meligethes aeneus* Fabricius (Coleoptera, Nitidulidae), and its specialized larval endoparasitoids (Hymenoptera, Ichneumonidae). The expansion of oilseed rape cultivation since the 1970s was accompanied by increased outbreaks of rape pest populations, so that farmers are recommended to use insecticides almost yearly. Analysing trophic interactions at multiple spatial scales may provide new insights on organism-space interactions and the potential of biological pest control.

9.2 Analyzing Spatial Landscape Pattern and Field Experiments

The landscapes in Southern Lower Saxony in Germany represent a continuum from extremely simple and structurally poor landscapes (>95% annual crops) to complex

Table 9.1 Proportions of habitat types in circular landscape sectors of 1.5 km diameter based on digital thematic maps of Southern Lower Saxony, Germany

Land use type	Mean ± SD	Minimum	Maximum
Arable	74.6 ± 17.1	50.9	98.3
Grass	12.9 ± 9.8	0	36.6
Forests	9.6 ± 11.2	0	29.3
Hedgerows	0.2 ± 0.4	0	1.4
Gardens	0.1 ± 0.1	0	0.5
Settlement	2.2 ± 3.3	0	12.7

and structurally rich ones with up to 50% uncropped habitats such as field margins, hedges, fallows and grassland. These landscapes have a land use history of >1,000 years and were formerly almost completely covered with forests. With human settlement emerged land use forms with arable land in flat areas with deep soil, forests and low intensity grassland in mountainous areas with shallow soil, and high intensity grasslands in wet areas along rivers (Table 9.1). Structurally simple and complex landscapes are geographically interdispersed in the region according to climate conditions, topography and natural soil fertility. Arable land is dominated by cereals (71% of crop fields), sugar beet (12%), oilseed rape (8%) and maize (4%), with a mean field size of 5.3 ha ± 2.6 S.D. ($N = 166$ fields).

Analyses of landscapes are concerned with a variety of potential methods and measures to describe and quantify landscape structure (for a review, see Gustafson 1998). We analysed two commonly used groups of landscape parameters, (i) metrics of landscape composition such as the proportion of habitat types (%) and habitat type diversity (after Shannon; H_S), and (ii) metrics of landscape configuration such as the perimeter-to-area ratio of arable fields (a measure of field edge density; P/A) and habitat isolation (a negative exponential weighting function measuring area and distance of study sites from semi-natural habitats; $I_{i;j}$; see Thies and Tscharntke 1999).

The percentage arable land turned out to be a simple predictor of landscape structural complexity due to its close negative correlation with H_S and P/A, and close positive correlation with $I_{i;j}$ (all P-values <0.001) (Roschewitz et al. 2005). Thus, low proportions of arable land are typical for landscapes with a high diversity of habitat types, smaller sizes of arable fields, higher numbers of field edges, and low degrees of isolation of study sites from semi-natural habitats. In structurally complex landscapes, the probability of semi-natural habitats to function as sources for the colonisation of crop fields is enhanced, especially for those species with low dispersal ability.

Moreover, agricultural landscapes are characterized by a temporal dimension of changes in landscape composition, owing to the rotation of crop species and changing political constraints. Oilseed rape is usually grown in the same field at intervals of 3–4 years, mainly as a consequence of accumulation of populations of soil pathogens and nematodes during crop growing, thereby resulting in changes of spatial landscape composition between years. We quantified this spatio-temporal

Fig. 9.1 Landscapes sectors in southern Lower Saxony, Germany, illustrating differences in landscape composition. (**a**) A structurally complex landscape with a high proportion (>50%) of semi-natural habitats such as grasslands, fallows, field margins and hedges. (**b**) A structurally simple landscape with a high proportion (>95%) of arable land. (**c**) A landscape with a high proportion (>15%) of rape crops, which changes (mainly into winter wheat) in the course of annual crop rotation (Photographs: Carsten Thies)

dynamic by calculating the percent difference of rape crop area cultivated in two consecutive years ($\%_{t+1} - \%_t = \Delta\%$), showing distinct interannual changes in landscape composition, with 7.2% maximum expansion of rape crop area and 8.2% maximum reduction of rape crop area from year-to-year in circular landscape sectors of 1.5 km diameter. The percentage of arable land was not correlated with the interannual changes of rape crop ($P = 0.290$). Thus, our landscape metrics were statistically independent; they were distributed over the full range of potential values, thereby meeting basic criteria to allow estimates of the role of spatio-temporal changing landscapes for trophic interactions (Fig. 9.1).

The pollen beetle, *M. aeneus*, is one of the economically most important pests on oilseed rape (Hoffmann and Schmutterer 1999, Alford et al. 2003, see also Williams Chapter 1 this volume). The adults feed on pollen in rape flowers as well as in young rape buds, causing the latter to drop leaving podless stalks and thereby reducing pod and seed development. The larvae develop in rape flowers; there are two larval instars. The larvae are attacked by three specialized univoltine endoparasitoids, *Tersilochus heterocerus* Thomson, *Phradis interstitialis* (Thomson) and *P. morionellus* (Holmgren) (Nilsson 2003, see also Ulber et al. Chapter 2 this volume). *Phradis interstitialis* can oviposit into the eggs and first instar larvae whereas *P. morionellus* and *T. heterocerus* attack mostly second instar larvae. These parasitoid species kill their host after the host larvae drop to the ground before pupation in the soil (Jourdheuil 1960). Parasitism was analyzed during spring rape flowering in June. Rates of parasitism were measured by dissection of the second instar larvae of the pollen beetle. Parasitoid species identification was based on egg shell characteristics, which were either black (*T. heterocerus*) or white (*Phradis* spp.) (Osborne 1960). Young beetle larvae (<3 mm long) were discarded from this measurement, because our earlier studies showed that their parasitism is very low. Plant damage caused by adult beetle feeding was quantified at peak ripeness of the spring rape in August. The number of pods, as well as the number of destroyed buds which had not developed into pods and appeared as podless stalks, were assessed.

Here, we present results of two experiments on trophic interactions between the pollen beetle and its parasitoids using 15 agricultural landscapes differing in (i) structural complexity and (ii) interannual changes of rape crop area, respectively. In the very centre of each landscape a patch of rape plants was exposed in a grassy field margin strip for standardized measurement. The inherent problem of confounding variables in landscape comparisons that many local variables change simultaneously, was met by exposing potted rape plants. These experimental patches had the same soil type, nutrient and water availability, and were planted with the same crop variety. We analysed the effects of landscape complexity and interannual changes of rape crop area, respectively, on trophic interactions at multiple spatial scales in two steps: Firstly, we examined a spatial scale of Ø 1.5 km, which is known to influence this specific plant-herbivore-parasitoid system (Thies and Tscharntke 1999). Secondly, we examined the effects of the landscape context at four further spatial scales ranging from Ø 0.5–Ø 3 km, because it was not a priori clear at which spatial scale the landscape context has the strongest effect (Fig. 9.2).

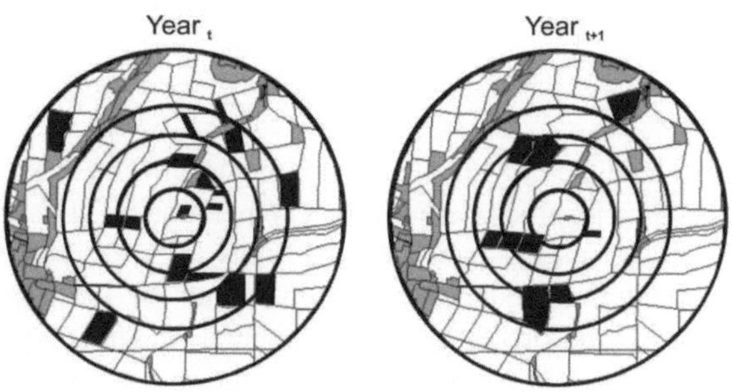

Fig. 9.2 Crop-dominated landscape sectors (Ø 1.5 km: 97.7% annual crops; *black* + *white areas*) in a digital map illustrating the rotation of oilseed rape (*black areas*) at five spatial scale in two consecutive years (Ø 1.5 km and year$_t$: 11.1% rape crops; Ø 1.5 km and year$_{t+1}$: 16.3% rape crops). Grey areas indicate semi-natural habitats. The *circles* show the five circular sectors (Ø 0.5, Ø 1, Ø 1.5, Ø 2, Ø 3 km), representing a nested set of landscape sectors at five spatial scales

9.3 Biological Control Across Landscape Complexity Gradients

The analyses across a gradient of landscape complexity showed the importance of the landscape context for local plant-herbivore-parasitoid interactions, thereby explaining why the same experimental design can result in a positive or a negative effect on biocontrol. The percentage of parasitism decreased and the percentage of destroyed buds increased as the percentage of arable land in the surrounding landscape (Ø 1.5 km) increased (Fig. 9.3a, b). High pollen beetle mortality due to parasitism and low plant damage caused by the pollen beetle were only found in structurally complex landscapes with a high percentage of semi-natural habitats and relatively little area converted to annual crop fields. Thus, semi-natural habitats appeared to support large parasitoid populations greatly contributing to the reduction of populations of the pollen beetle (Thies et al. 2003).

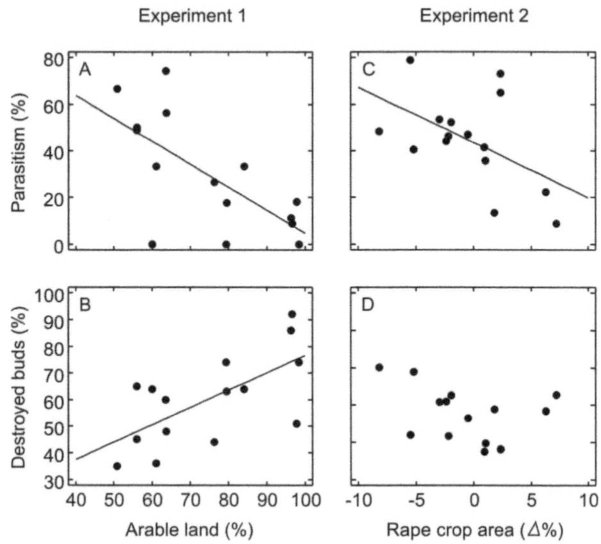

Fig. 9.3 Dependence of plant–herbivore–parasitoid interactions on (**a**, **b**) the percentage of arable land (experiment 1), and (**c**, **d**) the interannual changes of the percentage of rape crop area (Δ%) (experiment 2) in 15 agricultural landscapes at a spatial scale of Ø 1.5 km. For statistics, see Table 9.2

The interannual changes of landscape composition (Ø 1.5 km) owing to crop rotation also influenced trophic interactions. The percentage of parasitism decreased following rape crop expansion, and increased following rape crop contraction, indicating interannual dilution and concentration effects of the higher trophic level populations. In contrast, herbivory by the pollen beetle did not respond to these interannual landscape changes (Fig. 9.3c, d). These dilution and concentration effects on percent parasitism owing to crop rotation support the idea that populations of higher trophic levels are more sensitive to disturbance and ecological change (Holt et al. 1999). They also suggest that the regional population pool may be more important for biological control than local management such as timing of insecticide applications, reducing tillage and introducing field margins (Thies et al. 2008).

When semi-natural habitat area dropped below a value of ~20% of the landscape, or when the expansion of rape crop area between years exceeded ~5% of the landscape, respectively, percent parasitism dropped below a threshold value of about 32–36%, below which success in classical biological control has never been reported (Hawkins and Cornell 1994). In such landscape sectors pollen beetles appeared to be released from natural biological control. The conservation and/or creation of more semi-natural habitats in agricultural landscapes in combination with low spatial dynamics of rape crop rotation may therefore enhance populations of natural enemies, which immigrate into crop fields, attack pest insects and contribute to the reduction of pest populations below an economic threshold.

Accompanying inspections of rape crops support these findings. In structurally simple landscapes (with <20% semi-natural habitats), parasitism of the pollen beetle significantly decreased from the edge to the centre of the crop. In contrast, in structurally complex landscapes (with >20% semi-natural habitats), parasitism was generally higher and did not decrease from the edge to the centre (Tscharntke et al. 2002). Hence, local habitat management to conserve natural enemies such as the creation of field margins or flower strips (Landis et al. 2000) is less effective in complex landscapes than in structurally simple landscapes.

9.4 Identifying 'Functional Spatial Scales' for Landscape Management

The identification of ecological processes changing across spatial and temporal scales is a central challenge in landscape ecology. However, there is no single scale to describe an ecological system. Interacting communities are made up of species with different spatial strategies (Kareiva 1990, Holt 1996, With et al. 1999), and the perception of landscape complexity facilitating or impeding movement among habitat patches is species specific (Taylor et al. 1993, Wiens et al. 1997). The pollen beetle and its parasitoids are widely distributed in agricultural landscapes as they also use alternative hosts and host plants in semi-natural habitats. Different life history traits make the understanding of trophic interactions and population dynamics even more difficult. The experiments showed that parasitism was a negative and herbivory a positive function of the proportion of arable land with most significant results at landscape sectors of Ø 1.5 km. Moreover, parasitism responded negatively to interannual rape crop expansion, and positively to interannual rape crop contraction at this landscape scale. We tested how the plant-herbivore-parasitoid interactions responded at smaller and larger landscape sectors to explore how the species perceive their environment. The predictive power of the proportion of arable land for parasitism and herbivory as well as the predictive power of the interannual changes of rape crop area for parasitism differed in dependence on the spatial scale considered, indicating scale-dependence for both plant-herbivore and herbivore-parasitoid interactions. Herbivory and parasitism showed the best correlations with both the percentage of arable land and percentage of interannually changing rape crops at the same spatial scales, i.e., landscape sectors of 1–2 km diameter, thereby suggesting that this 'functional spatial scale' indicates their dispersal abilities (Table 9.2). The hypothesis that higher trophic levels experience their environment at a larger spatial scale was not supported (Holt 1996). Nonetheless, the predictive power of non-crop area changed only slightly for herbivory, but greatly with respect to parasitism as scales increased from Ø 0.5 to Ø 1.5 km and from Ø 1.5 to Ø 3 km, which may indicate a higher spatial susceptibility of parasitoids. Furthermore, the slopes of the regression lines tended to be steeper in parasitism than herbivory suggesting a greater effect of landscape changes on parasitoids. This is in support of the general idea that higher trophic levels should be more susceptible to disturbance.

Table 9.2 Dependence of plant-herbivore-parasitoid interactions on the percentage of arable land (experiment 1) and the interannual changes of the percentage of rape crop area ($\Delta\%$) (experiment 2) in 15 agricultural landscapes at five spatial scales. Percentage of parasitism due to *T. heterocerus* + *Phradis* spp. Percentage of destroyed rape buds (herbivory by *M. aeneus*). F-values, correlation coefficients (R) and levels of significance are from linear regressions ($N = 15$ landscapes)

Response variable	F	R		Regression model
Dependence on percentage of arable land (%)				
Parasitism (%)				
Scale: 0.5 km	2.2	−0.38		Y = 65.6 - 0.5X
Scale: 1 km	4.9	−0.52	*	Y = 76.7 - 0.7X
Scale: 1.5 km	6.5	−0.58	*	Y = 82.1 - 0.9X
Scale: 2 km	5.0	−0.53	*	Y = 78.2 - 0.8X
Scale: 3 km	2.7	−0.41		Y = 67.6 - 0.7X
Destroyed buds (%)				
Scale: 0.5 km	8.1	0.62	*	Y = 18.7 + 0.5X
Scale: 1 km	9.0	0.64	*	Y = 19.4 + 0.5X
Scale: 1.5 km	9.6	0.65	**	Y = 18.6 + 0.5X
Scale: 2 km	6.2	0.57	*	Y = 22.3 + 0.5X
Scale: 3 km	5.5	0.54	*	Y = 23.4 + 0.5X
Dependence on interannual changes of percentage of rape crop area ($\Delta\%$)				
Parasitism (%)				
Scale: 0.5 km	0.1	−0.102		Y = 41.4 - 0.1X
Scale: 1 km	6.5	−0.579	*	Y = 42.3 - 1.0X
Scale: 1.5 km	5.1	−0.530	*	Y = 45.3 - 1.7X
Scale: 2 km	3.5	−0.458		Y = 40.6 - 1.9X
Scale: 3 km	2.1	−0.371		Y = 40.7 - 2.1X
Destroyed buds (%)				
Scale: 0.5 km	0.1	0.070		Y = 37.5 - 0.1X
Scale: 1 km	0.4	0.177		Y = 37.4 - 0.1X
Scale: 1.5 km	2.0	−0.366		Y = 41.4 - 0.4X
Scale: 2 km	1.4	−0.316		Y = 37.2 - 0.5X
Scale: 3 km	< 0.1	−0.052		Y = 37.4 - 0.1X

* $P < 0.05$; ** $P < 0.01$.

9.5 Conclusions

Our results show that the spatio-temporally changing landscape context significantly influences plant-herbivore-parasitoid interactions. Therefore, understanding of local trophic interactions and biological control profits from a landscape perspective. The interaction between the first and the second trophic level (herbivory) as well as the second and the third trophic level (parasitism) were scale-dependent, suggesting 'functional spatial scales' for species-specific landscape management. The mortality rates of the pollen beetle due to parasitim decreased linearly with decreasing proportion of semi-natural habitats, with no critical threshold beyond which parasitism drastically dropped. Thus, this type of ecological functioning continuously decreased with the loss of semi-natural habitat (for theoretical models, see Kareiva and Wennergren 1995, With and Crist 1995).

Semi-natural habitats provide important resources for both pests and their natural enemies. In particular, wild Brassicaceae are known as alternative host plants of *M. aeneus* and its parasitoids (Frenzel and Brandl 1998), and several other *Meligethes* species feeding on non-cruciferous herbs, such as *Lamium* spp. and *Symphytum* spp., are also hosts of the parasitoids attacking *M. aeneus* (Horstmann 1981). These alternative host plants mainly occur on fallows, field margins, extensively managed grasslands and hedges in structurally complex landscapes. In addition, the parasitoids hibernate in the soil and are known to be negatively affected by ploughing (Nilsson 1985, see also Nilsson Chapter 11 this volume). Undisturbed areas should better enable parasitoid populations to build up over years, whereas a high proportion of annual crops should disadvantage them. Complex landscapes also provide more nectar resources due to a larger cover of flowering plants (Steffan-Dewenter et al. 2002), thereby extending parasitoid lifetimes and enhancing parasitism rates (for reviews, see Powell 1986, Wratten and Van Emden 1995).

The inter-annually changing mosaics of rape crops appeared to affect parasitoid populations more strongly than their herbivorous hosts. Lower dispersal rates of specialist parasitoids compared to their host might explain this process. Pollen beetle parasitoids remain for overwintering in the upper soil layer in or adjacent to the rape crop and emerge in May of the subsequent year. In contrast, new generation pollen beetles emerge in June/July of the current year, disperse to wild and cultivated Brassicaceae, and many other flowering plants, to feed on pollen before further dispersing to hibernating sites in the humus layer of nearby forests, hedges, river banks, and similar habitats (Hoffmann and Schmutterer 1999). The pollen beetles thereby can counterbalance differences in rape crop area between landscapes, whereas the parasitoids appear to disperse at smaller landscape scales, making them more susceptible to management changes at small spatial scales. Thus, the spatial distance between rape crops in two consecutive years due to crop rotation may often be too large to be covered by parasitoids, thereby releasing pollen beetles from natural biological control. Such an increased susceptibility of natural enemies to changing landscape composition of crops can be generally hypothesized to occur in weakly dispersing and specialized parasitoids such as *T. heterocerus* and *Phradis* spp., but not in high dispersal and generalist natural enemies. Dispersal processes of interacting organisms at the landscape scale have not been explicitly quantified so far, but appear to be essential to understanding functional connectivity between habitats (Holt et al. 1999, Cronin and Reeves 2005, Thies et al. 2005, Rand et al. 2006). The landscape context is related to local processes via dispersal, which is often related to species traits such as body size, foraging range, resource specialization, population size variability and trophic position (Tscharntke and Brandl 2004, Tscharntke et al. 2007).

In summary, community structures and dynamics in agricultural landscapes are spatially heterogeneous, exhibiting considerable variability in time and space. Our results indicate strong links between the landscape context and local ecological processes. Spatially explicit approaches may therefore provide a perspective to identify 'functional spatial scales' at which species experience their environment, with implication for future resource and landscape management.

Acknowledgements This work was supported by the German Research Foundation (Deutsche Forschungsgemeinschaft) and the German Ministry for Research and Education (Bundesministerium für Bildung und Forschung).

References

Alford DV, Nilsson C, Ulber B (2003) Insect pests of oilseed rape crops. In: Alford DV (ed.) *Biocontrol of oilseed rape pests*. Blackwell, Oxford, UK.

Altieri MA, Cure JR, Garcia MA (1993) The role and enhancement of parasitic hymenoptera biodiversity in agroecosystems. In: LaSalle J, Gauld ID (eds.) *Hymenoptera and biodiversity*. CAB International, London.

Bianchi FJJA, Booij CJH, Tscharntke T (2006) Sustainable pest regulation in agricultural landscapes: A review on landscape composition, biodiversity and natural pest control. *P Roy Soc Lond B Bio* 273: 1715–1727.

Burel F, Baudry J (1995) Farming landscapes and insects. In: Glen DM, Greaves MP, Anderson HP (eds.) *Ecology and integrated farming systems*. Wiley, New York, USA.

Cronin JT, Reeves JD (2005) Host-parasitoid spatial ecology: A plea for a landscape-level synthesis. *P Roy Soc Lond B Bio* 272: 2225–2235.

Frenzel M, Brandl R (1998) Diversity and composition of phytophagous insect guilds on Brassicaceae. *Oecologia* 113: 391–399.

Gustafson EJ (1998) Quantifying landscape spatial pattern: What is the state of art? *Ecosystems* 1: 143–156.

Halaj J, Wise DH (2001) Terrestrial trophic cascades: How much do they trickle? *Am Nat* 157: 262–281.

Hawkins BA, Cornell HV (1994) Maximum parasitism rate and successful biological control. *Science* 262: 1886.

Hoffmann GM, Schmutterer H (1999) *Krankheiten und Schädlinge an landwirtschaftlichen Kulturpflanzen*, 2nd edn. Ulmer Verlag, Stuttgart, Germany.

Holt RD (1996) Food webs in space: An island biogeographic perspective. In: Polis GA, Winemiller KO (eds.) *Food webs – Integration of patterns and dynamics*. Chapman & Hall, New York, USA.

Holt RD, Lawton JH, Polis GA, Martinez ND (1999) Trophic rank and species-area relationship. *Ecology* 80: 1495–1504.

Horstmann K (1981) Revision der Europäischen Tersilochinen II (Hymenoptera: Ichneumonidae). *Spixiana* Suppl 4: 1–76.

Jourdheuil P (1960) Influence de quelques facteurs écologiques sur les fluctuations de population d'une biocénose parasitaire. Etude à quelques Hymenoptère parasites de divers Coléoptères inféodès aux Crucifères. *Ann Epiphyties* 11: 445–539.

Kareiva P (1990) Population dynamics in spatial complex environments: Theory and data. *Philos T Roy Soc B* 330: 175–190.

Kareiva P, Wennergren U (1995) Connecting landscape pattern to ecosystem and population processes. *Nature* 373: 299–302.

Landis DA, Wratten SD, Gurr GM (2000) Habitat management to conserve natural enemies of arthropod pests in agriculture. *Annu Rev Entomol* 45: 175–201.

Matson PA, Parton WJ, Power AG, Swift MJ (1997) Agricultural intensification and ecosystem properties. *Science* 277: 504–509.

McCauley E, Wilson WG, de Roos AM (1993) Dynamics of age-structured and spatially structured predator-prey interactions: Individual-based models and population-level formulations. *Am Nat* 142: 412–442.

Menalled FD, Marino PC, Gage SH, Landis DA (1999) Does agricultural landscape structure affect parasitism and parasitoid diversity? *Ecol Appl* 9: 634–641.

Nilsson C (1985) Impact of ploughing on emergence of pollen beetle parasitoids after hibernation. *Z Angew Entomol* 100: 302–308.

Nilsson C (2003) Parasitoids of pollen beetles. In: Alford DV (ed.) *Biocontrol of oilseed rape pests*. Blackwell, Oxford, UK.

Osborne P (1960) Observations on the natural enemies of *Meligethes aeneus* (F.) and *Meligethes viridescens* (F.) (Coleoptera Nitidulidae). *Parasitology* 50: 91–110.

Östman Ö, Ekbom B, Bengtson J (2001) Farming practice and landscape heterogeneity influence biological control. *Basic Appl Ecol* 2: 365–371.

Pickett STA, Cadenasso ML (1995) Landscape ecology: Spatial heterogeneity in ecological systems. *Science* 269: 331–334.

Pimm SL (1991) *The balance of nature*. The University of Chicago Press, Chicago.

Powell W (1986) Enhancing parasitoid activity in crops. In: Waage J, Greathead W (eds.) *Insect parasitoids*. Academic Press, London, UK.

Rand TA, Tylianakis JM, Tscharntke T (2006) Spillover edge effects: The dispersal of agriculturally subsidized insect predators into adjacent natural habitats. *Ecol Lett* 9: 603–614.

Ricketts TH (2001) The matrix matters: Effective isolation in fragmented landscapes. *Am Nat* 158: 87–99.

Roschewitz I, Thies C, Tscharntke T (2005) Are landscape complexity and farm specialisation related to land-use intensity of annual crop fields? *Agr Ecosyst Environ* 105: 87–99.

Rosenzweig ML (1995) *Species diversity in time and space*. Cambridge University Press, Cambridge, UK.

Steffan-Dewenter I, Münzenberg U, Bürger C, Thies C, Tscharntke T (2002) Scale-dependent effects of landscape structure on three pollinator guilds. *Ecology* 83: 1421–1432.

Taylor PD, Fahrig L, Hennein K, Merriam G (1993) Connectivity is a vital element of landscape structure. *Oikos* 68: 571–573.

Thies C, Roschewitz I, Tscharntke T (2005) The landscape context of cereal aphid-parasitoid interactions. *P Roy Soc Lond B Bio* 272: 203–210.

Thies C, Steffan-Dewenter I, Tscharntke T (2003) Effects of landscape context on herbivory and parasitism at different spatial scales. *Oikos* 101: 18–25.

Thies C, Steffan-Dewenter I, Tscharntke T (2008) Interannual landscape changes influence plant-herbivore-parasitoid interactions. *Agr Ecosyst Environ* 125: 266–268.

Thies C, Tscharntke T (1999) Landscape structure and biological control in agroecosystems. *Science* 285: 893–895.

Tscharntke T, Bommarco R, Clough Y, Crist TO, Kleijn D, Rand T, Tylianakis JM, van Nouhuys S, Vidal S (2007) Conservation biological control and enemy diversity on a landscape scale. *Biol Control* 43: 294–309.

Tscharntke T, Brandl R (2004) Plant-insect interactions in fragmented landscapes. *Annu Rev Entomol* 49: 405–430.

Tscharntke T, Klein A-M, Kruess A, Steffan-Dewenter I, Thies C (2005) Landscape perspectives on agricultural intensification and biodiversity-ecosystem service management. *Ecol Lett* 8: 857–874.

Tscharntke T, Steffan-Dewenter I, Kruess A, Thies C (2002) Contribution of small habitat fragments to conservation of insect communities of grassland-cropland landscapes. *Ecol Appl* 12: 354–363.

Van Driesche RG, Bellows TS (1996) *Biological control*. Chapman & Hall, New York, USA.

Vollhardt IMG, Tscharntke T, Wäckers FL, Bianchi FJJA, Thies C (2008) Diversity of cereal aphid parasitoids in simple and complex landscapes. *Agr Ecosyst Environ* 126: 289–292.

Wiens JA, Schooley RL, Weeks RD (1997) Patchy landscapes and animal movements: Do beetles percolate? *Oikos* 78: 257–264.

Wiens JA, Stenseth NC, van Horne B, Ims RA (1993) Ecological mechanisms and landscape ecology. *Oikos* 66: 369–380.

With KA, Cadaret SJ, Davis C (1999) Movement responses to patch structure in experimental fractal landscapes. *Ecology* 80: 1340–1353.

With KA, Crist TO (1995) Critical thresholds in species' response to landscape structure. *Ecology* 76: 2446–2459.

Wratten SD, Van Emden HF (1995) Habitat management for enhanced activity of natural enemies of insect pests. In: Glen DM, Greaves MP, Anderson HM (eds.) *Ecology and integrated farming systems*. Wiley, New York, USA.

Chapter 10
Insect Pests and Spiders in Oilseed Rape and Their Response to Site and Landscape Factors

Thomas Frank, Thomas Drapela, Dietmar Moser, and Johann G. Zaller

Abstract The abundance of the insect pests: pollen beetle (*Meligethes aeneus*), stem weevils (*Ceutorhynchus napi, C. pallidactylus*) and brassica pod midge (*Dasineura brassicae*), pest damage, species richness and activity density of spiders, and density, body size and offspring of the wolf spider, *Pardosa agrestis*, in oilseed rape fields relative to site and landscape factors were investigated. Abundances of pollen beetles and stem weevils were significantly positively correlated with soil quality and negatively related to oilseed rape area in the surroundings. Generally, abundances of all groups were positively related to woody areas. Damage by pollen beetle and pod midge was negatively correlated with rape area, damage by the stem weevils responded positively to soil index. Spider richness was positively related to woody areas at small spatial scale, spider density increased with length of road-side strips at large scale. Also, body size of *P. agrestis* was best explained by length of road-side strips and number of offspring increased as distance to woody areas decreased. Non-crop areas surrounding rape fields promoted both spider fitness and assemblages in rape fields, thus underlining the importance of these habitats for biological pest control. This may become particularly significant as future plans to boost biofuel production will drastically reduce non-crop areas in agricultural landscapes. Our finding that the spatial configuration of non-crop habitats favours predators indicates that landscape management strategies should focus on interspersing non-crop habitats within the matrix of arable fields in a way that distances between refuge or source habitats and arable fields are kept short.

10.1 Introduction

The importance of oilseed rape (OSR, *Brassica napus* L., canola) as a source for industrial and nutritional oil has been increasing worldwide during the last decades; however in many regions this rising acreage is accompanied by a dramatic disproportionate increase in pesticide applications to this crop (Gianessi and

---

T. Frank (✉)
Department of Integrative Biology and Biodiversity Research, Institute of Zoology, University of Natural Resources and Applied Life Sciences Vienna, Vienna, A-1180, Austria
e-mail: thomas.frank@boku.ac.at

I.H. Williams (ed.), *Biocontrol-Based Integrated Management of Oilseed Rape Pests*, DOI 10.1007/978-90-481-3983-5_10,

Marcelli 2000). Prevalent pest problems are often attributed to the transformation of formerly heterogeneous landscapes with high proportions of semi-natural non-crop habitats to more monotonous landscapes mainly dominated by arable land and its local decrease of biodiversity (Kareiva 1990, Pickett and Cadenasso 1995) and detrimental consequences for beneficial organisms that might control various pest species (Polis et al. 1997, Roland and Taylor 1997, Menalled et al. 1999, Thies and Tscharntke 1999, Östman et al. 2001b).

Although in agroecological research it has been appreciated that plant-insect interactions depend on scales much larger than a single habitat (Wiens 1989, Levin 1992, Schneider 1994), crop-pest interactions have mainly been studied on single pest species by focusing either on the impact of site factors or on landscape structure but only rarely included both factors (Östman et al. 2001a). Here we focus on how major insect pests of oilseed rape and spiders respond to site and landscape characteristics at various spatial scales.

Both insect pests and predators in oilseed rape fields have recently been reviewed (Alford 2003). Thus, rather than just providing another review on this topic, the aim of this chapter is to briefly summarise the findings of a series of recent studies of pest-spider-landscape interactions in 29 landscapes (Drapela et al. 2008, 2009, Moser et al. 2009, Zaller et al. 2008a, b, 2009). Novel aspects of the research were the (i) differentiation between various types of non-crop habitats, (ii) inclusion of several pest species and (iii) investigation of the combined influence of site and landscape factors on insect pests, insect pest damage and spiders. This research was conducted in an agricultural region near Vienna (Austria) in winter oilseed rape fields that were embedded in differently structured landscapes, forming a complexity gradient ranging from structurally poor to structurally complex. Prior to the study year, the OSR fields were fertilised and treated with herbicides, fungicides, and insecticides following common agricultural practice, but, in the study year, a 1-ha-area of each OSR field was excluded from pesticide applications. The surrounding landscape of each study field was analysed at eight circular landscape sectors within radii between 250 and 2,000 m (Fig. 10.1). Within these sectors we

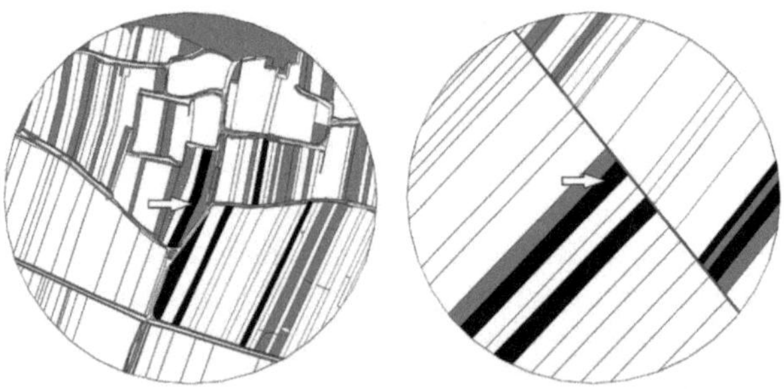

Fig. 10.1 Two of the study landscapes representing a structurally complex (*left*) and a structurally poor landscape (*right*). *Black* with *arrow*: studied OSR field (= study site). *Black*: other OSR fields. *Grey*: non-crop areas, i.e. fallows, road-side strips and woody areas. *White*: non-OSR crop fields and roads. Radius = 500 m (Moser unpublished)

calculated the area of non-crop habitats, OSR fields, fallows and woody habitats, the total length of road-side strips and of hedges, isolation of OSR fields and landscape diversity (Shannon-Wiener index). In addition, the site variables soil index, soil cultivation intensity, nitrogen fertiliser input, pesticide use, OSR ground cover and stand density were measured.

10.2 Insect Pests

In Europe, the most important insect pests in OSR are the cabbage stem flea beetle (*Psylliodes chrysocephala* L., Chrysomelidae), the pollen beetle (*Meligethes aeneus* (Fabricius), Nitidulidae), the cabbage seed weevil (*Ceutorhynchus assimilis* (Paykull) syn. *C. obstrictus* (Marsham), Curculionidae), the rape stem weevil (*C. napi* Gyllenhall), the cabbage stem weevil (*C. pallidactylus* (Marsham)) and the brassica pod midge (*Dasineura brassicae* (Winnertz), Cecidomyiidae) (Alford et al. 2003). Although the oilseed rape crop has been shown to compensate considerably after insect damage (Free and Williams 1978, 1979), yield losses up to 80% have been reported when insecticide spraying ceased (Hansen 2003).

Crop-pest interactions have usually been studied on single pest species by focusing either on the impact of within-field or on landscape parameters but only rarely included effects of both factors on more pest species (Östman et al. 2001a). Here we present findings of studies where we investigated how abundance and damage of three major groups of pest species in OSR, the pollen beetle, the two stem weevils and the brassica pod midge, are related to site and landscape factors (more details can be found in Zaller et al. 2008a, b). These species infest different parts of the crop at different developmental stages: (i) the stem weevils lay eggs in leaf petioles or midribs of OSR plants, hatched larvae tunnel in the stems; (ii) pollen beetle adults feed on pollen damaging flowering structures particularly during the green to yellow bud stages resulting in podless peduncles; and (iii) brassica pod midge lay eggs into pods where the hatched larvae feed on the inner wall of the pod and cause the pods to split prematurely (Alford et al. 2003). Because pest species must migrate to colonize the crop we hypothesized that the spatial patterns of OSR fields and potential overwintering sites will affect their abundance and the damage they cause. With the exception of the pollen beetle (Thies and Tscharntke 1999, Thies et al. 2003) these pest species have not been studied in a landscape context before.

10.2.1 Pest Abundance in Relation to Site and Landscape Factors

10.2.1.1 Pest Abundance and Site Characteristics

Univariate regression analyses showed that only pollen beetle and the stem weevils but not pod midge were positively related to soil quality but remained unrelated to nitrogen fertilization levels (for more details see Zaller et al. 2008a). This indicates that for pollen beetle and the stem weevils soil characteristics described by the soil index (e.g., soil texture, water holding capacity, clay content) seemed to be more important for their abundance than nitrogen fertilizer inputs. Because, in spring, pollen beetle and stem weevils immigrate into the OSR fields from overwintering habitats, soil quality of the OSR field they colonise is unlikely to have any direct effect on their current abundance. However, it is most likely that effects of soil quality on the nutritional quality or on the canopy microclimate of OSR could have influenced the searching efficiency of pollen beetle and stem weevils and thus indirectly altered their abundance (Walters et al. 2003). In fact, glucosinolates and their catabolites in OSR have been found to be important cues to host selection of cruciferous pests, aiding both orientation to and recognition of the host plant (Bartlet 1996). A curvilinear shape of the relationship between pollen beetle and stem weevil abundance and soil index with a maximum at average levels (details are shown below) indicates that OSR quality at low soil index was not suitable for pests; on the other hand at a higher soil quality the crop could have exhibited a better ability to protect itself from herbivores via the production of secondary compounds (Cipollini and Bergelson 2002). It remains to be tested experimentally, whether soil quality could also have affected the colour and odour of OSR plants and thereby altered the attractiveness for pest species.

10.2.1.2 Pest Abundance and Landscape Characteristics

Pollen beetle was the pest species that showed the most pronounced relations to landscape variables with a negative relationship to OSR area and positive relations with the isolation of OSR in the landscape, landscape diversity, and the proportions of non-crop and of woody areas (Fig. 10.2a, d, g, j, m). Stem weevil larvae were negatively related to OSR area, but positively related to OSR isolation and the proportion of woody areas (Fig. 10.2b, e, n). Abundance of pod midge larvae was positively related to landscape diversity and the proportion of woody areas (Fig. 10.2i, o).

We hypothesized that specialist pest species respond more to the proportion and isolation of OSR than to non-OSR elements in the landscape (Holt et al. 1999). This was confirmed for pollen beetle, partly confirmed for the stem weevils but not for pod midge. Pollen beetle and the stem weevils were negatively associated with OSR area in the landscape revealing that they were more abundant in OSR fields in landscapes with less OSR area and less abundant when more OSR area was available. This is in contrast to findings by Thies et al. (2003), but can be explained by the different methodological approaches used. While potted OSR plants located in old field margins were used in the former study we studied pollen beetle abundance on OSR plants growing in the field. Because of the annual generation cycles of the studied species these patterns could be expected to mirror the situation regarding the areas of OSR and overwintering sites of the preceding year that enabled the buildup of a certain landscape pest pool that was then dispersed among available OSR area in the current year (Hokkanen 2000). However, since in the study region OSR cropping history and non-crop areas had remained fairly constant for several years prior to the study because agri-environmental programmes provided subsidies to promote both OSR cropping and extensification in the region (BMLFUW 2006), we assume that pest migration processes are most likely responsible for these findings.

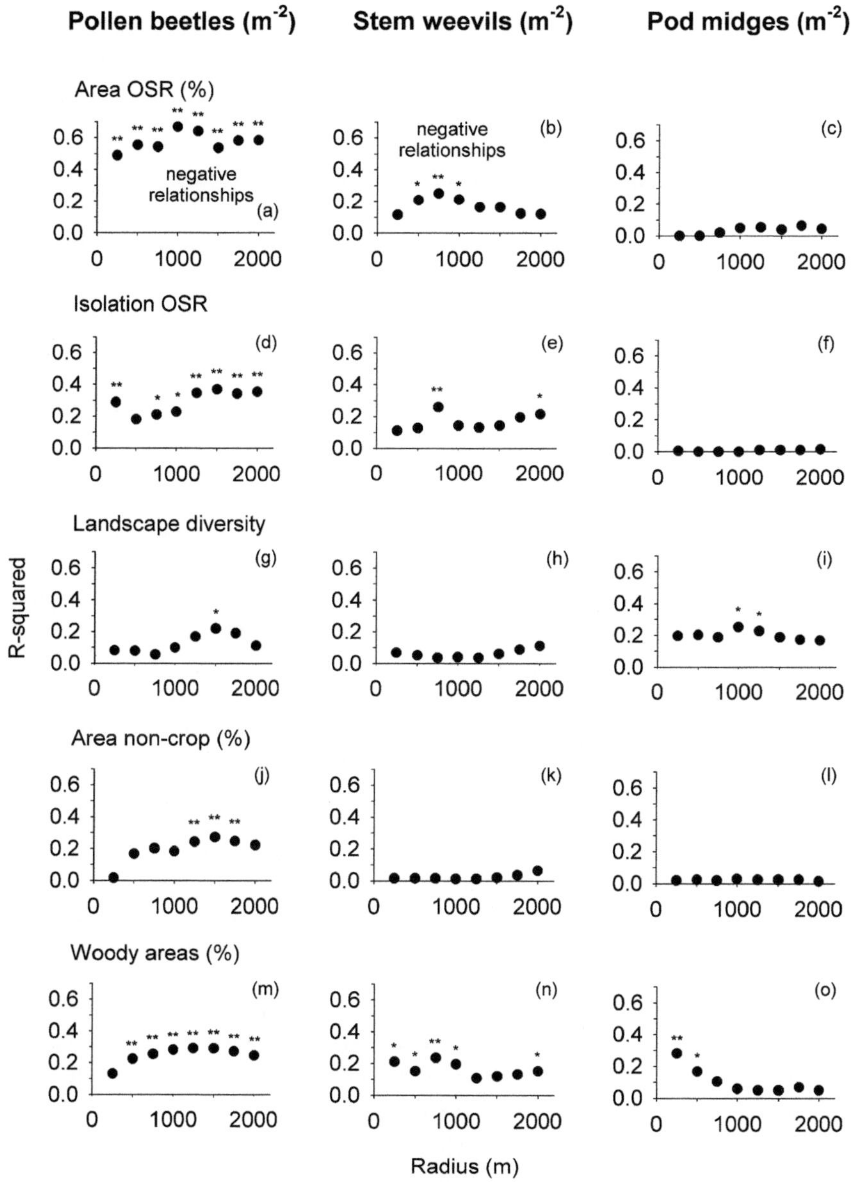

Fig. 10.2 Relationships between abundance of pollen beetle (*M. aeneus*) adults, stem weevil (*C. napi*, *C. pallidactylus*) larvae and pod midge (*D. brassicae*) larvae, and landscape variables across the investigated radii derived from univariate ordinary least square regression analyses. *Asterisks* denote statistical significance: $* \ P < 0.05$, $** \ P < 0.01$. With the exception of graphs **a** and **b** all relationships are positive (after Zaller et al. 2008a)

Pollen beetle abundance and non-crop area were positively related suggesting that more complex landscapes support a greater variety of alternative host plants for the beetle and therefore complex landscapes may have enhanced pest populations (Frenzel and Brandl 1998). However, this finding again appears to be in contrast to that of others (Thies and Tscharntke 1999, Thies et al. 2003) who showed a negative relationship between pollen beetle damage (i.e., number of podless stalks) and non-crop area. Because pollen beetle abundance and pollen beetle damage correlated well (Zaller et al. 2008b) several explanations could be considered for this discrepancy. Firstly, the parameter non-crop area per se does not adequately describe landscapes because non-crop usually encompasses various habitats (e.g., fallows, meadows, woody areas) that can individually have contrasting effects on pest abundance. For instance, in the data presented here non-crop was mainly comprised of grassy fallows and woody areas, however only the latter showed a significant effect on pest abundance. Moreover, grassland elements were mainly composed of grassy fallows while, in the former studies, pastures and meadows formed the grassland elements (Thies and Tscharntke 1999, Thies et al. 2003). Secondly, contrasting findings might again result from the fundamentally different methodological approaches used (see above). The positive relations between pollen beetle and non-crop area also show that natural enemies of pests inhabiting these non-crop areas (e.g., hymenopterous parasitoids – Thies and Tscharntke 1999, carabid beetles – Östman et al. 2001a, spiders – Schmidt et al. 2005, Drapela et al. 2008) had no detrimental influence on pests in the region investigated. Indeed, we observed that parasitoids of pollen beetle seemed to play only a minor role in the study region (Zaller et al. 2009) and there is evidence that, beside beneficial organisms, pests are also supported by non-crop structures (Thies et al. 2005).

Pod midge abundance was unrelated to OSR variables but positively related to landscape diversity and woody areas. This finding seems somewhat surprising because pod midge adults are reported to emerge in the spring from the OSR field of the previous year and then to migrate to current OSR fields (Alford et al. 2003). However, it also has been observed that second and third generations of these polyvoltine pests overwinter in stands of other Brassicaceae (Paul 2003). Thus, our data suggest that for this species a more diverse landscape with more woody areas increases the likelihood of finding alternative host plants. However, our understanding of landscape effects on pod midge is still too scarce to be able to further interpret these findings.

Two overall trends for the three species groups could be seen in our data: (i) pest abundance was consistently positively related to woody areas, and (ii) there was no relation between pest abundance and grassy fallows. For pollen beetle, the stimulating effect of woody areas can be explained by the choice of these habitats as overwintering sites. For the other species that are thought to overwinter in non-woody areas, it is possible that woody areas in the vicinity of the actual overwintering habitats could have climatically influenced these habitats. Therefore, the regression analysis perhaps revealed the indirect contribution of woody areas on non-woody overwintering sites (e.g., reduced wind exposure or increased relative humidity). No detrimental effects of grassy fallows on populations of OSR pests could be identified although these sites have frequently been shown to be important overwintering habitats for natural enemies of OSR pests (Östman et al. 2001b, Thies et al. 2003). The lack of a relation between pests and grassy fallows also suggests that the widely-held concern, namely that fallows provide refuge for insect pests (Lethmayer et al. 1997), was not substantiated by our data.

10.2.1.3 Multivariate Responses to Site and Landscape Factors

Pollen beetle abundance was best explained by the proportion of OSR at 1,000 m radius and soil quality with a distinct negative response to OSR area and a curvilinear relationship to soil index. The plot of the partial effects indicated that pollen beetle abundance was affected by the soil index mainly below 4% OSR area in the landscape (Fig. 10.3a). Above 4% OSR area soil index played only a marginal role in modulating the primary correlation of pollen beetle abundance to OSR area. Maximum pollen beetle abundance is predicted by this model at lowest levels of OSR area and average soil index values. The final model for stem weevil abundance also consisted of two variables: soil index and proportion of woody areas at a radius of 250 m. The plot of the partial effects showed that abundance is highest when soil index values were slightly above average and proportion of woody areas is high (Fig. 10.3b). While abundance showed a linear response to woody area, the relation to soil index is curvilinear which resembles a saturation curve. For pod midge abundance beside the proportion of woody areas no second variable could significantly enhance the predictive power of the model.

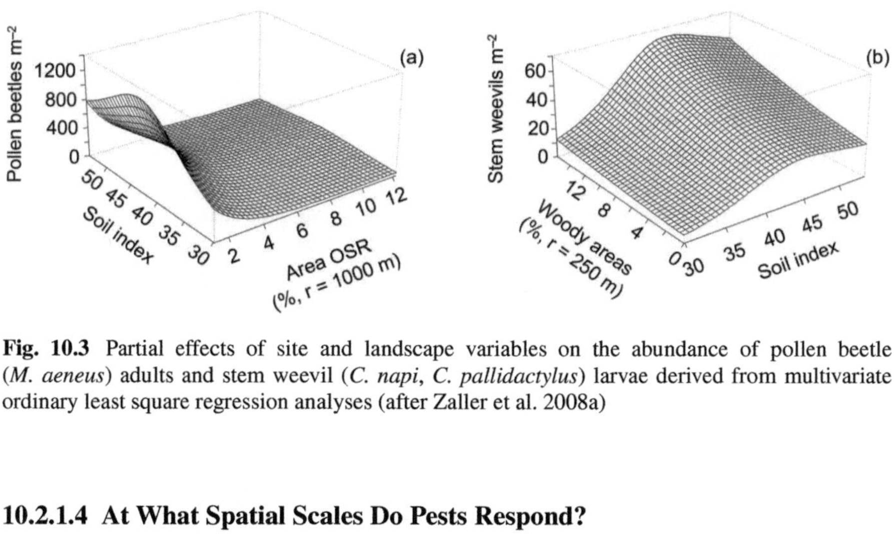

Fig. 10.3 Partial effects of site and landscape variables on the abundance of pollen beetle (*M. aeneus*) adults and stem weevil (*C. napi*, *C. pallidactylus*) larvae derived from multivariate ordinary least square regression analyses (after Zaller et al. 2008a)

10.2.1.4 At What Spatial Scales Do Pests Respond?

For most pest species responses to landscape variables showed clear maxima at certain scales. Abundance of pod midge larvae was explained by proportion of woody areas up to 500 m only but no relation to woody areas at greater radii (Fig. 10.2o). Additionally, pod midge abundance was significantly associated with landscape diversity at 1,000 and 1,250 m radius but was unrelated at other scales (Fig. 10.2i). Pollen beetle abundance was negatively related to OSR area across all tested scales (Fig. 10.2a), and strongly related to OSR isolation at short (250 m) and longer radii ($\geq$ 1,250 m; Fig. 10.2d). Additionally, pollen beetle abundance was significantly related to landscape diversity at 1,500 m (Fig. 10.2g), non-crop area between 1,250–1,750 m radius (Fig. 10.2j) and woody areas above a radius of 250 m (Fig. 10.2m). Abundance of stem weevil larvae was significantly explained by OSR area between 500 and 1,000 m (Fig. 10.2b), by the isolation of OSR at 750 and 2,000 m radius (Fig. 10.2e) and by woody areas between 250 and 1,000 m radius and at 2,000 m radius (Fig. 10.2n). The data indicate that the scale at which each insect group is influenced is related to each group's dispersal capacity. Whereas pollen beetle showed effects at medium to large scales (1,000–2,000 m; all variables except grassy fallows), pod midge and the stem weevils responded to landscape factors at small to medium scales (pod midge: 250–500 m; woody areas. 1,000–1,250 m landscape diversity. Stem weevil: 500–1,000 m; OSR area, isolation OSR, woody areas). Limitation of available OSR food at certain scales seemed unlikely to be responsible for these relationships; however some dispersal limitations might be the reason for these patterns (Kruess and Tscharntke 1994, Tscharntke and Kruess 1999, With et al. 1999, Kruess and Tscharntke 2000). Pod midge is known to be a weak flier with females usually dispersing no more than a few hundred meters from their emergence sites, which might explain its response to woody areas at the very smallest scale. In contrast, pollen beetle can easily cross distances greater than those considered in the current study (Fritzsche 1957) and are therefore affected by landscape characteristics on medium to large scales. Information on the mobility of the stem weevils is generally scarce but they are reported to prefer making short flights (Schmutterer 1956). However, this did not seem to hinder them reacting to landscape effects on small to medium scales.

10.2.2 Pest Damage in Relation to Site and Landscape Factors

Multiple regression analyses showed that a significant proportion of variation for pollen beetle and pod midge damage was accounted for by a negative relationship with OSR area being the only explanatory variable (Fig. 10.4a, c). Pollen beetle and pod midge damage was best explained by a negative relationship to OSR

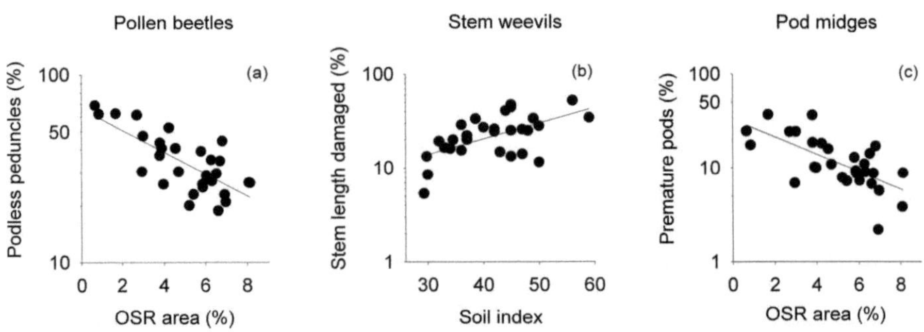

Fig. 10.4 Relations between species-specific damage measures in 29 winter oilseed rape fields and best-explaining site and landscape variables shown by multiple regression analyses (after Zaller et al. 2008b)

area in the landscape indicating that damage was greater in landscapes where less OSR area was available and *vice versa*. An analogous concentration effect has been reported for other insect species in fragmentation studies (Debinski and Holt 2000, Tischendorf et al. 2005); however we were among the first who could show this relationship for OSR pests. In contrast to our findings, Thies et al. (2003) found no relationship between pollen beetle damage and OSR area in the landscape. An explanation for this discrepancy appears to be the differential methodological approaches used: potted OSR plants distributed in the landscape were used in the former study while field-grown plants were investigated in the current study.

Stem weevil damage was significantly positively related to the site variables soil index, development of OSR and stand density (Fig. 10.4b), while other site and landscape variables did not explain significant amounts of variability. Stem weevil damage was mainly affected by site parameters while damage by pollen beetle and pod midge were responsive to OSR area in the landscape. For the stem weevils, it is unlikely that site characteristics would have directly affected damage because these species invaded the study fields from non-crop overwintering sites. However, soil index, OSR development and stand density may have affected stem weevil damage indirectly via (i) influencing the nutritional quality of OSR plants, e.g., through the production of glucosinolates (Bartlet 1996), (ii) altering the odour intensity of OSR plants and thereby changing the attractiveness for this pest species (Cook et al. 2006) and (iii) altering the OSR canopy microclimate and thus influencing the searching efficiency of pest species (Walters et al. 2003).

Assuming that the distribution of pest damage reflects the foraging range and dispersal ability of the studied pest species (Kruess and Tscharntke 1994, 2000) it was surprising to see that variables characterizing the distribution of OSR fields in the landscape (e.g., distance between OSR fields) were unrelated to pest damage in the multivariate analysis.

10.3 Spiders

Spiders are among the most abundant and species-rich groups of predators in arable fields and they can play an important role in natural pest control (Symondson et al. 2002, Nyffeler and Sunderland 2003). Arthropod predators in annual crop fields depend strongly on the surrounding landscape because annual crops are ephemeral habitats characterised by periodical disturbances such as soil cultivation, pesticide applications and harvesting and have to be recolonised by much of the arthropod fauna each year at least once (Wissinger 1997). A spider species' ability to colonise arable fields from perennial habitats depends mainly on its life cycle, its dispersal mode and the distribution of suitable source habitats in the landscape (Topping and Sunderland 1994, Topping 1999). For species with limited dispersal capabilities the nearby landscape is supposed to be of greater importance than for very mobile species for which the wider surrounding landscape is more relevant (Topping 1999, Schmidt and Tscharntke 2005b). Thus, we hypothesized that spider assemblages in OSR fields will be affected by the structure of the surrounding landscapes.

In addition to spider assemblages we also studied in more detail the response of *Pardosa agrestis* Westring (Lycosidae), the dominant wolf spider species in most arable fields in Central Europe (Samu and Szinetár 2002, Nyffeler and Sunderland 2003). Knowledge of the factors that support this species is desirable in order to sustain viable populations that can reduce pest populations. As an agrobiont species, *P. agrestis* is adapted to agricultural practices by a life-cycle synchronized to the disturbance regime and a high dispersal capacity (Richter 1970, Kiss and Samu 2005). Small- to medium-sized juveniles of *P. agrestis* occur in spring and autumn and are the main colonizer stages as they are able to cover large distances by ballooning (Richter 1970), while the exclusively cursorial larger instars and adults have limited dispersal abilities.

Non-crop habitats within agricultural landscapes may not only have a strong influence on abundance of generalist predators such as *P. agrestis*, but may also influence fitness related traits like body size and fecundity. Adult body size in spiders is to a large degree determined by food supply during pre-adult development (e.g., Uetz et al. 2002). Furthermore, female body size is a major determining factor for fecundity (Beck and Connor 1992) which in turn affects population build-up.

Even if several studies consistently showed that the surrounding landscape, and in particular semi-natural habitats, have important effects on spiders in arable fields (Weibull et al. 2003, Prasifka et al. 2004, Clough et al. 2005, Schmidt and Tscharntke 2005b, Schmidt et al. 2005, Schweiger et al. 2005, Isaia et al. 2006, Öberg et al. 2007), not much is known of multi-scale influences. Only very few of the above-mentioned studies integrated both landscape and site factors, and even fewer aimed at estimating the actual spatial scale of the landscapes' influence (Prasifka et al. 2004, Schmidt and Tscharntke 2005b). Additionally, most studies did not differentiate between types of semi-natural or perennial habitats, but simply distinguished crop versus non-crop area. We studied the influence of site and landscape factors at different radii on spider assemblages and on *P. agrestis* (for more details see Drapela et al. 2008, 2009); to our knowledge these topics have never been investigated before.

10.3.1 Spider Assemblages in Relation to Site and Landscape Factors

Spider assemblages were related to different types of semi-natural habitats in the surrounding landscapes. Species richness was positively related to the proportion of non-crop areas, woody areas, fallows and landscape diversity at several radii (Fig. 10.5a, c, e, i). Except for fallows, the highest values of explained variance were reached at the 500 m radius (Fig. 10.5a, c, i). These relations diminished with increasing radius but remained significant up to 1,250 m. Woody areas within a 500 m radius were the most important explanatory variable for species richness (Fig. 10.5c). This may be due to the large number of vagrant species with low activity densities that originated from different non-crop habitats. Since spider assemblages in arable fields differ much more from assemblages in woody habitats

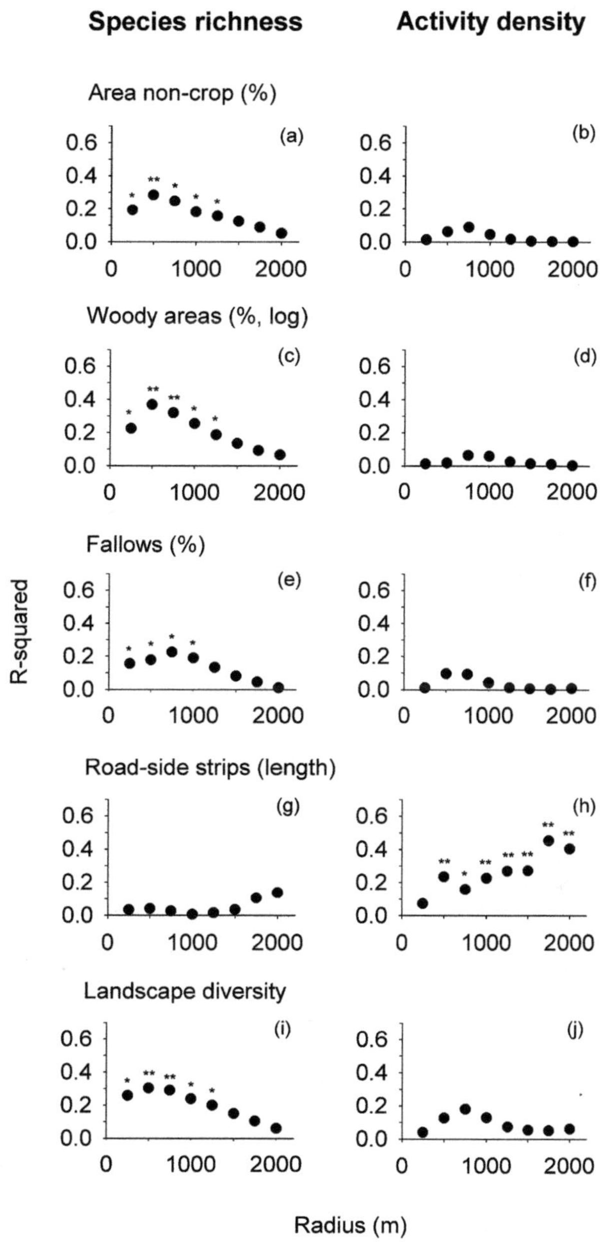

Fig. 10.5 Relationships between spider species richness and activity density, and landscape variables across the investigated radii derived from univariate ordinary least square regression analyses. *Asterisks* denote statistical significance: $* P < 0.05$, $** P < 0.01$. All relationships are positive (after Drapela et al. 2008)

than from assemblages in fallows (Bolaños 2003), species arriving in OSR fields from woody habitats are much more likely to add a new species to a spider assemblage than species arriving from fallows. Agrobiont species were shown to be not or negatively related to woody areas, but in most cases they were positively related

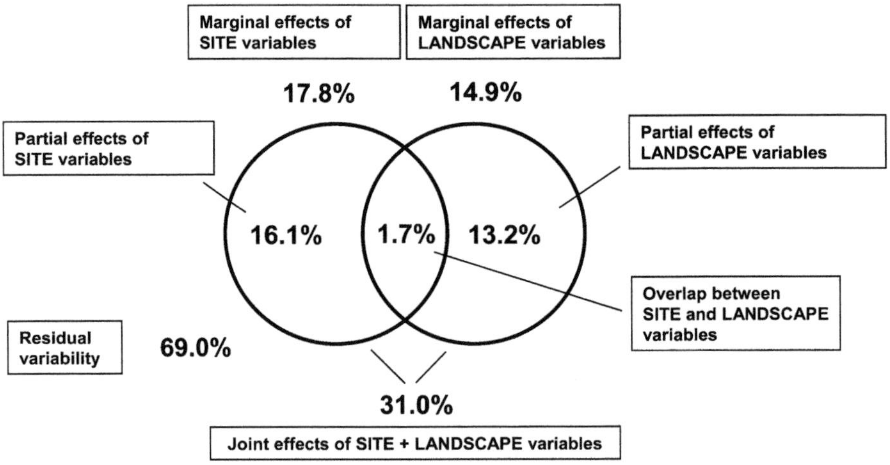

Fig. 10.6 Results of redundancy analyses (RDA) showing the influence of site and landscape variables as well as their overlap on spider assemblages in OSR fields. RDA with forward selection; all marginal and partial effects: $P < 0.002$; P-values are from Monte Carlo permutation tests with 499 permutations (Drapela unpublished)

to fallows. This indicates that fallows are of great importance for these species, perhaps as overwintering sites or refuge habitats during or after times of disturbances within arable fields, e.g., soil cultivation, pesticide spraying or harvest, or as sources from which arable fields can periodically be recolonised (Halley et al. 1996, Pfiffner and Luka 2000, Schmidt and Tscharntke 2005a). From fallows and other perennial semi-natural habitats arable fields are periodically recolonised (Thomas and Jepson 1997, Samu et al. 1999).

Redundancy analysis models (RDA) for landscape and site variables calculated to assess the relationships of spider assemblages with landscape and site variables explained about the same amount of variability with little overlap of only 1.7% between these two variable sets (Fig. 10.6). The proportions of woody areas and fallows at the 500 m radius were the most important explanatory variables in the landscape RDA model. Many of the spider species explained well by this model were positively correlated with the proportion of woody areas. Agrobiont species were either negatively correlated with woody areas (e.g., *Zelotes mundus, Drassyllus pusillus*) or closely related to fallows (e.g., *Xysticus kochi, Trochosa ruricola, P. agrestis*).

Differences in dispersal abilities and modes may be the reason for different patterns of Linyphiidae and Lycosidae, which contained by far the most agrobiont spiders, in the landscape RDA models. Contrary to the lycosid species (*Pardosa prativaga, P. agrestis, T. ruricola*), the three most frequent linyphiid spiders (*Oedothorax apicatus*, *Erigone dentipalpis*, *Meioneta rurestris*) were rather negatively related to fallows in the landscape RDA model. In spring, OSR fields are colonised by adult or subadult Lycosidae on the ground, while Linyphiidae balloon frequently throughout their life cycles (Bell et al. 2005). Therefore, fallows near the OSR fields (radius 500 m) seem to be of less importance for Linyphiidae than for Lycosidae. Site variables representing farming management and structural properties of the OSR vegetation had a significant influence on spider species composition (Fig. 10.6). It is known that habitat structure (e.g., structural complexity) is among the fundamental factors influencing spider assemblages (Uetz 1991, Langellotto and Denno 2004). As weeds were very scarce in the studied OSR fields, the density of OSR plants was the most important feature determining structural complexity within the study fields. Furthermore, stand density may have influenced species composition by altering microclimatic conditions within the fields. In contrast to many other arable fields, winter OSR fields usually have a considerable vegetation cover during winter, which is known to be an important factor for overwintering arthropods (Pfiffner and Luka 2000, Frank and Reichhart 2004). Although, there are no reports on spiders overwintering in OSR fields, it is very likely that some spiders may indeed overwinter in these fields, especially when the crop has a certain vegetation cover in autumn and winter offering shelter from harsh weather conditions.

Spider activity density was positively related to the total length of road-side strips (Fig. 10.5h). Road-side strips or field-margins might become particularly important in intensively-used landscapes where they can serve as corridors between other non-crop elements and often are the only remaining perennial habitats. The fact that the maximum response was found at large scales indicates that dispersal processes may be involved. Road-side strips may be especially important for species that are cursorial dispersers or have limited periods of aerial dispersal. Surprisingly, none of the site variables were significantly related to species richness or activity density. This emphasizes the importance of landscape characteristics for the occurrence of spiders in oilseed rape fields.

Overall, results of our spider study showed that (i) spider species richness was higher in more diverse landscapes with more non-crop area, (ii) activity density responded positively to road-side strips and non-crop area and (iii) fallows specifically supported agrobiont spiders, while woody areas enhanced overall species richness. Such aspects are important for the role of spiders in natural pest control, as more diverse predator communities may have a higher potential for suppressing pest species populations (Marc and Canard 1997, Losey and Denno 1999). Moreover landscape effects observed were shown to be scale-dependent due to differences in dispersal power between species. The high dispersal potential of ballooning species made them relatively independent of landscape characteristics in the closer surroundings of the OSR fields. Finally, it was found that site factors had a greater influence on spider species composition than on spider species richness and activity density.

10.3.2 Pardosa agrestis

Our studies demonstrated that non-crop habitats in the surrounding landscape were not only important for spider assemblages but also affected activity density, body size and fecundity of *P. agrestis* in OSR fields. Activity density of *P. agrestis* was positively related to length of road-side strips at radius 1,750 m and close to the nearest fallow (Drapela et al. 2009). Although *P. agrestis* is very abundant and dominant mainly in arable fields (Samu and Szinetár 2002), semi-natural non-crop habitats are additionally used as overwintering sites and refuges (Hänggi et al. 1995, Pfiffner and Luka 2000). The finding that road-side strips at a large radius have higher explanatory power than nearby fallows suggests that the former constitute a network of habitats embedded in the agricultural landscape that is probably attractive for *P. agrestis* and that may facilitate the colonisation of arable fields. Such a network function may be important because *P. agrestis* is restricted to cursorial movement except for young instars which are capable of aerial dispersal by ballooning (Richter 1970, Wissinger 1997). Our study area is generally characterised by relatively small fields and a rather dense network of road and farm tracks which are predominantly accompanied by road-side strips (ca. 50–100 cm wide vegetated margins). These road-side strips are the only semi-natural, perennial habitats present throughout the whole study area, while field margins between neighbouring fields are scarce. Not much is known about the role of road-side strips for epigeic spiders, but their ecological function in the landscape is similar to field margins in many respects. The positive effects of field margins on epigeic arthropods in arable land have been demonstrated elsewhere (reviewed in Sunderland and Samu 2000, Benton et al. 2003, Bianchi et al. 2006), but there is little information on relations at the landscape scale.

Road-side strips were also the most important explanatory variables for male and female body size: the more road-side strips in the surrounding landscape the smaller the spiders (Fig. 10.7a). Mean body size of males and females was also negatively correlated with activity density. Road-side strips could have enhanced colonisation of study fields which led to stronger intraspecific competition, mainly for limited food resources. Studies on other lycosid species showed that enhanced intraspecific competition during pre-adult development due to prey shortage can result in decreased adult body size (Uetz et al. 2002, Balfour et al. 2003). For both

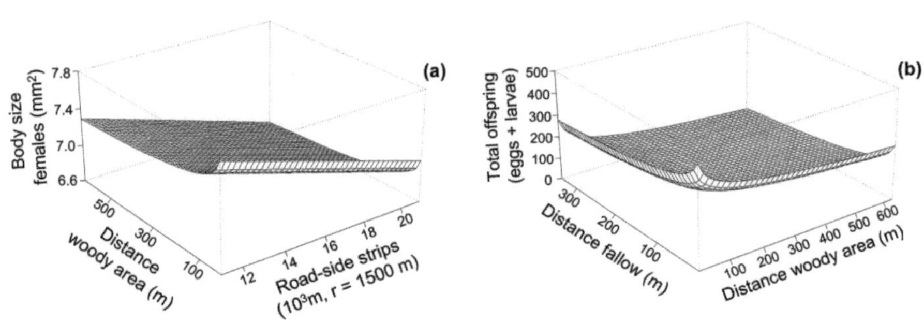

Fig. 10.7 Partial effects of landscape variables on mean body size of *P. agrestis* females and between total offspring of *P. agrestis* derived from multivariate ordinary least square regression analyses (after Drapela et al. 2009)

sexes, body size was positively related to woody areas in the surrounding landscape (Fig. 10.7a). This might seem surprising because *P. agrestis* is a species of open habitats (Hänggi et al. 1995). In the current study the category woody areas comprised, besides forests and copses, also hedges and shrublands – habitats where *P. agrestis* was found in other studies (Hänggi et al. 1995). Such habitats together with associated structures (e.g., forest edges, grassy margins) may constitute important overwintering sites and refuge habitats with abundant prey (Pywell et al. 2005). Woody habitats in the vicinity may also have indirectly affected *P. agrestis* by altering habitat conditions within the study fields (e.g., prey availability).

Total offspring and offspring per female were positively related to woody areas and fallows close to the study fields (Fig. 10.7b). Similar to the relations with body size, woody areas and fallows may serve as permanently available habitats or they may improve within-field conditions. Furthermore clutch size was not related to landscape factors but to female body size, and body size was negatively related to activity density which in turn was enhanced by fallows close to the study fields. All these relations suggest that population density influenced fecundity indirectly by a decreased size of adult females and consequently smaller clutches (Wise and Wagner 1992), while the positive effect of non-crop areas in the surrounding landscape was probably based on increased numbers of cocoon-carrying females but not on bigger clutches.

Taken collectively, it was observed that (i) non-crop habitats in the surrounding landscape increased *P. agrestis* activity density; grass-dominated habitats were more important for activity density than woody habitats but the latter contributed significantly to explaining body size and fecundity; (iii) there is no simple positive relation between body size and non-crop areas; (iv) clutch size, in contrast to total offspring and offspring per female, was not affected by landscape factors but only correlated with female body size. Road-side strips obviously played an important role as explanatory variable for *P. agrestis*. These appear to be the first data highlighting the importance of road-side strips for an epigeic arthropod predator at the landscape level. The results suggest that the importance of road-side strips may have been underestimated and future research in their role in networking perennial habitats in agricultural landscapes may provide valuable new insights in distribution and dispersal dynamics of epigeic arthropod predators in arable fields.

10.4 Implications for Farm and Landscape Management

The species-specific and spatially-explicit responses of the studied pest species to site and landscape characteristics paired with a lack of data from other regions should be carefully considered when trying to formulate general management recommendations. Also, it has to be kept in mind that management measures tailored to minimize OSR pest abundance might not be appropriate for pests in other crops because they might respond differently to site and landscape characteristics. From a farmer's perspective, the pollen beetle seems to be the most important species because it showed the most pronounced negative impact on yields (Zaller et al.

2008b). There may be scope to manipulate the placement and distribution of oilseed rape crops on large farms to optimize the spatial distribution of pests and thereby also influence mobile beneficial species, especially hymenopteran parasitoids (Landis et al. 2000, Walters et al. 2003). Our results showed that combined effects of soil quality and landscape structure were responsible for high abundance of pollen beetle and the stem weevils. While crop rotation schemes are commonly practised to reduce the carry-over of OSR pest and disease problems, these could be expanded by a landscape perspective by avoiding OSR cropping in landscapes with more woody habitats because these favored all three pest species in our study. Our results also suggest that stem weevil damage and yield loss would be reduced if OSR was cropped on fields with below-average soil quality combined with lower stand density and delayed crop development. Since it was shown that pollen beetle and pod midge damage was greater when less OSR area was available, dramatic fluctuations in OSR area between years should be avoided. Against common expectations grassy fallows seemed to play only a negligible role in promoting pest populations.

However, we could demonstrate that these non-crop areas surrounding OSR fields (e.g., fallows, road-side strips) promoted both spider fitness and assemblages in OSR fields, thus underlining the importance of these landscape elements for conservation biocontrol. This aspect may become even more significant as future plans to boost biofuel production are expected to drastically reduce non-crop areas in our landscapes. Our findings that not only the type and quantity of non-crop habitats are important but also their spatial configuration and distribution in the landscape indicate that landscape management strategies should also focus on interspersing these non-crop elements within the matrix of arable fields in a way that distances between refuge or source habitats and arable fields are kept short. In this regard, relatively small but interspersed and interconnected habitats like road-side strips or field margins become especially important in intensively-managed agricultural landscapes as exemplified in our studies for *P. agrestis*.

Clearly, for more general recommendations on management strategies aiming to improve natural control and reduce reliance on pesticides, a much better understanding of the population dynamics and dispersal behaviour of pests and their antagonists is necessary. The results summarized here represent a first attempt to understand how the abundance of different pest species might be affected by landscape characteristics and at which spatial scale this occurs, however much more research is necessary to be able to elucidate the underlying processes particularly regarding pest migration patterns and dispersion between metapopulations of pests in a landscape. There is a great need for more multi-scale and multi-species assessments conducted in different regions to adequately pinpoint how habitat features and their spatial configuration can affect migration patterns and dispersion between metapopulations of pests and their antagonists in the landscape.

References

Alford DV (ed.) (2003) *Biocontrol of oilseed rape pests*. Blackwell, Oxford, UK.

Alford DV, Nilsson C, Ulber B (2003) Insect pests of oilseed rape crops. In: Alford DV (ed.) *Biocontrol of oilseed rape pests*. Blackwell, Oxford, UK.

Balfour RA, Buddle CM, Rypstra AL, Walker SE, Marshall SD (2003) Ontogenetic shifts in competitive interactions and intra-guild predation between two wolf spider species. *Ecol Entomol* 28: 25–30.

Bartlet E (1996) Chemical cues to host-plant selection by insect pests of oilseed rape. *Agr Zool Rev* 7: 89–116.

Beck MW, Connor EF (1992) Factors affecting the reproductive success of the crab spider *Misumenoides formosipes* – the covariance between juvenile and adult traits. *Oecologia* 92: 287–295.

Bell JR, Bohan DA, Shaw EM, Weyman GS (2005) Ballooning dispersal using silk: World fauna, phylogenies, genetics and models. *B Entomol Res* 95: 69–114.

Benton TG, Vickery JA, Wilson JD (2003) Farmland biodiversity: Is habitat heterogeneity the key? *Trends Ecol Evol* 18: 182–188.

Bianchi FJJA, Booij CJH, Tscharntke T (2006) Sustainable pest regulation in agricultural landscapes: A review on landscape composition, biodiversity and natural pest control. *P Roy Soc Lond B Bio* 273: 1715–1727.

BMLFUW (ed.) (2006) *Grüner Bericht 2006*. Bundesministerium für Land- und Forstwirtschaft, Umwelt und Wasserwirtschaft, Vienna, Austria.

Bolaños A (2003) Spider assemblages and habitat bindings in Central Europe. *Verlag Agrarökologie, Bern.*

Cipollini DF, Bergelson J (2002) Interspecific competition affects growth and herbivore damage of *Brassica napus* in the field. *Plant Ecol* 162: 227–231.

Clough Y, Kruess A, Kleijn D, Tscharntke T (2005) Spider diversity in cereal fields: Comparing factors at local, landscape and regional scales. *J Biogeogr* 32: 2007–2014.

Cook SM, Smart LE, Martin JL, Murray DA, Watts NP, Williams IH (2006) Exploitation of host plant preferences in pest management strategies for oilseed rape (*Brassica napus*). *Entomol Exp Appl* 119: 221–229.

Debinski DM, Holt RD (2000) A survey and overview of habitat fragmentation experiments. *Conserv Biol* 14: 342–355.

Drapela T, Moser D, Zaller JG, Frank T (2008) Spider assemblages in winter oilseed rape affected by landscape and site factors. *Ecography* 31(2): 254–262.

Drapela T, Moser D, Zaller JG, Frank T (2009) Landscape affects activity density, body size and fecundity of the wolf spider *Pardosa agrestis* (Araneae, Lycosidae) in winter oilseed rape. *Agric For Entomol* (in revision).

Frank T, Reichhart B (2004) Staphylinidae and Carabidae overwintering in wheat and sown wildflower areas of different age. *B Entomol Res* 94: 209–217.

Free JB, Williams IH (1978) A survey of the damage caused to crops of oil-seed rape (*Brassica napus* L.) by insect pests in south-central England and their effect on seed yield. *J Agr Sci* 90: 417–424.

Free JB, Williams IH (1979) The distribution of insect pests on crops of oil-seed rape (*Brassica napus* L.) and the damage they cause. *J Agr Sci* 92: 139–149.

Frenzel M, Brandl R (1998) Diversity and composition of phytophagous insect guilds on Brassicaceae. *Oecologia* 113: 391–399.

Fritzsche R (1957) Zur Biologie und Ökologie der Rapsschädlinge aus der Gattung *Meligethes*. *J Appl Entomol* 40: 222–280.

Gianessi LP, Marcelli MB (2000) *Pesticide use in U.S. crop production: 1997*. National Center for Food and Agricultural Policy, Washington, DC, USA.

Halley JM, Thomas CFG, Jepson PC (1996) A model for the spatial dynamics of linyphiid spiders in farmland. *J Appl Ecol* 33: 471–492.

Hänggi A, Stöckli E, Nentwig W (1995) *Habitats of Central European Spiders. Characterisation of the Habitats of the Most Abundant Spider Species of Central Europe and Associated Species*. Centre suisse de cartographie de la faune, Neuchâtel.

Hansen LM (2003) A model for determination of the numbers of pollen beetles (*Meligethes aeneus* F.) (Col., Nitidulidae) per plant in oil-seed rape crops (*Brassica napus* L.) by estimating the percentage of plants attacked by pollen beetles. *J Appl Entomol* 127: 163–166.

Hokkanen HMT (2000) The making of a pest: Recruitment of *Meligethes aeneus* onto oilseed Brassicas. *Entomol Exp Appl* 95: 141–149.
Holt RD, Lawton JH, Polis GA, Martinez ND (1999) Trophic rank and the species-area relationship. *Ecology* 80: 1495–1504.
Isaia M, Bona F, Badino G (2006) Influence of landscape diversity and agricultural practices on spider assemblage in Italian vineyards of Langa Astigiana (northwest Italy). *Environ Entomol* 35: 297–307.
Kareiva P (1990) Population-dynamics in spatially complex environments – theory and data. *Philos T Roy Soc Lond Ser B* 330: 175–190.
Kiss B, Samu F (2005) Life history adaptation to changeable agricultural habitats: Developmental plasticity leads to cohort splitting in an agrobiont wolf spider. *Environ Entomol* 34: 619–626.
Kruess A, Tscharntke T (1994) Habitat fragmentation, species loss, and biological control. *Science* 264: 1581–1584.
Kruess A, Tscharntke T (2000) Effects of habitat fragmentation on plant insect communities. In: Ekbom B, Irwin ME, Robert Y (eds.) *Interchanges of insects between agricultural and surrounding landscapes*. Kluwer Academic Publishers, Dordrecht, The Netherlands.
Landis DA, Wratten SR, Gurr GM (2000) Habitat management to conserve natural enemies of arthropod pests in agriculture. *Annu Rev Entomol* 45: 175–201.
Langellotto GA, Denno RF (2004) Responses of invertebrate natural enemies to complex-structured habitats: A meta-analytical synthesis. *Oecologia* 139: 1–10.
Lethmayer C, Nentwig W, Frank T (1997) Effects of weed strips on the occurrence of noxious coleopteran species (Nitidulidae, Chrysomelidae, Curculionidae). *J Plant Dis Prot* 104: 75–92.
Levin SA (1992) The problem of pattern and scale in ecology. *Ecology* 73: 1943–1967.
Losey JE, Denno RF (1999) Factors facilitating synergistic predation: The central role of synchrony. *Ecol Appl* 9: 378–386.
Marc P, Canard A (1997) Maintaining spider biodiversity in agroecosystems as a tool in pest control. *Agr Ecosyst Environ* 62: 229–235.
Menalled FD, Marino PC, Gage SH, Landis DA (1999) Does agricultural landscape structure affect parasitism and parasitoid diversity? *Ecol Appl* 9: 634–641.
Moser D, Drapela T, Zaller JG, Frank T (2009) Interacting effects of wind direction and resource distribution on rape pest species abundance. *Basic Appl Ecol* 10: 208–215.
Nyffeler M, Sunderland KD (2003) Composition, abundance and pest control potential of spider communities in agroecosystems: A comparison of European and US studies. *Agr Ecosyst Environ* 95: 579–612.
Öberg S, Ekbom B, Bommarco R (2007) Influence of habitat type and surrounding landscape on spider diversity in Swedish agroecosystems. *Agr Ecosyst Environ* 122: 211–219.
Östman Ö, Ekbom B, Bengtsson J (2001a) Landscape heterogeneity and farming practice influence biological control. *Basic Appl Ecol* 2: 365–371.
Östman Ö, Ekbom B, Bengtsson J, Weibull A-C (2001b) Landscape complexity and farming practice influence the condition of polyphagous carabid beetles. *Ecol Appl* 11: 480–488.
Paul VH (2003) *Raps. Krankheiten – Schädlinge – Schadpflanzen.* Verlag Th. Mann, Gelsenkirchen-Buer.
Pfiffner L, Luka H (2000) Overwintering of arthropods in soils of arable fields and adjacent semi-natural habitats. *Agr Ecosyst Environ* 78: 215–222.
Pickett STA, Cadenasso ML (1995) Landscape ecology – spatial heterogeneity in ecological systems. *Science* 269: 331–334.
Polis GA, Anderson WB, Holt RD (1997) Toward an integration of landscape and food web ecology: The dynamics of spatially subsidized food webs. *Ann Rev Ecol Syst* 28: 289–316.
Prasifka JR, Heinz KM, Minzenmayer RR (2004) Relationships of landscape, prey and agronomic variables to the abundance of generalist predators in cotton (*Gossypium hirsutum*) fields. *Landscape Ecol* 19: 709–717.

Pywell RF, James KL, Herbert I, Meek WR, Carvell C, Bell D, Sparks TH (2005) Determinants of overwintering habitat quality for beetles and spiders on arable farmland. *Biol Conserv* 123: 79–90.

Richter CJJ (1970) Aerial dispersal in relation to habitat in eight wolf spider species (*Pardosa*, Araneae, Lycosidae). *Oecologia* 5: 200–214.

Roland J, Taylor PD (1997) Insect parasitoid species respond to forest structure at different spatial scales. *Nature* 386: 710–713.

Samu F, Sunderland KD, Szinetár C (1999) Scale-dependent dispersal and distribution patterns of spiders in agricultural systems: A review. *J Arachnol* 27: 325–332.

Samu F, Szinetár C (2002) On the nature of agrobiont spiders. *J Arachnol* 30: 389–402.

Schmidt MH, Roschewitz I, Thies C, Tscharntke T (2005) Differential effects of landscape and management on diversity and density of ground-dwelling farmland spiders. *J Appl Ecol* 42: 281–287.

Schmidt MH, Tscharntke T (2005a) The role of perennial habitats for Central European farmland spiders. *Agr Ecosyst Environ* 105: 235–242.

Schmidt MH, Tscharntke T (2005b) Landscape context of sheetweb spider (Araneae: Linyphiidae) abundance in cereal fields. *J Biogeogr* 32: 467–473.

Schmutterer H (1956) Zur Lebensweise und Bekämpfung des Großen Rapsstengelrüsslers *Ceutorhynchus napi* (Gyll.). *Z Angew Entomol* 39: 302–315.

Schneider DC (1994) *Quantitative ecology: Spatial and temporal scaling*. Academic Press, New York.

Schweiger O, Maelfait JP, Van Wingerden W, Hendrickx F, Billeter R, Speelmans M, Augenstein I, Aukema B, Aviron S, Bailey D, Bukacek R, Burel F, Diekötter T, Dirksen J, Frenzel M, Herzog F, Liira J, Roubalova M, Bugter R (2005) Quantifying the impact of environmental factors on arthropod communities in agricultural landscapes across organizational levels and spatial scales. *J Appl Ecol* 42: 1129–1139.

Sunderland K, Samu F (2000) Effects of agricultural diversification on the abundance, distribution, and pest control potential of spiders: A review. *Entomol Exp Appl* 95: 1–13.

Symondson W, Sunderland K, Greenstone M (2002) Can generalist predators be effective biocontrol agents? *Annu Rev Entomol* 47: 561–594.

Thies C, Roschewitz I, Tscharntke T (2005) The landscape context of cereal aphid-parasitoid interactions. *P Roy Soc Lond B Bio* 272: 203–210.

Thies C, Steffan-Dewenter I, Tscharntke T (2003) Effects of landscape context on herbivory and parasitism at different spatial scales. *Oikos* 101: 18–25.

Thies C, Tscharntke T (1999) Landscape structure and biological control in agroecosystems. *Science* 285: 893–895.

Thomas C, Jepson P (1997) Field-scale effects of farming practices on linyphiid spider populations in grass and cereals. *Entomol Exp Appl* 84: 59–69.

Tischendorf L, Grez A, Zaviezo T, Fahrig L (2005) Mechanisms affecting population density in fragmented habitat. *Ecol Soc* 10: 7 [online] URL: http://www.ecologyandsociety.org/vol10/iss11/art17/

Topping C (1999) An individual-based model for dispersive spiders in agroecosystems: Simulations of the effects of landscape structure. *J Arachnol* 27: 378–386.

Topping CJ, Sunderland KD (1994) A spatial population-dynamics model for *Lepthyphantes tenuis* (Araneae, Linyphiidae) with some simulations of the spatial and temporal effects of farming operations and land-use. *Agr Ecosyst Environ* 48: 203–217.

Tscharntke T, Kruess A (1999) Habitat fragmentation and biological control. In: Hawkins B and Cornell H (eds.) *Theoretical approaches to biological control*. Cambridge University Press, Cambridge, UK.

Uetz GW (1991) Habitat structure and spider foraging. In: Bell SS, McKoy ED, Mushinsky HR (eds.) *Habitat structure: The physical arrangement of objects in space*. Chapman and Hall, London.

Uetz GW, Papke R, Kilinc B (2002) Influence of feeding regime on body size, body condition and a male secondary sexual character in *Schizocosa ocreata* wolf spiders (Araneae, Lycosidae): Condition-dependence in a visual signaling trait. *J Arachnol* 30: 461–469.
Walters KFA, Young JEB, Kromp B, Cox PD (2003) Management of oilseed rape pests. In: Alford DV (ed.) *Biocontrol of oilseed rape insect pests*. Blackwell, Oxford, UK.
Weibull A-C, Östman Ö, Granqvist A (2003) Species richness in agroecosystems: The effect of landscape, habitat and farm management. *Biodivers Conserv* 12: 1335–1355.
Wiens JA (1989) Spatial scaling in ecology. *Funct Ecol* 3: 385–397.
Wise DH, Wagner JD (1992) Evidence of exploitative competition among young stages of the wolf spider *Schizocosa ocreata*. *Oecologia* 91: 7–13.
Wissinger SA (1997) Cyclic colonization in predictably ephemeral habitats: A template for biological control in annual crop systems. *Bio Control* 10: 4–15.
With KA, Cadaret SJ, Davis C (1999) Movement responses to patch structure in experimental fractal landscapes. *Ecology* 80: 1340–1353.
Zaller JG, Moser D, Drapela T, Schmöger C, Frank T (2008a) Insect pests in winter oilseed rape affected by field and landscape characteristics. *Basic Appl Ecol* 9: 682–690.
Zaller JG, Moser D, Drapela T, Schmöger C, Frank T (2008b) Effect of within-field and landscape factors on insect damage in winter oilseed rape. *Agr Ecosyst Environ* 123: 233–238.
Zaller JG, Moser D, Drapela T, Schmöger C, Frank T (2009) Parasitism of stem weevils and pollen beetles in winter oilseed rape is differentially affected by site and landscape characteristics. *Biocontrol* 54: 505–514.

Chapter 11
Impact of Soil Tillage on Parasitoids of Oilseed Rape Pests

Christer Nilsson

Abstract Some of the parasitoids of important oilseed rape pests over-winter in the soil within the pupal chambers prepared by the host larva. Thus post-harvest soil tillage can have a great impact on the survival of these insects and the parasitization of oilseed rape pests the following year. Conventionally, soil tillage following harvest of oilseed rape was usually by ploughing but, more recently, methods of reduced tillage and direct drilling have become increasingly common. Experiments have shown that ploughing almost always reduces the number of parasitoids emerging the following year by 50% or more, while other kinds of tillage have a varying but lesser effect. Reduced, non-inversion tillage can be part of an integrated strategy to control insect pests in oilseed rape.

11.1 Introduction

Most of the parasitoids of the major insect pests of oilseed rape belong to the Hymenopteran family Ichneumonidae, subfamily Tersilochinae, but a few, mainly parasitoids of brassica pod midge, belong to the Chalcidoidea (Table 11.1); most develop during the larval stage of their hosts (see also Ulber et al. Chapter 2 this volume). The major part of the development and pupation of the parasitoid takes place in the soil chamber that the host prepares before pupation. The parasitoids hatch in the soil, but only the polyvoltine species e.g., *Omphale clypealis*, also leave the soil and eventually also the rape field that year. Univoltine species stay in the soil until the following spring, usually emerging when the oilseed rape of that year is at the appropriate stage for their hosts to be present. The fact that they stay in the soil after harvest of the oilseed rape crop, makes them vulnerable to post-harvest tillage and may be affected differently by different tillage, practised to establish the succeeding crop, usually cereals e.g., winter wheat or spring barley.

C. Nilsson (✉)
Swedish University of Agricultural Sciences, Alnarp SE-230 53, Sweden
e-mail: ncnilsson@telia.com

I.H. Williams (ed.), *Biocontrol-Based Integrated Management of Oilseed Rape Pests*,
DOI 10.1007/978-90-481-3983-5_11, © Springer Science+Business Media B.V. 2010

Table 11.1 Parasitoids of key oilseed rape pests that emerge from last year's rape field

Pest	Parasitoid
	Family Ichneumonidae
Psylliodes chrysocephala Cabbage stem flea beetle	*Tersilochus microgaster* (Szépligeti)
Ceutorhynchus napi Rape stem weevil	*Tersilochus fulvipes* (Gravenhorst)
Ceutorhynchus pallidactylus Cabbage stem weevil	*Tersilochus obscurator* Aubert
Meligethes aeneus Pollen beetle	*Tersilochus heterocerus* Thomson *Phradis morionellus* (Holmgren) *Phradis interstitialis* (Thomson)
	Superfamily Chalcidoidea
Dasineura brassicae Brassica pod midge	*Platygaster subuliformis* (Kieffer) *Omphale clypealis* (Thomson)

Most of these parasitoid species have no means of discriminating between a host already containing a parasitoid egg of the same or another parasitoid species and multiple oviposition is more the rule than the exception. Thus the number of parasitoids migrating into an oilseed rape field is important for the final parasitization level. As was shown by Jourdheuil (1960), the pattern of oviposition follows a Poisson distribution and the percentage of parasitized pest larvae is a curvilinear function of the quotient of parasitoid to host. A parasitization level in the pollen beetle (*Meligethes* sp.) of 50% needs 0.7 parasitoids for every host and a parasitisation level close to 100% would be possible only when there are more than three parasitoids for every host. Evidently, if soil tillage can influence the number of important parasitoids in the oilseed rape crop, this can be of great importance to the mortality of some of the more important insect pests and hence the need for chemical control or other control measures in subsequent years.

11.2 Tillage Systems

A couple of decades ago, soil tillage, following harvest of oilseed rape and before the sowing of winter or spring cereals, was usually by ploughing. However, methods of reduced tillage and direct drilling have recently become increasingly common, mainly due to the development of new machine concepts. Seeders with coulters (Fig. 11.1) that cut a furrow in the soil for the seed have, on many farms, replaced seeders with shoe coulters, the exception being when a seeder with a shoe coulter is used together with a rotary cultivator. In many areas, the soil structure after a winter oilseed rape crop is good enough to allow the establishment of winter wheat without ploughing and a much reduced tillage is often used.

Modern reduced-tillage seed drills perform tillage while preparing a seedbed for sowing. Disc coulters, clod crushers and pressing rollers are usually mounted on the

Fig. 11.1 Seed drill coulter (Photo: C. Nilsson)

same machine as the drill and row spacing is about 0.12 m. The first experiments (1981–1983) that showed an effect of tillage on parasitoids (Nilsson 1985) were done with direct drills (Bettinson) that had a row spacing of 0.18 m and any impact on the soil was even less than that of present day direct drills.

In conventional tillage, the turning of the soil will put the pupal chambers at a depth of 15–25 cm, as the larvae pupate no deeper that 5 cm. Modern ploughs can work at any depth from 5 to 25 cm. Very shallow ploughing will hardly affect hibernating insects in the soil. The following harrowing, compaction and seed bed preparation will have more impact when ploughing has been less than 10 cm deep. At deeper ploughing, the insects are buried and can have problems working their way up to surface. Brassica pod midge can stay in the soil for several years (Nilsson et al. 2004) and will thus be ploughed up again the second and fourth autumn. The cocoons are spun encrusted with soil particles and will probably withstand different kinds of soil tillage quite well and so will the parasitoids within. The parasitoids will, as far as is known, not stay in the soil more than one winter and will have the same problem reaching the soil surface as other parasitoids, whereas the pod midge can survive for at least 5 years in the soil which increases its chances of survival.

In reduced tillage and tillage to prepare a seedbed after ploughing, a variety of other machines can be used, such as, different kinds of cultivators (tines), ducksfoot harrows, rotary harrows or disc harrows (Figs. 11.2 and 11.3). These machines break and sort soil aggregates. Rollers and cultipresses also compact the soil. Seeders for reduced tillage will also treat the soil partly in the same way. Pupal chambers can be broken and aggregates containing pupal chambers can be brought up to soil surface and dry out or be exposed to predators. It is not possible to foresee the effects in a specific situation. In Fig. 11.4 the resulting aggregate distribution from different machines in two fields in central Sweden, not far from each other, is shown (Arvidsson et al. 2004). As can be seen, the proportion of bigger aggregates can vary widely, more between fields than between machines. This investigation also shows that the soil water content, soil texture and type of machine are the most important parameters determining aggregate distribution and size.

Fig. 11.2 Cultivator (Photo: C. Nilsson)

Fig. 11.3 Ducksfoot cultivator (Photo: C. Nilsson)

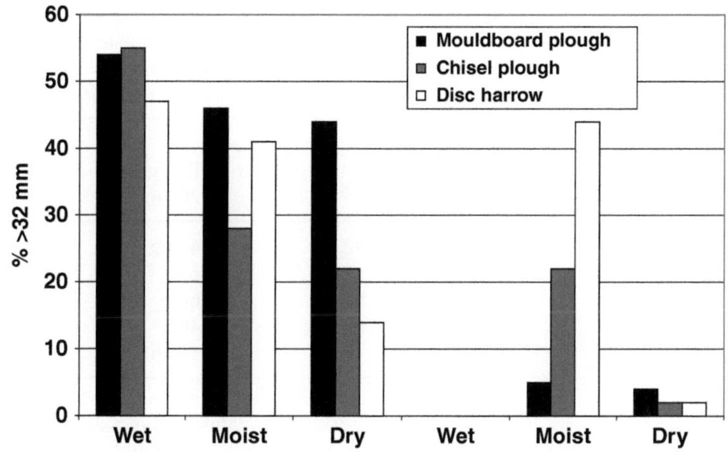

Fig. 11.4 Proportion of aggregates > 32 mm after tillage with 3 different machines at wet, moist and dry soil conditions at two locations: *left*, Ultuna (30–44% clay) and *right*, Saby (14–25% clay), both near Uppsala in central Sweden

11.3 Effects of Tillage on Parasitoids

Experiments have been made in Sweden, Finland, Germany and UK during a 30-year period, with different machine systems. The first experiments in Sweden and Finland (Nilsson 1985, Hokkanen et al. 1988) showed that the over-wintering of parasitoids, mainly of pollen beetles, was about four times higher from fallow or when the winter wheat was sown with direct drill compared to ploughing or other forms of tillage (Table 11.2).

In the UK, a similar experiment, but with very low emergence of parasitoids was also done during 2005 (Ferguson et al. 2007). Here ploughing reduced emergence by an average of 30%, for some species more, whereas a shallow cultivation without

Table 11.2 Numbers of pollen beetle parasitoids (m^{-2}) and the proportions (% relative to fallow) emerging from plots with different tillage systems. Comparisons followed by the same letter are not significantly different. Field experiments in Southern Sweden (Nilsson 1985) and Finland (Hokkanen et al. 1988)

	Sweden					Finland	
	Fallow (no. m^{-2})	Fallow	Direct drill (%)	Disc harrow (%)	Plough + harrow (%)	Fallow (%)	Plough + harrow (%)
P. morionellus	6.9	100 a	145 a	26 b	54 b	100 c	23 d
T. heterocerus	28.2	100	77	19	18	–	–
P. interstitialis	10.8	100	127	24	17	–	–

Table 11.3 Numbers of parasitoids (m^{-2}) and proportion (% relative to fallow) emerging from plots with different tillage systems. Comparisions followed by the same letter are not significantly different. Female parasitoids, males and a few *T. morionellus* are included in the sum. Field experiments from Weende, near Göttingen, Germany (Klingenberg and Ulber 1994)

	Fallow (no. m^{-2})	Fallow	Chisel plough 15 cm (%)	Ploughed 20 cm (%)	Ploughed + disc harrow 10 cm on stubble (%)
T. microgaster	43	100 a	12 b	42 ab	19 b
T. obscurator	193	100 a	42 a	63 a	34 a
T. heterocerus	26	100 a	81 ab	60 b	27 b
P. interstitialis	25	100 a	28 a	48 a	32 a
All Tersilochinae	478	100 a	40 b	55 ab	29 b

Table 11.4 Number of parasitoids m^{-2} and proportion (% relative to direct drill) emerging from plots with different tillage systems. Numbers followed by the same letter are not significantly different. Field experiments in 1996 from Wendelsbreite (Nitzsche and Ulber 1998) and from Reinshof (Wahmhoff et al. 1999) near Göttingen, Germany

	Direct drill (no. m^{-2})	Direct drill	Chisel plough 8 cm (%)	Rotary harrow 5 cm (%)	Ploughed 25 cm (%)	Ploughed 25 cm + spade roller harrow 5 cm (%)
Wendelsbreite						
T. heterocerus	16.0	100	71	32	59	72
P. interstitialis	32.1	100	135	72	85	53
All Tersilochinae	48.1	100 a	114 a	59 b	75 ab	47 b
Reinshof						
T. obscurator	12.8	100 a	–	144 a	50 b	–
T. heterocerus	11.0	100 a	–	65 b	9 c	–
P. interstitialis	14.2	100 a	–	50 b	18 c	–
All Tersilochinae	38.0	100	–	86	26	–

ploughing reduced the parasitoid populations less and was not significantly different from the emergence from fallow. Similar results were obtained in an experiment in Poland (Klukowski pers. comm.)

In Germany, several experiments (Tables 11.3 and 11.4) with different machines have also given a strong reduction of the emerging parasitoid population when ploughing was used, amplified by a stubble treatment after the harvest of the preceding crop. At this time in August, parasitoids start to develop into adults, but stay in their cocoons in the soil and are probably very sensitive to soil cultivation. The overall effect is 25 and up to 90% reduction of parasitoid numbers. In these experiments, the effects of a rotary harrow or a chisel plough are often as severe as ploughing, but, in the experiments shown in Table 11.4, especially that in Reinshof, the effects on the parasitoids are clearly less than those of ploughing.

These experiments show that direct drilling and reduced, non-inversion cultivation of soil after harvest of oilseed rape can be recommended to farmers as part of an integrated strategy to actively enhance parasitoid populations and thereby at least sometimes improve biological control of economically-important pests of oilseed rape. It is however not possible to predict the results of a specific tillage method in a specific situation. The use of this method to increase biological control would be much greater if we had a more detailed knowledge of which factors of post-harvest soil cultivation have the most significant impact on the winter survival and spring emergence of these parasitoids.

References

Arvidsson J, Keller T, Gustafsson K (2004) Specific draught for mouldboard plough, chisel plough and disc harrow at different water contents. *Soil Till Res* 79: 221–231.

Ferguson AW, Holdgate R, Mason NS, Clark SJ, Williams IH (2007) Non-inversion tillage to conserve functional biodiversity for biocontrol of oilseed rape pests. *Proc XVI Int Plant Protection Cong*, 2007, 15–18 October 2007, Glasgow. *BCPC* 2: 818–819.

Hokkanen H, Husberg GB, Söderblom M (1988) Natural enemy conservation for the integrated control of the rape blossom beetle *Meligethes aeneus* F. *Ann Agr Fenn* 27: 281–294.

Jourdheuil P (1960) Influence de quelques facteurs écologiques sur les fluctuations de population d'une biocénose parasitaire: Étude relative á quelques Hyménoptères (Ophioninae, Diospilinae, Euphorinae) parasites de divers Coléoptères inféodés aux Crucifères. *Ann Épiphyties* 11: 445–658.

Klingenberg A, Ulber B (1994) Untersuchungen zum Auftreten der Tersilochinae (Hym., Ichneumonidae) als Larvalparasitoide einiger Rapsschädlinge im Raum Göttingen 1990 und 1991 und zu deren Schlupfabundanz nach unterschiedlicher Bodenbearbeitung. *J Appl Entomol* 117: 287–299.

Nilsson C (1985) Impact of ploughing on emergence of pollen beetle parasitoids after hibernation. *J Appl Entomol* 100: 302–308.

Nilsson C, Vimarlund L, Gustafsson G (2004) Long term survival of Brassica Pod Midge (*Dasineura brassicae*) populations. *IOBC/wprs Bull* 27(10): 299–305.

Nitzsche O, Ulber B (1998) Einfluss differenzierter Bodenbearbeitungssysteme nach Winterraps auf die Mortalität einiger Parasitoiden des Rapsglanzkäfers (*Meligethes* spp.). *Z PflKrankh PflSchutz* 105(4): 417–421.

Wahmhoff W, Hedke K, Tiedemann AV, Nitzsche O, Ulber B (1999) Zum Einfluss von Fruchtfolge und Bodenbearbeitung auf die Entwicklung wichtiger Schaderreger des Winterrapses. *Z PflKrankh PflSchutz* 106(1): 57–73.

Chapter 12
Chemical Control of Insect Pests and Insecticide Resistance in Oilseed Rape

Thomas Thieme, Udo Heimbach, and Andreas Müller

Abstract Public concerns about environmental hazards and widespread resistance in pollen beetle populations on oilseed rape in Europe are threatening the availability of a variety of insecticidal modes of action for pest control on the crop. For a sustainable use of insecticides any overuse has to be avoided to minimize risk of resistance development. Pollen beetles are present in the crop for a long period during the season and are therefore exposed to most insecticide applications carried out, irrespective of any need for control above threshold values. To avoid resistance development active substances with several modes of action are needed to ensure sustainable control of pollen beetle. Insecticide resistance is an example of an evolutionary process in which insecticide resistant genotypes are selected in treated populations. The available insecticides, their modes of action, methods for identifying resistance and both, factors affecting the dynamics of insecticide resistance in the field as well as management strategies to prevent or delay the evolution of resistance, are reported.

12.1 Introduction

Pest insects in oilseed rape, *Brassica napus* L., among them the pollen beetle, *Meligethes aeneus* (Fabricius), are one of the main target groups for development of new insecticides by the ever-shrinking number of pesticide manufacturers.

Some 40 years ago, the insecticides dominating the pollen beetle control market were mostly cyclodiene organochlorines, DDT, carbamates and organophosphates (OPs); pyrethroids were introduced into this market from about 1980. Insecticides in the former groups were usually not systemic, were relatively persistent, most were highly toxic to users, and not just to the target pests but also to many beneficial insects; many have since been withdrawn under pressure from environmental and human health aspects. The EU regulation of pesticides has accepted only 43 out

T. Thieme (✉)
BTL Bio-Test Labor GmbH Sagerheide, Sagerheide 18184, Germany
e-mail: tt@biotestlab.de

I.H. Williams (ed.), *Biocontrol-Based Integrated Management of Oilseed Rape Pests*,
DOI 10.1007/978-90-481-3983-5_12,

of 218 insecticidal active substances that had been used in the EU countries before 1993. Since 1993, only 11 new insecticidal active substances have been accepted (January 2009). Pyrethroid insecticides have replaced more and more the other groups, but showing relatively persistent activity they also lack systemic activity. Similar to OPs they have also broad-spectrum effects on many non-target insects but, as a very positive aspect for pollen beetle control, pyrethroids are relatively bee safe and less toxic for users and mammals compared to OPs.

Increasing public concerns about environmental hazards and widespread resistance in pest populations are threatening the availability of a variety of insecticidal modes of action. For the effectiveness of insecticides to continue, there is a need to avoid any overuse to minimize risk of resistance development. In oilseed rape, several pest insects need to be controlled within a few weeks resulting in a high selection pressure for resistance. Pollen beetles in particular are present in the crop for a long period from early spring to summer and are therefore exposed to most insecticide applications carried out in spring including those on other target pests such as the weevils (*Ceutorhynchus napi, C. pallidactylus* and *C. obstrictus* syn. *assimilis*) and the brassica pod midge (*Dasineura brassicae*), irrespective of any need for control because of appearance in numbers above threshold values. In the last 10 years, the demand for safer insecticides for human health and the environment has stimulated the development of some novel groups. Some have properties that are ideal for pollen beetle control, including the neonicotinoids, thiacloprid and acetamiprid, both of which are bee safe and quite safe for several other environmental aspects.

To avoid resistance development several active substances with different modes of action are needed to ensure sustainable control of the pollen beetle and other pest insects in oilseed rape. Pyrethroids are an important tool for pest management because they show high efficacy for all pests in oilseed rape even at lower temperature, they have a moderate price and are quite user-friendly. However any overuse leads to selection for resistance.

Insecticide resistance is an example of an evolutionary process in which mutations conferring protection against insecticides are selected in treated populations. During the last 25 years, advances have been made in the characterization and understanding of such adaptations. These have delivered valuable information about the origin and nature of selection and evolution in farm land.

The evolution of insecticide resistance has contributed to overall increases in the application of chemicals to crops. Besides seed treatment with insecticides, oilseed rape in Europe is sprayed with insecticides on average about twice but, in some cases, up to five times, with large differences between regions (Richardson 2008). Despite this, resistant insects, such as the pollen beetle, continue to affect agricultural productivity. It is proving impossible to fight back resistance by being focused only on application of insecticides. The development of a new insecticide takes eight to 10 years at a cost of € 15–35 million, and the rate of discovery of new molecules, unaffected by cross resistance, appears to be declining. Only by monitoring, characterizing and predicting the appearance and spread of resistant pollen beetles can we hope to use chemical tools in a sustainable manner.

12.2 Chemical Control of Insect Pests

The pollen beetle is a serious pest of oilseed rape throughout Europe and a major target pest for control with insecticides (Williams Chapter 1 this volume). It is polyphagous and damages many agricultural and horticultural crops; the adults are generalist feeders on pollen of oilseed rape and many other plants, including cauliflower, rose, blackberry, chamomile and lime trees; the presence of beetles has an impact on the market value of horticultural crops. On oilseed rape, the pollen beetle causes primary damage by adult feeding on pollen within the flower buds, thereby destroying them. Damage occurs also by larval feeding in the buds and flowers. However, where populations are potentially exposed to intensive applications of insecticides with the same mode of action, there is a high level of concern regarding the increased potential for selection pressure and resistance development.

12.2.1 Choice of Insecticide

In cases where a pest reaches the damage threshold farmers need to choose an effective insecticide. Table 12.1 lists the range of insecticides approved for use against the pollen beetle in oilseed rape in Germany in 2009 (https://portal.bvl.bund.de/psm/jsp/ListeMain.jsp?page=1&ts=1237033472771). This list differs from that for other countries, but it does illustrate the choice of available insecticides these days.

Pyrethroids have been the tower of strength for pollen beetle control as well as for the control of other oilseed rape pest insects for nearly the last 20 years in Europe. However, neonicotinoids are now also approved for pollen beetle control

Table 12.1 List of insecticides approved to control the pollen beetle on oilseed rape in Germany as notified in the German Pesticide Guide (BVL, 2009)

			Control activity on populations	
Insecticide	Class	Properties	Sensitive to pyrethroids	Resistant to pyrethroids
Chlorpyrifos-ethyl*	OP	Cont	***	***
Chlorpyrifos-methyl*	OP	Cont	***	***
Bifenthrin	Pyrethroid	Cont	****	**
Alpha-cypermethrin	Pyrethroid	Cont	****	*
Beta-cyfluthrin	Pyrethroid	Cont	****	*
Lambda-cyhalothrin	Pyrethroid	Cont	****	*
Zeta-cypermethrin	Pyrethroid	Cont	***	*
Deltamethrin	Pyrethroid	Cont	****	*
Esfenvalerate	Pyrethroid	Cont	***	*
Etofenprox	Pyrethroid	Cont	***	**
Tau-fluvalinate	Pyrethroid	Cont	**	**
Thiacloprid	Neonicotinoid	Cont, Syst	***	***

*Approved only for emergency actions with high populations densities of pollen beetles for 120 days in 2009 for a limited area; cont, contact; syst, systemic.

in many countries, but the use of many active substances with different modes of action showing some control action of pollen beetle (e.g., pymetrozine, spinosad, indoxacarb) is still restricted to the control of other pests on other crops.

Neonicotinoids were first discovered in the early 1970s, but they were not developed for use in agriculture until 1991, when imidacloprid (Elbert et al. 1990, Altmann and Elbert 1992) was introduced to the market. Imidacloprid had the required photostability, insecticidal activity, and residual persistence to be marketed for a wide range of uses. It is a systemic broad-spectrum insecticide, with translaminar activity and with contact and stomach action. Other insecticides developed within this group include thiacloprid (Elbert et al. 2000, Jeschke et al. 2001), and acetamiprid (Takahashi et al. 1992). All have pollen beetle control properties, but some are more active than others. Among the neonicotinoids, only thiacloprid and acetamiprid have low bee toxicity, though their effectivity on pollen beetles and some other pests in oilseed rape is limited.

12.2.2 Modes of Action

To understand the differences in properties of the novel insecticides now being used for pollen beetle control, their modes of action are recapitulated here.

12.2.2.1 Carbamates and Organophosphates

Carbamates and organophosphates are systemic insecticides with contact and stomach action. Both are acetylcholinesterase (AchE) inhibitors. They interrupt the transmission of nerve impulses across the synaptic gap between two nerve cells by preventing the breakdown of the neurotransmitter, acetylcholine (Tomizawa and Casida 2003). This results in paralysis that destroys the ability of animals to respond to external stimuli. Both groups of insecticides tend to have high toxicity to honey bees, fish and mammals (Plapp 1991).

12.2.2.2 Pyrethroids

Pyrethroids prevent the voltage-gated sodium channel from functioning, so that no nerve impulses travel along the nerve axon. Pyrethroids have high toxicity to fish but low mammalian toxicity. To have maximum effect they need to be applied just at the point when beetle numbers exceed threshold values. But it is very difficult to establish threshold values and they differ quite markedly between European countries and regions (see Williams Chapter 1 this volume). Additionally, it is quite difficult to monitor the number of pollen beetles with the required precision.

12.2.2.3 Neonicotinoids

Neonicotinoids act as antagonists by binding to post-synaptic nicotinic acetylcholine-receptors (nAChR), affecting the synapses in the insect central nervous system (Ishaaya and Horowitz 1998, Nauen et al. 2001, 2003, Tomizawa and Casida 2003). These substances are xylem-mobile and therefore suitable for seed treatment and soil application. They have low mammalian toxicity due to differences between the nAChRs of insects and those of mammals (Tomizawa and Casida 2003), but some have high toxicity to birds (Anonymous 1993) and honey bees.

Imidacloprid and acetamiprid, applied topically to the upper surface of leaves, showed translaminar (movement across leaf layers) and acropetal (movement towards leaf margins) activity against pests (Bucholz and Nauen 2002). They also suppress feeding of pollen beetle when treated with sublethal concentrations, even though the beetles are not killed (Nauen *pers comm*).

The long persistence of imidacloprid may be explained by the slow degradation on or in plants and the toxicity of its metabolites (Nauen et al. 1998). Most of the imidacloprid was metabolized within 97 days (Westwood et al. 1998). This neonicotinoid is metabolized by loss of the nitro group, hydroxylation, hydrolysis and formation of conjugates.

Thiacloprid is an acute contact stomach poison with systemic properties. It has a short half-life with a good safety for bees (Elbert et al. 2002), making it ideal for spray application. Its mode of action is similar to that of the other neonicotinoids.

12.2.2.4 Adjuvants for Use Against Resistant Populations

The efficacy of some insecticides can be improved by the use of adjuvants in the mix. For example, piperonyl butoxide (PBO) improves the efficacy of lambda-cyhalothrin and other pyrethroids against the pollen beetle, especially if they have pyrethroid resistance. This is attributed to a degradation of monooxygenase in the targeted beetles. But PBO also causes some mortality of pyrethroid resistant beetles in the laboratory if applied without any insecticide though interestingly PBO is always reported as a non-toxic adjuvant.

12.2.3 Application

All insecticides for control of the pollen beetle are applied as sprays, often in response to the economic threshold. However, the only forecasting scheme to guide such usage in Germany is presented by proPlant (http://www.proplant.de/german/pp_frameset/frs_home/pp_home_data.htm) (see also Johnen et al. Chapter 15 this volume). It is difficult to estimate how many farmers make use of threshold values, as usage of insecticides has tended to remain constant or even increase from year to year, irrespective of the pest pressure. The use of pyrethroids has increased since fungicide use during flowering of oilseed rape was introduced and made the use of tank mixtures with insecticides attractive. The reduction of cost for the application of tank mixtures made insecticide use more economic for farmers and also increased the prophylactic use of insecticides which was additionally supported by the low price of pyrethroids. A major problem for good insecticide coverage of plants seems to be induced by the tendency to reduce the amount of water for application. When farmers observe that many pollen beetles are still present after spraying of an insecticide they usually argue that the product is failing even if good agricultural practice was not followed. This is the case if the amount of water sprayed was to low for proper coverage of plants or spraying took place at an inappropriate time.

12.2.4 Thresholds for Control

Thresholds for control of the pollen beetle are useful to encourage rational use of pesticides. However, the applied thresholds were calculated many years ago and differ quite drastically between countries and regions, the crop health conditions and crop growth stage (Richardson 2008, Williams Chapter 1 this volume). Their use is therefore hampered by the lack of information about their relevance to changed modern cropping systems, e.g., reduction in the number of plants per m^2 and use of cultivars which start flowering much earlier and have a much stronger ability to compensate loss of flower buds than the older cultivars.

12.3 Insecticide Resistance in the Pollen Beetle

The economic importance of the pollen beetle as a pest has resulted in intensive selection by insecticides and led to the evolution of resistance mechanisms. Here, we focus on their characterization by bioassay, the molecular basis of resistance, their impact on insecticide efficacy and, where known, the dynamics of resistance in pollen beetle populations and the factors driving them.

With the expansion of oilseed rape in Europe more farmers became aware of pest problems in their crop (Kirk-Spriggs 1996, Hansen 2003) which resulted in an increasing number of insecticide applications. In the late 1980s, resistance to insecticides was identified in populations of pollen beetles in Poland (Lakocy 1977). Since the late 1990s, further insecticide resistance has been found in different countries and is still increasing (Hansen 2003, Ballanger et al. 2003, 2007, Derron et al. 2004, Heimbach 2005, Nauen 2005, Heimbach and Müller 2006, Heimbach et al. 2006, Thieme et al. 2006, Wegorek and Zamojska 2006). In 2007, an EPPO workshop on pollen beetle resistance took place in which the relevance for all European countries was discussed (EPPO Bulletin 38, 2008).

More than 40 years ago, DDT resistant pollen beetles were documented in Poland (Table 12.2). By switching to insecticides with other mode of action this resistance was broken.

The first documented case of reduced pyrethroid susceptibility in the pollen beetle was reported in 1999 in the Champagne region in North-Eastern France (Table 12.2). Anecdotal reports on pyrethroid resistance development date back to 1997, although it should be noted that a number of these reports have not been validated. A similar phenomenon was observed in the Scandinavian countries, such as Denmark and Sweden. First rumours of reduced pyrethroid efficacy in Germany came in 2001 in the federal state of Rhineland-Palatinate. These rumours were confirmed in 2002, and since then pyrethroid resistance has spread all over Germany; since 2006, it has affected more than 50% of the winter oilseed rape acreage (Nauen

Table 12.2 Insecticide resistant pollen beetles reported from oilseed rape in Europe (after IRAC 2006)

Compounds resisted		Resistance first recorded
Insecticide	MOA	Year and country
Carbaryl	Carbamates	1974 PL[a]
Dioxacarb	Carbamates	1974 CZ[b], 1985 PL[a]
Propoxur	Carbamates	1974 PL[a]
Aldrin	Cyclodiene organochlorines	1974 PL[a]
Dieldrin	Cyclodiene organochlorines	1974 PL[a]
Endosulfan	Cyclodiene organochlorines	1974 PL[a], 1984 CZ[c]
Kelevan	Cyclodiene organochlorines	1974 CZ[b], 1985 PL[a]
Toxaphene	Cyclodiene organochlorines	1974 PL[a]
DDT	DDT	1965 PL[d], 1974 CZ[b]
Methoxychlor	DDT	1974 PL[a]
HCH-gamma	Cyclodiene organochlorines	1974 PL[a], 1974 CZ[b]
Cartap	Cartap	1974 CZ[b]
Bromophos	Organophosphates	1974 PL[a]
Chlorfenvinphos	Organophosphates	1974 PL[a]
Diazinon	Organophosphates	1974 PL[a]
Dimethoate	Organophosphates	2003 DK[e]
Methamidophos	Organophosphates	1974 PL[a]
Methidathion	Organophosphates	1974 PL[a]
Parathion-methyl	Organophosphates	1974 PL[a]
Phoxim	Organophosphates	1974 PL[a]
Pirimiphos-methyl	Organophosphates	1974 PL[a]
Lambda-cyhalothrin	Pyrethroids	2003 DK[e], 2005 D[f]
Alpha-cypermethrin	Pyrethroids	2004 PL[g]
Esfenvalerate	Pyrethroids	2003 DK[e]
Tau-fluvalinate	Pyrethroids	2003 DK[e]

[a]Lakocy (1974).
[b]Anonymous (1974).
[c]No entry FAO.
[d]Wegorek (1965).
[e]Hansen (2003).
[f]Heimbach et al. (2006).
[g]Wegorek (2005).

2007). In 2008, more than 80% of German populations showed high or very high resistance.

Periodically, pyrethroid resistance can escalate, as in 2007, when samples taken from a range of crops contained high proportions of resistant beetles. This apparent instability most probably reflected selection gradually acting through the response of growers to up-to-date advice on resistance management. In more recent years, high pyrethroid resistance has become more common in EU field samples, and has now also been found in the UK. Variation in susceptibility to pyrethroids has been documented previously among geographically-distinct pollen beetle populations and annual evaluations of susceptibility to pyrethroids have been conducted for insects collected from oilseed rape for many years.

12.3.1 Identification of Resistance by Bioassay

Although many laboratory bioassay methods have been developed to detect resistance, most of them are limited to defining phenotypes and provide little or no information on the underlying mechanisms. Nonetheless, bioassays remain the indispensable tower of strength of most large-scale resistance monitoring programs and are essential prior to the development of alternative diagnostic techniques based on the knowledge of the mechanism of resistance.

The phenotypic expression of resistance is assessed in small-scale laboratory bioassays by exposing pollen beetles to different applications of insecticides. The response of potentially insecticide-resistant populations is compared with known insecticide-susceptible standards. The only way of clearly identifying resistant populations needs baseline studies to be carried out prior to the use of the active substance. Such baseline studies are nowadays required by the pesticide regulation procedure (EPPO PP 213(2)) and hopefully will be available for new compounds. With pyrethroids and pollen beetles no such studies were carried out before about 20 years of use. Really sensitive pollen beetle populations may be extremely rare and therefore, in the majority of instances, susceptibility is a relative rather than an absolute concept.

The increase in resistant populations of pollen beetle is demonstrated by the response to lambda-cyhalothrin (applied at the recommended field rate in a adult vial test) for pollen beetles collected each year from 2002 to 2007 from a locality in Germany subjected to extensive use of insecticides (Fig. 12.1) and by results of a monitoring of pollen beetle resistance in Germany organised by the Julius Kühn-Institute (Table 12.3).

Susceptibility of these beetles to lambda-cyhalothrin decreased from 2005 to 2007, hence they have to be classified as being resistant in the last 3 years, according to the definition of EPPO (Heimbach et al. 2006). These analyses also showed that the resistance to lambda-cyhalothrin of pollen beetles from other locations increased

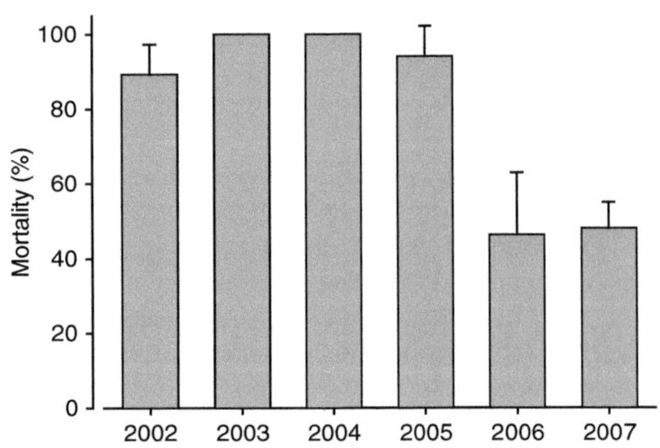

Fig. 12.1 Susceptibility to lambda-cyhalothrin of pollen beetles from a location with extensive use of insecticides in Germany

Table 12.3 Frequency of resistance classes of pollen beetle populations collected in Germany from 2006 to 2008 (from class 1 = very sensitive to class 5 = very resistant)

Resistance class	2006 (*n*=103)	2007 (*n*=281)	2008 (*n*=297)
1	7.0	3.2	0.3
2	8.0	5.3	2.4
3	18.0	19.9	15.8
4	41.0	38.8	40.4
5	26.0	32.7	41.1

up to 2007, even in large fields where there had been no previous application of insecticides (Thieme et al. 2008). This indicates that pollen beetles are very good at dispersing and makes it difficult to use field experiments for characterizing resistance. It is hardly possible to identify the recolonization of a field by insects that are such active flyers as pollen beetles.

The classification of beetles for resistance can be dependent on sampling time. Their great mobility allows fast mixing with beetles from other locations. Therefore, it is difficult to implicate a relationship between the application of an insecticide and resistance. This was demonstrated in an experiment on a 10 ha field sprayed with a pyrethroid. The application took place in the evening, the number of pollen beetles on plants and their susceptibility to lambda-cyhalothrin were assessed just before spraying, the next morning and until up to 10 days after application (Fig. 12.2). Because of immigration of beetles from other areas, the resistance identified directly after spraying was dramatically different from that on the second day after application. Therefore, it is essential to judge the susceptibility/resistance of a population of mobile insects by combining field results with laboratory analyses.

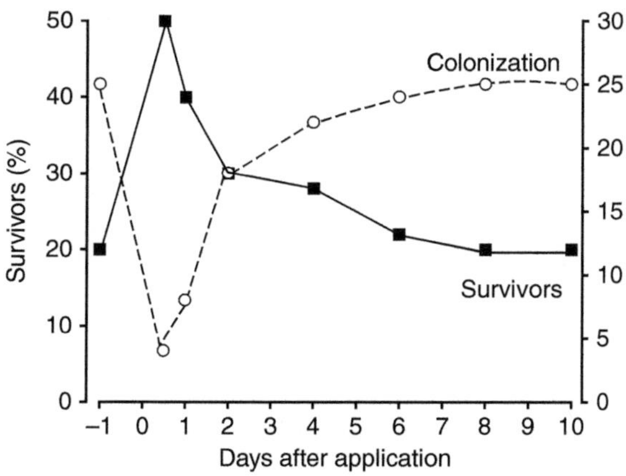

Fig. 12.2 Colonization of oilseed rape by pollen beetles and their survival after 5 h exposure to 0.075 μg lambda-cyhalothrin/cm^2 applied to glass tubes (Assessments of 20 plants per time and sampling were done in April 2007 on a 10 ha field in North-East Germany)

12.3.1.1 Problems with Bioassays

The response of pollen beetles to the active ingredient may differ from their response to the formulated product (Thieme et al. 2008). This is especially important when determining the sensitivity for products that seem to fail in the field.

Often, it is easier to run bioassays with pure technical substances. This has the advantage that they are not influenced by any formulation change which often occurs with products. Whereas technical substances can be dissolved in acetone, formulated products usually need at least some water which increases the time of drying test tubes. It is often argued that all insecticides with technical substances belonging to the same mode of action group should be cross-resistant. Thus, one might expect that, when Karate Zeon® is applied in the field, because of the many years of selection for resistance to lambda-cyhalothrin in a pollen beetle population, these beetles must be cross-resistant to all other pyrethroids. This is supported by bioassays using pure active substances, but not when formulated products are applied. Whereas formulated lambda-cyhalothrin shows resistance, formulated beta-cyfluthrin shows less resistance (Fig. 12.3a). For one sample of pollen beetles, a dosage of 1.41 g/ha lambda-cyhalothrin induced the same mortality (50%) as 1.24 g a.i./ha of the formulated product. However, for the same sample, a dosage of 1.74 g/ha beta-cyfluthrin induced the same mortality (50%) as 0.62 g a.i./ha of the formulated product. Similar results were obtained for etofenprox and its formulation (Fig. 12.3b). Therefore, results obtained using a formulated product and its active substance should not be generalized to other pyrethroids.

In bioassays suitable for testing insecticides that have best efficacy if consumed by feeding (e.g., Biscaya®), complete inflorescences or excised flower buds of oilseed rape that have been dipped into the insecticide can be used. Although more labour intensive and time consuming (in terms of preparation), bioassays using this

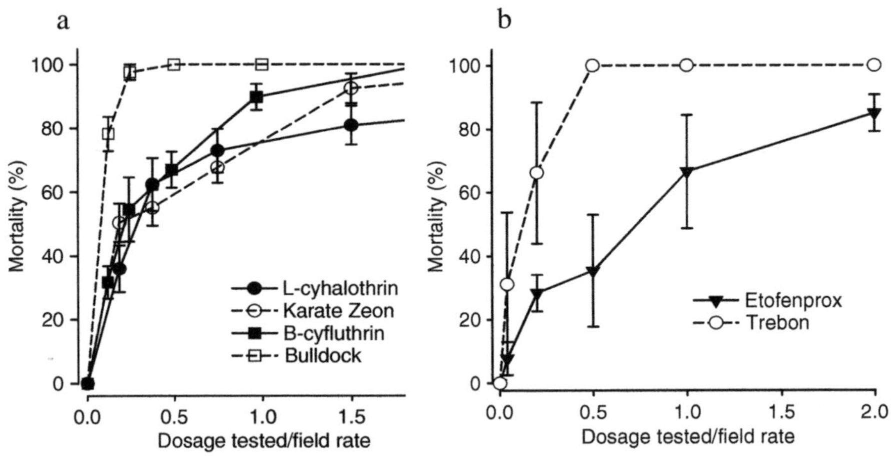

Fig. 12.3 a, b Mortality of pollen beetles induced by different dosages of pyrethroids (active substances compared with formulated products)

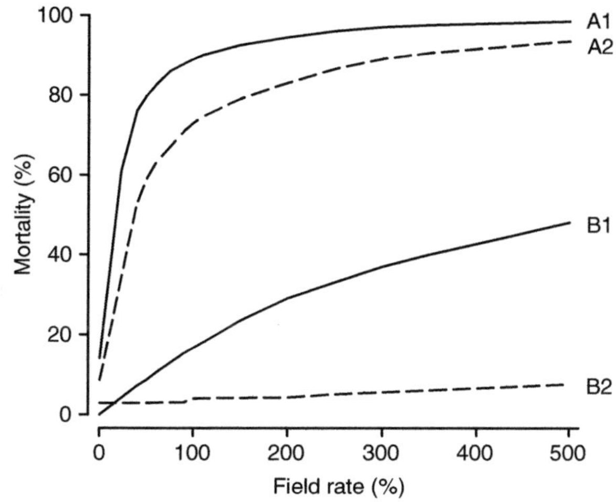

Fig. 12.4 Mortality of pollen beetles induced by 24 h exposure to Biscaya® (A) and formulated lambda-cyhalothrin (B) applied either to excised flower buds (1) or to a complete inflorescence (2) of oilseed rape. The beetles were collected in May 2007 near Dommritz/Saxony (Germany)

method allow oral application of formulated products. However, results can differ with the oilseed rape material to which the insecticide is applied as demonstrated for a pollen beetle population collected from rape in 2007 in Saxony. Although highly resistant to Karate Zeon®, these beetles were susceptible to Biscaya® (Fig. 12.4). But, for both products, exposure to excised and dipped buds resulted in significantly higher mortality than exposure to dipped complete inflorescences. For Biscaya® the hypothesis for equality ($\chi^2 = 31.76$; d. f. = 2; P = 0.000) and also the hypothesis of parallelism ($\chi^2 = 9.06$; d. f. = 2; P = 0.002) were rejected. Also for Karate Zeon® the hypothesis for equality was rejected ($\chi^2 = 48.49$; d. f. = 2; P = 0.000) as well as the hypothesis of parallelism ($\chi^2 = 7.96$; d. f. = 2; P = 0.005).

These differences may be caused by the larger size of container needed for the complete inflorescence, allowing the beetles to disperse more in space (which may result in a reduced consumption of treated plant material). The differences may also be caused by a stronger concentration of volatiles in the smaller glass tubes containing the excised buds. These volatiles are attractive to the pollen beetle and are not sex related (Cook et al. 2002). Further experiments will show if beside attractive floral volatiles, damage volatiles too are olfactory cues inducing increased consumption of treated plant material.

12.3.1.2 Do We Have to Know the Species?

Pest monitoring surveys have shown that *Meligethes* spp. differ in their susceptibility to insecticides (Derron et al. 2004). Currently insecticide resistance is determined under the assumption of testing only the true and most-commonly found species of pollen beetle, namely *M. aeneus*. But it is strikingly evident that samples of the genus *Meligethes* collected in the field can include several different species. Ignoring this may result in apparent susceptibility. The occurrence of different species in both winter and spring oilseed rape is not new (Reitter 1871, Friederichs

1921, Kauffmann 1925, Bollow 1950, Nolte and Fritzsche 1952, Scherney 1953, Fritzsche 1955, Kirk-Spriggs 1996, Marczali and Keszthelyi 2003). The problem is that a time-consuming dissection of male and female genitalia is necessary to distinguish between most species and, in monitoring surveys often, if at all, only those species are recorded that can be distinguished by colour or by morphometric characters that are easy to recognize (as in *M. viridescens*). Analyses have shown that in Germany and the UK, species of *Meligethes* other than *M. aeneus* occur in samples collected from oilseed rape fields (Thieme et al. 2006; Figs. 12.5 and 12.6).

As already reported by Derron et al. (2004), *M. viridescens* is not resistant and is present in variable proportions in relation to collection time and origin. Insecticide

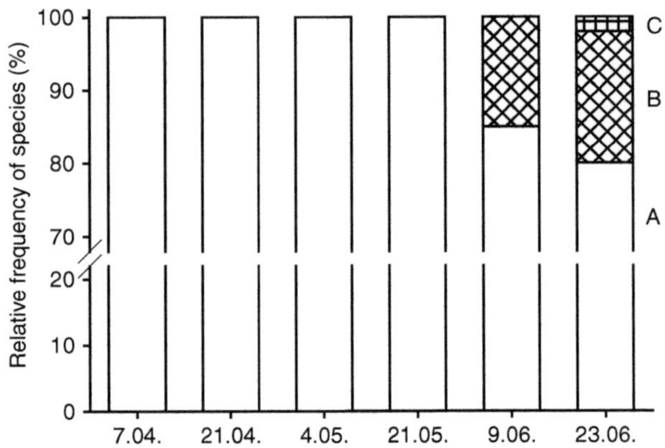

Fig. 12.5 Frequency of *Meligethes* species in samples collected in 2004 near Schwerin/Germany (A – *M. aeneus*; B – *M. viridescens*; C – *M. nigrescens, M. coeruleovirus and M. coracinus*)

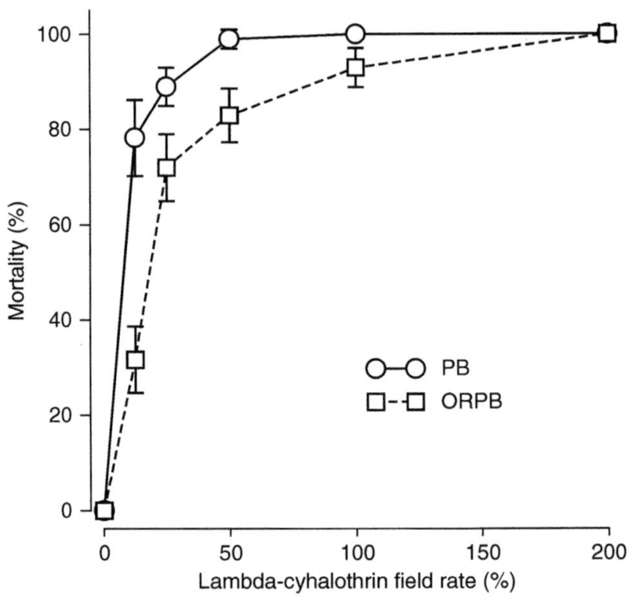

Fig. 12.6 Mortality induced by lambda-cyhalothrin to a mixture of different pollen beetle species (PB) and to *M. aeneus* (ORPB) collected in July 2007 near Rothamsted, UK

resistance should be determined only for the true oilseed rape pollen beetle *M. aeneus*. Ignoring this distorts bioassay results.

Additionally, other aspects can influence the outcome of bioassays such as time of beetle collection (see below), the location within a field, any pyrethroid application prior to the collection, the fitness of the beetles, as well as their weight (Tiilikainen and Hokkanen 2008).

12.3.1.3 Relevance of Bioassays to Field Control

In addition to quantifying resistance in bioassays, it is important to establish whether resistance quantified in this manner is of practical importance in the field. One way to do this is to apply insecticides in a more realistic manner under field conditions. These approaches ensure that bioassays are correlated with potential field control problems.

Field experiments have been used to study the performance of various established insecticides applied at recommended field rates against the pollen beetle carrying resistance to insecticides. To decide if the population is resistant or not they need to be combined with bioassays with adequate sampling of beetles shortly before and/or after the spraying. Bioassay results need then to be compared to known resistant or susceptible strains. Unfortunately, it is not yet possible to culture the pollen beetle for several generations in the laboratory. Therefore, it is impossible to compare bioassay results of well-defined strains that may support the development of biochemical and molecular characterization.

12.3.2 Biochemistry and the Molecular Basis of Resistance

The information provided by the use of bioassay techniques is essential. However, attention is being focused increasingly on the development of more sophisticated diagnostics that not only offer greater precision and throughput, but also identify the specific mechanism(s) present and even the genotype of the resistant insect. This is exemplified by research on aphids where significant advances have led to rapid and precise methods for the detection of different resistance mechanisms in individual insects.

But one of the disadvantages of the pollen beetle is that it is not yet possible to culture characterized strains in the laboratory.

In Europe, the pollen beetle possesses at least two resistance mechanisms: (i) metabolic resistance, caused by overproduced monooxygenase conferring resistance to pyrethroids (Nauen 2007), and (ii) target site resistance (kdr) to pyrethroids (Kristensen pers comm, Nauen pers comm, http://www.jki.bund.de/fileadmin/dam_uploads/_A/resistenz/insektizide_akarizide/5te_Kurzprotokoll zur 5. Sitzung des Fachausschuss Pflan.pdf). The latter mechanism confers strong resistance to virtually all available pyrethroids. It is now possible to diagnose both mechanisms in beetles using an immunoassay for the overproduced monooxygenase, and a molecular diagnostic for the *kdr* allele.

12.3.2.1 Metabolic Resistance Mechanism

In recent years, the majority of insecticides used to control the pollen beetle on oilseed rape have belonged to the pyrethroid class, which is an ester. Their bonds are particularly prone to cleavage and resistance to these active substances is commonly mediated by the enhanced hydrolysis of ester bonds. An association between resistance in bioassays and enhanced esterase activity was established in aphids (*M. persicae*) over 30 years ago (Needham and Sawicki 1970). This discovery was central to the understanding of this type of resistance in aphid species.

Biochemical studies were performed by Nauen (2007) with pyrethroid-susceptible and pyrethroid-resistant pollen beetle populations. No differences were found in the overall esterase activity and glutathione S-transferase activity but monooxygenase levels differed significantly. This suggested a metabolic mechanism of resistance contributing to the differences in pyrethroid susceptibility. The over-produced monooxygenase sequester or degrade insecticide esters before they reach their target sites in the nervous system and confer strong resistance to pyrethroids.

12.3.2.2 Target Site Resistance Mechanism

A target site mechanism, termed knockdown resistance (kdr), has been identified as the main threat of resistance to pyrethroid insecticides (Martinez-Torres et al. 1999). Target site resistance is conferred by mutations in the voltage-gated sodium channel gene (Soderlund and Knipple 2003). Initial recognition of the kdr mechanism was dependent on a molecular biological approach in which a highly conserved point mutation in the sodium channel gene was identified based on studies in kdr houseflies, aphids and other pest species with the same mechanism (Martinez-Torres et al. 1997, 1999, Devonshire et al. 1998). A replacement of leucine by phenylalanine in domain II of transmembrane segment II-S6 is the most common amino acid substitution in the *kdr* allele. This homologous mutation has been documented in many insects (Dong 1997, Park et al. 1997, Jamroz et al. 1998, Martinez-Torres et al. 1998) and was initially particularly well-studied in the house fly (Williamson et al. 1993, 1996a, b). Molecular diagnostic methods continue to be essential in understanding the role of this mechanism in resistant aphids (Guillemaud et al. 2003).The kdr mechanism was found recently in pollen beetle populations in Denmark (see above).

The biochemical and molecular genetic diagnostics described will enable a greater understanding of the incidence and effect of the different resistance mechanisms in the pollen beetle.

12.3.3 Factors Affecting the Dynamics of Insecticide Resistance in the Field

12.3.3.1 Selection Pressures

Insecticide Application Factors

The use of reduced rates, poor application techniques, or too early applications before most of the beetles arrive, will lead to substantial survival and population increase. It causes selection of resistant beetles when the insecticide is present at concentrations lower than those expected at the time of correct treatments. Such conditions also arise from the natural decay of the insecticide with time, the application at unfavorable conditions (e.g., pyrethroids show reduced efficacy applied at high temperatures) or the application of reduced rates, either to cut cost or to target other pests. This demonstrates that variation in susceptibility is the first evolutionary step towards resistance. In Germany at least, studies were recently launched to analyze the fitness costs or advantages of maintaining a fluctuating polymorphism of susceptible and resistant pollen beetle populations despite prolonged and often intense selection by insecticides.

Ecological Factors

The diversity of resistance mechanisms known to date leads to the assumption that no insecticide is immune to the appearance of genes conferring resistance. The probability of these achieving detectable frequencies depends instead on a suite of ecological and genetic factors and how these interact with insecticide usage patterns. As a consequence, the same product can face very different resistance risks with different pest species and even within the same species in different cropping systems (Foster et al. 2007). So selection pressure and resistance risk seem to be much higher in spring oilseed rape crops than in winter oilseed rape crops (Hansen 2008).

One of the basic principles of resistance management is that resistant forms are less fit than the susceptible genotype in the absence of insecticides; otherwise, the former would probably be present at higher frequencies prior to selection by insecticides. Some of the best evidence of pleiotropic effects of resistance genes comes from recent work on aphids where field and laboratory studies suggest the existence of adverse selection in the form of poor winter survival, maladaptive behavior, and reduced reproductive fitness imposed primarily during times of stress. Significant selection pressure from synthetic insecticides dates back only 40 years, but the intensity of usage already has imposed extremely strong selection.

Nevertheless, susceptible beetles persist and could increase in proportion within a population when insecticide selection pressure is reduced, for example, over winter months when pollen beetles have settled within hibernation places and are untreated. The fitness of resistant forms in the absence of insecticides has not been studied. Monitoring of pollen beetle populations in Germany did not show a fall in the frequencies of resistant beetles during the winter months in bioassays (Thieme et al. 2008, Fig. 12.7); the same level of resistance was found in pollen beetles caught during infestation of oilseed rape crops in spring/summer as in those caught in the previous autumn.

In Germany, a large field experiment was undertaken to show the consequences of different treatment regimes for the build-up of metabolic resistance. To check if any reduction of selection pressure has effects on the sensitivity of the populations present, a region in North-East Germany (diameter about 6–8 km) with strong resistance problems in 2006 was selected in which no pyrethroids were used on oilseed rape crops after 2007 during periods when pollen beetles were present. No clear change in sensitivity (Fig. 12.8) was detected; this may have been caused either by too high mobility of the beetles, too short a period for the analysis, or a learning effect of near-by non-regulated farmers also avoiding the use of pyrethroids. The

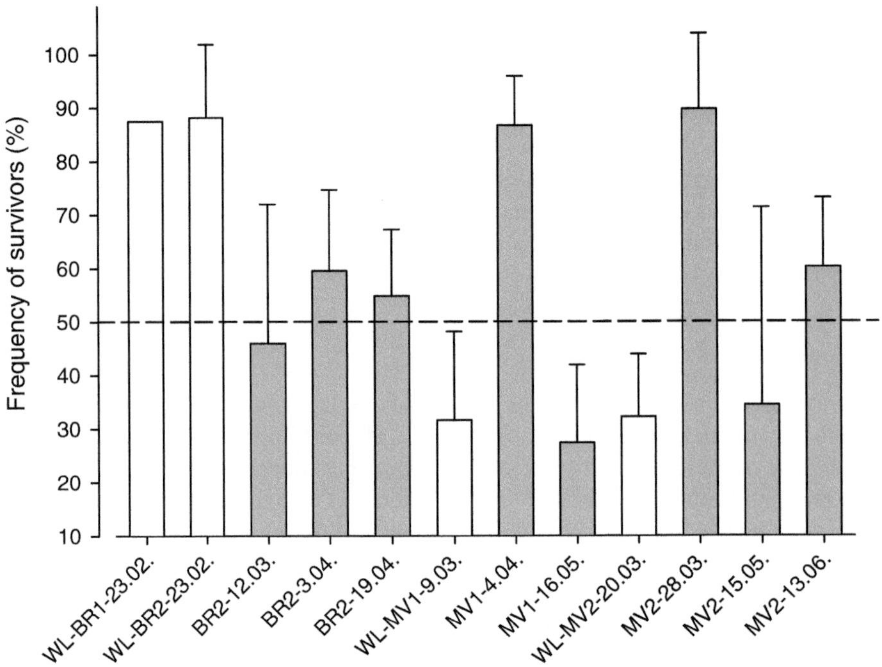

Fig. 12.7 Mean percentage survivors (± SD) of pollen beetles collected at the same location during different periods in Brandenburg (BR) and Mecklenburg-West Pomerania (MV) in hibernation places (WL) and in the field after 5 h exposure to 0.075 μg lambda-cyhalothrin/cm^2 in an adult vial test (*dashed line*: threshold for resistance class 'highly resistant' according Heimbach et al. 2006)

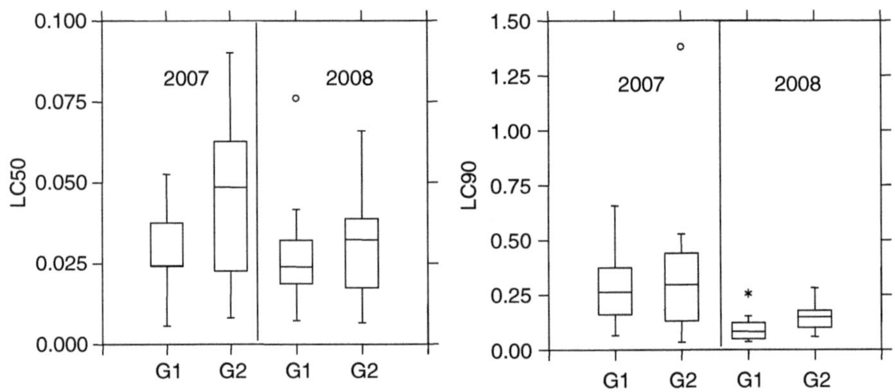

Fig. 12.8 LC50 and LC90 (μg a.i./cm^2) values for pollen beetles exposed to lambda-cyhalothrin in an adult vial test. The beetles belonging to populations sampled at different times in 2007 and 2008 in a German region without pyrethroid use (G1) and a non-regulated region nearby (G2)

latter may be relevant because, in 2008, an improvement in the resistant status of pollen beetles was detected in North East Germany in contrast to the situation in Germany as a whole where resistance is still increasing. This might support the idea that a reduction of the selection pressure will result in less resistant populations and that replacement of pyrethroids seems to be an efficient resistance management strategy.

12.3.4 Resistance Management Strategies

The possibility of resistance highlights the need to develop and implement resistance management strategies to prevent or delay the evolution of resistance to new insecticides. These strategies are dependent on the development of effective resistance monitoring programs capable of early detection of resistance that will allow implementation of appropriate management decisions in a timely manner. The initial steps in implementing such programs include development of appropriate bioassay techniques and establishment of baseline susceptibility data among populations across the geographic range of the target species. With this information, potential population susceptibility changes in response to selection with a distinct insecticide can be identified.

Variation in susceptibility to pyrethroids has been documented previously among geographically-distinct *M. aeneus* populations and annual evaluations of susceptibility to pyrethroids are conducted for insects collected from oilseed rape. The objective of different studies was to establish a baseline of susceptibility to pyrethroids from geographically-distinct populations of *M. aeneus* collected from oilseed rape, emphasizing areas where there is likely to be exposure to insecticides.

The variation in baseline susceptibility to insecticides may reflect differences in vigour among parental populations, attributes that are the product of genotype, the maternally-determined nutritional status of the egg, age, size and health status. Variation in susceptibility to insecticidal toxins has been reported from repeated bioassays against larvae from single strains of Colorado potato beetle and diamond-back moth (Robertson et al. 1995) and European corn borer (Marcon et al. 1999). Intrapopulation variation in response to insecticides is a common phenomenon when any bioassay is repeated. The extents of both inter- and intra-population variation in susceptibility to a given insecticide should be investigated before biologically-important changes can be identified with any certainty. This should be done before the product is used commercially rather than after resistance is already widespread.

Resistance strategies should be implemented before a resistance problem becomes acute. This was apparently not the case for pyrethroid-resistant pollen beetles. Now, in several countries, resistance groups have been established. For example, in Germany, resistance strategies for insecticidal use in oilseed rape have been discussed and published since 2006, to try to ensure effective control of the pollen beetle as well as at the same time to slow down the development of resistance (http://www.jki.bund.de/no_cache/de/startseite/fachinformationen/pflanzenschutz/

pflanzenschutzresistenz/insektizide-akarizide.html). Such strategies need support by all relevant parties including pesticide regulators, industry, pesticide merchants, advisors and last, but not least, farmers. The German strategy still relies on the use of pyrethroids because of the limited availability of other control options. OPs are implemented but show only low control options on stem weevils and can only be used clearly before flowering because of bee safety issues. Bee-safe neonicotinoids are less effective on relevant oilseed rape pests than pyrethroids and not sufficient under high pollen beetle pressure. The use of some pyrethroid active substances, such as tau-fluvalinate, etofenprox and bifenthrin, which sometimes still seem to be less affected by resistance (e.g., Hansen 2008), is the only chance, though the risk of further resistance selection is very high and, in Denmark, kdr resistance has already been detected (see above) after the use of such active substances for some years. The same is true for additives such as PBO, which have not been accepted for use in Germany. There is a strong need for new modes of action with good efficacy on several pest insects of oilseed rape, no cross resistance, bee safety and a competitive price.

Resistance management strategies should also more and more implement new growing schemes for oilseed rape such as border rows with early flowering varieties or other attractive crops (Cook and Denholm 2008) as well as biological control options (Hokkanen 2008). But there is only limited experience with such growing forms and often they are associated with more economic risk for farmers. So, at the moment, only chemical control is accepted by farmers.

In 2007, an 'Insecticide Resistance Action Committee' (IRAC) Pollen Beetle Working Group was established to bring together expertise from agrochemical companies and independent researchers in order to monitor the development of insecticide resistance in oilseed rape pests and to provide guidance and advice on the best practices to monitor and to prevent further insecticide resistance development. The monitoring showed the wide distribution of resistant populations in Europe (http://www.irac-online.org/documents/pdf_pb_monitoring_%202008_v2%202.pdf).

All resistance groups and participants of the EPPO workshop in 2006 (Zlof 2008) have concluded that an efficient strategy to reduce the resistance level or to slow down resistance development requires at least 2 additional modes of action with no cross resistance to pyrethroids, the correct use of the full dose with the correct equipment, and insecticide application only in those situations in which relevant pest damage is expected. Such strategies need also to take account of other insect pests of oilseed rape. In Germany, many populations of other pests (*Ceutorhynchus* spp. weevils), have been monitored with no clear incidence of any resistance yet (Müller et al. 2008), although there seems to be an increase in their survival rate from 2005 to 2008 (Fig. 12.9).

Development of baseline susceptibility data represents the first step toward the development of a monitoring program designed to detect changes in susceptibility that may result from repeated and prolonged exposure to a given insecticide. These data also may provide information that will allow development of diagnostic bioassays that would be more efficient in detection of resistant populations. However, it

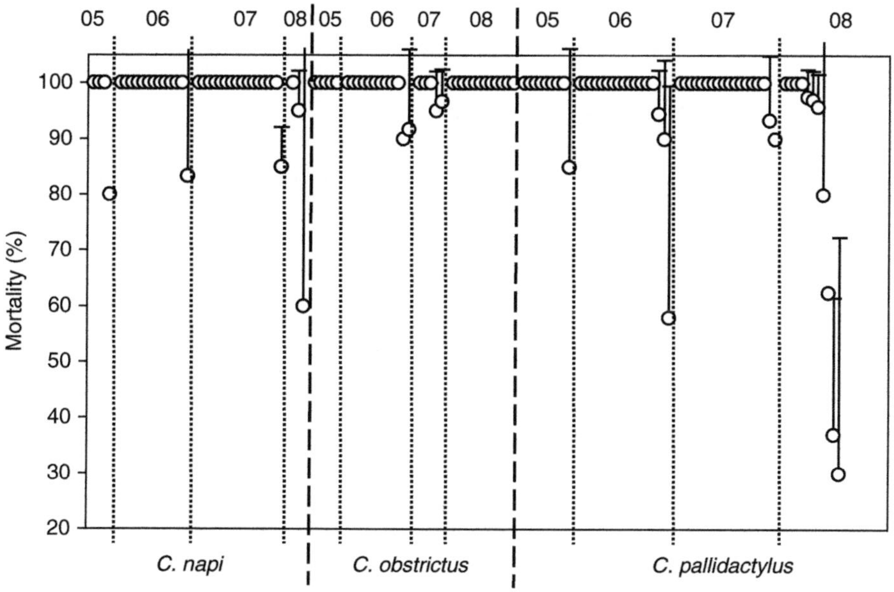

Fig. 12.9 Mortality of populations of different *Ceutorhynchus* spp. 5 h after exposure to 0.015 μg/cm^2 lambda-cyhalothrin in adult vial tests. The beetles were collected in Germany between 2005 and 2008. Each *dot* represents a different population

should be noted that pollen beetle populations are exposed to multiple applications of pyrethroids for a long time. As a result, designation of a shift in susceptibility to pyrethroids may be difficult to achieve because of the lack of field populations not being exposed to these insecticides.

12.4 Conclusions

The battle to control the pollen beetle and other pest insects of oilseed rape will continue as long as these pests compete with us for food resources. The new insecticides such as neonicotinoids with novel modes of action, will provide more effective control in the future. Their use will achieve less exposure of potentially risky chemicals to non-target organisms and farmers compared to older insecticides from the last century. In future, conventional breeding programs for resistance or genetic modified plants may provide more options for control. But there is also a need to develop oilseed rape growing systems which reduce pollen beetle pressure and reduce the need to control. Integrated pest management protocols for oilseed rape pollen beetle problems are needed. Sustainable control of pest insects on oilseed rape needs to alternate the use of different active substances having no cross resistance and needs a selection of optimal active substances with good efficacy for the different pest species and climatic condition including bee safety aspects. Such an optimal choice is not possible at the moment in Europe with only one or two types of active substances being available in many countries.

As a consequence, insecticidal control of oilseed rape pests continues to select for resistance and is clearly a case of co-evolution demonstrating how such processes generate biological diversity. Over the past 15 years, scientists have made great progress in monitoring and characterizing insecticide resistance and in understanding some of the factors that affect the speed of its development. Regardless of the progress that has been made, the competition between insect evolution and human inventiveness will continue to present major challenges. In order to meet these, it will be necessary to increase understanding of the processes that mediate the development of resistance in pollen beetles. There is a strong need for empirical research on the mechanisms conferring resistance to new chemical groups, and an understanding of the width of resistance that these mechanisms confer. There is also a lack of information on the ecological factors that mediate resistance development – fitness costs for resistant beetles, and the effects of dispersal and migration.

Acknowledgements Thanks should be given to many persons from different countries for the collection of pollen beetles and using the test kits supplied by the JKI as well as for technical support by K. Burlak, K. Gloyna, I. Rogotzki, L. Bansemir, R. Bull and U. Hoffmann.

References

Altmann R, Elbert A (1992) Imidacloprid – ein neues Insektizid für die Saatgutbehandlung in Zuckerrüben, Getreide und Mais. *Mitt Dtsch Ges Allg Angew Ent* 8: 212–221.

Anonymous (1974) Cited in Georgiou, GP and Lagunes-Tejeda L (1991) *The Occurrence of Resistance to Pesticides in Arthropods: An Index of Cases Reported through 7989.* FAO, Rome, Italy.

Anonymous (1993) *Evolution on Imidacloprid.* Disclosure Document for Imadacloprid in the Products 'Gaucho' and 'Zelmone'. Advisory Committee on Pesticides Issue No. 118. Stationary Office, London, UK.

Ballanger Y, Détourné D, Delorme R, Pinochet X (2003) Difficulties to control pollen beetle (*Meligethes æneus* F.) in France revealed by unusual high level infestations in winter rape fields. *Proc GCIRC 11th Int. Rapeseed Congress, Copenhagen, 6–10 July 2003*, 3: 1048–1050.

Ballanger Y, Détourné D, Delorme R, Pinochet X (2007) France, difficulties to manage insect pests of winter oilseed rape (*Brassica napus* var. *oleifera*): Resistances to insecticides. *Proc GCIRC, 12th Int Rapeseed Congress, Wuhan, 26–30 March 2007,* 4: 276–279.

Bollow H (1950) Vorkommen verschiedener *Meligethes*-Arten an Raps in Bayern. *Z Pflbau und PflSchutz* 2: 86–93.

Bucholz A, Nauen R (2002) Translocation and translaminar bioavailability of two neonicotinoid insecticides after foliar application to cabbage and cotton. *Pest Manage Sci* 58: 10–16.

Cook SM, Bartlet E, Murray DA, Williams IH (2002) The role of pollen odour in the attraction of pollen beetles to oilseed rape flowers. *Entomol Exp Appl* 104: 43–50.

Cook SM, Denholm I (2008) Ecological approaches to the control of pest insects in oilseed rape. *EPPO Bull* 38: 110–113.

Derron JO, Clech EL, Bezençon N, Goy G (2004) Résistance des méligèthes du colza aux pyréthrinoïdes dans le bassin lémanique. *Revue Suisse d'Agriculture* 36: 237–242.

Devonshire AL, Field LM, Foster SP, Moores GD, Williamson MS, Blackman RL (1998) The evolution of insecticide resistance in the peach-potato aphid, *Myzus persicae*. *Philos T Roy Soc B* 353: 1677–1684.

Dong K (1997) A single amino acid change in the para sodium channel protein is associated with knockdown-resistance (*kdr*) to pyrethroid insecticides in German cockroach. *Insect Biochem Mol Biol* 27: 93–100.

Elbert A, Ebbinghaus D, Maeyer L, Nauen R, Comparini S, Pitta L, Brinkmann R (2002) CalypsoReg., a new foliar insecticide for berry fruit. *Acta Hort* 585: 337–341.

Elbert A, Erdelen C, Kuhnhold J, Nauen R, Schmidt HW, Hattori Y (2000) Thiacloprid, a novel neonicotinoid insecticide for foliar application. *Proc Brighton Crop Protection Conf, Pests and Diseases* 1: 21–26.

Elbert A, Overbeck H, Iwaya K, Tsuboi S (1990) Imidacloprid, a novel systemic nitromethylene analogue insecticide for crop protection. *Proc Brighton Crop Protection Conf, Pests and Diseases, November 1990* 1: 21–28.

Foster SP, Devine G, Devonshire AL (2007) Insecticide resistance. In: vanEmden HF, Harrington R (eds.) *Aphids as crop pests*. CABI, Wallingford.

Friederichs K (1921) Untersuchungen über den Rapsglanzkäfer in Mecklenburg. *Z Angew Entomol* 7: 1–36.

Fritzsche R (1955) Zur Morphologie von *Meligethes aeneus* Fabr., *M. viridescens* Fabr., *M. coracinus* Sturm und *M. picipes* Sturm. *Beit Entomol* 5: 309–333.

Guillemaud T, Brun A, Anthony N, Sauge M-H, Boll R, Delorme R, Fournier D, Lapchin L, Vanlerberghe-Masutti F (2003) Incidence of insecticide resistance alleles in sexually-reproducing populations of the peach-potato aphid *Myzus persicae* (Hemiptera: Aphididae) from southern France. *B Entomol Res* 93: 289–297.

Hansen LM (2003) Insecticide-resistant pollen beetles (*Meligethes aeneus* F.) found in Danish oilseed rape (*Brassica napus* L) fields. *Pesticide Manage Sci* 59: 1057–1059.

Hansen LM (2008) Occurrence of insecticide resistant pollen beetles (*Meligethes aeneus* F.) in Danish oilseed rape (*Brassica napus* L.) crops. *EPPO Bull* 38: 95–98.

Heimbach U (2005) Ausschuss für Resistenzfragen- Insektizide und Akarizide, Bericht über das erste Treffen im Februar 2005 in der BBA in Braunschweig. *Nach Dtsch PflSchutz* 57: 172–173.

Heimbach U, Müller A (2006) Achtung: Resistente rapsschädlinge. *DLZ Agrarmagazin* 2: 40–43.

Heimbach U, Müller A, Thieme T (2006) First steps to analyse pyrethroid resistance of different oil seed rape pests in Germany. *Nach Dtsch PflSchutz* 58: 1–5.

Hokkanen HMT (2008) Biological control methods of pest insects in oilseed rape. *EPPO Bull* 38: 104–109.

IRAC (2006) IRAC Susceptibility Test Methods Series, Method No: 11 Pollen Beetle, *Meligethes aeneus,* adults, synthetic pyrethroids. http://irac-online.org/documents/method11.pdf, May 2007.

Ishaaya I, Horowitz AR (1998) Insecticides with novel modes of action: An overview. In: Ishaaya I, Degheele D (eds.) *Insecticides with novel modes of action. Mechanisms and application.* Springer, Berlin, Germany.

Jamroz RC, Guerrero FD, Kammlah DM, Kunz SE (1998) Role of the *kdr* and *super-kdr* sodium channel mutations in pyrethroid resistance: Correlation of allelic frequency to resistance level in wild and laboratory populations of horn flies (*Haematobia irritans*). *Insect Biochem Mol Biol* 28: 1031–1037.

Jeschke P, Moriya K, Lantzsch R, Seifert H, Lindner W, Jelich K, Gohrt A, Beck ME, Etzel W (2001) Thiacloprid (Bay YRC 2894) a new member of the chloronicotinyl insecticide (CNI) family. *PflSchutz Nach Bayer* 54: 147–160.

Kauffmann O (1925) Beobachtungen und Versuche zu Fragen der Überwinterung und Parasitierung von Ölfruchtschädlingen aus den Gattungen *Meligethes*, *Phyllotreta*, *Psylliodes* und *Ceutorrhynchus*. *Arb Biol Reichsanst Land- und Forstwirtsch* 12: 109–169.

Kirk-Spriggs AH (1996) *Pollen Beetles. Coleoptera: Kateretidae and Nitidulidae: Meligethinae.* Handbooks for the Identification of British Insects, Vol. 5 (6a). Royal Entomological Society of London, London, UK.

Lakocy A (1974) *Communication to FAO*. Cited in IRAC 2006.

Lakocy A (1977) The influence of some biological and ecological factors on the development of resistance to insecticides and on the course of chemical control of *Meligethes aeneus* F in the Voivodships of Poznan and Wroclaw. *Prace Naukowe Institutu Ochrony Roslin* 19(1): 123–181.

Marcon PCHG, Young LJ, Steffey KL, Siegfried BD (1999) Baseline susceptibility of European corn borer (Lepidoptera: Crambidae) to *Bacillus thuringiensis* toxins. *J Econ Entomol* 92: 279–285.

Marczali Z, Keszthelyi S (2003) A study on *Meligethes* species in Keszthely, 2002. *J European Agr* 4: 237–244.

Martinez-Torres D, Chandre F, Williamson MS, Darriet F, Berge JB, Devonshire AL, Guillet P, Pasteur N, Pauron D (1998) Molecular characterization of pyrethroid knockdown resistance (*kdr*) in the major malaria vector *Anopheles gambiae* s.s. *Insect Mol Biol* 7: 179–184.

Martinez-Torres D, Devonshire AL, Williamson MS (1997) Molecular studies of knockdown resistance to pyrethroids: Cloning of domain II sodium channel gene sequences from insects. *Pestic Sci* 51: 265–270.

Martinez-Torres D, Foster SP, Field LM, Devonshire AL, Williamson MS (1999) A sodium channel point mutation is associated with resistance to DDT and pyrethroid insecticides in the peach-potato aphid, *Myzus persicae* (Sulzer) (Hemiptera: Aphididae). *Insect Mol Biol* 8: 1–8.

Müller A, Heimbach U, Thieme T (2008) Pyrethroid sensitivity monitoring in Germany of oilseed rape pest insects other than pollen beetle. *EPPO Bull* 38: 85–90.

Nauen R (2005) Insecticide resistance in European agriculture: Research instead of rumours. *Proc Brighton Crop Protection Conf – Crop Science & Technology 2005* 3: 123–130.

Nauen R (2007) Pyrethroid resistance and its management in European populations of pollen beetles, *Meligethes aeneus,* in winter oilseed rape. *Proc XVI Int Plant Protection Cong 2007*: 522–523.

Nauen R, Ebbinghaus-Kintscher U, Salgado VL, Kausmann M (2003) Thiamethoxam is a neonicotinoid precursor converted to clothianidin in insects and plants. *Pestic Biochem Physiol* 76: 55–69.

Nauen R, Ebbinghaus-Kintscher U, Schmuck R (2001) Toxicity and nicotinic acetylcholine receptor interaction of imidacloprid and ist metabolites in *Apis mellifera* (Hymenoptera: Apidae). *Pest Manage Sci* 57: 577–586.

Nauen R, Tietjen K, Wagner K, Elbert A (1998) Efficacy of plant metabolites of imidacloprid against *Myzus persicae* and *Aphis gossypii* (Homoptera: Aphididae). *Pestic Sci* 52: 53–57.

Needham PH, Sawicki RM (1970) Diagnosis of resistance to insecticides in *Myzus persicae. Nature* 230: 125–126.

Nolte HW, Fritzsche R (1952) Untersuchungen über das Vorkommen verschiedener *Meligethes*-Arten auf Raps. *Beit Entomol* 2: 434–448.

Park Y, Taylor MFJ, Feyereisen R (1997) A valine421 to methionine mutation in IS6 of the *hscp* voltage-gated sodium channel associated with pyrethroid resistance in *Heliothis virescens* F. Biochem. *Biophys Res Commun* 239: 688–691.

Plapp FW (1991) The nature, modes of action, and toxicity of insecticides. In: Pimentel D (ed.) *CRC handbook of pest management in agriculture*. CRC Press, Boston, USA.

Reitter E (1871) *Revision der europäischen Meligethes-Arten.* Verhandlungen des naturforschenden Vereins in Brünn, Bd. IX, 133 pp; Brünn.

Richardson DM (2008) Summary of findings from a participant country pollen beetle questionnaire. *EPPO Bull*38: 68–72.

Robertson JL, Preisler HK, Ng SS, Hickle LA, Celernter WD (1995) Natural variation: A complicating factor in bioassays with chemical and microbial pesticides. *J Econ Entomol* 88: 1–10.

Scherney F (1953) Zur Biologie der an Raps vorkommenden *Meligethes* Arten. *Z PflBau PflSchutz* 4: 154–176.

Soderlund DM, Knipple DC (2003) The molecular biology of knockdown resistance to pyrethroid insecticides. *Insect Biochem Mol Biol* 33: 563–577.

Takahashi J, Mitsui J, Takakusa N, Matsuda M, Yoneda H, Suzuki J, Ishimitsu K, Kishimoto T (1992) NI-25, a new type of systemic and broad spectrum insecticide. *Proc Brighton Crop Protection Conf Pests and Diseases November 1992* 1: 89–96.

Thieme T, Drbal U, Gloyna K, Hoffmann U (2008) Different methods of monitoring susceptibility of oilseed rape beetles to insecticides. *EPPO Bull* 38: 114–117.

Thieme T, Hoffmann U, Mühlschlegel F (2006) Susceptibility of pollen beetles to insecticides on oilseed rape. *CD-ROM Proc Int Symp 'Integrated Pest Management of Oilseed Rape Pests', Göttingen, Germany, 3–5 April 2006.*

Tiilikainen TM, Hokkanen HMT (2008) Pyrethroid resistance in Finnish pollen beetle (*Meligethes aeneus*) populations – is it around the corner? *EPPO Bull* 38: 99–103.

Tomizawa M, Casida JE (2003) Selective toxicity of neonicotinoids attributable to specificity of insect and mammalian nicotinic receptors. *Annu Rev Entomol* 48: 339–364.

Wegorek W (1965) In: FAO 1967. Cited in IRAC 2006.

Wegorek P (2005) Preliminary data on resistance appearance of Pollen beetle PB (*Meligethes aeneus* F.) to selected pyrethroids, organophosphorous and chloronicotynyls insecticide, in 2004 year in Poland. *Resistance Pest Manage Newsl* 14: 19–21.

Wegorek P, Zamojska J (2006) Resistance of pollen beetle (*Meligethes aeneus* F.) to pyrethroids, chloronicotynyls and organophosphorous insecticides in Poland. *IOBC/wprs Bull* 29(7): 135–140.

Westwood F, Bean KM, Dewar AM, Bromilow RH, Chamberlain K (1998) Movement and persistence of [14C]imidacloprid in sugar-beet plants following application to pelleted sugar-beet seed. *Pestic Sci* 52: 97–103.

Williamson MS, Denholm I, Bell CA, Devonshire AL (1993) Knockdown resistance (*kdr*) to DDT and pyrethroid insecticides maps to a sodium channel gene locus in the house by (*Musca domestica*). *Mol Gen Genet* 240: 17–22.

Williamson MS, Martinez-Torres D, Hick CA, Castells N, Devonshire AL (1996a) Analysis of sodium channel gene sequences in pyrethroid-resistant houseflies. In: Brown TM (ed.) *Molecular genetics and evolution of pesticide resistance*. American Chemical Society, Washington, USA.

Williamson MS, Martinez-Torres D, Hick CA, Devonshire AL (1996b) Identification of mutations in the housefly para-type sodium channel gene associated with knockdown resistance (*kdr*) to pyrethroid insecticides. *Mol Gen Genet* 252: 51–60.

Zlof V (2008) Recommendations and conclusions of the Ad hoc EPPO Workshop on insecticide resistance of *Meligethes* spp. (pollen beetle) on oilseed rape. *EPPO Bull* 38: 65–67.

Chapter 13
Impact of Insecticides on Parasitoids of Oilseed Rape Pests

Bernd Ulber, Zdzisław Klukowski, and Ingrid H. Williams

Abstract Insecticide application for control of insect pests on oilseed rape may impact parasitoid activity and the levels of parasitism. Field and laboratory studies show that the insecticide product, the dose rate and the timing of application can affect parasitoid abundance and levels of parasitism differentially. Insecticides applied during flowering have most potential to threaten parasitoid populations. This knowledge will help in the future to provide recommendations to the agricultural industry on the selection of less harmful insecticides and the optimization of the time of application, in order to minimize detrimental side effects on the beneficial entomofauna.

13.1 Introduction

The assessment of the side-effects of insecticide application on non-target organisms is a major pre-requisite for developing integrated pest management (IPM) systems. One of the main objectives in IPM is the optimization of insecticide use as far as product selection and the timing of application are concerned, while minimizing their negative impact on beneficial arthropods (Flint and van den Bosch 1981).

Minimizing the effects of insecticides on non-target organisms is especially important in oilseed rape, where frequent applications of broad-spectrum, non-selective insecticide compounds, mainly synthetic pyrethroids, are commonly sprayed each year for the control of various economically-important insect pests in autumn, spring and summer (Pruszynski and Mrowczynski 1990, Williams 2004, Thieme et al. Chapter 12 this volume). These are often applied prophylactically without regard to pest control thresholds, thereby reducing the economic profitability of the crop. Further, abundant treatments have recently caused a widespread resistance of pollen beetle populations, particularly to pyrethroid

B. Ulber (✉)
Department of Crop Sciences, Plant Pathology and Crop Protection Division, Georg-August University Göttingen, Göttingen D-37077, Germany
e-mail: bulber@gwdg.de

I.H. Williams (ed.), *Biocontrol-Based Integrated Management of Oilseed Rape Pests*, DOI 10.1007/978-90-481-3983-5_13,

insecticides in many European countries, including in Denmark (Hansen 2003), in France (Ballanger et al. 2007), in Germany (Heimbach et al. 2006) and in Poland (Wegorek and Zamojska 2006); for a review see Thieme et al. (Chapter 12 this volume). In autumn, seed treatment and area sprays are used to control winter pests, such as the cabbage stem flea beetle, *Psylliodes chrysocephala* (L.). Two to three insecticide treatments are applied commonly in continental Europe, targeted against major pests attacking the crop in spring and summer, namely the rape stem weevil, *Ceutorhynchus napi* Gyllenhal, the cabbage stem weevil, *Ceutorhynchus pallidactylus* (Marsham) (syn. *C. quadridens* Panzer) at stem elongation, the pollen beetle, *Meligethes aeneus* (Fabricius) at green bud, and the cabbage seed weevil, *Ceutorhynchus obstrictus* Marsham (syn. *C. assimilis* Paykull), and the brassica pod midge, *Dasineura brassicae* Winnertz, at flowering (Alford et al. 2003, Menzler-Hokkanen et al. 2006, Kirch and Basedow 2008, Buerger and Gerowitt 2009). Until the 1980s, the insecticides used on oilseed rape were mainly organochlorines, organophosphates, and carbamates but then pyrethroids began to be used and now dominate.

Insecticides may also kill the natural enemies of pests (Jepson 1989). In the past, studies focussed on their effects on predators. Carabid beetles and other ground-dwelling predators have potential to decrease pest populations (Goltermann 1994, 1995, Warner et al. 2000, Büchs 2003, Williams et al. Chapter 4 this volume) and side-effects of insecticide application on their abundance and predation efficiency in arable crops including oilseed rape have been investigated in Germany (Ulber et al. 1990), Poland (Klukowski et al. 2003, 2006) and the UK (Vickerman 1992, Sunderland 1992). More recently, there has been increased focus on the effects of pesticides on the hymenopteran parasitoids which we now know can exert substantial natural control of the major pests on oilseed rape (Ulber et al. Chapter 2 this volume). Thus, integration of chemical and conservation biological control into the pest management system is a major challenge (Nitzsche 1998, Ulber 2003, Veromann et al. 2006, Williams 2004, 2006, Nilsson Chapter 16 this volume).

This chapter reviews the literature and recent results of the EU project MASTER on the effects of chemical insecticides on parasitoids attacking pests on oilseed rape.

13.2 Effects of Insecticides on Parasitoids

The effects of insecticides on parasitoid populations within the crop depend on many factors including mode of action (see also Thieme et al. Chapter 12 this volume), concentration, persistency and its temporal and spatial application. Parasitoids can be exposed to insecticides through direct contact with spray droplets, through chemical residues taken from the plant surface when seeking for food or hosts, and through feeding on contaminated nectar or honeydew (Longley and Jepson 1996). They can also be affected during their development within the host. Consequently, the effects of insecticides have been assessed on the abundance of adult parasitoids and the level of parasitism in both field trials and laboratory bioassays. Most studies have focussed on lethal effects, while only a few have addressed sublethal effects.

13.2.1 Lethal Effects

Although pyrethroids are now the main group of insecticides used to control insect pests on oilseed rape, some earlier assessments of the effects of organochlorines and organophosphates on parasitoids, and of the effects of these earlier-used insecticides compared with pyrethroids are reported in the literature, and so are reviewed here.

13.2.1.1 Effects of Organochlorines

In the UK, marked reductions in the abundance and level of parasitism of parasitoids attacking pollen beetle larvae were reported with the organochlorine insecticides Dieldrin and DDT which were used for control of pollen beetle around 50 years ago (Winfield 1963). Following three field applications in 1960 and 1961, smaller numbers of *Isurgus morionellus* Holmgren (syn. *Phradis morionellus*) and *Isurgus heterocerus* Thomson (syn. *Tersilochus heterocerus*), two key parasitoids of the pollen beetle, were caught by sweep-netting and percentage parasitism was very low compared to untreated plots. Earlier sprays at the green bud stage were reported to affect *I. morionellus* more severely, while sprays during early flowering had stronger effects on *I. heterocerus*.

In another study, in France, the contact toxicity of four insecticides, the organochlorines: lindane, endosulfan and toxaphene, and the organophosphate: diethion, applied for control of the cabbage seed weevil and brassica pod midge in the past, was tested on the pod midge parasitoid *Prosactogaster* (syn. *Platygaster*) *oebalus* Walker (Hym.: Platygastridae) in the laboratory (Coutin and Coulon 1966). The test insects were obtained by rearing from field-collected host larvae. Adults of *P. oebalus* were exposed to dried residues of various dilutions either for 40 min or for 10 min, followed by 30 min of observation in untreated vials. Similar results were obtained from both methods: based on the LC50, the relative toxicities of the insecticides lindane, endosulfan, toxaphene and diethion were 1.0, 9.0, 1581 and 1990, respectively. The toxicities of the insecticides to *P. oebalus* ranged within the same order as the toxicities to adult pod midge.

13.2.1.2 Effects of Organophosphates and Pyrethroids Compared

The initial contact toxicity of two pyrethroids (deltamethrin, alpha-cypermethrin) and an organophosphate (fenitrothion) for the pollen beetle parasitoid *P. morionellus* was tested in the laboratory in Finland (Hokkanen et al. 1988). The pyrethroids proved to be slightly less harmful to this parasitoid than fenitrothion. Survival of adults in treatments with pyrethroids at full and reduced (50%) dose rates was 70–80% after 4 h, while in treatments with fenitrothion all individuals were killed within 1.5 h. However, no parasitoid survived for up to 12 h in any insecticide treatment.

Only a few studies have investigated the effects of insecticides on parasitoid species attacking the larvae of the cabbage seed weevil. In the UK, commercial application of the organophosphate triazophos in winter rape at the end of flowering substantially reduced the level of parasitism by *Trichomalus perfectus* (Walker) (Hym.: Pteromalidae), the key larval ectoparasitoid of cabbage seed weevil (Murchie et al. 1997). At that time, *T. perfectus* was most actively searching for host larvae for oviposition. The pyrethroid alpha-cypermethrin was less harmful because it was sprayed during flowering, before the main immigration period of *T. perfectus*. Moreover, application of alpha-cypermethrin post-flowering was less detrimental to *T. perfectus* which led to the conclusion that this pyrethroid is less toxic to *T. perfectus* than triazophos. The replacement of triazophos by a widespread use of pyrethroids in 1993 and 1994 coincided with a substantial increase of the level of seed weevil parasitism in the UK (Alford et al. 1995, 1996).

13.2.1.3 Effects of Pyrethroids

Two pyrethroids, tau-fluvalinate and lambda-cyhalothrin are widely used for pest control in many European countries (Thieme et al. Chapter 12 this volume) and both have been tested for their effects on various parasitoids of oilseed rape pests. Because of its remarkably low bee toxicity, tau-fluvalinate was formerly used as an acaricide for control of Varroa mite in bee-hives and has been registered only recently for control of oilseed rape pests in Europe. Tau-fluvalinate is known to degrade faster in soil and is less toxic to mammals and other animals than lambda-cyhalothrin. Data on the mammalian toxicity and ecotoxicology of both chemicals have been summarized by the Environmental Protection Agency (2005) and Tomlin (2006).

The effect of the pyrethroid tau-fluvalinate on the abundance of adult pollen beetle parasitoids and parasitization of pollen beetle larvae was estimated in field trials in Sweden (Haldén 2004). Two flowering crops of spring rape, 10 ha each, were treated in one part with tau-fluvalinate (Mavrik) at the recommended dose rate of 0.2 and 0.3 l/ha, respectively, in June (GS 60–62), while the other part was left untreated. The abundance of adult parasitoids on the crops was assessed by sweep-netting during the days before and after the application. In the plots sprayed with Mavrik, parasitoid catches, consisting of 90% *P. morionellus and* 5% *Diospilus capito,* were lower than in the untreated plots: the number of parasitoids was reduced by approximately 50% compared to untreated. In one of the two crops, the level of parasitism of pollen beetle larvae was considerably lower in the plot sprayed with Mavrik than in the unsprayed plot, while there was no difference between treated and untreated plots in the other field trial.

The direct contact toxicity of the two pyrethroid insecticides, tau-fluvalinate (Mavrik 2F) and lambda-cyhalothrin (Karate Zeon 050 CS) on the percentage mortality of adult pollen beetle parasitoids was compared in laboratory bioassays in Poland (Jackowski et al. 2009). To compare the effects of full and reduced dose rates, various dilutions of the commercial formulated insecticide products (100, 75, 50% of the recommended field dose) in water, equivalent to spraying of 400 l of water per ha, were prepared and equal volumes pipetted onto filter paper placed in glass petri dishes. In another series of experiments, leaf discs of oilseed rape were dipped into the insecticide dilutions and then air-dried before placing on the bottom of the petri dishes. Four hours after the insecticide application, 10 field-collected adults of *Phradis interstitialis* (Thomson) and *P. morionellus* (Holmgren), in approximately equal proportions, were transferred to the dried residual deposit in petri dishes (10 replicates per treatment). Flowers of oilseed rape were added as food. The percentage survival of adults was recorded every 30–60 min for 7 and 12 h, respectively. The tests were conducted in the laboratory at temperatures ranging from 25.0 to 27.5°C.

On filter paper and leaf discs treated with water, only 49 and 66% of the test individuals (*Phradis* spp.), respectively, survived after 6.5 h (Table 13.1). The lethal effect of the insecticides on adult parasitoids was higher on filter paper than on leaf discs. Mavrik was less toxic than Karate: at 100% of the recommended dose rate, the insecticidal effect of Mavrik on parasitoid mortality was 6 and 8%, respectively, lower compared to Karate (filter paper, leaf discs). When only 50% of the recommended doses of Mavrik and Karate were applied, survival of *Phradis* spp. was 9% on treated filter paper and 31% (Karate) and 35% (Mavrik) on treated leaf discs (Table 13.1). The lower mortality rates on treated leaf discs may result from penetration of the active ingredients into the wax layer of the leaf surface. Furthermore, there was only a small difference between the mortality caused by 100 and 50% of Mavrik which means that the reduction of the dose rate applied for control of a specific pest is less important than with Karate. However, more laboratory studies are needed aimed particularly at species of *Phradis* or *Tersilochus*.

The effects of the same two pyrethroid insecticides, tau-fluvalinate (Mavrik 2F) and lambda-cyhalothrin (Karate 25 EC), on the abundance of a wider range of key parasitoids of oilseed rape pests were studied extensively in field trials in 2 years (2002, 2004) during the EU MASTER project in Poland (Klukowski 2006).

The field trials were laid out in a randomized block design with four replicated plots. In the first experiment, in 2002, the recommended full dose rate of the insecticides was applied in the bud stage and in mid-flowering at growth stages 54–56 and

Table 13.1 Product-limit survival analysis S(T) of adult *Phradis* spp. after 6.5 h of exposure to dried residual deposits of insecticides either on filter paper or on leaf discs in the laboratory. Applications were 100, 75, and 50% of the recommended field dose

	Cumulative Survival S(T) (mean ± S.E.)	
Treatment	on filter paper	on leaf discs
Untreated (water)	0.491 ± 0.033	0.659 ± 0.033
Mavrik 100%	0.089 ± 0.028	0.262 ± 0.049
Mavrik 75%	–	0.269 ± 0.059
Mavrik 50%	0.088 ± 0.028	0.308 ± 0.061
Karate 100%	0.031 ± 0.015	0.178 ± 0.046
Karate 75%	–	0.208 ± 0.058
Karate 50%	0.089 ± 0.025	0.350 ± 0.064

65–68, respectively. In the second experiment, in 2004, half dose rates were applied in addition to full dose rates at both growth stages in separate plots. Untreated plots were used as a reference. Plot size was 30 × 30 m in 2002 and 25 × 15 m in 2004. Sweep net samples were collected from the centre of each plot to evaluate the direct effect of insecticide applications on the activity density of adult parasitoids of stem weevils, pollen beetle and cabbage seed weevil. Sampling was done 3 days and 1 day before pyrethroid application, and 6–7 times with increasing intervals up to 15 days after the applications.

The effects of the insecticide treatments on adult parasitoids in 2002 and 2004 are presented in Figs. 13.1, 13.2, 13.3, and 13.4; the number of insects caught in treated plots is expressed as a percentage of the mean number caught in the untreated plots. The insecticides were applied in the period just before the main immigration of parasitoids into the crop, when the number of insects captured increased.

In 2002, the application of tau-fluvalinate (Mavrik 2F) and lambda-cyhalothrin (Karate 25 EC) on 24 April (GS 54) caused significant reductions in the numbers of *Tersilochus fulvipes*, a parasitoid of the rape stem weevil, and of *Tersilochus obscurator*, a parasitoid of the cabbage stem weevil, caught in sweep nets (Fig. 13.1). Even 2 weeks after the application of tau-fluvalinate and lambda-cyhalothrin the total numbers of *T. obscurator* and *T. fulvipes* caught were reduced by about 39 and 58%, respectively, compared to the untreated plots. The pteromalid parasitoids of the cabbage seed weevil showed only a weak response to the second application of either pyrethroid on 15 May (Fig. 13.2).

In 2004, the first application of tau-fluvalinate and lambda-cyhalothrin, both at full and reduced dose rate, was carried out at growth stage 56 on 28 April, 3–4 days

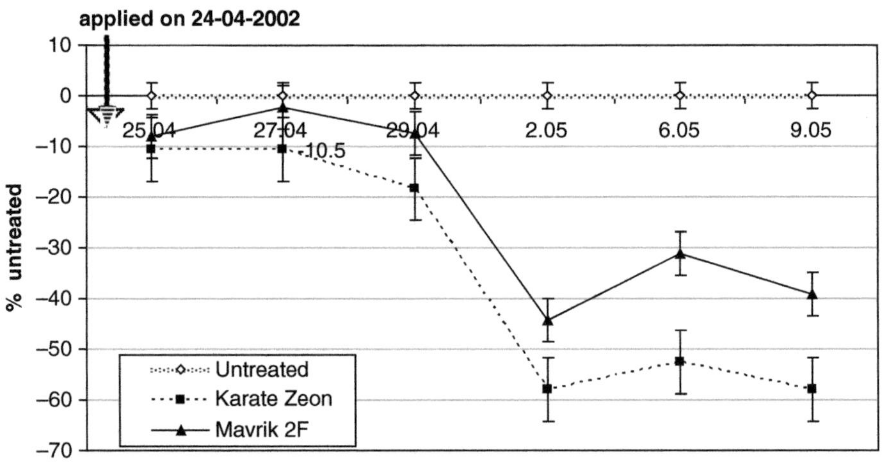

Fig. 13.1 Impact of the application of the pyrethroid insecticides Karate Zeon and Mavrik 2F at full dose rate to plots of winter oilseed rape on the abundance of the parasitoids *Tersilochus obscurator* and *Tersilochus fulvipes*. Data is expressed as percentages of the mean number caught by sweep net in untreated plots and is standardized according to the different abundances recorded in treated and untreated plots before insecticide application

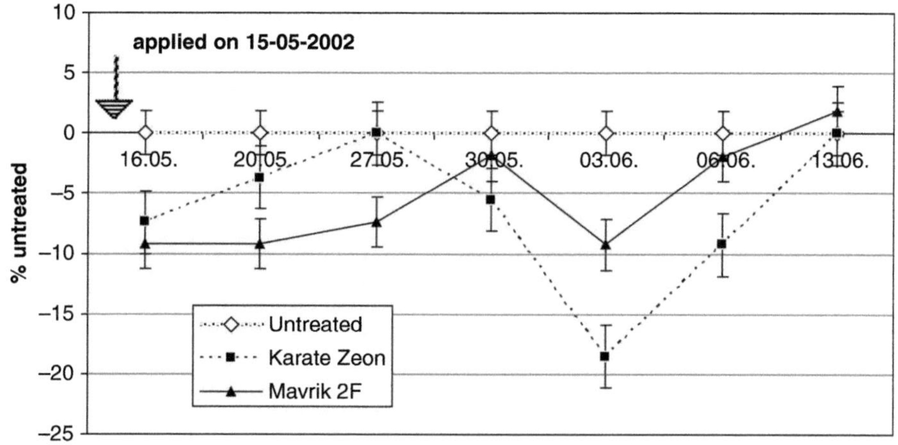

Fig. 13.2 Impact of the application of the pyrethroid insecticides Karate Zeon and Mavrik 2F at full dose rate to plots of winter oilseed rape on the abundance of Pteromalidae. Data is expressed as percentages of the mean number caught by sweep net in untreated plots and is standardized according to the different abundances recorded in treated and untreated plots before insecticide application

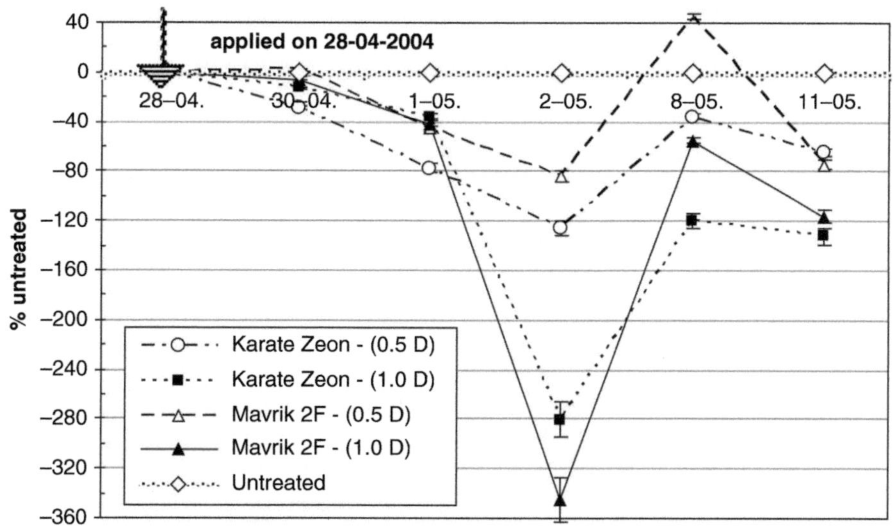

Fig. 13.3 Impact of the application of the pyrethroid insecticides Karate Zeon and Mavrik 2F at full (1.0D) and half (0.5D) dose rates to plots of winter oilseed rape on the abundance of *Phradis morionellus*. Data is expressed as percentages of the mean number caught by sweep net in untreated plots and is standardized according to the different abundances recorded in treated and untreated plots before insecticide application

before the main migration of *P. interstitialis* into the crop started. No significant difference was found between the numbers of adult *Phradis* sp. caught in different plots immediately after application (Fig. 13.3). The first seven specimens of *P. morionellus* were caught on 28 April, but the maximum number of this species

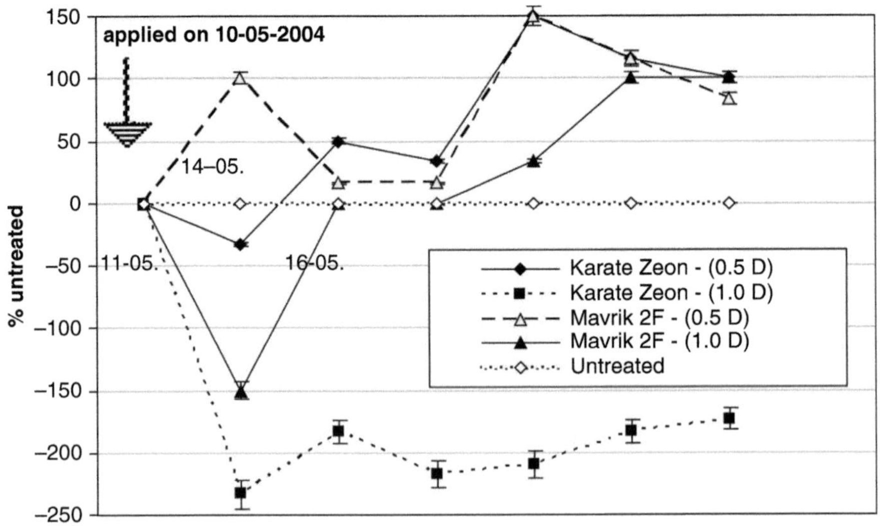

Fig. 13.4 Impact of the application of the pyrethroid insecticides Karate Zeon and Mavrik 2F at full (1.0D) and half (0.5D) dose rates on the abundance of *Phradis interstitialis*. Data is expressed as percentages of the mean number caught by sweep net in untreated plots and is standardized according to the different abundances recorded in treated and untreated plots before insecticide application

was recorded in sweep net samples on 2 May (71 specimens in untreated plots). On this date, the full dose rate of both pyrethroids caused a very high mortality. At half-dose rates, this effect was less, the more so with tau-fluvalinate (Fig. 13.3).

Following the second application on 10 May 2004, the residual effect of the full dose of tau-fluvalinate reduced the number of *P. interstitialis* only for 3 days, while in plots treated with the full dose of lambda-cyhalothrin the effect persisted until the end of the activity period of *P. interstitialis* (Fig. 13.4).

In both years, the effect of lambda-cyhalothrin and tau-fluvalinate on the mean larval abundance of major pests and their level of parasitism was studied as well (Table 13.2). Sampling of larvae was conducted twice during the field seasons, at the yellow bud stage (BBCH 54–56) and at mid-flowering or briefly after. Of each the three pest species, 40–50 of full-grown larvae/plot were collected from trays placed below the crop canopy or through pod dissection.

In the field trials of 2002 and 2004, the level of parasitism of *Meligethes* larvae was generally higher in plots treated with tau-fluvalinate than in plots treated with lambda-cyhalothrin. Application of tau-fluvalinate at mid-flowering had no significant effect on percentage parasitism of pollen beetle larvae, while it was significantly reduced by lambda-cyhalothrin. Similarly, following the application of tau-fluvalinate in the bud stage and particularly in flowering in 2002, the level of parasitism of cabbage stem weevil larvae was higher than following usage of lambda-cyhalothrin. In 2002, the parasitism of cabbage stem weevil with tau-fluvalinate was significantly higher (9.3%) than with lambda-cyhalothrin, although

Table 13.2 Effect of application of lambda-cyhalothrin and tau-fluvalinate to winter oilseed rape on the mean percentage parasitism (%) of larvae of the pollen beetle (*Meligethes* sp.), the cabbage stem weevil (*Ceutorhynchus pallidactylus*) and the cabbage seed weevil (*Ceutorhynchus obstrictus* syn. *C. assimilis*)

		Meligethes sp.			*C. pallidactylus*			*C. obstrictus*		
Treatment	BBCH growth stage	Mean (%)	SD	Dunnett test*	Mean (%)	SD	Dunnett test*	Mean (%)	SD	Dunnett test*
Field experiment 2002										
Untreated	54	24.6	4.8		13.8	4.6		7.5	5.6	n.s
Lambda-cyhalothrin		14.9	3.4	0.010	7.3	1.8	0.021	10.2	2.4	n.s
Tau-fluvalinate		16.9	3.0	0.033	9.3	1.3	0.101	10.1	4.8	n.s
Untreated	65	31.0	1.4		12.3	2.1		10.0	2.0	n.s
Lambda-cyhalothrin		20.9	4.7	0.009	13.1	9.6	0.998	11.0	1.7	n.s
Tau-fluvalinate		23.9	4.6	0.053	21.4	5.9	0.078	12.2	1.6	n.s
Field experiment 2004										
Untreated	56	31.1	3.4		21.5	2.6		5.3	2.2	n.s
Lambda-cyhalothrin		15.9	1.8	<0.001	11.1	3.7	<0.001	4.2	2.4	n.s
Tau-fluvalinate		19.3	1.5	<0.001	13.7	2.7	<0.001	7.9	3.2	n.s
Untreated	68	32.7	2.7		Not applicable at this growth stage			7.0	0.4	n.s
Lambda-cyhalothrin		22.7	2.0	<0.001				6.2	1.9	n.s
Tau-fluvalinate		27.3	2.0	0.064				3.8	1.4	n.s

*Dunnett test was applied when a significant difference was found by ANOVA. n.s. = not significantly different from Untreated.

this difference was only 2%. The tested pyrethroid compounds and the application dates had only minor effects on the level of parasitism of cabbage seed weevil in both years. However, the insecticide sprays were applied relatively early, and effects on percent parasitism of seed weevil larvae by the key parasitoid *T. perfectus* could not be analysed properly.

13.2.2 Sublethal Effects

Most studies on the impact of insecticides on parasitoids have focussed on lethal effects. Only a few experiments address sublethal effects, and most of these concentrate on life history parameters, such as longevity, fecundity and searching rate (Desneux et al. 2007).

Sub-lethal effects on the aphid parasitoid *Diaretiella rapae* (M'Intosh) (Hym.: Braconidae) and on a population of the aphid *Myzus persicae* were studied on oilseed plants that had been treated with the pyrethroid deltamethrin. One, two, seven and 14 day-old insecticide residuals had no effect on the reproduction of parasitoid females or on aphid population growth (Desneux et al. 2005).

Host location by hymenopteran parasitoids is mainly based on olfactory cues (see Williams and Cook Chapter 4 this volume). Volatiles emitted by infested host plants or host larvae might be particularly important for detecting plant-mining host larvae which are hidden within stems or petioles. To study the effect of insecticide spray deposits on the host-finding success of *T. obscurator*, the key parasitoid of the cabbage stem weevil, insecticide-treated (thiacloprid 'Biscaya'; lambda-cyhalothrin 'Karate Zeon') and untreated leaves, both infested by larvae of cabbage stem weevil, were offered to *T. obscurator* females for parasitisation in dual-choice experiments (Neumann et al. 2010). Females spent less time foraging on insecticide-treated leaves compared to untreated leaves or even avoided treated leaves. Further, on insecticide-treated leaves they performed less ovipositor probings than on untreated leaves. Coupled EAG-GC/MS analyses provided evidence that *T. obscurator* females could discriminate between odours released from treated and untreated leaves.

Further studies are needed to determine whether insecticide spray deposits inhibit host finding by *T. obscurator* females by changing plant volatile emission or by masking specific plant volatiles needed for host location. Direct repellent effects of insecticide residuals might also reduce the level of parasitism by *T. obscurator*.

Similar results have been obtained with the aphid parasitoid *D. rapae*. When females foraged on Brussels' sprouts plants that were treated with sublethal doses of the insecticides pirimicarb, permethrin and malathion, they spent less time on sprayed plants than on untreated plants, and on treated plants tended to concentrate their activity to areas not covered by spray residuals (Jiu and Waage 1990).

13.3 Conservation of Parasitoid Populations in Integrated Pest Management Systems

Appropriate selection of the insecticide to be used on the oilseed rape crop and the dose rate at which it is applied can help to conserve parasitoids on the crop. Improved standard methods for regulatory testing of pesticide effects on parasitoids in the oilseed rape crop are needed and would further aid this selection. In addition, insecticide application should be targeted to pest density, both in time and space. This can be achieved by using pest monitoring, control thresholds (see Williams Chapter 1 this volume) and decision support systems (see Johnen et al. Chapter 15 this volume).

13.3.1 Selection of Product and Dose Rate

Choice of an insecticide that is less harmful to parasitoids, applied at an appropriate dose rate, can help to conserve parasitoids. Results obtained recently in the MASTER project, suggest that the application of tau-fluvalinate is less harmful to parasitism of pollen beetle than application of lambda-cyhalothrin, and that the application of pyrethroids at half dose rates has potential to conserve parasitoids more effectively than the choice of the active ingredient.

Differential toxicity of various pyrethroids and a relatively low toxicity of tau-fluvalinate to parasitoids occurring in other crops have been also reported in other studies. Among six insecticides tested for their toxicity to beneficial non-target invertebrates in winter wheat fields, tau-fluvalinate was the least toxic pyrethroid compared to zeta-cypermethrin and deltamethrin, but more toxic than pirimicarb (Moreby et al. 2001). When exposed to dried residues of pyrethroids in the laboratory, racemat fluvalinate was less toxic to the aphid parasitoid *Aphidius rhopalosiphi* (Hym.: Aphidiidae) than lambda-cyhalothrin, bifenthrin, cyfluthrin, cypermethrin and deltamethrin on both glass plates and maize leaves (Jansen 1996).

13.3.2 Regulatory Testing of Pesticide Effects on Parasitoids

Following the initiative of the International Organisation for Biological Control about 25 years ago, research groups and agrochemical companies started to investigate the effects of pesticides on beneficial organisms, developing protocols for laboratory and field test methods and for regulatory testing requirements in Europe (Sterk et al. 1999). The application of agreed protocols for testing is of crucial importance to the environmentally-acceptable use of pesticides and to the further development of IPM systems.

Improved standard methods for regulatory testing of pesticide effects on beneficial insects on oilseed rape are required in Europe. The procedure should include laboratory and field tests on key parasitoid species which are active in the oilseed rape crop during the time of insecticide application. Knowledge of relative toxicities of different insecticides to parasitoids would aid selection of the appropriate product and dose rate for use in IPM systems that aim to enhance biocontrol of pests.

13.3.3 Temporal Targeting of Insecticides

Targeting of insecticide application to the crop with due regard to pest incidence and parasitoid phenology on the crop has potential to minimize mortality of parasitoids.

The combined parasitoid complex in oilseed rape is active over a long period during the spring/summer; individual species also have long periods of activity and may have several peaks of migration. To avoid killing any parasitoids, insecticide applications would need to be applied outside this period of activity which is impractical for control of the pests. The use of phenological models to find spray windows that allow insecticide application consistent with good pest control but that minimize effects on parasitoids is reviewed by Johnen et al. (Chapter 15 this volume).

The period of main activity of the tersilochine parasitoids (except *T. microgaster*) on crops of winter rape persists from the late bud stage to the end of flowering. Peak activity of all tersilochine parasitoids is mainly confined to the period of full flowering of the crop (Fig. 13.5). Consequently, insecticide sprays applied just before or during flowering (Fig. 13.5, insecticide applications 2 and 3) potentially have the most adverse effects on parasitoid populations (Johnen and Ulber 2004, Johnen et al. 2006). The avoidance of insecticide application during the flowering

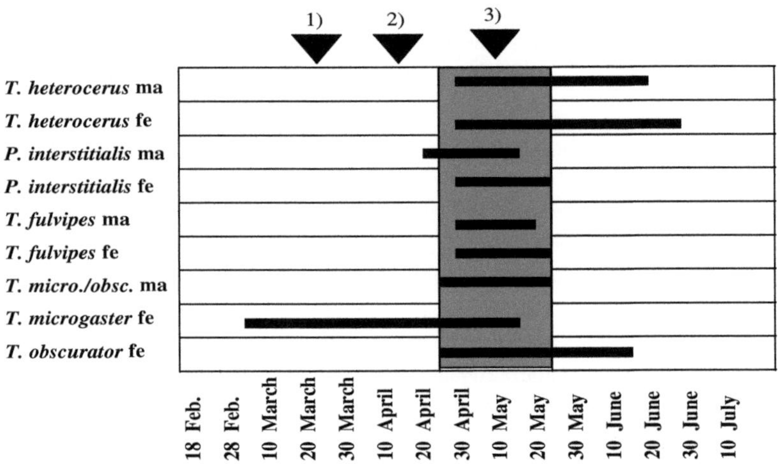

Fig. 13.5 Period of activity of tersilochine parasitoids (*black bars*) in relation to flowering of oilseed rape and insecticide application. The *grey area* encloses the period of flowering, the *triangles* (▼) indicate potential dates of insecticide applications. ma = males, fe = females

period could reduce negative effects on natural control of pest populations in IPM systems.

The initiation of dispersal flights of tersilochine parasitoids and their phenologies within the crops show a species-specific sequence (Ulber and Nitzsche 2006). In Germany, females of *T. microgaster* immigrate during the early bud stage of winter rape in March. However, in years with extended frost temperatures, emergence of *T. microgaster* can be delayed until early April. Females of *T. obscurator, T. fulvipes* and *P. interstitialis* colonize crops from mid-April onwards. Females of *T. heterocerus* seldom occur before the end of April or early May, i.e., usually after the beginning of flowering (Fig. 13.5). These results are consistent with previously reported observations on the phenologies of *T. microgaster* and *T. obscurator* in Czechoslovakia (Šedivý 1983), Germany (Lehmann 1965, Klingenberg and Ulber 1994) and the UK (Barari et al. 2005, Williams et al. 2006) and with reports on the seasonal occurrence of *P. interstitialis* and *T. heterocerus* in Germany (Lehmann 1965, Klingenberg and Ulber 1994), Southern Sweden (Nilsson 2003, Jönsson et al. 2004) and the UK (Ferguson et al. 2003a). See also Williams and Ferguson (Fig. 8.15 this volume) for parasitoid phenologies in the UK.

Studies of the phenology of the cabbage seed weevil and its key parasitoid *T. perfectus* on winter rape crops in the UK showed dissociation between the immigration flights of pest and parasitoid, the former arriving before the latter (Murchie et al. 1997). This provides an opportunity for the temporal targeting of pyrethroids before the main migration period of the parasitoids. Avoidance of treatments post-flowering, when the parasitoids are most active, conserves natural populations of *T. perfectus* and has been shown to be effective and economically-viable as a method for controlling cabbage seed weevil populations in the UK (Alford et al. 1996).

Thus avoidance of application of insecticides during flowering provides potential for the safe integration of insecticide use in IPM systems with conservation of key tersilochine parasitoids while avoidance of application of insecticides post-flowering helps conserve parasitoids of the cabbage seed weevil.

More information on the diel periodicities and behaviour patterns of pests and parasitoids within the crop may help define times of day when parasitoids are less exposed than pests to insecticides. Ferguson et al. (2010) found that, during flowering, peak flight activity of the pollen beetle and its parasitoid *P. interstitialis*, and of *T. obscurator*, a parasitoid of the cabbage stem weevil, was around midday and that few insects were caught before 10.00 h.

13.3.4 Spatial Targeting

Insecticides are usually applied to the whole area of the oilseed rape crop even though pests are usually irregularly distributed within the crop. This suggests scope for reducing insecticide inputs by spatially targeting it to areas of high pest density only, if these areas could be determined, thereby conserving parasitoid populations in unsprayed areas.

There is a tendency for most of the major pests to be strongly edge-distributed during their immigration phase to the crop. This has been demonstrated for the cabbage stem weevil (e.g., Free and Williams 1979, Klukowski 2006), the pollen beetle (Free and Williams 1979), the cabbage seed weevil (eg., Free and Williams 1979, Murchie et al. 1997, Ferguson et al. 2000) and the brassica pod midge (eg., Free and Williams 1979, Ferguson et al. 2003b, 2004). Application of insecticide only to crop edges during the immigration phase would kill a large proportion of the adult pests, while, at the same time, would avoid killing any parasitoids in the central unsprayed areas of the crop.

Most parasitoid species have been shown to have close spatial associations with their hosts on the oilseed rape crop, eg., *P. interstitialis* and the pollen beetle (Ferguson et al. 2003a), *T. microgaster* and the cabbage stem flea beetle (Ferguson et al. 2006), *T. obscurator* and the cabbage stem weevil (Ferguson et al. 2006), *T. perfectus* and the cabbage seed weevil (Murchie et al. 1999; Ferguson et al. 2000) and *P. subuliformis* and *O. clypealis* and the brassica pod midge (Ferguson et al. 2004); for a review see Williams and Ferguson Chapter 8 this volume. In these species there is little potential for parasitoid conservation through spatial targeting of insecticide to high pest densities except through border spraying. However, the within-crop distributions of some parasitoid species do not always coincide completely with those of their hosts. Thus, in a study in the UK, *T. heterocerus* was found to be more evenly spread across the crop than its host, the pollen beetle, so any spatial targeting of insecticide to areas of high beetle density, such as to crop borders, would conserve those parasitoids in areas of low beetle density (Ferguson et al. 2003a). Further information about the environmental and behavioural factors affecting parasitoid distributions on the crop is needed to enable areas of high parasitoid densities to be predicted.

There may also be potential for spatial targeting of insecticide to kill pests but conserve parasitoids in association with the use of a trap crop as part of a so-called 'push-pull' IPM strategy (see also Williams and Cook Chapter 7 this volume). 'Push-pull' strategies exploit behaviour-modifying stimuli, particularly those associated with host plant location, to manipulate the distribution of pests and their parasitoids on a crop (Cook et al. 2007b). The push-pull strategy being developed for oilseed rape utilises turnip rape (*Brassica rapa* L.) as the trap crop to 'pull' pests away from the main oilseed rape crop (Nilsson 2004, Cook et al. 2006, 2007a). Field trials have shown that a turnip rape trap crop border can reduce the abundance of cabbage stem flea beetle (Barari et al. 2005) and pollen beetle (Cook et al. 2006) in the oilseed rape crop and reduce the need for insecticide.

The effect of pyrethroid treatment on pest incidence and the parasitism of their larvae was determined in field experiments in the UK which included 6 m-wide turnip rape (*Brassica rapa*) borders as a trap crop around plots of winter rape (Barari et al. 2005). Water trap samples were used to record the incidence of adult pests and parasitoids, and plant samples were collected to determine plant infestation by pest larvae and percentage larval parasitism. Treatment of the turnip rape borders with lambda-cyhalothrin in early October 2001 and mid April 2002 reduced the incidence of cabbage stem flea beetle and cabbage stem weevil larvae in the turnip rape but not in the oilseed rape plots. There was no significant difference between numbers of *T. obscurator* caught neither in sprayed and unsprayed turnip rape borders nor in the adjacent oilseed rape plot centres. Application of lambda-cyhalothrin within the turnip rape trap crop had no significant effect on the level of parasitism of cabbage stem weevil in both the treated trap crop border and the untreated oilseed rape centres. This can be explained by the phenology of *T. obscurator* which showed peak activity during flowering, about 8 weeks after the application of the insecticide. Percentage parasitism of cabbage stem flea beetle weevil was on a low level in sprayed (5.6%) and unsprayed (7.7%) turnip rape borders, and no parasitized larvae were found in the oilseed rape plot centres.

13.3.5 Filter Effect of the Crop Canopy

The flowering canopy of the oilseed rape crop can act as a filter for insecticide sprays applied during flowering; a large proportion of the spray is deposited on the upper level of flowers and pods and does not penetrate to the lower levels of the crop (Goltermann 1994).

The vertical distribution of parasitoids within the crop varies with species depending on their foraging/host-finding behaviour; species may therefore differ in the extent to which they are affected by the application of insecticide. Yellow water traps placed in the crop at the ground level caught larger numbers of *T. obscurator* and *T. microgaster*, parasitoids of the cabbage stem weevil and the cabbage stem flea beetle larvae, respectively, than traps placed on top of the crop canopy (Ulber and Nitzsche 2006). In contrast, more *T. heterocerus*, a parasitoid of pollen beetle larvae, were caught in traps at flower level of the crop canopy than at ground level. Adults of *T. fulvipes,* a parasitoid of the rape stem weevil larvae, and *P. interstitialis*, another parasitoid of the pollen beetle larvae, were found in similar numbers at ground level and on the top of the crop canopy.

13.4 Conclusions

Within IPM strategies, the application of insecticides may be necessary if pest numbers exceed their control threshold levels. In past decades, the selection of insecticide for use in oilseed rape has been restricted largely to synthetic pyrethroids which are effective against the pests but also harm their parasitoids. In recent years, a number of new active substances of the insecticide class neonicotinoids, such as thiacloprid and acetamiprid, have been developed and registered for control of pests on oilseed rape.

The effect of the pyrethroid application on the level of parasitism of pests on oilseed rape can show significant differences mainly on the level of the different dose rates applied. Reducing the dose rate to 50% can conserve parasitoids more effectively than selecting active ingredients of higher selectivity. In many field experiments, the effect of the tested pyrethroids on parasitism of pest larvae is not significantly different. Both at early and late dates of application, there appears to exist a constant tendency that tau-fluvalinate can ensure slightly higher levels of parasitisation than lambda-cyhalothrin.

For regulatory testing procedures of pesticide effects on beneficial insects occurring on oilseed rape, appropriate methods are required which include a standard set of key biocontrol species. These methods should be used for integration into the guidelines of pesticides risk assessment in Europe. The testing procedure should include a set of laboratory and field experiments orientated on non-target species most active at the time of spraying in the crop.

More research is needed to assess the effect of various insecticide groups on natural enemies of insect pests on oilseed rape and to evaluate the potential for their integration. IPM programs focused on conservation of parasitoid populations and on greater implementation of biological control of pests will reduce growing public concern about the negative effects of pesticides on consumers and the environment.

Acknowledgements The project MASTER (QLK5-CT-2001-01447) was co-funded by the EU under its Framework 5 Quality of Life and Management of Living Resources programme. We thank all scientists and students who contributed to the project. Preparation of this review was supported by the Estonian Targeting Financing Project SF0170057s09.

References

Alford DV, Murchie AK, Williams IH (1995) Observations on the impact of standard insecticide treatments on *Trichomalus perfectus*, a parasitoid of seed weevil on winter oilseed rape in the UK. *IOBC/wprs Bull* 18(4): 122–126.

Alford DV, Nilsson C, Ulber B (2003) Insect pests of oilseed rape crops. Chapter 2. In: Alford DV (ed.) *Biocontrol of oilseed rape pests*. Blackwell, Oxford, UK.

Alford DV, Walters KFA, Williams IH, Murchie AK (1996) A commercially viable low-cost strategy for the management of seed weevil populations on winter oilseed rape in the UK. *Brighton Crop Prot Conf – Pests and Diseases* -1996: 609–614.

Ballanger Y, Détourné D, Delorme R, Pinochet X (2007) France, difficulties to manage insect pests of winter oilseed rape (*Brassica napus* var. *oleifera*): Resistances to insecticides. *Proc GCIRC, 12th Int Rapeseed Congress, Wuhan, 26–30 March 2007,* 4: 276–279.

Barari H, Cook SM, Clark SJ, Williams IH (2005) Effects of a turnip rape (*Brassica rapa*) trap crop on stem-mining pests and their parasitoids in winter oilseed rape *(Brassica napus*). *BioControl* 50: 69–86.

Buerger J, Gerowitt B (2009) Anwendungsmuster von Pflanzenschutzmitteln in Winterweizen und Winterraps. *Gesunde Pflanzen* 61: 11–17.

Büchs W (2003) Predators as biocontrol agents of oilseed rape pests. In: Alford DV (ed.) *Biocontrol of oilseed rape pests*. Blackwell, Oxford, UK.

Cook SM, Smart LE, Martin JL, Murray DA, Watts WP, Williams IH (2006) Exploitation of host plant preferences in crop protection strategies for oilseed rape (*Brassica napus*). *Entomol Exp Appl* 119: 221–229.

Cook SM, Khan ZR, Pickett JA (2007a) The use of 'push-pull' strategies in integrated pest management. *Annu Rev Entomol* 52: 357–400.

Cook SM, Rasmussen HB, Birkett MA, Murray DA, Pye BJ, Watts NP, Williams IH (2007b) Behavioural and chemical ecology underlying the success of turnip rape (*Brassica rapa*) trap crops in protecting oilseed rape (*Brassica napus*) from the pollen beetle (*Meligethes aeneus*). *Arthropod-Plant Interact* 1: 57–67.

Coutin R, Coulon J (1966) Toxicités comparées de plusieurs insecticides vis-a-vis de *Dasyneura brassicae* Winn. et de son parasite principal *Prosactogaster oebalus* Walk. *Phytiatrie-Phytopharmacie* 15: 105–112.

Desneux N, Fauvergue X, Dechaume-Moncharmont F-X, Kerhoas L, Ballanger Y, Kaiser L (2005) *Diaretiella rapae* limits *Myzus persicae* populations after applications of deltamethrin in oilseed rape. *J Econ Entomol* 98: 9–17.

Desneux N, Decourtye A, Delpuech J-M (2007) The sublethal effects of pesticides on beneficial arthropods. *Annu Rev Entomol* 52: 81–106.

Environmental Protection Agency (2005) *Registration Eligibility Decision for tau-fluvalinate.* United States Environmental Protection Agency, Case no. 2295, 78 pp.

Ferguson AW, Klukowski Z, Walczak B, Perry JN, Mugglestone MA, Clark SJ, Williams IH (2000) The spatio-temporal distribution of adult *Ceutorhynchus assimilis* in a crop of winter oilseed rape in relation to the distribution of their larvae and that of the parasitoid *Trichomalus perfectus*. *Entomol Exp Appl* 95: 161–171.

Ferguson AW, Campbell JM, Warner DJ, Watts NP, Schmidt JEU, Williams IH (2003a) Spatio-temporal distributions of *Meligethes aeneus* and its parasitoids in an oilseed rape crop and their significance for crop protection. *Proc 11th Int Rapeseed Congress, 6–10 July 2003, Copenhagen, Denmark*: 1057–1059.

Ferguson AW, Klukowski Z, Walczak B, Clark SJ, Mugglestone MA, Perry JN, Williams IH (2003b) Spatial distribution of pest insects in oilseed rape: Implications for integrated pest management. *Agric Ecosyst Environ* 95: 509–521.

Ferguson AW, Campbell JM, Warner DJ, Watts NP, Schmidt JEU, Williams Ingrid H (2004) Phenology and spatial distributions of *Dasineura brassicae* and its parasitoids in a crop of winter oilseed rape: Implications for integrated pest management. *IOBC/wprs Bull* 27(10): 243–252.

Ferguson AW, Barari H, Warner DH, Campbell JM, Smith ET, Williams IH (2006) Distributions and interactions of the stem miners, *Psylliodes chrysocephala* and *Ceutorhynchus pallidactylus*, and their parasitoids in a crop of winter oilseed rape *(Brassica napus.)*. *Entomol Exp Appl* 119: 81–92.

Ferguson AW, Holdgate R, Mason NS, Clark SJ, Williams IH (2010) Phenologies and diel periodicities of within-crop flight by pests and parasitoids in oilseed rape. *IOBC/wprs Bull* (in press).

Flint ML, van den Bosch R (1981) *Introduction to pest management*. Plenum, New York, USA.

Free JB, Williams IH (1979) The distribution of insect pests on crops of oil-seed rape (*Brassica napus* L.) and the damage they cause. *J Agric Sci* 93: 139–149.

Goltermann S (1994) *Das Auftreten von Laufkäfern (Col., Carabidae) auf Winterrapsfeldern und deren Einfluss auf den Massenwechsel von Meligethes aeneus F. (Col., Nitidulidae)*. PhD thesis University of Rostock, Germany.

Goltermann S (1995) Nützlingsschonender Insektizideinsatz im Winterraps? *Raps* 13: 153–154.

Haldén P (2004) *Influence of tau-fluvalinate on parasitoid abundance and parasitisation rate of pollen beetles in spring oilseed rape* (in Swedish). MSc thesis, Dept Crop Science SLU, Alnarp, Sweden.

Hansen LM (2003) Insecticide-resistant pollen beetles (*Meligethes aeneus* F.) found in Danish oilseed rape (*Brassica napus* L) fields. *Pesticide Manage Sci* 59: 1057–1059.

Heimbach U, Müller A, Thieme T (2006) First steps to analyse pyrethroid resistance of different oil seed rape pests in Germany. *NachrBl Dtsch PflSchutz* 58: 1–5.

Hokkanen HMT, Husberg G-B, Söderblom M (1988) Natural enemy conservation for the integrated control of the rape blossom beetle *Meligethes aeneus* F. *Ann Agriculturae Fenniae* 27: 281–294.

Jackowski J, Klukowski Z, Irzykowicz M (2009) The effect of τ-fluvalinate and λ-*cyhalothrin* on two parasitic species of *Phradis* spp. (Hymenoptera, Ichneumonidae, *Tersilochinae*). *Pestycydy/Pesticides* 2008(3–4): 79–92.

Jansen JP (1996) Side effects of insecticides on *Aphidius rhopalosiphi* (Hym: Aphidiiae) in laboratory. *Entomophaga* 41: 37–43.

Jepson PC (1989) The temporal and spatial dynamics of pesticide side-effects on non-target invertebrates. In: Jepson PC (ed.) *Pesticides and non-target invertebrates.* Intercept, Andover, UK.

Jiu GD, Waage JK (1990) The effect of insecticides on the distribution of foraging parasitoids, *Diaretiella rapae* (Hym.: Braconidae) on plants. *Entomophaga* 35: 49–56.

Jönsson M, Nilsson C, Anderson P (2004) Occurrence of pollen beetle parasitoids in the south of Sweden. *IOBC/wprs Bull* 27(10): 239–242.

Johnen A, Ulber B (2004) Perspektiven der Nützlingsschonung im Winterraps durch die Entwicklung von phänologischen Modellen. *Mitt Biol Bundesanst Land-Forstwirtsch* 396: 278.

Johnen A, Williams IH, Ferguson AW, Büchs W, Klukowski Z, Luik A, Nilsson C, Ulber B (2006) MASTER: Construction of phenological models of key parasitoids in Europe and prospects for spray windows compatible with their conservation in winter oilseed rape. *Proc Int Symp Integrated Pest Management in Oilseed Rape, 3–5 April 2006, Göttingen, Germany*: 137–139.

Kirch G, Basedow T (2008) The importance and control of insect pests of winter wheat and winter oilseed rape in Schleswig-Holstein 1999–2001, and the trends of insecticide use in 1999–2004. *Mitt Dtsch Ges Allg Angew Ent* 16: 327–331.

Klingenberg A, Ulber B (1994) Untersuchungen zum Auftreten der Tersilochinae (Hym., Ichneumonidae) als Larvalparasitoide einiger Rapsschädlinge im Raum Göttingen 1990 und 1991 und zu deren Schlupfabundanz nach unterschiedlicher Bodenbearbeitung. *J Appl Entomol* 117: 287–299.

Klukowski Z (2006) The impact of insecticide application and timing on parasitoid activity and levels of parasitism. *Proc Int Symp Integrated Pest Management in Oilseed Rape, 3–5 April 2006, Göttingen, Germany.*

Klukowski Z, Twardowski J, Irzykowicz M (2003) [Preliminary results of subsequent effects on the pyrethroids Karate and Mavrik to ground beetle (Coleoptera, Carabidae) activity in winter rapeseed crop agrocenosis] In Polish. *Rośliny Oleiste/Oilseed Crops* 24(1): 269–279.

Klukowski Z, Twardowski J, Irzykowicz M (2006) Subsequent effect of tau-fluvalinate (Maverik) and lambda-cyhalotrine (Karate) pyrethroids on the activity of carabid beetles (Coleoptera: Carabidae) in winter oilseed rape. *IOBC/wprs Bull* 29(7): 125–130.

Lehmann W (1965) Einfluss chemischer Bekämpfungsmassnahmen auf einige Rapsschädlinge und ihre Parasiten. II. Knospen- und Stengelschädlinge. *Arch Pflanzenschutz* 1: 209–219.

Longley M, Jepson PC (1996) Effects of honeydew and insecticide residue on the distribution of foraging aphid parasitoids under glasshouse and field conditions. *Entomol Exp Appl* 81: 189–198.

Menzler-Hokkanen I, Hokkanen HMT, Büchs W, Klukowski Z, Luik A, Nilsson C, Ulber B, Williams IH (2006) Insect problems in European oilseed rape cultivation, and how to deal with them: The OSR farmers' perspective. *IOBC/wprs Bull* 29(7): 91–94.

Moreby SJ, Southway S, Barker A, Holland JM (2001) A comparison of the effect of new and established insecticides on nontarget invertebrates of winter wheat fields. *Environ Toxicol Chem* 20: 2243–2254.

Murchie AK, Williams IH, Alford DV (1997) Effects of commercial insecticide treatments to winter oilseed rape on parasitism of *Ceutorhynchus assimilis* Paykull (Coleoptera: Curculionidae) by *Trichomalus perfectus* (Walker) (Hymenoptera: Pteromalidae). *Crop Protect* 16: 199–202.

Murchie AK, Williams IH, Perry J (1999) Edge distributions of *Ceutorhynchus assimilis* (Paykull) and its parasitoid *Trichomalus perfectus* (Walker) in a crop of winter oilseed rape (*Brassica napus* L.). *Biocontrol* 44: 79–90.

Neumann N, Schütz S, Eisenwiener U, Ulber B (2010) Influence of insecticide application on host finding of *Tersilochus obscurator* (Hym.: Ichneumonidae). *IOBC/wprs Bull* (in press).

Nilsson C (2003) Parasitoids of the pollen beetles. In: Alford DV (ed.) *Biocontrol of oilseed rape pests*. Blackwell, Oxford, UK.

Nilsson C (2004) Trap plants to avoid insecticide application against pollen beetles in oilseed rape. *IOBC/wprs Bull* 27(10): 217–224.

Nitzsche O (1998) *Auftreten und Effizienz von Parasitoiden als natürliche Gegenspieler von Schadinsekten im Winterraps unter besonderer Berücksichtigung unterschiedlicher Bodenbearbeitungsmassnahmen nach Winterraps*. PhD thesis Georg-August-University of Göttingen, Göttingen, Germany.

Pruszynski S, Mrowczynski M (1990) Trial results on use of lambda- cyhalothrin for pest control in agricultural crops in poland. *Brighton Crop Prot. Conf. – Pest & Diseases*: 715–720.

Šedivý J (1983) Tersilochinae as parasitoids of insect pests of winter rape (Hymenoptera: Ichneumonidae). *Contributions Amer Entomol Inst* 20: 266–276.

Sterk G, Hassan SA, Baillod M, Bakker F, Bigler F, Blumel S, Bogenschutz H, Boller E, Bromand B, Brun J et al. (1999) Results of the seventh joint pesticide testing programme carried out by the IOBC/WPRS-Working Group 'Pesticides and Beneficial Organisms'. *Biocontrol* 44: 99–117.

Sunderland K (1992) Effects of pesticides on the population ecology of polyphagous predators. *Aspects Appl Biol* 31: 19–28.

Tomlin CDS (ed.) (2006) *The pesticide manual*. BCPC Publications, Farnham, UK.

Ulber B (2003) Parasitoids of Ceutorhynchid stem weevils. In: Alford DV (ed.) *Biocontrol of oilseed rape pests*. Blackwell Publishing, Oxford, UK.

Ulber B, Nitzsche O (2006) Phenology of parasitoids (Hym., Ichneumonidae-Tersilochinae) of oilseed rape pests in northern Germany in 1995–1997. *IOBC/wprs Bull* 29(7): 173–179.

Ulber B, Stippich G, Wahmhoff W (1990) Möglichkeiten, Grenzen und Auswirkungen des gezielten Pflanzenschutzes im Ackerbau: III. Auswirkungen unterschiedlicher Intensitäten des chemischen Pflanzenschutzes auf epigäische Raubarthropoden in Winterweizen, Zuckerrüben und Winterraps. *Z PflKrankh PflSchutz* 97: 263–283.

Veromann E, Tarang T, Kevväi R, Luik A, Williams I (2006) Insect pests and their natural enemies on spring oilseed rape in Estonia: Impact of cropping systems. *Agr Food Sci* 15: 61–72.

Vickerman GP (1992) The effects of different pesticide regimes on the invertebrate fauna in winter wheat. In: Greg-Smith P, Frampton GK, Hardy AR (eds.) *Pesticides, cereal farming and environment, the boxworth project*. HMSO, London, UK.

Warner DJ, Allen-Williams LJ, Ferguson AW, Williams IH (2000) Pest-predator spatial relationships in winter rape: Implications for Integrated Crop Management. *Pest Manage Sci* 56: 977–982.

Wegorek P, Zamojska J (2006) Resistance of pollen beetle (*Meligethes aeneus* F.) to pyrethroids, chloronicotinyls and organophosphorous insecticides in Poland. *IOBC/wprs Bull* 29(7): 135–140.

Williams IH (2004) Advances in insect pest management of oilseed rape in Europe. In: Horowitz AR, Ishaaya I (eds.) *Insect pest management – field and protected crops*. Springer-Verlag, Heidelberg, Germany.

Williams IH (2006) Integrating parasitoids into management of pollen beetle on oilseed rape. *Agronomy Research 4 (Special Issue)*: 465–470.

Williams IH, Barari H, Frearson D (2006) Upwind anemotaxis by the parasitoid *Tersilochus obscurator* (Hym., Ichneumonidae) during its migration flights to oilseed rape. *IOBC/wprs Bull* 29(7): 181–186.

Winfield AL (1963) A study on the effects of insecticides on parasites of larvae of blossom beetles (*Meligethes aeneus* F., Coleoptera: Nitidulidae). *Entomol Exp Appl* 6: 309–318.

Chapter 14
Oilseed Rape, Bees and Integrated Pest Management

Marika Mänd, Ingrid H. Williams, Eneli Viik, and Reet Karise

Abstract As a major mass-flowering crop producing an abundance of nectar and pollen, oilseed rape is very attractive to honey bees, bumblebees and solitary bees. It provides a food resource of considerable value in sustaining bee populations in agroecosystems at a time when bees are in decline. Although the flowers are self-fertile, they are entomophilous, and pollination studies, both in the glasshouse and in the field, suggest that bee foraging activities on the crop have many beneficial effects for the grower, including improving both the quantity and quality of the seed produced. However, bees foraging on the crop are vulnerable to the effects of insecticides, mostly pyrethroids applied to the crop, particularly when these are applied during flowering to control inflorescence pests. Effects may be lethal or sub-lethal; the latter have been little studied but there is growing evidence that insecticides affect many aspects of bee behaviour and physiology, such as division of labour, foraging and orientation, reproduction and respiration. Husbandry practices on the crop must therefore seek to minimise the use of insecticides on the crop, particularly during flowering, in order to sustain and not diminish bee populations foraging on the crop. Bees may even have a role in integrated pest management strategies incorporating biocontrol through their capacity to vector entomopathogenic fungal spores to the flowering canopy of oilseed rape to kill inflorescence pests.

14.1 Introduction

Oilseed rape (*Brassica napus* L.) is an oil crop of increasing importance worldwide. It is the major oilseed crop grown in northern and central Europe with over 5 million ha grown and a production of over 15 million tonnes in 2006 (Eurostat 2009, see also Williams Chapter 1 this volume). The flowers of oilseed rape yield abundant nectar and pollen and are very attractive to bees, which consequently are often abundant on flowering rape crops. The growing of mass-flowering oilseed rape

M. Mänd (✉)
Institute of Agricultural and Environmental Sciences, Estonian University of Life Sciences, Tartu, 51014, Estonia
e-mail: marika.mand@emu.ee

I.H. Williams (ed.), *Biocontrol-Based Integrated Management of Oilseed Rape Pests*,
DOI 10.1007/978-90-481-3983-5_14,

crops thus greatly enhances nectar and pollen resource availability in agricultural areas and, when appropriately managed, have potential to promote the abundance as well as the fitness of bee populations (Westphal et al. 2009). Many beekeepers move their honey bee colonies to crops of oilseed rape during flowering; honey is therefore an important by-product of the crop (Williams 1980, Williams and Cook 1982, Williams et al. 1993).

Although commonly considered to be a self-pollinating species, oilseed rape has entomophilous flowers capable of both self- and cross-pollination and there is substantial evidence that seed quantity and quality can be improved by the foraging activities of bees on the crop.

On the other hand, bees foraging on the crop are vulnerable to the toxic effects of pesticides applied to the crop and this may contribute to the decline of wild bees as well as honey bees (Corbet et al. 1991, Miranda et al. 2003). Frequent applications of broad-spectrum, non-selective insecticide compounds, mainly synthetic pyrethroids, are commonly applied to rape crops throughout Europe each year for the control of economically-important insect pests in autumn, spring and summer (see Williams et al. Chapter 1, Ulber et al. Chapter 13, this volume); some applications are made during flowering when bees may be foraging on the crop and are particularly vulnerable to their toxic effects. Further, insecticides are often applied in tank-mixes with fungicides; this may change the effects of both products on non-target organisms; the toxicity of the insecticide may be greater when applied in a tank-mix (Muranjan et al. 2006). Despite research data indicating severe mortality effects on beneficial insects, less attention has been paid to sub-lethal effects (Gels et al. 2002, Thompson 2003). There is increasing concern amongst beekeepers that sub-lethal doses of pesticides may have a significant impact on the behaviour of honey bees (Pajot 2001) and there is growing evidence that they also affect their physiology.

The intensification of agriculture has lead to a rapid decline in the species-richness of farmland (Benton et al. 2003). General and widespread shortage of bee-pollinators is predicted in agricultural areas of America (Kremen et al. 2004), Asia (Klein et al. 2003) and Europe (Williams et al. 1991, Williams 1996). Bees are important pollinators not only of agricultural ecosystems but of almost all terrestrial ecosystems because they provide a vitally important ecosystem service as pollinators for a wide range of agricultural, horticultural and wild plants (Corbet et al. 1991, Williams 1994, 1996, Klein et al. 2007). The decline of bee populations is therefore currently giving cause for great concern (Williams 1996, Biesmeijer et al. 2006, Gabriel and Tscharntke 2007).

Several factors have been suggested as possible contributors to this decline, including changes in climate and the effects of predators and parasites (Williams 1986). However, the principal factor is likely to have been the loss and degradation of habitats and of critical food resources due to changes in land-use and agricultural practice (Osborne et al. 1991, Williams et al. 1993, Mänd et al. 2002, Sepp et al. 2004, Goulson et al. 2005, Öckinger and Smith 2007). The supply of nectar and pollen is now often insufficient in European agricultural landscapes to support healthy bee populations (Goulson et al. 2005, Öckinger and Smith 2007). Oilseed rape, as a mass flowering crop, provides highly rewarding resources of both nectar and pollen for bees and therefore promotes colony growth and bee abundance (Westphal et al. 2003, 2009). Thus it is vital that husbandry of the oilseed rape crop helps to sustain and not to diminish bee populations.

It is essential therefore to consider bee populations, their interactions with the oilseed rape crop as well as their importance to the wider environment when developing pest management strategies for the crop. Alternatives to chemical insecticides for pest management are needed to reduce pesticide applications to the crop and thereby minimize the pressure on beneficial insects such as bees and parasitoids (see also Ulber et al. Chapter 13 this volume). Due to their morphological and behavioural characteristics, bees may even be used to aid pest management on the crop. Their hairy bodies are adapted for carrying pollen grains but they can also be used to vector antagonistic micro-organisms, such as entomopathogenic fungi for the control of inflorescence pests. Development of bee-mediated biological control vector-technology has great potential in integrated pest management strategies for crop protection (Williams 2004, Williams et al. 2005).

This review analyses the importance of oilseed rape as a food resource for bees, describes its pollination requirements, discusses the vulnerability of bees to pesticides applied to the crop and examines the potential for use of bees as entomovectors within integrated pest management strategies for the control of inflorescence pests of oilseed rape.

14.2 Oilseed Rape as a Source of Forage for Bees

14.2.1 The Flower

Oilseed rape is a typical cruciferous plant with yellow (or in some cultivars, white) flowers arranged in elongated terminal racemes. Each flower has four sepals, four petals and, usually, six stamens, four of which are longer and two shorter than the style. The flower bears four partly-hidden nectar glands (nectaries) at the base of the six stamens, two at the inner bases of the short stamens and two outside the ring of stamens (Hasler and Maurizio 1950, Eisikowitch 1981).

The flowers may open at any time of the day, but usually begin to open early in the morning and most are fully open by 9.00 h. They remain open for up to 3 days, closing slightly at night, but opening fully again the next morning; winter rape flowers are open for 1–3 days, whereas flowers of spring rape open for 1–2 days. Flowering extends from 22 to 45 days (Radchenko 1964, Free 1993, Delaplane and Mayer 2000) depending on weather conditions. When the weather is cold and damp, the flowers are open for longer time than in warmer and drier weather (Williams 1985).

14.2.2 Nectar Production

Oilseed rape flowers yield abundant nectar. Nectar volume can vary greatly from 0.2 μl per flower up to 6 μl per flower (Free 1993, Davis et al. 1994, Pierre et al. 1999). Pierre et al. (1999) tested 71 cultivars of winter oilseed rape for floral nectar volume and found that on average a flower secretes about 2 μl. Nectar volume per flower may be affected by genotype (Pierre et al. 1999), cultivar (Davis et al. 1994), flower age (Williams 1980) and local environmental conditions (Williams 1985, Rathcke 1992). Nectar production has been reported to decrease towards the end of the day (Radchenko 1964), and to be greater in the morning and early afternoon than midday (Williams 1985). The flowers are able to replenish the level of nectar completely within 30 min of being emptied which makes them very attractive to bees. Nectar production even increases if bee density is high, and flowers are visited more than three times per day (Williams 1985). In a given genotype, nectar secretion can fluctuate from one- to three-fold depending on the time of day (Williams 1985, Pierre et al. 1999).

Nectar production in the two types of nectaries varies within a single flower. Inner nectaries begin to secrete nectar before the flowers are fully open and produce much more nectar than the two outer nectaries (Hasler and Maurizio 1950, Eisikowitch 1981), whereas, the outer nectaries are more accessible to pollinators than the inner ones, particularly towards the end of flowering (Davis et al. 1994, Pierre et al. 1999).

Due to the significant heterosis for seed yield, in addition to the conventional cultivars, hybrid cultivars of oilseed rape were evolved (Riaz et al. 2001). Hybrid composites consisting of a male-sterile component and a male-fertile component have been widely used in the EU. However, the male-sterile lines did not secrete enough nectar for pollinators. Pierre et al. (1999) demonstrated a clear difference in nectar production between male-sterile lines and their isogenic male-fertile counterparts. Mesquida and Renard (1979) showed that 68% of male-sterile flowers had only two of the four nectaries present, 20% had only one nectary and 12% had none. The remaining nectaries of male sterile flowers were small, with the consequence that male-sterile flowers secreted ten times less nectar than male-fertile ones. Under different environmental conditions, five male-sterile cybrid (hybrid composite) lines of 'Darmor' produced from 50% up to 90% less nectar than male-fertile lines (Mesquida et al. 1991). By contrast, Pierre et al. (1999) showed that nectar production of some of the male-sterile lines, compared with male-fertile genotypes, was generally not all that low. For example, male sterile 'Fu58 Darmor' produced 2.83 μl of nectar per flower which was greater than the average amount of nectar produced by male-fertile genotypes.

More recently, composite hybrid cultivars have been replaced with restored hybrid cultivars (Pinochet and Bertrand 2000). Unlike the male-sterile lines of composite hybrid cultivars, the nectar quantities produced by restored lines are similar to those produced by male-fertile oilseed rape cultivars (Pierre et al. 1999). However, for the breeding of restored hybrid cultivars and seed production for commercial growing, the combination of male-sterile and male-fertile lines is still necessary (Steffan-Dewenter 2003).

The nectar of oilseed rape flowers contains carbohydrates, such as sucrose, glucose, fructose and ribose (Hasler and Maurizio 1950, Pierre et al. 1999). The sugar concentration in the nectar is highest at the beginning of the flowering period (30.24 g/100 ml) and decreases towards the end (10.64 g/100 ml) (Pernal and Currie 1998). The same temporal trend was observed in different cultivars (Pierre et al. 1999). Similarly, during the life of a flower, sugar concentration of the nectar is greatest when the flower opens, and lowest before it withers (Radchenko 1964). Nectar production is greatest at the beginning of the day, while the sugar concentration increases toward the end of the day (Meyerhoff 1958, Radchenko 1964). Most (95%) of the total nectar carbohydrate per flower is secreted by the inner pair of glands, because the inner nectaries are directly supplied with phloem alone, whereas the outer glands, which are poor nectar yielders, lack any vascular supply or are barely innervated by phloem (Davis et al. 1994).

Climatic factors influence nectar sugar concentration of many plants, including that of oilseed rape (Corbet et al. 1979); these include temperature, rainfall, relative humidity of air and sunshine, as well as edaphic factors (Mesquida et al. 1991). For example, at high relative humidity (80–90%) nectar from the inner and outer nectaries has the same sugar concentration (22–33%), but at a lower range of relative humidity the outer nectaries, which are relatively exposed, have a higher sugar concentration (Eisikowitch 1981).

14.2.3 Pollen Production

Oilseed rape flowers produce a lot of pollen. For example, the number of pollen grains produced per flower of the spring oilseed rape cultivar 'Drakkar' averaged 125×10^3 (Pertl et al. 2002). Pollen contains proteins, lipids, carbohydrates, starch, sterols, vitamins, and minerals (Herbert 1992, Day et al. 1990). All are important nutrients for brood rearing and development of young worker bees, particularly the protein content (Winston 1987, Hrassnigg and Crailsheim 1998). Pollens from different plant species differ in amino acid composition, concentration or both, and pollens with high proportions of essential amino acids are assumed to be of greater nutritional value. Oilseed rape pollen is rich in the amino acids most essential for bees, i.e., leucine, valine and isoleucine (Cook et al. 2003).

14.3 Pollination Requirements of Oilseed Rape

The flowers of oilseed rape are self-fertile (autogamous). Before the corolla fully expands, the four long stamens dehisce and release pollen outward the flower. Anthers on the two short stamens release pollen below the stigma which lengthens during flowering to reach the height of the anthers of the long stamens. When the flower is old, the long stamens bend towards the flower centre so that they become directed towards the stigma, and self-pollination can occur. Thus the morphology and behaviour of the oilseed rape flower encourage cross-pollination at first, but self-pollination later (Eisikowitch 1981, Williams 1985, Free 1993, Bell and Cresswell 1998, Delaplane and Mayer 2000).

Pollination studies (Williams 1978, 1984, Williams et al. 1986, 1987) have shown that oilseed rape cultivars set equally well whether self- or cross-pollinated; cultivars differed in the proportion of seed set from cross-pollination (up to 40%) (Williams 1985). However, in the case of cross-pollination more pollen can reach the stigmas, particularly pollen from the short stamens (Free 1993). Cross-pollination with pollen from short stamens is significantly superior to that from long stamens, and gives a 14% greater weight of seed per pod (Free 1993, Steffan-Dewenter 2003). Moreover, in a normal population there are individual plants which are self-sterile or prefer foreign pollen (Rives 1957, Williams et al. 1987, Williams and Simpkins 1989, Becker et al. 1992, Free 1993).

Overall, most authors agree that pollen vectored by wind, insects or gravity is necessary for seed production in oilseed rape (Williams 1978, Eisikowitch 1981, Free 1993, Westcott and Nelson 2001). However, the proportion of pollen vectored by wind and insects and over what distance, is still debated (Timmons et al. 1995, Ramsay et al. 2003, Devaux et al. 2008).

Plants grown in the still air of a glasshouse have poor seed set (Eisikowitch 1981, Mesquida and Renard 1982, Mesquida et al. 1988); shaking plants to simulate movement by wind improves seed set (Williams et al. 1986). Pollination studies in the field have shown that plots exposed to wind but caged to exclude insects often yield at least as well as open-pollinated plots (Williams 1978, 1984, Williams et al. 1987). Wind has been even suggested to be a primary pollen vector of oilseed rape (Timmons et al. 1995, Wilkinson et al. 2003). Under field conditions, the movement of plants by wind could increase the self-pollination of cultivars that auto-pollinate poorly. Pollen grains may be carried over long distance: from 400 m up to 3,000 m (Scheffler et al. 1995, Hall et al. 2000, Rieger et al. 2002, Beckie et al. 2003, Devaux et al. 2008). Thus, wind not only causes self-pollination of flowers by moving them, but also causes cross-pollination by transporting considerable quantities of pollen. But Rieger et al. (2002) have questioned the efficacy of wind and others have shown that wind alone is insufficient to attain maximum seed set (Williams 1978, Eisikowitch 1981, Free 1993, Cresswell et al. 2002, 2004, Ramsay et al. 2003). Oilseed rape has entomophilic pollen grains, which cannot be transferred by wind alone; anthers when flicked by insects or artificially under dry conditions behave like catapults raising a cloud of pollen grains (Eisikowitch 1981). So, additional pollination by insects may be necessary.

14.4 Bees as Pollinators of Oilseed Rape

Oilseed rape is visited by honey bees, bumblebees and solitary bees, including species of *Andrena, Halictus* and *Megachile*. Honey bees are usually the most abundant visitors. Rape flowers produce such abundant nectar and at a time when there are few other cultivated food plants available for them, that honey bees visit rape crops from a distance of 3.5–4 km from their hives and neglect fruit trees in favour of rape (Free 1993). Furthermore, many beekeepers move their colonies onto or near to oilseed rape crops to benefit from the nectar and pollen it produces (Williams 1980, Williams and Cook 1982, Williams et al. 1993, Carreck et al. 1997). Although many species of bumblebee and solitary bee may visit a crop, their proportion is often quite low (Free 1993, Varis 1995, Karise et al. 2004). All bee species successfully transfer rape pollen from anthers to stigmas.

Earlier studies have shown that insect pollination of oilseed rape can lead to higher seed set and yield (Williams 1978, Williams and Simpkins 1989, Westcott and Nelson 2001). According to Free and Nuttall (1968), plants caged with bees produced 13% more seed than plants caged without bees. Recorded benefit from bee pollination ranges from 13 to 64% more seeds per pod (Williams 1985). But there are still some questions about the degree of benefit to seed production from insect pollinators. Positive effects are dependent on cultivar, environmental growing conditions, and the compensatory capacity of the crop (Williams and Free 1979, Williams et al. 1987) and include shortening of the flowering period, reduction of raceme production, acceleration of ripening (Mesquida and Renard 1981, Williams 1984, Mesquida et al. 1988), and increases in seed germination rate (Frediani et al. 1987, Kevan and Eisikowitch 1990) and seed oil content (Radchenko 1964, Mishra and Kaushic 1992).

The influence of honey bees on oilseed rape flowering may be explained by the fact that flowers are visited early in their development. Such early visiting is immediately followed by deposition of abundant pollen on the receptive stigmas. Consequently flowers pollinated in this way wither more quickly. Flower life is strongly reduced, flowering is shorter, and is more uniform and coordinated than for plants that are not insect pollinated (Mesquida et al. 1988).

Mesquida and Renard (1979) found that bee pollination slightly increased the final yield of the male-sterile plants, but significantly increased the yield of the male-fertile plants. Sabbahi et al. (2005) showed an improvement in rape seed yield of 46% in the presence of three honey bee hives per hectare, compared with the absence of hives. This suggests that supplemental pollination may increase set of early flowers, evenness of ripening, and ease of harvest (Williams 1978), therefore the plant would produce fewer flowers (Free 1993), and the flowering period and vegetative growth would shorten (Mesquida et al. 1988, Free 1993). It increases the number of seeds per pod, the number of seeds per plant (Steffan-Dewenter 2003), the evenness of ripening, thus reducing seed loss at harvesting (Free 1993). Altogether the seed yield of oilseed rape could be higher by up to 25–46% (Delaplane and Mayer 2000, Sabbahi et al. 2005).

14.5 Toxicity to Bees of Insecticides Applied to Oilseed Rape

Bees are especially vulnerable to the toxic effects of insecticides applied during flowering when they are foraging on the crop. They may be exposed through direct contact with spray droplets, through chemical residues left on the plant surface, and through feeding on contaminated nectar or pollen, either as adults or larvae. Effects may be lethal or sub-lethal; most studies have assessed lethal effects, while only a few have addressed sub-lethal effects. The effects of pesticides on non-target organisms have been studied extensively. It is obligatory for chemical companies to provide mortality data for their products for all larger organism groups. But, despite research data indicating the severe mortality rate on bees, less attention has been paid to the sub-lethal effects. In recent years, this has been an increasing area of study and a subject of discussion between scientists and regulatory authorities (Thompson and Maus 2007).

In addition to deficient information of the sub-lethal effects of insecticides, there exists the problem of extrapolating data from honey bees to bumblebee and other pollinating bees. Pesticide risk assessments for honey bees are based on hazard ratios which rely on application rates and toxicity data that are unlikely to be appropriate for bumblebees. The latter are active at different times and on different crop species and, therefore, are likely to have different exposure profiles. Unlike honey bees, deaths of bumblebees due to pesticides are unlikely to be reported, since the bees are not kept domestically and die in small numbers (Thompson and Hunt 1999). The information on pesticide toxicity on non-*Apis* bees is scarce, and limited to species managed for crop pollination (Ladurner et al. 2003).

14.5.1 Lethal Effects

In conventional farming, application of many insecticides (e.g., pyrethroids) considered to be safe for honey bees, is permitted to the oilseed rape crop while it is in flower. Despite this, 57 out of 117 honey bee poisoning incidents in UK during 1994–2003 resulted from spray applications to flowering crops; 17 of these incidents were through approved use of the products (Barnett et al. 2007). Pyrethroids, most often sprayed on flowering oilseed crops, have been reported to be repellent to honey bees (Thompson 2001), although this is still in question in the field situation. Karise et al. (2007) found no repellency of alpha-cypermethrin to honey bees on oilseed rape under field conditions but found that flower visitation depended on the density of flowers present. If any repellency does occur with respect to this insecticide, the attractiveness of the flower resource is likely to override it.

In organic farming, pesticides are also needed and many botanical insecticides are permitted for use in controlling pests. The main ingredient of Neem extracts, azadirachtin, is considered to be safe for honey bees (Zehnder et al. 2007), but has been found to cause changes in the foraging behaviour in bumblebees (Karise et al. 2006). Pyrethrins are toxic to bees; quassia and rotenone do not harm bees (Zehnder et al. 2007). The toxicity of botanical compounds to bees tends to be lower than that of synthetic compounds because their degradation time is shorter and timing of application helps to minimize harmful effects on beneficial insects (Kühne 2008).

14.5.2 Sub-lethal Effects

Studying the sub-lethal effects of pesticides is complicated due to difficulties in measuring the effects. Results obtained in the laboratory may not match with those obtained in the field (Thompson and Maus 2007). Under certain circumstances, sub-lethal effects may cause more harm than lethal doses since they affect the survival of the brood and colony. Systemic compounds have been considered safe for pollinators when not applied to the flowers. However, the residues of the compounds still contaminate nectar and pollen in sub-lethal doses via both active and passive transport (Thompson 2001, Cutler and Scott-Dupree 2007). Contamination may occur after application of the compounds to other parts of plants (Ferguson 1987), to the soil (Jaycox 1964) or on seeds (Dikshit et al. 2002, Sur and Stork 2003). Contaminated nectar and pollen poses a potential danger not only to forager bees but also to bees in the hive and to brood. The toxicity of pesticides to brood has been investigated far less than toxicity for adults (Alix and Vergnet 2007).

14.5.2.1 Effects on Division of Labour

Division of labour plays an important role in colonies of social insects. Workers have specific, often age-dependent tasks. Treatment of honey bees with juvenile hormone analogues (synthetic hormone-like compounds used as insecticides), results in a decreasing ability of young emerging bees to feed larvae, due to the early degeneration of the hypopharyngeal glands and precocious foraging ability (Tasei 2001). Changes in the division of labour of honey bees, such as decreased house cleaning abilities, delayed onset and duration of foraging and handling of nectar, have also been recorded (reviewed by Thompson 2003). These changes affect both honey yield and the overwintering of colonies (Thompson et al. 2005).

14.5.2.2 Effects on Foraging and Orientation

Foraging depends on the bee's ability to discriminate odours, to learn, to communicate, and to orientate within its environment; altering these systems may result in a decrease in foraging. The bees' orientation and communication ability have been found to be affected by sub-lethal doses of organophosphorus insecticides (Schricker and Stephen 1970), synthetic pyrethroids (Cox and Wilson 1984, Vandame et al. 1995) and neonicotinoids (Bortolotti et al. 2003, Yang et al. 2008). Pyrethroids and neonicotinoids have also been shown to affect both foraging activity (Thompson 2003) and learning capacities (Decourtye et al. 1999, 2003, Guez et al. 2001, Ramirez-Romero et al. 2005). Pyrethroids may also affect thermoregulation (Jagers op Akkerhuis et al. 1999b, Belzunces et al. 2001); in cooler climates, this can lead to decreased flying ability. The decrease in foraging and in returning foragers reduces brood production (Thompson 2003), which in turn may weaken a colony's potential to survive the winter.

14.5.2.3 Effects on Reproduction

All classes of insecticides affect the reproductive behaviour of bees (reviewed by Thompson 2003). Reduction of brood may have more damaging consequences for honey bees than simply the moderate loss of foragers (Haynes 1988, Thompson et al. 2007). Thompson et al. (2005) have reported 40–95% egg mortality over 2 weeks after diflubenzuron application and 45–60% egg mortality over 2 weeks after fenoxycarb application. The insect growth regulator fenoxycarb has caused the death of almost all larvae or developing malformed pupae (Van der Steen and de Ruijter 1990, Aupinel et al. 2007).

Besides killing brood, insecticides can cause changes in the development of the larvae. Contamination of the food by insect growth regulators (Tasei 2001) can increase development time and cause malformations. In solitary bees, pyrethroids (Tasei et al. 1988) and in honey bees, pyrethroids (Tasei et al. 1988) and neonicotinoids (Schmuck et al. 2001) have been found to affect their fecundity. Some organophosphates, pyrethroids and neonicotinoids have affected the honey bee queen's status or have interfered with a colony's ability to requeen itself (Stoner et al. 1985, Thompson et al. 2005). Organophosphates have decreased the longevity of honey bees (Johansen and Mayer 1990). Neonicotinoids (Tasei et al. 2000) and organophosphates (Johansen and Mayer 1990) have decreased brood production in the bumblebee.

14.5.2.4 Effects on Respiration

Better understanding of the effects of insecticides in the field benefits from insight into their effects on different physiological functions, for example, on respiration. In the case of bees, it is difficult to examine the effects of insecticides on respiration patterns because there is little data on their normal respiration patterns. However, this has been an area of increasing interest during the past decade.

Since water is a key element in every living organism, most insects have probably evolved mechanisms to prevent excessive water loss (Klowden 2002). Resting insects often exhibit discontinuous gas exchange cycles (DGC), a function of which may be the reduction of respiratory water loss (Levy and Schneiderman 1966, Lighton 1994) through the large inner surface of the tracheal system.

According to Lighton (1994, 1996), in the state of discontinuous gas exchange, the spiracles are closed most of the time. At low oxygen rates inside the trachea the spiracular valves flutter, allowing oxygen to enter the tracheal system. As larger amounts of carbon dioxide accumulate in the tracheae and haemolymph (Wobschall and Hetz 2004), the spiracles open and allow the gas to escape. Thus, as compared with continuous respiration, loss of carbon dioxide along with evaporated water occurs only discontinuously during the brief open phases of the spiracles. There are different views about the origin of DGC, as reviewed by Chown (2002) and Chown et al. (2006). There are also hypotheses that DGC serves as an adaptation for coping with hypercapnia and/or hypoxia in soil-living insects (Lighton 1998, Vogt and Appel 2000, Lighton et al. 2004) or protection against the oxidative damage during the periods with low metabolic cost (Hetz and Bradley 2005).

The existence and the precise pattern of DGC depend on the species (Lighton 1994, 1996, Slama 1999, Chown et al. 2006), individual characteristics (Marais and Chown 2003, Gibbs and Johnson 2004, Karise et al. 2010), life stage of the individuals (Beekman and van Stratum 1999, Mänd et al. 2005, 2006) and environmental conditions like temperature (Lighton and Lovegrove 1990, Lighton 1996, Vogt and Appel 2000, Kovac et al. 2007), relative humidity (Duncan et al. 2002, Lighton 2007, Slama et al. 2007) and the amount of oxygen or carbon dioxide in the air (Lighton 1998, Vogt and Appel 2000, Lighton et al. 2004).

DGC patterns have been used to characterize the physiological state of an insect, as several stress factors, including chemical ones, can affect them (Kestler 1991). Although knowledge about the sub-lethal effects of pesticides on insect physiology is scarce, it is known that treatments of arthropods with pyrethroids cause neurotoxic effects in parts of the nervous system, including the central nervous system and sensory, motor or neurosecretory neurons (Corbett 1974, Jagers op Akkerhuis et al. 1995). Because the closing and opening of spiracular valves is controlled by the nervous system, the neurotoxic effects may also include interference by DGC. In pupae of cabbage butterfly *Pieris brassicae*, after the treatment with original pyrethrum, the DGCs disappeared and metamorphosis was disrupted (Harak et al. 1999, Jõgar et al. 2008).

Pyrethroids, as well as many other insecticides, can induce increased water loss in arthropods (Gerolt 1976, 1983), due to production of diuretic hormones (Jagers op Akkerhuis et al. 1999a). This process could be reversible if the insect could replenish its water reserves. Since the pyrethroids often affect motion as well, causing the knockdown effect, death may come through desiccation (Jagers op Akkerhuis et al. 1995, 1999a, Thompson 2003).

14.6 Bees as Vectors of Entomopathogenic Fungi for Pest Control on Oilseed Rape

Bees are covered in an abundance of branched body hair, specially adapted to trap and transport pollen grains from flowers back to the colony or nest site to feed to brood (Free and Williams 1972). These hairs can also trap and transport the spores of bacteria and fungi (Batra et al. 1973, Sandu and Waraich 1985). This ability has been utilized in the development of biocontrol strategies to control various plant pests and diseases on a variety of crops. For example, Thomson et al. (1990) showed that honey bees could be used to carry spores of the bacteria *Pseudomonas fluorescens* (Trevisan) and *Erwinia herbicola* (Brown) to the flowers of apple to control fireblight disease caused by the bacterium *Erwinia amylovora* (Burrill). Similarly, honey bees have vectored spores of the fungus *Gliocladium roseum* (Bainier) to strawberry (Peng et al. 1992) and to raspberry (Yu and Sutton 1997) flowers to control growth of the grey mould fungus *Botrytis cinerea* Pers. More recently, the bumblebee, *Bombus impatiens* (Cresson) has been used in the glasshouse, to transport spores of the entomopathogenic fungus *Beauveria bassiana* (Balsamop-Crivelli) Vuillemin to sweet pepper flowers to control two insect pests, the plant bug, *Lygus lineolaris* (Palisot de Beauvois) and the thrips, *Frankliniella occidentalis* (Pergande) (Al-mazra'awi et al. 2006).

Bees have similarly been shown able to deliver entomopathogenic fungal spores to oilseed rape flowers to infect and kill insect pests living within the flowering canopy of the oilseed rape crop (Butt et al. 1998, Carreck et al. 2006). The pollen beetle, *Meligethes aeneus* (Fabricius) and the cabbage seed weevil, *Ceutorhynchus assimilis* (Paykull) are major inflorescence pests of oilseed rape throughout Europe (Williams Chapter 1 this volume); the latter is also a major pest in North America

(Dosdall and Mason Chapter 6 this volume). The pollen beetle feeds, as an adult, on pollen in the buds and flowers of the crop, and lays its eggs in the buds. Its larvae also feed on pollen in the buds and flowers, usually lying alongside the filaments of the stamens. The larvae are mobile, moving up the flowering inflorescence to younger flowers as they grow (Williams and Free 1978). On maturity, second instar larvae drop to the ground to pupate in the soil. The cabbage seed weevil also feeds on pollen in the flowers as well as on young buds, shoots and pods. The females lay their eggs singly in young pods on the flowering racemes. The seed weevil larva feeds within the pod on the growing seeds and on maturity, bores an exit hole through the pod wall and drops to the soil to pupate (Williams and Free 1978).

Honey bees foraging from hives, fitted with inoculum dispensers at their entrances (Fig. 14.1), have been shown to effectively deliver conidia of the entomopathogenic fungus, *Metarhizium anisopliae* (Metschnikoff) Sorokin, to the flowers of oilseed rape plots enclosed in field cages (Fig. 14.2, Butt et al. 1998, Carreck et al. 2006). Pollen beetles and seed weevils, sampled from the treated plots, both picked up lethal doses of the conidia from the flowers. When incubated in the laboratory, the fungus caused infection and mortality of both adult and larval pollen beetles, as well as of adult seed weevils (Figs. 14.3 and 14.4). Pod infestation by seed weevil larvae was too low to determine whether they were also infected by the fungus. After death, the bodies of many of the pest insects showed external conidiation of the fungus, confirming infection by *M. anisopliae* (Fig. 14.5).

Conidia of *M. anisopliae* disseminated initially from an inoculum source to rape flowers by honey bees in this way, would probably be further disseminated horizontally within the crop canopy by other insects, such as by bumblebees, foraging on the flowers. The inoculum would also be disseminated to the soil below the crop, as inoculated flowers shed their petals. How long the conidia can survive on petals is not known, but they occur naturally, albeit at a low level, and persist well in soil (Vanninen et al. 2000, Hokkanen et al. 2003). Laboratory and pot experiments have

Fig. 14.1 Honey bee hive fitted at its entrance with an inoculum dispenser containing the entomopathogenic fungus *Metarhizium anisopliae* for dissemination by the bees to the flowering canopy of oilseed rape (Photo: Ingrid Williams)

Fig. 14.2 Oilseed rape plots enclosed in field cages for the study of bee-mediated dissemination of the entomopathogenic fungus *Metarhizium anisopliae* (Photo: Ingrid Williams)

shown that mature larvae of the pollen beetle are susceptible to the fungus, not only when directly exposed to an inoculum, but also when the inoculum is applied to soil before the insects pupate in it (Husberg and Hokkanen 2000). However, in the semi-field experiments, described above, Carreck et al. (2006) found no effect on the numbers of new generation pollen beetle and seed weevil adults that emerged from pupation, following dissemination of inoculum by honey bees to the flowering crop canopy.

The effects of *M. anisopliae* on bees need further investigation as extrapolating risk from laboratory tests to bees in the field may be misleading (Alves et al. 1996). Butt et al. (1994) showed, in laboratory studies, that the honey bee was susceptible to *M. anisopliae* V245; when inoculated and then incubated at 30°C, the mean LT_{50} was 8.5 days. However, they also showed, in the laboratory, that isolates vary in

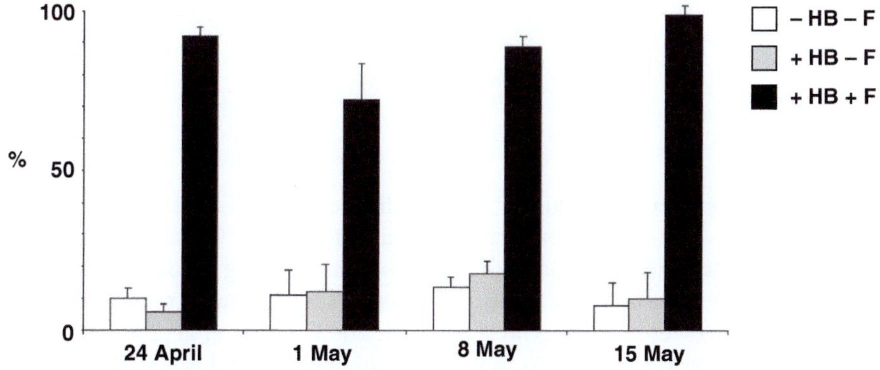

Fig 14.3 Percentage mortality of adult pollen beetles (*Meligethes aeneus*) on plots of winter oilseed rape when exposed to honey bees (HB) with and without dispensers containing the entomopathogenic fungus *Metarhizium anisopliae* (F) at their hive entrances (modified after Carreck et al. 2006)

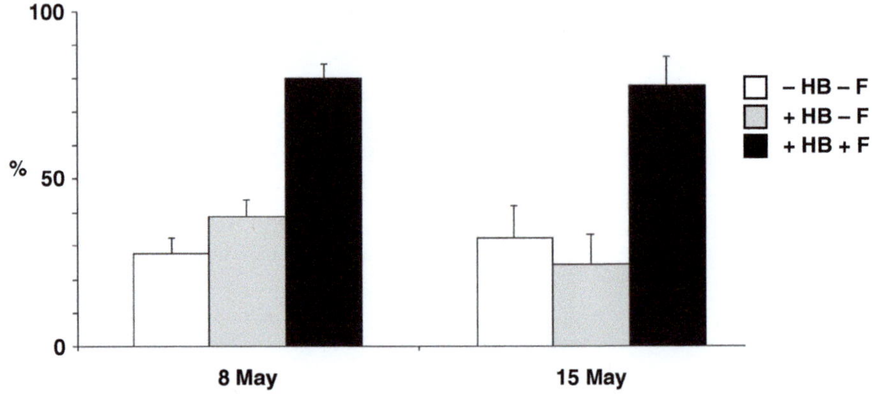

Fig. 14.4 Percentage mortality of adult seed weevils (*Ceutorhynchus obstrictus* syn. *C. assimilis*) on plots of winter oilseed rape when exposed to honey bees (HB) with and without dispensers containing the entomopathogenic fungus *Metarhizium anisopliae* (F) at their hive entrances (modified after Carreck et al. 2006)

Fig. 14.5 Body of the pollen beetle, *Meligethes aeneus*, showing external conidiation of the fungus, confirming infection by *Metarhizium anisopliae* (Photo: Ingrid Williams)

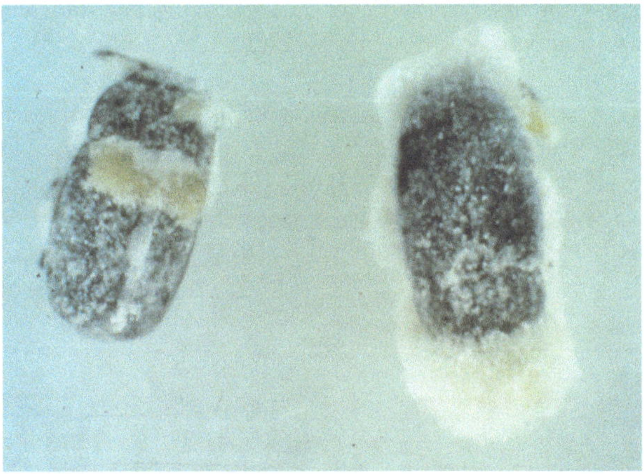

their temperature tolerances and so may have different effects on honey bees within the brood nest, where the temperature is maintained at ca. 35°C, than on foragers at outside temperatures. Some isolates of *M. anisopliae* are being tested for the biological control of the parasitic mite *Varroa destructor* Anderson and Trueman in honey bee colonies (Shaw et al. 2002, Davidson et al. 2003, Kanga et al. 2003, Lodesani et al. 2003). Carreck et al. (2006) found that, although in field cages where honey bees were disseminating *M. anisopliae* inoculum some of the bees that died showed external conidiation when incubated, declines in colony population size appeared to be unrelated to fungal infection, as they were no greater in colonies disseminating fungal inoculum than in control plots with bees but no inoculum. Population decline is usual in honey bee colonies, particularly large ones, when they are confined in field cages with limited forage (Pinzauti 1994). Further, if this strategy were to be used for pest control in oilseed rape crops some loss of honeybee foragers may be acceptable as their colonies are managed and to some extent therefore replaceable by beekeepers. The effect of *M. anisopliae* on bumblebees foraging on the crop remains to be investigated.

The effect of *M. anisopliae* on key parasitoids of the inflorescence pests of oilseed rape needs further investigation as Husberg and Hokkanen (2000) found that the hymenopterous larval endoparasitoids of the pollen beetle, *Phradis morionellus* (Holmgren) and *Diospilus capito* (Nees), both key agents in conservation biocontrol of the beetle (Ulber Chapter 2 this volume), were also susceptible, although to different extents, to spray treatment with the fungus.

14.7 Implications for Biocontrol-Based Integrated Management of Insect Pests of Oilseed Rape

Oilseed rape, as a widespread mass-flowering crop of agroecosystems of northern and central Europe, as well as in North America and other regions of the world, provides an abundant resource of pollen and nectar for bees. Many beekeepers move their honey bee colonies to oilseed rape crops during flowering; honey is a valuable by-product from the crop. Loss of food resources for bees in arable landscapes is probably a major cause of their decline over recent decades in many regions. The foraging activities of bees on the crop have been shown to improve both the quality and quantity of seed produced. Husbandry practices on the rape crop should therefore seek to sustain and not diminish bee populations.

Currently crop protection on oilseed rape, particularly against inflorescence pests such as the pollen beetle and the cabbage seed weevil, relies heavily on the application of pyrethroid insecticides. These kill beneficial insects, such as bees and parasitoids (Ulber et al. Chapter 13 this volume) foraging on the crop, particularly when applied during flowering. They also cause sub-lethal effects, although these have been little studied. The recent widespread development in many European countries of resistance to pyrethroids in the pollen beetle (Thieme et al. Chapter 12 this volume) has increased the urgency of developing integrated pest management strategies that minimise the use of insecticides on the crop, particularly during flowering. Further development of biocontrol strategies incorporating parasitoids and predators is essential to achieve this.

Honey bees also have potential for employment in biocontrol strategies. Their ability to vector the entomopathogenic fungus *M. anisopliae* for the control of inflorescence pests could perhaps be further enhanced by using an early-flowering cultivar of oilseed rape or turnip rape as a trap crop to concentrate both pest and honey bee populations. This would facilitate both concentration and horizontal transfer of the inoculum to its target pest populations before they move onto the oilseed rape main crop (Cook et al. 2006). However, to be of use in integrated pest management, any entomopathogenic fungus to be used should be benign both to bees, needed to pollinate the crop, and to parasitoids of the pests which contribute to their biocontrol.

Acknowledgements This work was funded by Estonian targeted financing project No. SF 0170057s09 and by Estonian Science Foundation grants No 6722 and 7391.

References

Al-mazra'awi MS, Shipp L, Broadbent B, Kevan PG (2006) Biological control of *Lygus lineolaris* (Hemiptera: Miridae) and *Frankliniella occidentalis* (Thysanoptera: Thripidae) by *Bombus impatiens* (Hymenoptera: Apidae) vectored *Beauveria bassiana* in greenhouse sweet pepper. *Biol Control* 37: 89–97.

Alix A, Vergnet C (2007) Risk assessment to honey bees: A scheme developed in France for non-sprayed systemic compounds. *Pest Manag Sci* 63: 1069–1080.

Alves SB, Marchini LC, Pereira RM, Baumgratz LL (1996) Effects of some insect pathogens on the Africanized honey bee, *Apis mellifera* L. (Hym., Apidae). *J Appl Entomol* 120: 559–564.

Aupinel P, Fortini D, Michaud B, Marolleau F, Tasei JN, Odoux JF (2007) Toxicity of dimethoate and fenoxycarb to honey bee brood (*Apis mellifera*), using a new in vitro standardized feeding method. *Pest Manag Sci* 63: 1090–1094.

Barnett EB, Charlton AJ, Fletcher MR (2007) Incidents of bee poisoning with pesticides in the United Kingdom, 1994–2003. *Pest Manag Sci* 63: 1051–1057.

Batra LR, Batra SWT, Bohart GE (1973) The mycoflora of domesticated and wild bees (Apoidea). *Mycopathol Mycol Appl* 49: 13–44.

Becker HC, Damgaard C, Karlsson B (1992) Environmental variation for outcrossing rate in rapeseed (*Brassica napus*). *Theor Appl Genet* 84: 303–306.

Beckie HJ, Warwick SI, Nair H, Séguin-Swartz G (2003) Gene flow in commercial fields of herbicide-resistant canola (*Brassica napus*). *Ecol Appl* 13: 1276–1294.

Beekman M, van Stratum P (1999) Respiration in bumblebee queens: Effect of life phase on the discontinuous ventilation cycle. *Ent Exp Appl* 92: 295–298.

Bell SA, Cresswell JE (1998) The phenology of gender in homogamous flowers: Temporal change in the residual sex function of flowers of oilseed rape *Brassica napus*. *Funct Ecol* 12: 298–306.

Belzunces LP, Vandame R, Xingfa Gu (2001) Joint effects of pyrethrinoid insecticides and azole fungicides on honeybee thermoregulation. In: Belzunces LP, Pelissier C, Lewis GB (eds.) *Hazards of pesticides to bees*. INRA, France.

Benton TG, Vickery JA, Wilson JD (2003) Farmland biodiversity: Is habitat heterogeneity the key? *Trends Ecol Evol* 18(4): 182–188.

Biesmeijer JC, Roberts SPM, Reemer M, Ohlemüller R, Edwards M, Peeters T, Schaffers AP, Potts SG, Kleukers R, Thomas CD, Settele J, Kunin WE (2006) Parallel declines in pollinators and insect-pollinated plants in Britain and the Netherlands. *Science* 313: 351–354.

Bortolotti L, Montanari R, Marcelino J, Medrzycki P, Maini S, Porrini C (2003) Effects of sub-lethal imidacloprid doses on the homing rate and foraging activity of honey bees. *Bull Insectol* 56: 63–67.

Butt TM, Carreck NL, Ibrahim L, Williams IH (1998) Honey bee-mediated infection of pollen beetle (*Meligethes aeneus* Fab.) by the insect-pathogenic fungus, *Metarhizium anisopliae*. *Biocontrol Sci Techn* 8: 533–538.

Butt TM, Ibrahim L, Ball BV, Clark SJ (1994) Pathogenicity of the entomogenous fungi *Metarhizium anisopliae* and *Beauveria bassiana* against crucifer pests and the honey bee. *Biocontrol Sci Techn* 4: 207–214.

Carreck NL, Butt TM, Clark SJ, Ibrahim L, Isger EA, Pell JK, Williams IH (2006) Honey bees can disseminate a microbial control agent to more than one inflorescence pest of oilseed rape. *Biocontrol Sci Techn* 17: 179–191.

Carreck NL, Williams IH, Little DJ (1997) The movement of honey bee colonies for crop pollination and honey production by beekeepers in Great Britain. *Bee World* 78: 67–77.

Chown SL (2002) Respiratory water loss in insects. *Comp Biochem Phys A* 133: 791–804.

Chown SL, Gibbs AG, Hetz SK, Klok CJ, Lighton JRB, Marais E (2006) Discontinuous gas exchange in insects: A clarification of hypotheses and approaches. *Physiol Biochem Zool* 79: 333–343.

Cook SM, Awmack CS, Murray DA, Williams IH (2003) Are honey bees' foraging preferences affected by pollen amino acid composition? *Ecol Entomol* 28: 622–627.

Cook SM, Smart LE, Martin JL, Murray DA, Watts NP, Williams IH (2006) Exploitation of host plant preferences in pest management strategies for oilseed rape (*Brassica napus*). *Entomol Exp Appl* 119: 221–229.

Corbet SA, Williams IH, Osborne JL (1991) Bees and the pollination of crops and wild flowers in the European community. *Bee World* 72: 47–59.

Corbet SA, Willmer PG, Beament JW, Unwin DM, Prys-Jones OE (1979) Post secretory determinants of sugar concentration in nectar. *Plant Cell Environ* 2: 293–308.

Corbett JR (1974) *The biochemical mode of action of pesticides*. Academic Press, London, New York.

Cox RL, Wilson WT (1984) Effects of permethrin on the behaviour of individually tagged honey bees, *Apis mellifera* L. (Hymenoptera: Apidae). *Environ Entomol* 13: 375–378.

Cresswell JE, Davies TW, Patrick MA, Russell F, Pennel C, Vicot M, Lahoubi M (2004) Aerodynamics of wind pollination in a zoophilous flower, *Brassica napus*. *Funct Ecol* 18: 861–866.

Cresswell JE, Osborne L, Bell SA (2002) A model of pollinator-mediated gene flow between plant populations with numerical solutions for bumblebees pollinating oilseed rape. *Oikos* 98: 375–384.

Cutler GC, Scott-Dupree CD (2007) Exposure to clothianidin seed-treated canola has no long-term impact o honey bees. *Ecotoxicology* 100: 765–772.

Davidson G, Phelps K, Sunderland KD, Pell JK, Ball BV, Shaw K, Chandler D (2003) Temperature growth interactions of entomopathogenic fungi with potential for control of *Varroa destructor* (Acari: Mesostigmata), using a nonlinear model of poikilotherm development. *J Appl Microbiol* 94: 816–825.

Davis AR, Sawhney VK, Fowke LC, Low NH (1994) Floral nectar secretion and ploidy in *Brassica rapa* and *B. napus* (Brassicaceae). I. Nectar size and nectar carbohydrate production and composition. *Apidologie* 25: 602–614.

Day S, Beyer R, Mercer A, Ogden S (1990) The nutrient composition of honey bee-collected pollen in Otago. New Zealand. *J Apic Res* 29: 138–146.

Decourtye A, Lacassie E, Pham-Delègue MH (2003) Learning performances of honeybees (*Apis mellifera* L) are differentially affected by imidacloprid according to the season. *Pest Manag Sci* 59: 269–278.

Decourtye A, Le Metayer M, Pottiau H, Tisseur M, Odoux JF, Pham-Delegue MH (1999) Impairment of olfactory learning performances of *Apis mellifera* L. by long term ingestion of imidacloprid. In: Belzunces LP, Pélissier C, Lewis GB (eds.) *Hazards of pesticides to bees. 7th Int Symp ICPBR Bee Protection Group*. INRA, Avignon, France, pp 113–117.

Delaplane KS, Mayer DF (2000) *Crop pollination by bees*. CAB International, Wallingford, UK.

Devaux C, Klein EK, Lavigne C, Sausse C, Messéan A (2008) Environmental and landscape effects on cross-pollination rates observed at long distance among French oilseed rape *Brassica napus* commercial fields. *J Appl Ecol* 45: 803–812.

Dikshit AK, Lal OP, Sinha SR, Srivastava YN (2002) Safety evaluation, persistence and bioefficacy of Imidacloprid and beta cyfluthrin on okra. Part II. *Pestology* 26: 30–37.

Duncan FD, Krasnov B, McMaster M (2002) Novel case of a tenebrionid beetle using discontinuous gas exchange cycle when dehydrated. *Physiol Entomol* 27: 79–83.

Eisikowitch D (1981) Some aspects of pollination of oilseed rape (*Brassica napus* L.). *J Agric Sci* 96: 321–326.

Eurostat (2009) http://epp.eurostat.ec.europa.eu
Ferguson F (1987) Long term effects of systemic pesticides on honey bees. *Austr Beekeeper* 89: 49–54.
Frediani D, Pinzauti M, Caterini B (1987) Influence of the entomophilous pollination on the oilseed rape production (*Brassica napus* L. spp. *oleifera*). *Apicoltura* 3: 63–73.
Free JB (1993) *Insect pollination of crops*, 2nd edn. Academic Press, London, UK.
Free JB, Nuttall PM (1968) The pollination of oilseed rape (*Brassica napus*) and the behaviour of bees on the crop. *J Agric Sci* 71: 91–94.
Free JB, Williams IH (1972) The transport of pollen on the body hair of honeybees (*Apis mellifera* L.) and bumblebees (*Bombus* spp. L.). *J Appl Ecol* 9: 609–615.
Gabriel D, Tscharntke T (2007) Insect pollinated plants benefit from organic farming. *Agr Ecosyst Environ* 118: 43–48.
Gels JE, Held DW, Potter DA (2002) Hazards of insecticides to the bumble bees *Bombus impatiens* (Hymenoptera: Apidae) foraging on flowering white clover in turf. *J Econ Entomol* 95: 722–728.
Gerolt P (1976) The mode of action of insecticides: Accelerated water loss and reduced respiration in insecticide-treated *Musca domestica* L. *Pestic Sci* 7: 604–620.
Gerolt P (1983) Insecticides: Their route to entry, mechanisms of transport and mode of action. *Biol Rev* 58: 233–274.
Gibbs AG, Johnson RA (2004) The role of discontinuous gas exchange in insects: The chtonic hypothesis does not hold water. *J Exp Biol* 207: 3477–3482.
Goulson D, Hanley ME, Darvill B, Ellis JS, Knight ME (2005) Causes of rarity in bumblebees. *Biol Conserv* 122: 1–8.
Guez D, Suchail S, Gauthier M, Maleszka R, Belzunces LP (2001) Contrasting effects of imidacloprid on habituation in 7- and 8-day-old honeybees (*Apis mellifera*). *Neurobiol Learn Mem* 76: 183–191.
Hall L, Topinka K, Huffman J, Davis L, Good A (2000) Pollen flow between herbicide-resistant *Brassica napus* is the cause of multiple-resistant *B. napus* volunteers. *Weed Sci* 48: 688–694.
Harak M, Lamprecht I, Kuusik A, Hiiesaar K, Metspalu L, Tartes U (1999) Calorimetric investigations of insect metabolism and development under the influence of a toxic plant extract. *Thermochim Acta* 333: 39–48.
Hasler A, Maurizio A (1950) Ueber den Einfluss verschiedener Nährstoffe auf Blütenansatz, Nektarsekretion und Samenertrag von honigenden Pflanzen, speziell von Sommerraps (*Brassica napus* L.). *Schweiz Landw Mh* 6: 201–211.
Haynes KF (1988) Sublethal effects of neurotoxic insecticides on insect behaviour. *Annu Rev Entomol* 33: 149–168.
Herbert EW (1992) Honey bee nutrition. In: Graham JM (ed.) *The hive and the honey bee*. Dadant & Sons, Hamilton, Illinois, USA.
Hetz SK, Bradley TJ (2005) Insects breathe discontinuously to avoid oxygen toxicity. *Nature* 433: 516–519.
Hokkanen HMT, Menzler-Hokkanen I, Butt TM (2003) Pathogens of oilseed rape pests. In: Alford DV (ed.) *Biocontrol of oilseed rape pests*. Blackwell, Oxford, UK.
Hrassnigg N, Crailsheim K (1998) The influence of brood on the pollen consumption of worker bees (*Apis mellifera* L.). *J Insect Physiol* 44: 393–404.
Husberg G-B, Hokkanen HMT (2000) Effects of *Metarhizium anisopliae* on the pollen beetle *Meligethes aeneus* and its parasitoids *Phradis morionellus* and *Diospilus capito*. *Biocontrol* 46: 261–273.
Jagers op Akkerhuis GAJM, Kjær C, Damgaard C, Elmegaard N (1999b) Temperature-dependent, time-dose-effect model for pesticide effects on growing, herbivorous arthropods: Bioassays with dimethoate and cypermethrin. *Environ Toxicol Chem* 18: 2370–2378.
Jagers op Akkerhuis GAJM, Seidelin N, Kjaer C (1999a) Are we analysing knockdown in the right way? How independence of the knockdown-recovery process from mortality may affect measures for behavioural effects in pesticide bioassays. *Pestic Sci* 55: 62–68.

Jagers op Akkerhuis GAJM, Westerhof R, Straalen NMvan, Koeman JH (1995) Water balance, respiration and immobilisation in relation to deltamethrin poisoning and physical conditions in the epigeal spider *Oedothorax apicatus*. *Pestic Sci* 44: 123–130.

Jaycox E (1964) Effect on honey bees of nectar from systemic insecticides treated plants. *J Econ Entomol* 57: 31–35.

Johansen CA, Mayer DF (1990) *Pollinator protection: A bee and pesticide handbook*. Wicwas Press, Cheshire, Conn.

Jõgar K, Kuusik A, Metspalu L, Hiiesaar K, Grishakova M, Luik A (2008) Effects of Neem EC on gas exchange, tracheal ventilation, and water loss in diapausing pupae of *Pieris brassicae*. *Ent Exp Appl* 2: 165–173.

Kanga LHB, Jones WA, James RR (2003) Field trials using the fungal pathogen, *Metarhizium anisopliae* (Deuteromycetes: Hyphomycetes) to control the ectoparasitic mite *Varroa destructor* (Acari: Varroidae) in honey bee, *Apis mellifera* (Hymenoptera: Apidae) colonies. *J Econ Entomol* 96: 1091–1099.

Karise R, Kuusik A, Mänd M, Metspalu L, Williams IH, Hiiesaar K, Luik A, Muljar R, Liiv K (2010) Gas exchange patterns of bumble bee foragers before and after exposing to lowered temperature. *J Insect Physiol* (in press).

Karise R, Mänd M, Koskor E, Bender A (2006) The effect of Neem EC on the pollen forage of the bumble bee *Bombus terrestris* L., *Proc Int Conf: Information Systems in Sustainable Agriculture, Agrienvironment and Food Technology, Volos, Greece, 2006*, pp 367–372.

Karise R, Mänd M, Viik E, Martin A-J, Lääniste P (2004) Flower visitors of spring oilseed rape in different cropping systems. *Latvian J Agronomy* 7: 6–11.

Karise R, Viik E, Mänd M (2007) Impact of alpha-cypermethrin on honey bees foraging on spring oilseed rape *Brassica napus* flowers in field conditions. *Pest Manag Sci* 63: 1085–1089.

Kestler P (1991) Cyclic CO_2 release as a physiological stress indicator in insects. *Comp Biochem Phys C* 100: 207–211.

Kevan PG, Eisikowitch D (1990) The effect of insect pollination on canola (*Brassica napus* L. cv. OAC Triton) seed germination. *Euphytica* 45: 33–41.

Klein AM, Steffan-Dewenter I, Tscharntke T (2003) Fruit set of highland coffee increases with the diversity of pollinating bees. *Proc R Soc B* 270: 955–961.

Klein A-M, Vaissiere BE, Cane JH, Steffan-Dewenter I, Cunningham SA, Kremen C, Tscharntke T (2007) Importance of pollinators in changing landscapes for world crops. *Proc R Soc B* 274: 303–313.

Klowden MJ (2002) *Physiological systems in insects*. Academic Press, San Diego, USA.

Kovac H, Stabentheiner A, Hetz SK, Petz M, Crailsheim K (2007) Respiration of resting honeybees. *J Insect Physiol* 53: 1250–1261.

Kremen C, Williams NM, Bugg RL, Fay JP, Thorp RW (2004) The area requirements on an ecosystem service: Crop pollination by native bee communities in California. *Ecol Lett* 7: 1109–1119.

Kühne S (2008) Prospects and limits of botanical insecticides in organic farming. *Agronomski Glasnik* 4: 377–382.

Ladurner E, Bosch J, Maini S, Kemp WP (2003) A method to feed individual bees (Hymenoptera: Apiformes) known amounts of pesticides. *Apidologie* 34: 597–602.

Levy RI, Schneiderman HA (1966) Discontinuous respiration in insects. 2. The direct measurement and significance of changes in tracheal gas composition during the respiratory cycle of silkworm pupae. *J Insect Physiol* 12: 83–104.

Lighton JRB (1994) Discontinuous ventilation in terrestrial insects. *Physiol Zool* 67: 142–162.

Lighton JRB (1996) Discontinuous gas exchange in insects. *Annu Rev Entomol* 41: 309–324.

Lighton JRB (1998) Notes from the underground: Towards ultimate hypotheses of cyclic, discontinuous gas-exchange in tracheate arthropods. *Am Zool* 38: 483–491.

Lighton JRB (2007) Why insects evolved discontinuous gas exchange. *Curr Biol* 17: 645–647.

Lighton JRB, Lovegrove BGL (1990) A temperature-induced switch from diffusive to convective ventilation in the honeybee. *J Exp Biol* 154: 509–516.

Lighton JRB, Schilman PE, Holway DA (2004) The hyperoxic switch: Assessing respiratory water loss rates in tracheate arthropods with continuous gas exchange. *J Exp Biol* 207: 4463–4471.

Lodesani M, Costa C, Nipoti P, Prodi A (2003) Use of three mitosporic fungi for the control of the honey bee parasitic mite *Varroa destructor* Anderson & Trueman. *Proc XXXVIIth Int Apic Cong, Ljubiana, Slovenia*, pp 422–423.

Marais E, Chown SL (2003) Repeatability of standard metabolic rate and gas exchange characteristics in a highly variable cockroach, *Perisphaeria* sp. *J Exp Biol* 206: 4565–4574.

Mesquida J, Pham-Delegue MH, Marilleau R, le Metayer M, Renard M (1991) The floral secretion in male-sterile cybrid of winter rapeseed (*Brassica napus* L.). *Agronomie* 11: 217–227 (in French).

Mesquida J, Renard M (1979) Preliminary results on the pollination of male sterile winter rape and methods for production of hybrid seed. *Bull Inf Tech CETIOM* 65: 3–14.

Mesquida J, Renard M (1981) Pollinisation du colza d'hiver mâle-fertile et mâle-sterile (*Brassica napus* L. *var. oleifera* Metzger) par i'abeille domestique (*Apis m. mellifera* L.) Effects sur la phenologie et le rendement. *Apidologie* 12: 345–362.

Mesquida J, Renard M (1982) Etude de la dispersion du pollen par le vent et de l'importance de la pollinisation anemophilie chez le colza (*Brassica napus* L. *var. oleifera* Metzger). *Apidologie* 13: 353–367.

Mesquida J, Renard M, Pierre JS (1988) Rapeseed (*Brassica napus* L.) productivity: The effect of honey bees (*Apis mellifera* L.) and different pollination conditions in cage and field tests. *Apidologie* 19: 51–72.

Meyerhoff G (1958) The foraging behaviour of honeybee on rape. *Leipzig Bienenztg* 72: 164–165 (in German).

Miranda JE, Navickiene HMD, Nogueira-Couto RH, De Bortoli SA, Kato MJ, Bolzani VdaS, Furlan M (2003) Susceptibility of *Apis mellifera* (Hymenoptera: Apidae) to pellitorine, an amide isolated from *Piper tuberculatum* (Piperaceae). *Apidologie* 34: 409–415.

Mishra RC, Kaushic HD (1992) Effect of cross pollination on yield and oil content of *Brassica* spp. and *Eruca sativa* with pollination efficiency of honeybees. *Ann Entomol* 10: 33–37.

Muranjan PN, Gandhale DN, Chaudhari CS, Patil BD, Pokharkar DS, Naik RL (2006) Toxocity of ready-mix formulations of pyrethroids and ad-mixed insecticides on forager honey bee, *Apis cerana indica* Fabricius. *Ann Plant Prot Sci* 14(1) Print ISSN: 0971–3573.

Mänd M, Kuusik A, Martin AJ, Williams IH, Luik A, Karise R, Metspalu L, Hiiesaar K (2005) Discontinuous gas exchange cycles and active ventilation in pupae of the bumblebee *Bombus terrestris*. *Apidologie* 36: 561–570.

Mänd M, Kuusik A, Martin AJ, Williams IH, Luik A, Karise R, Metspalu L, Hiiesaar K (2006) Regular periods of abdominal contractions recorded from larvae of the bumblebee, *Bombus terrestris* L. (Hymenoptera: Apidae). *Eur J Entomol* 103: 319–322.

Mänd M, Mänd R, Williams IH (2002) Bumblebees in the agricultural landscape of Estonia. *Agr Ecosyst Environ* 89: 69–76.

Öckinger E, Smith HG (2007) Semi-natural grasslands as population sources for pollinating insects in agricultural landscapes. *J Appl Ecol* 44: 50–59.

Osborne JL, Williams IH, Corbet SA (1991) Bees, pollination and habitat change in the European community. *Bee World* 72: 99–116.

Pajot S (2001) Dossier Gaucho. *Abielles et Fleurs* 616: 160–165.

Peng G, Sutton JC, Kevan PG (1992) Effectiveness of honey bees for applying the biological control agent *Gliocladium roseum* to strawberry flowers to suppress *B. cinerea*. *Can J Plant Pathol* 14: 117–129.

Pernal SF, Currie RW (1998) Nectar quality in open-pollinated, pol CMS hybrid and dominant SI hybrid oilseed summer rape. *Can J Plant Sci* 78: 79–89.

Pertl M, Hauser TP, Damgaard C, Jørgensen RB (2002) Male fitness of oilseed rape (*B. napus*), weedy *B. rapa* and their F_1 hybrids when pollinating *B. rapa* seeds. *Heredity* 89: 212–218.

Pierre J, Mesquida J, Marilleau R, Pham-Delegue MH, Renard M (1999) Nectar secretion in winter oilseed rape, *Brassica napus* – quantitative and qualitative variability among 71 genotypes. *Plant Breed* 118: 471–476.

Pinochet X, Bertrand R (2000) Oilseed rape grain yield productivity increase with hybrid varietal types: A first balance sheet with post registration tests in France and Europe. *OCL-Oleagineux Corps Gras Lipides* 7: 11–16.

Pinzauti M (1994) Pollinator behaviour and activity in colonies of *Apis mellifera* in confined environments. *Ethol Ecol Evol* 3: 101–106.

Radchenko TG (1964) The influence of pollination on the crop and the quality of the seed of winter rape. *Bdzhil'nitstvo* 1: 68–74 (in Ukrainian).

Ramirez-Romero R, Chaufaux J, Pham-Delègue MH (2005) Effects of Cry1Ab protoxin, deltamethrin and imidacloprid on the foraging activity and the learning performances of the honeybee *Apis mellifera*, a comparative approach. *Apidologie* 36: 601–611.

Ramsay G, Thompson C, Squire G (2003) *Quantifying Landscape-Scale Gene Flow in Oilseed Rape*. Department for Environment, Food and Rural Affairs, London, UK.

Rathcke BJ (1992) Nectar distributions, pollinator behaviour, and plant reproductive success. In: Hunter MD, Ohgushi T, Price RPW (eds.) *The effect of Resource Distribution on Animal-Plant Interactions*. Academic Press, London, UK.

Riaz A, Li G, Quresh Z, Swati MS, Quiros CF (2001) Genetic diversity of oilseed *Brassica napus* inbred lines based on sequence-related amplified polymorphism and its relation to hybrid performance. *Plant Breed* 120: 411–415.

Rieger M, Lamond M, Preston C, Powles SBN, Roush RT (2002) Pollen mediated movement of herbicide resistance between commercial canola fields. *Science* 296: 2386–2388.

Rives M (1957) Etudes sur la selection du Colza d'hiver. *Ann Amelior Plantes* 1: 61–107.

Sabbahi R, De Olivera D, Marceau J (2005) Influence of honey bee (Hymenoptera: Apidae) density on the production of canola (Crucifera: Brassicacae). *J Econ Entomol* 98: 367–372.

Sandu DK, Waraich MK (1985) Yeasts associated with pollinating bees and flower nectar. *Microbial Ecol* 11: 51–58.

Scheffler JA, Parkinson R, Dale PJ (1995) Evaluating the effectiveness of isolation distances for field plots of oilseed rape (*Brassica napus*) using a herbicide-resistance transgene as a selectable marker. *Plant Breeding* 114: 317–321.

Schmuck R, Schöning R, Stork A, Schramel O (2001) Risk posed to honeybees (*Apis mellifera* L, Hymenoptera) by an imidacloprid seed dressing of sunflowers. *Pest Manag Sci* 57: 225–238.

Schricker B, Stephen WP (1970) The effect of sublethal doses of parathion on honeybee behavior. I. Oral administration and the communication dance. *J Apicult Res* 9: 141–153.

Sepp K, Mikk M, Mänd M, Truu J (2004) Bumblebee communities as an indicator for landscape monitoring in the agri-environmental programme. *Landscape Urban Plann* 67: 173–183.

Shaw KE, Davidson G, Clark SJ, Ball BV, Pell JK, Chandler D, Sunderland K (2002) Laboratory bioassays to assess the pathogenicity of mitosporic fungi to *Varroa destructor*, an ectoparasitic mite of the honeybee, *Apis mellifera*. *Biol Control* 24: 266–276.

Slama K (1999) Active regulation of insect respiration. *Ann Entomol Soc Am* 92: 916–929.

Slama K, Sobotnik J, Hanus R (2007) Respiratory concerts revealed by scanning microrespirography in a termite *Prorhinotermes simplex* (Isoptera: Rhinotermitidae). *J Insect Physiol* 53: 295–311.

Steffan-Dewenter I (2003) Seed set of male-sterile and male-fertile oilseed rape (*Brassica napus*) in relation to pollinator density. *Apidologie* 34: 227–235.

Stoner A, Wilson WT, Harvey J (1985) Acephate (Orthene): Effects on honey bee queen, brood and worker survival. *Am Bee J* 125: 448–450.

Sur R, Stork A (2003) Uptake, translocation and metabolism of imidacloprid in plants. *Bull Insectol* 56: 35–40.

Tasei JN (2001) Effects of insect growth regulators on honey bees and non-*Apis* bees. *Apidologie* 32: 527–545.

Tasei JN, Carré C, Moscatelli B, Grondeau C (1988) Recherche de la DL50 de la deltaméthrine (Decis) chez: *Megachile rotundata* F., abeille pollinisatrice de la luzerne et des effets de doses infraléthales sur les adultes et les larves. *Apidologie* 19: 291–306.

Tasei JN, Lerin J, Ripault G (2000) Sub-lethal effects of imidacloprid on bumblebees, *Bombus terrestris* (Hymenoptera: Apidae), during laboratory feeding test. *Pest Manag Sci* 56: 784–788.

Thompson HM (2001) Assessing the exposure and toxicity of pesticides to bumblebees (*Bombus* sp.). *Apidologie* 32: 305–321.

Thompson HM (2003) Behavioural effects of pesticides in bees – their potential for use in risk assessment. *Ecotoxicology* 12: 317–330.

Thompson HM, Hunt LV (1999) Extrapolating from honeybees to bumble bees in pesticide risk assessment. *Ecotoxicology* 8: 147–166.

Thompson HM, Maus C (2007) The relevance of sublethal effects in honey bee testing for pesticide risk assessment. *Pest Manag Sci* 63: 1058–1061.

Thompson HM, Wilkins S, Battersby AH, Waite RJ, Wilkinson D (2005) The effects of four insect growth regulating (IGR) insecticides on honeybee (*Apis mellifera*) (L.) colony development, queen rearing and drone sperm production. *Ecotoxicology* 14: 757–769.

Thompson HM, Wilkins S, Battersby AH, Waite RJ, Wilkinson D (2007) Modelling long term effects of IGRs on honey bee colonies. *Pest Manag Sci* 63: 1081–1084.

Thomson SV, Shotwell KM, Vandenberg JD (1990) The distribution of antagonistic bacteria by honey bees for biological control of fire blight. *Phytopathol* 80: 1017.

Timmons AM, O'Brain ET, Chatters YM, Dubbels SJ, Wilkinson MJ (1995) Assessing the risks of wind pollination from fields of genetically modified *Brassica napus* ssp. *oleifera*. *Euphytica* 85: 417–423.

van der Steen JJM, de Ruijter A (1990) Test methods to determine the hazards of IGRs (insect growth regulators) on honey bee brood, using phenoxycarb as an example, *Proc 4th Int Symp Harmonization of Methods for Testing the Toxicity of Pesticides to Bees, May 15–18, Rez near Prague, Research Institute of Apiculture, Dol, Czechoslovakia 1990*, pp 88–91.

Vandame R, Meled M, Colin ME, Belzunces LP (1995) Alteration of the homing-flight in the honey bee *Apis mellifera* L. exposed to sublethal dose of deltamethrin. *Environ Toxicol Chem* 14: 855–860.

Vänninen I, Tyni-Juslin J, Hokkanen H (2000) Persistence of augmented *Metarhizium aniosopliae* and *Beauveria bassiana* in Finnish agricultural soils. *Biocontrol* 45: 201–222.

Varis A-L (1995) Abundance, species composition and daily pattern of bees visiting field bean, goat's rue and turnip rape in southern Finland. *Agricult Sci in Finland* 4: 473–478.

Vogt JT, Appel AG (2000) Discontinuous gas exchange in the fire ant, *Solenopsis invicta* Buren: Caste differences and temperature effects. *J Insect Physiol* 46: 403–416.

Westcott L, Nelson D (2001) Canola pollination: An update. *Bee World* 82: 115–129.

Westphal C, Steffan-Dewenter I, Tscharntke T (2003) Mass flowering crops enhanced pollinator densities at a landscape scale. *Ecol Lett* 6: 961–965.

Westphal C, Steffan-Dewenter I, Tscharntke T (2009) Mass flowering oilseed rape improves early colony growth but not sexual reproduction of bumblebees. *J Appl Ecol* 46: 187–193.

Wilkinson MJ, Elliot LJ, Allainguillaume J, Shaw W, Norris C, Welters R, Alexander M, Sweet J, Mason DC (2003) Hybridization between *Brassica napus* and *B. rapa* on a national scale in the United Kingdom. *Science* 302: 457–459.

Williams IH (1978) The pollination requirements of swede rape (*Brassica napus*) and turnip rape (*Brassica campestris*). *J. Agric Sci* 91: 343–348.

Williams IH (1980) Oilseed rape and beekeeping, particular in Britain. *Bee World* 61(4): 141–153.

Williams IH (1984) The concentrations of air-borne rape pollen over a crop of oil-seed rape (*Brassica napus* L.). *J Agric Sci* 103: 353–357.

Williams IH (1985) The pollination of swede rape (*Brassica napus* L.). *Bee World* 66: 16–22.

Williams PH (1986) Environmental change and the distributions of British bumble bees (*Bombus* Latr.). *Bee World* 67: 50–61.

Williams IH (1994) The dependence of crop production within the European Union on pollination by honeybees. *Agricult Sci Rev* 6: 229–257.

Williams IH (1996) Aspects of bee diversity and crop pollination in the European Union. In: Matheson A, Buchmann SL, O'Toole C, Westrich C, Williams IH (eds.) *Conservation of Bees*. Academic Press for IBRA and the Linnean Society, London, UK.

Williams IH (2004) Advances in insect pest management of oilseed rape in Europe. In: Horowitz AR, Ishaaya I (eds.) *Insect pest management: Field and protected crops*. Springer-Verlag, Berlin, Heidelberg.

Williams IH, Büchs W, Hokkanen H, Menzler-Hokkanen I, Johnen A, Klukowski Z, Luik A, Nilsson C, Ulber B (2005) MASTER – Integrating biological control within IPM for winter oilseed rape across Europe. *Proc BCPC Int Cong, Crop Sci Techn, Glasgow, 31 Oct–2 Nov 2005*, pp 1, 301–308.

Williams IH, Carreck N, Little DJ (1993) Nectar sources for honey bees and the movement of honey bee colonies for crop pollination and honey production in England. *Bee World* 74: 160–175.

Williams IH, Cook VA (1982) The beekeeping potential of oilseed rape. *BBJ* 110: 68–70.

Williams IH, Corbet SA, Osborne JL (1991) Beekeeping, wild bees and pollination in the European community. *Bee World* 72(4): 170–180.

Williams IH, Free JB (1978) The feeding and mating behaviour of pollen beetles (*Meligethes aeneus* Fab.) and seed weevils (*Ceutorhynchus assimilis* Payk.) on oilseed rape (*Brassica napus* L.). *J Agri Sci* 91: 453–459.

Williams IH, Free JB (1979) Compensation of oil-seed rape (*Brassica napus* L.) plants after damage to their buds and pods. *J Agr Sci* 92: 53–59.

Williams IH, Martin AP, Ferguson AW, Clark SJ (1987) The effect of insect pollination on plant development and seed production in winter oil-seed rape (*Brassica napus* L.). *J Agr Sci* 109: 135–139.

Williams IH, Martin AP, White RP (1986) The pollination requirements of oil-seed rape (*Brassica napus* L.). *J Agr Sci* 106: 27–30.

Williams IH, Simpkins J (1989) Honeybee pollination of the double low oil-seed rape line Ariana. *Asp Appl Biol* 23: 343–346.

Winston ML (1987) The biology of the honey bee. Harvard University Press, Cambridge, Massachusetts, USA.

Wobschall A, Hetz SK (2004) Oxygen uptake by convection and diffusion in diapausing moth pupae (*Attacus atlas*). *Int Cong Series* 1275: 157–164.

Yang EC, Chuang YC, Chen YL, Chang LH (2008) Abnormal foraging behaviour induced by sublethal dosage of imidacloprid in the honey bee (Hymenoptera: Apidae). *J Econ Entomol* 101: 1743–1748.

Yu H, Sutton JC (1997) Effectiveness of bumble bees and honey bees for delivering inoculum of *Gliocladium roseum* to raspberry flowers to control *B. cinerea*. *Biol Control* 10: 113–122.

Zehnder G, Gurr GM, Kühne S, Wade MR, Wratten SD, Wyss E (2007) Arthropod pest management in organic crops. *Annu Rev Entomol* 52: 57–80.

Chapter 15
The proPlant Decision Support System: Phenological Models for the Major Pests of Oilseed Rape and Their Key Parasitoids in Europe

Andreas Johnen, Ingrid H. Williams, Christer Nilsson, Zdzisław Klukowski, Anne Luik, and Bernd Ulber

Abstract The computer-based DSS proPlant incorporates weather-based phenological models for the six major pests of oilseed rape in Europe: the cabbage stem flea beetle, the rape stem weevil, the cabbage stem weevil, the pollen beetle, the cabbage seed weevil and the brassica pod midge. The models have now been validated for several European countries. The system predicts the start and course of pest infestation and provides site-specific crop protection treatment decisions, a selection of suitable chemicals, computes optimum dates and rates for application and evaluates the efficacy of past applications. Phenological models for migration of some key parasitoids of the pests have also been constructed recently; their use in finding spray windows to conserve parasitoids is described. Further work is needed to validate them before they can be integrated into a pest control strategy for farmers.

15.1 The proPlant DSS

The proPlant DSS is a commercially-available decision support system that provides plant protection advice for oilseed rape, as well as for cereals, potatoes and sugar beet (Epke et al. 1996, Frahm et al. 1996, Newe et al. 2003). The software was developed in Germany by The Agricultural Computer Science Institute, Münster, in cooperation with the Chamber of Agriculture, Westfalen-Lippe, and upgraded by proPlant GmbH. It can be either PC-based or internet-based (www.proPlantexpert.com). proPlant was first used in Germany in 2001 but is now also used in Austria, Belgium, Czech Republic, Finland, France, Hungary, The Netherlands, Sweden and Switzerland. In 2008, by mid-year, users had consulted the proPlant internet services 67,000 times and sought information from the website 410,000 times.

The system requires information inputs about the crop, about pest densities on the crop and about local daily weather data to predict the need for pest control

A. Johnen (✉)
proPlant GmbH, Münster 48147, Germany
e-mail: A.Johnen@proPlant.de

I.H. Williams (ed.), *Biocontrol-Based Integrated Management of Oilseed Rape Pests*,
DOI 10.1007/978-90-481-3983-5_15,

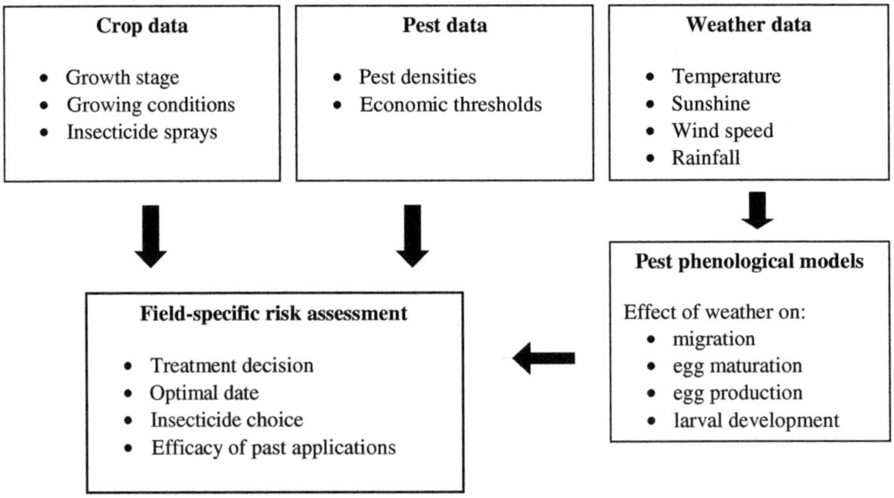

Fig. 15.1 Inputs required for a field-specific risk assessment and crop protection advice by the proPlant DSS for winter oilseed rape

(Fig. 15.1). It automatically imports regional weather data with a 3-day forecast either from the Internet or from a local weather station. Using these inputs the system produces a field-specific risk assessment of the need for insecticide application, provides advice on suitable chemicals and optimum dates and rates for their application, and evaluates the efficacy of past applications (Johnen and Meier 2000).

Recent developments within the EU-funded project MASTER (Williams et al. 2005, Williams 2006) include the validation in other countries of the proPlant pest phenological models developed for spring/summer pests in Germany (Johnen et al. 2006a) and the construction of phenological models for some of the most important parasitoids of the major pests (Johnen et al. 2006b).

15.2 Pest Phenological Models

The proPlant DSS for oilseed rape incorporates phenological models for the six major European insect pests: the cabbage stem flea beetle (*Psylliodes chrysocephala* (L.)), the pollen beetle (*Meligethes aeneus* (Fabricius)), the rape stem weevil (*Ceutorhynchus napi* Gyllenhal), the cabbage stem weevil (*Ceutorhynchus pallidactylus* (Marsham) syn. *C. quadridens* (Panzer)), the cabbage seed weevil (*Ceutorhynchus obstrictus* (Marsham) syn. *C. assimilis* (Paykull)), and the brassica pod midge (*Dasineura brassicae* Winnertz). The models were developed using data from the literature and from 8 years of crop monitoring in various regions of Germany. They predict pest infestation based on weather parameters; the parameters used are daily maximum, minimum and mean temperatures at 2 m, minimum temperature 20 cm above ground, sunshine hours, wind speed and rainfall.

When work on the models began, there was little information on the effects of weather on pest phenology. The strategy for pest control in oilseed rape in Germany, as in most other countries in Europe, was based solely on control thresholds (e.g., the numbers of adult beetles or weevils counted on plants or collected in yellow water traps). This strategy works only when the threshold value relates to the damaging stage of the pest, as for adult pollen beetles. Problems arise when it relates to a non-damaging stage, for e.g., when adults are monitored but the larvae cause the damage, as for cabbage stem flea beetle or cabbage stem weevil. Further problems arise when larval infestation cannot be measured or treatment against the larvae is impossible. Monitoring the number of adults then introduces uncertainties because the potential for reproduction is ignored. Assessment of the need for control and the timing of field inspections and pesticide applications can be improved only if the effects of weather on pest phenology are understood and incorporated.

Several key development stages for each pest are considered in the modelling. The models calculate the start and end of migration of each pest into the crop, daily conditions for flight, and provide an estimate of the percentage of adult pests that have migrated to the crop by a specific date. After arrival on the crop, they calculate the time needed for ovary maturation and the beginning, end and intensity of the egg-laying periods for the cabbage stem flea beetle and the stem weevils. For the cabbage stem flea beetle, the conditions for egg maturation, beginning of egg hatch and the rate of larval development are also calculated. Knowledge representation and processing by the system is rule-based. The rules define specific conditions that result in varying treatment decisions ('if then rules'). This method allows many parameters to be taken into consideration in the decision-making and new knowledge to be incorporated as it becomes available.

The derived data guide the decision about whether a treatment is needed by giving advice on the choice of dates for risk assessment, field inspection and insecticide application.

15.2.1 For Autumn Pests

In the autumn, the cabbage stem flea beetle is a major pest on winter rape. Because the larval stage is the most damaging, the likely larval density to establish, given a known adult pest density and known weather conditions, must be estimated. Previously, the number of adult beetles caught in yellow water traps had been used to predict larval density. However, a large catch of beetles does not necessarily lead to a high larval density (Büchs 1992, Hossfeld 1993, Nilsson 2002) because temperature influences both the number of eggs laid in September and October and the proportion of them that hatch (Bonnemaison and Jourdheuil 1954).

The influence of temperature on migration, egg-laying, egg maturation and larval density in autumn in two different years is shown in Fig. 15.2. The density of adult beetles was similar in both years. However, in 1993, low temperatures from October onwards limited egg-laying and delayed egg hatch so few eggs hatched in the autumn. By contrast, in 1994, the milder weather allowed egg-laying to continue

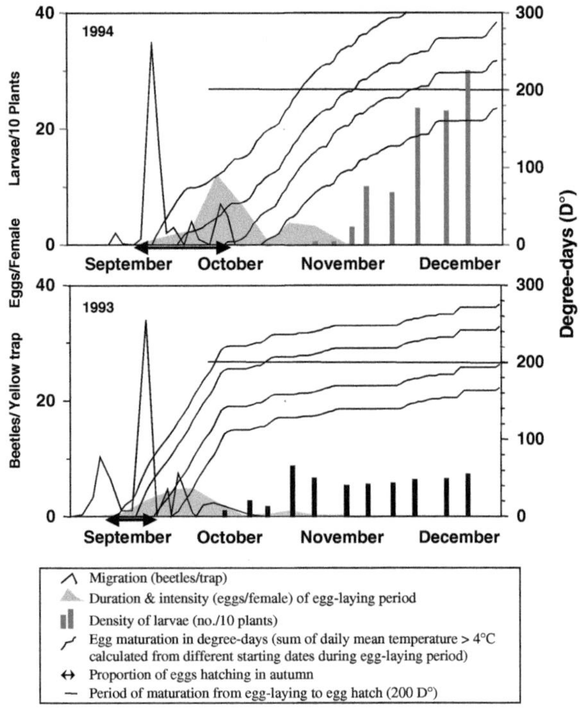

Fig. 15.2 Influence of autumn temperatures on the migration, egg-laying, egg maturation and larval density of the cabbage stem flea beetle (*Psylliodes chrysocephala*) in 1993 and 1994

until November, a greater percentage of them hatched and a greater larval density resulted.

Table 15.1 shows how proPlant uses weather data to identify years when crops are at risk from damage by cabbage stem flea beetle larvae. Temperatures in September and October are important for both migration to the crop and for egg-laying. Key points in development of this pest are the beginning of egg hatch and the dates of occurrence of the second and third larval instars. These are calculated from daily temperature sums; the appearance of the first larvae requires a sum of 200 degree-days (D°) of daily mean temperature > 4°C. Thus, in 2005 and 2006, for example, conditions were optimal for migratory flight during September, as well as for egg-laying and rapid egg and larval development during October, so crops were at risk of damage from this pest in those years. The dates of occurrence of later larval instars are important for decision-making as it is these larger larvae that are the most damaging and the efficiency of insecticides decreases as the larvae grow (Lane et al. 1995).

15.2.2 For Spring Pests

The spring pests (pollen beetle, rape stem weevil, cabbage stem weevil, cabbage seed weevil and brassica pod midge) migrate to winter rape over a long period of

Table 15.1 Weather conditions and the population dynamics of the cabbage stem flea beetle (*Psylliodes chrysocephala*)

Year	Difference between maximum temperature in September and long-term average (°C)	Difference between mean temperature in October and long-term average (°C)	Day degree sum of daily mean temperature above 4°C in October (°D)
1996	−3	0	190
1997	+1	−1	140
1998	0	−0.5	160
1999*	+4	0	190
2000*	0	+1.5	220
2001*	−3	+4	310
2002	0	−1	150
2003	+1	−3	100
2004	+0.5	−0.5	225
2005*	+1.5	+1.5	270
2006*	+3.5	+2	300
2007	−2	−1.5	175
2008	−2	−1.5	185

Years marked * indicate good conditions for migration and egg-laying (September and October temperatures) and/or for egg and larval development in autumn (day degree sum in October).

2–3 months (February/March–May) (Fig. 15.3). Their emergence from overwintering sites is based on accumulated temperature in spring. The daily average of the air temperature is used to calculate this with different base temperatures for the accumulation for each pest. The temperature sum for the last 5 days only is used to analyse periods with a continuous temperature increase which activates pests overwintering in the soil, and is used to analyse the risk for the first pest migration to the crop for a particular day.

Pest arrival on the crop is usually in sequence, first the stem weevils, then pollen beetle, cabbage seed weevil and finally brassica pod midge as they differ in their temperature requirements for emergence and flight. The program assumes that they have all left their over-wintering sites within a few days of favourable or optimal flight conditions.

In the phenological models, migration to the crop is defined by the start and the end of flight, i.e., by the first and last days of adult migration to the crop. Combinations of several weather parameters (daily maximum temperature, daily mean temperature, wind speed and sunshine hours) are used to calculate the suitability of weather conditions for migration; this is expressed as a flight index ranging from 0 when conditions make flight impossible to 3 when conditions are optimal for flight. Table 15.2 illustrates how the selected parameters of temperature and sunshine determine the probability of migration in the cabbage stem weevil. Wind speed is used in addition either as an excluding parameter (migration can be excluded even on days with temperature and sunshine conditions for optimal migration possibilities by higher wind speed) or a diminishing factor of the flight indices listed in Table 15.2. To calculate the end of migration, the system generates a maximum sum

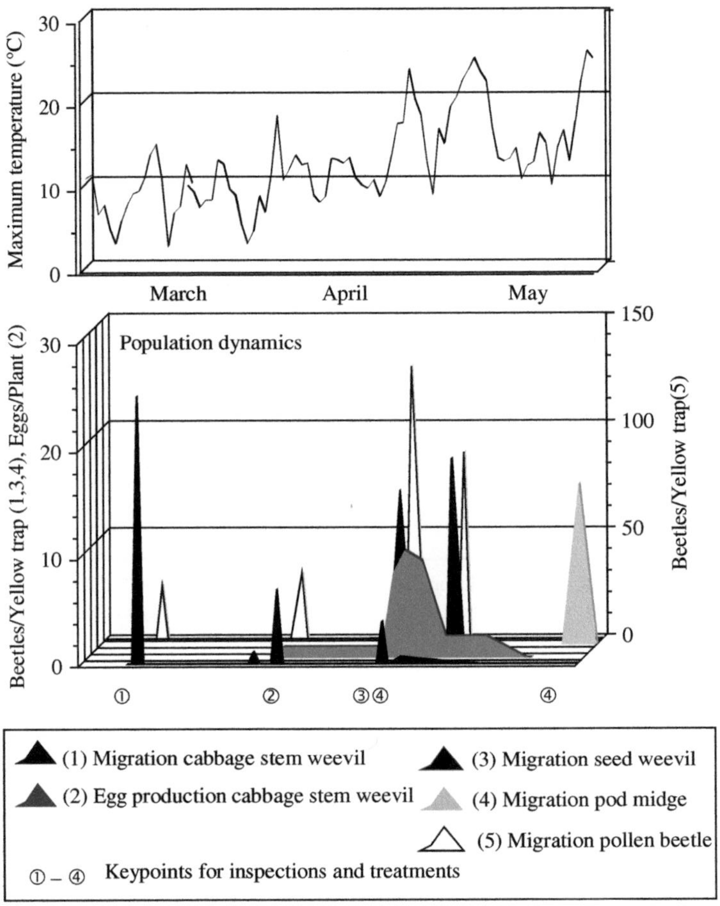

Fig. 15.3 Population dynamics of spring pests, in relation to the temperature in 1995, and important phenological dates for risk assessments, field inspections and treatments

of the flight indices for each pest. With this maximum summed index, it is possible to estimate the percentage of adults that have already migrated to the crop and the increase in pest density in the crop.

In most years, migration flights of early spring pests (pollen beetles, stem weevils) to the crop can be monitored, using yellow water traps, over the same period of time during March to April whereas flowering pests (cabbage seed weevil, brassica pod midge) migrate later during late April–May. Figure 15.3 indicates the important phenological dates for crop inspections and treatments in 1995. In this year, the stem weevils and the pollen beetle were the first pests to migrate to the crop during March (Fig. 15.3 point 1) although the main migration of the pollen beetle was not until mid-April when the maximum temperature exceeded 20°C (Fig. 15.3 point 3). The migration of the cabbage seed weevil and the brassica pod midge started in late April (Fig. 15.4 point 3/4).

Table 15.2 Examples of combinations of selected weather parameters that allow migration of the cabbage stem weevil (*Ceutorhynchus pallidactylus*) into winter oilseed rape

Weather parameters			
Maximum temperature (2 m) °C	Mean temperature (2 m) °C	Sunshine (h)	Flight index
≤ 10.5	–	–	0
> 10.5 ≤ 12	< 5	> 3 ≤ 7	0
	≥ 5	> 3 ≤ 7	1
	< 9	≤ 3	0
	≥ 9	≤ 3	1
>12 ≤ 15	< 5	> 3 ≤ 7	1
	≥ 5	> 3 ≤ 7	2
	< 9	≤ 3	0
	≥ 9	≤ 3	1
	–	> 7	2
> 15	–	≤ 3	2
	–	> 3	3

When the flight index (FI) = 0 migration does not occur. Conditions for migration are low when FI = 1, good when FI = 2 and optimal when FI = 3. Wind speed is also used for the migration forecast as an exclusion or diminishing factor.

After arrival on the crop, a period of time is needed for the ovaries of adult stem weevil females to mature before egg-laying can begin. This period is determined largely by daily minimum and maximum temperatures. On cold nights, weevils may leave the plants, becoming active again on days with higher temperatures. Cabbage stem weevil females, for example, require 100 degree-days of the daily maximum temperature before they are ready to lay eggs. This accumulated temperature takes into account only days with a difference between minimum and maximum temperature > 9°C (e.g., if the minimum temperature at night is −2°C, the maximum day temperature must be ≥ 11°C). For the intensity of egg-laying, daily indices are generated that are comparable with the flight indices. The total egg-laying potential is then calculated as a summed index.

The stem weevils damage plants as eggs or larvae, and, because pyrethroid insecticides do not kill larvae in plant stems, treatment must prevent egg-laying by the adults. The length of the ovary maturation period and the intensity of egg production depend on temperature and daily sunshine hours. Cool, cloudy weather can delay egg-laying for several weeks after the adults arrive in the crop (Fig. 15.3 point 3).

15.2.3 Validation in Other Countries

The proPlant phenological models developed for spring/summer pests in Germany were recently validated for use in other European countries differing in climate by comparing the phenology of pest activity in the field with that forecast by proPlant (Johnen et al. 2006a). Pest activity in crops of winter oilseed rape was monitored

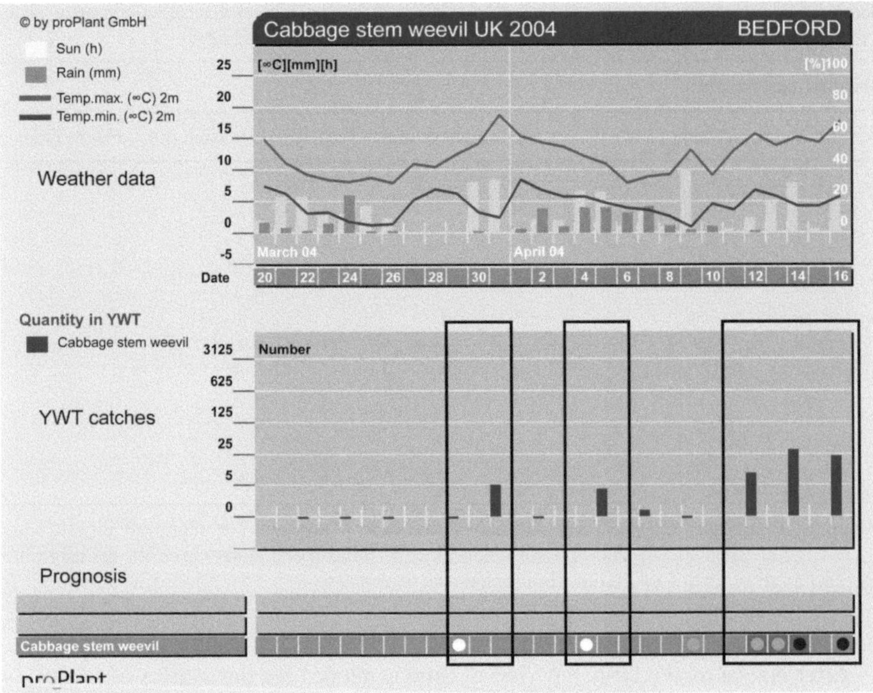

Fig. 15.4 Computer screen display of the test version of the proPlant DSS used for the validation of the existing phenological pest models. The example shows a comparison of the catches for cabbage stem weevil in the UK in 2004 with the prognosis of the proPlant DSS. The *upper graph* displays weather data between 20 March and 16 April 2004: sunshine (h), light *grey bars*; rain (mm), *dark grey bars*; maximum temperature at 2 m (°C), *upper line* and minimum temperature at 2 m (°C), *lower line*. The lower graph shows actual numbers of the weevil caught in *yellow* water traps. The prognosis indicates days when weather conditions were computed to be good for migration (*lighter dots*) or optimal for migration (*darker dots*) or at least allowed some migration (*white dots*). Days when migration was possible do not show in *greyscale*

for 4 years (2002–2005) in Germany and Sweden, and for 3 years (2003–2005) in Poland and the UK, using yellow water traps placed and maintained at canopy level in the crop. Traps were emptied three times each week. The datasets provided information on the phenology of pest migration into crops and their activity densities within them. A total of 7–13 data sets for each pest were obtained for evaluation (Table 15.3); less data was available for the rape stem weevil which occurs in Germany and Poland, but not in Sweden or the UK, and, in some years, some pests were caught in such low numbers that these data sets were omitted from the evaluation.

The pest activity data obtained from the field were compared with migration forecasts computed by proPlant. Daily weather data were imported from official weather stations located nearest to the trial site (Germany–Göttingen, Sweden–Falsterbo, Poland–Wroclaw II, UK–Bedford). A test version of proPlant expert.classic was

Table 15.3 Numbers of yellow water trap year data sets and of each pest species (in brackets) obtained in each country for validation of proPlant pest models

	Germany	Sweden	UK	Poland	Total
Rape stem weevil	4 (6,772)	–	–	3 (1,695)	7 (8,467)
Cabbage stem weevil	4 (5,713)	3 (1,724)	3 (1,332)	3 (3,319)	13 (12,088)
Pollen beetle	3 (2,049)	4 (18,790)	3 (43,915)	3 (2,32,221)	13 (2,96,975)
Cabbage seed weevil	3 (3,946)	3 (1,120)	3 (3,399)	3 (30,244)	12 (38,709)
Brassica pod midge	2 (1,623)	4 (6,747)	3 (9,441)	3 (5,308)	12 (23,119)

developed which showed recent weather data together with predicted migration and pest catch at the trial sites. By importing the local weather data and the yellow water trap catches for all trial sites it was possible to compare the results of the proPlant phenological models with the actual catches directly and thereby check the accuracy of the models. As an example, Fig. 15.4 shows the screen display from the test version for the cabbage stem weevil.

proPlant accurately forecast the beginning of migration for four of the five pests in three of the four countries. In Germany, UK and Poland, the dates that proPlant predicted for the beginning of migration of rape stem weevil (only in Germany and Poland), cabbage stem weevil, pollen beetle, and cabbage seed weevil into winter rape matched exactly the dates when the first insects were caught in the yellow traps, in all years. In Sweden however, temperatures were very low in two of the 4 years, and cabbage stem weevils and pollen beetles were caught in the yellow water traps earlier than the dates predicted by proPlant for the beginning of migration. The reasons for this remain unclear; it may be that the weather station chosen did not represent the trial region well. By contrast, the beginning of cabbage seed weevil migration in Sweden, as indicated by trap catches, did match the proPlant forecast in all years.

proPlant predictions as to periods of main migration and activity as well as the end of migration, based on yellow water trap catches, were accurate for all pests in all countries and for each year. The only exception was for brassica pod midge, for which they misjudged migration to a certain extent. They did not forecast all of the periods of migration and activity indicated by yellow trap catches. This was attributed to the fact that the system ruled out pod midge migration for periods when the maximum temperature was low ($< 20°C$) while average temperatures were high. These rules have now been modified, significantly improving the accuracy of forecasts for midge migration.

In conclusion, the proPlant system requires only minimum adjustment to adapt it for field use in other countries. These trials confirmed that the weather-based rules proPlant uses to predict migration of major pests into winter oilseed rape produce accurate results, not only for Germany, but also for the climates that prevail in Sweden, Poland and the UK. Further evaluations in France have verified the system for the warmer climate there, and the French rape growers' organisation, Cetiom, now offers advice, based on the proPlant phenological models, via the internet to its members (www.cetiom.fr).

Table 15.4 proPlant decision tree for optimised pest control in spring/summer

Key point 1: Start of pollen beetle and stem weevil immigration
Does pollen beetle infestation exceed threshold? i.e., 2 beetles/plant at G.S 51– 8 beetles/plant at G.S. 59 in normal crops. In thin or backward crops the values can be halved
No ⇩ Yes ⇨ TREATMENT
Does cabbage stem flea beetle larval infestation exceed threshold? i.e., > 80% of the leaf petioles infested in normal crops, > 60% in thin or backward crops
No ⇩ Yes ⇨ TREATMENT
Key point 2: Start of egg-laying by the stem weevils
Does pollen beetle infestation exceed threshold? i.e., 2 beetles/plant at GS 51 (green bud) to 8 beetles/plant at GS 59 (late yellow bud) in normal crops. In thin or backward crops the values can be halved
No ⇩ Yes ⇨ TREATMENT
Do stem weevil infestations exceed thresholds? i.e., >10 rape stem weevils or > 30 cabbage stem weevils per yellow water trap in 3 days
No ⇩ Yes ⇨ TREATMENT
Key point 3: Before the egg-laying peak of the stem weevils
Do stem weevil infestations exceed thresholds? i.e., >10 rape stem weevil/yellow water trap, or > 30 cabbage stem weevil per yellow water trap
No ⇩ Yes ⇨ TREATMENT
Key point 4: Start of immigration of the cabbage seed weevil
Is infestation by cabbage seed weevil high this year and/or was infestation by brassica pod midge high last year? i.e., 1 weevil/plant (this year) + small midge infestation (last year) . 0.5 weevil/plant (this year) + high midge infestation (last year)
No ⇩ Yes ⇨ TREATMENT
NO TREATMENT

15.3 Field-Specific Risk Assessment

The proPlant DSS integrates the computed data from the pest phenological models with input data on pest densities and crop growth to produce the field-specific risk assessment with advice on the need for insecticide application (Fig. 15.1).

In the autumn, pest density of the cabbage stem flea beetle on the crop is first established by monitoring using yellow water traps. If very few adults are caught in the traps (fewer than a total of 50 beetles/trap captured during the whole migration period starting with the first migration), a decision is made even before egg-laying has begun and no treatment is recommended, because, even with good weather conditions for egg production and larval development during autumn, a risk for damaging larval densities can be excluded in most years (Hossfeld 1993). Only in years with abnormally high temperatures in winter (e.g., 2006/2007), is an additional control of the larval attack needed when there are fewer than 50 beetles/trap for validation. If moderate numbers are caught (sum of 50–100 beetles per trap), no decision can yet be made, and future conditions for egg-laying must be considered. If very large numbers of beetles are caught in the traps (a total of more than 100 beetles per trap), an application of insecticide is advised since, even under unfavourable conditions for egg-laying, a larval density sufficient to cause economic damage would be expected.

After this first analysis, the weather conditions for egg-laying and egg maturation by the cabbage stem flea beetle are evaluated as described above. A summed index is calculated to describe the egg-laying period and this is used as a basis for the treatment decision. For this sum, the daily calculated indexes for egg laying based on weather data are used. Days with weather conditions not allowing or limiting egg laying (e.g., days with mean temperatures < 10°C or days with higher temperatures but rainfall without sunshine hours) have an index '0' or '1', whereas days with good or optimal conditions are valued by the system with '2' or '3'. In September and the first 2 weeks of October, the temperatures are normally high enough not to limit egg laying and the main reason for the differences in the summed index between the years are the temperatures in the second half of October. For example, in 1993, when low temperatures limited egg-laying and delayed egg maturation (Fig. 15.2) application of an insecticide was computed to be unnecessary. By contrast, in 1994, when milder conditions allowed egg-laying to continue into November but most of the eggs did not hatch in the autumn (Fig. 15.2) and application of an insecticide was recommended in the spring for fields with normal crops and, in autumn only, in fields with thin or backward crops. Thus, depending on site and season, if the larvae are still in their first or second instar stages in early autumn, the system will recommend that the application of insecticide is delayed in the autumn or, even until early spring.

In the spring, the system assesses the need for treatment of every pest, again from the weather-derived pest index, crop data and pest density data, taking into account established control thresholds for each pest (Fig. 15.1). It connects the pest-specific treatment decisions with the treatment dates to reach an optimised decision.

The spring risk assessment defines four phenological key points for crop inspections and treatments (Fig. 15.3 and Table 15.4).

Key point 1 marks the start of immigration of the stem weevils and the pollen beetle. At this point, early treatment against the pollen beetle is recommended only if their numbers per plant exceed the control thresholds or if numbers of cabbage stem flea beetle larvae are high. The stem weevils do not yet require control as their ovaries are immature (see below) and they are not ready to lay eggs.

Keypoint 2 marks the start of egg-laying by the stem weevils. At this point, treatment may be required against the stem weevils if they exceed their threshold values in yellow water traps. The rape stem weevil starts egg-laying earlier because the process of maturation runs more quickly than for cabbage stem weevil and is more damaging than the cabbage stem weevil, as deposition of only one rape stem weevil egg per plant can cause severe stem deformation. It therefore has a lower control threshold and requires treatment earlier, before the start of egg-laying. Pollen beetle may also require control if their numbers on the crop exceed the threshold.

Keypoint 3 is before the peak egg-laying period of the stem weevils. At this point, the stem weevils must be controlled at the latest if their numbers exceed threshold levels in yellow water traps. If daily maximum temperatures > 20°C then this may co-incide with the main migration of pollen beetles into the crop and they will also be killed by any treatment applied against the stem weevils that is also effective against the pollen beetle. If egg-laying of stem weevils is delayed until late April, keypoint 3 may co-incide with keypoint 4.

Key point 4 marks the start of immigration of the cabbage seed weevil. At this point, the cabbage seed weevil/ brassica pod midge complex may require control if the seed weevil threshold is breached or if the crop is at risk of midge attack. It occurs after the main migration of the pollen beetle and during the period of egg-laying by the stem weevils, but by this time it is too late to control them.

The prolonged period of migration of spring pests into the crop has meant that several applications of insecticide are often made to the crop. However, the proPlant system aims at delaying the first spring treatment for as long as possible, until later migrating pests have arrived, thereby often reducing the number of insecticide applications needed. For example, if cabbage stem flea beetle infestation is not severe in the spring, and weather conditions are not yet conducive to egg-laying by stem weevils, insecticide treatment can be delayed until required to control pollen beetle and cabbage seed weevil. In many years, if the first critical egg-laying period of the cabbage stem weevil, related to temperatures > 20°C, occurs in April (Fig. 15.3 point 3/4), it is possible to control this weevil simultaneously with the main migration of pollen beetle adults and the start of seed weevil migration.

This 'delaying tactic' does not work in every year. In some circumstances, when an early treatment is recommended by the program, a second treatment cannot always be avoided, notably when (i) pollen beetle adults and cabbage stem flea beetle larvae need to be treated early in the season, (ii) the rape stem weevil is the dominant stem weevil, (iii) stem weevil egg production starts soon after first emergence (e.g., if March temperatures are high) but migration of pollen beetle and cabbage seed weevil does not occur until April.

Decisions vary with crop parameters. Figure 15.5 shows how the outcome can vary with crop growth stage (GS) when all other parameters are the same. In this example, the monitored density of rape stem weevil was between 10–30 weevils per trap (medium). The weather-based phenological pest models calculated that the majority (>70%) of weevils had migrated to the crop so that no further large-scale migration to increase pest density on the crop would occur, and that conditions for egg-laying were optimal (egg-laying index >5). The assessments and the decisions

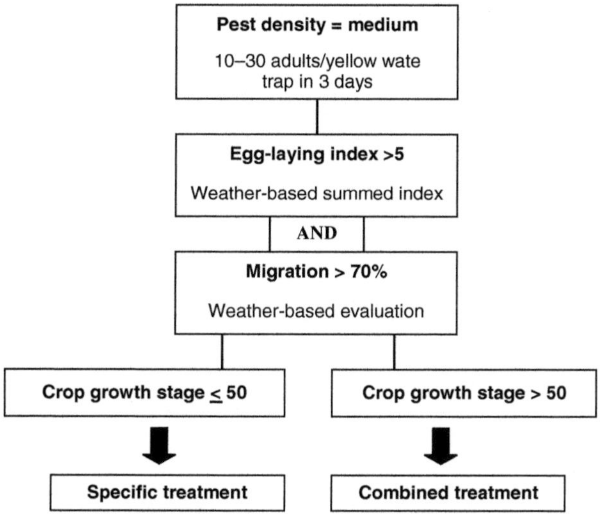

Fig. 15.5 Effect of crop growth stage (GS) on proPlant decision on the field-specific need for treatment against the rape stem weevil (*Ceutorhynchus napi*). GS 50 = flower buds present, still enclosed by leaves (Meier 2001)

differ according to the GS. Early egg-laying in young plants increases the risk of damaging attacks. Therefore, at GS $\leq$ 50 a specific treatment is advised, whereas at GS > 50 a combined treatment is advised, i.e., treatment is recommended only if a specific treatment is also advised for another pest.

15.4 Parasitoid Models

Data on the phenologies of occurrence, migration and activity of the key hymenopteran parasitoids of oilseed rape pests were collected over a period of 4 years (2002–2005) in Germany and Sweden and for a period of 3 years (2003–2005) in Poland and the UK, using the same yellow water traps as used for validation of the pest models (Johnen et al. 2006b). The traps were placed and maintained at canopy height in winter oilseed rape crops and emptied three times each week. Sufficient data were collected to develop models for six parasitoid species, based on a total of 35,794 individual parasitoids caught (see Table 15.5).

Daily weather data from the official meteorological services nearest to the trapping location were used to analyze the weather conditions during the migration periods of these parasitoids to winter rape crops. To define the relationships between weather parameters initiating migration of parasitoids, the same system of 'if then rules' was used, as used to develop the migration models for oil seed pests in the proPlant DSS. These rules described combinations of weather parameters (maximum temperature, mean temperature, wind speed and sunshine hours) and assigned them to an flight index.

Table 15.5 Numbers of yellow water trap year data sets and of individuals of each species of parasitoid (in brackets) used to develop phenological models for parasitoids of oilseed rape pests

Pest	Parasitoid	Germany	Sweden	UK	Poland	Total
Cabbage stem flea beetle	*Tersilochus microgaster*	3 (182)	1 (183)	1 (27)	0	5 (392)
Cabbage stem weevil	*Tersilochus obscurator*	4 (444)	3 (410)	1 (157)	2 (160)	10 (1,171)
Pollen beetle	*Tersilochus heterocerus*	1 (41)	4 (9,099)	3 (7,379)	2 (110)	10 (16,629)
	Phradis interstitialis	2 (63)	3 (3,114)	3 (822)	2 (160)	10 (4,319)
Cabbage seed weevil	*Trichomalus perfectus*	3 (423)	1 (179)	3 (575)	2 (180)	9 (1,357)
Brassica pod midge	*Platygaster subuliformis*	1 (56)	4 (2,734)	2 (8,963)	1 (173)	8 (11,926)

As with the pests, the order in which their key parasitoids were first caught in the yellow water traps in winter rape was identical in every country and in every year. The first parasitoid to be caught in winter oilseed rape was the cabbage stem flea beetle parasitoid *Tersilochus microgaster* (Szépligeti) (Hymenoptera, Ichneumonidae) in the second half of March and April. The cabbage stem weevil parasitoid *Tersilochus obscurator* Aubert (Hymenoptera, Ichneumonidae) followed in late March in some years, but more normally in the second half of April. The pollen beetle parasitoids *Phradis interstitialis* (Thomson) (Hymenoptera, Ichneumonidae) and *Tersilochus heterocerus* Thomson (Hymenoptera, Ichneumonidae) migrated into crops, just before or at the start of flowering, followed a little later by the pod midge parasitoid *Platygaster subuliformis* (Kieffer) (Hymenoptera, Platygastridae). The last species to migrate into crops was the cabbage seed weevil parasitoid *Trichomalus perfectus* (Walker) (Hymenoptera, Pteromalidae) which was caught in large numbers in June. As all of these parasitoids, except *T. perfectus* overwinter in the soil, these differences in their emergence times reflect the different soil temperatures they require for emergence.

However, direct measurement of soil temperature is problematical, as it can vary widely with soil depth and type, so it was not used as a parameter in developing the parasitoid phenological models. Instead, the maximum temperature at 2 m above ground was used. Numbers of parasitoids caught in the yellow water traps increased during periods of increasing daily maximum temperature (often in combination with intensive radiation) but decreased during periods when daily maximum temperatures remained constant. From a biological point of view this was an indirect effect. Extended periods of increasing air temperatures and sunshine warm the soil, stimulating parasitoid emergence.

Each parasitoid species was first classified to a temperature category conducive to and optimal for its migratory flight to winter rape (see Fig. 15.6). Ambient maximum temperatures of 14°C (optimum temperature for *T. microgaster*) to 23°C (optimum temperature for *T. perfectus*) were found to be adequate for all species to migrate.

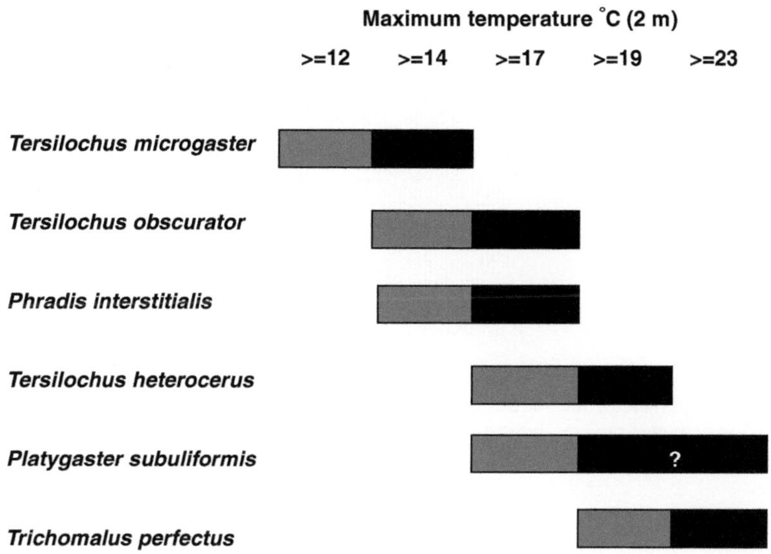

Fig. 15.6 Classification of parasitoids to temperature categories conducive to (*grey*), or optimal (*black*), for migration

Maximum temperatures for flight of *T. obscurator* and *P. interstitialis* were similar. Migration of *T. obscurator* normally starts a little earlier than that of *P. interstitialis* but the periods and the peaks of migration are the same. Catch data indicated that the migration periods of *T. heterocerus* and *P. subuliformis* were similar, although respective peak activities could vary in some years, peak catch of *P. subuliformis* sometimes occurring later than that of *T. heterocerus*. The reason for these years with later peaks is not yet clear, but further field data may help refine the temperature categories for *P. subuliformis*.

Variability in climate in different regions of Europe affects the start dates of the main migration of parasitoids into winter rape crops in different countries. The earliest and latest dates in 2003–2005 on which there was a strong increase in the numbers of parasitoids of each species caught in the yellow water traps, indicating the start of the main migration period, in Germany, Poland and the UK were compared with those in Sweden where the climate is colder in the spring. This clearly showed the effect of climatic variability on parasitoid phenology, the later warming of the soil in spring in Sweden delaying the emergence of parasitoids (Fig. 15.7).

In Germany, Poland and the UK, maximum daily temperatures exceeded 14°C, the optimum temperature for *T. microgaster* flight, in the second half of March enabling migration to winter rape to start. By contrast, in Sweden where temperatures were still lower in March/April, the start of migration by *T. microgaster* did not occur until later, in mid April/early May. The start of migration of *T. obscurator* and *P. interstitialis* occurred almost a month later than that of *T. microgaster*

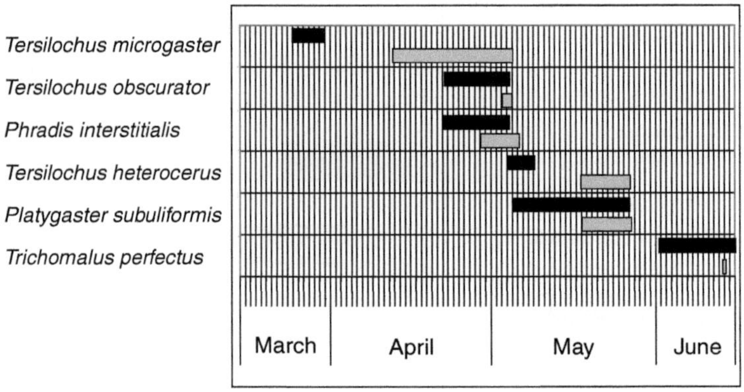

Fig. 15.7 Comparison of the start of migration (earliest and latest dates) by different species of parasitoid into winter rape, as indicated by the increasing numbers caught in yellow water trap catches in 2003–2005, in Germany, Poland and the UK (*black*) compared with Sweden (*grey*)

in Germany, Poland and the UK and not until late April/early May and early May, respectively, in Sweden; their optimum temperature for flight is > 17°C. The captured number of *T. heterocerus* and *P. subuliformis* increased first in yellow water traps in May when temperatures exceeded 19°C, while that of *T. perfectus* increased later in June when temperatures exceeded 23°C, all three species migrating later in Sweden than in Germany, Poland and the UK. Differences in the dates of first flights of parasitoids between Sweden (with later and cooler spring temperature increase) and the other countries were consistently greater for the early-migrating parasitoids, e.g., *T. microgaster*, than for the later-migrating species, e.g., *T. perfectus*. This was probably because temperature differences between Sweden and the other countries gradually decreased from spring to summer.

To compute the start of migration in the phenological models, it was important to consider maximum temperatures over a 3-day period: the previous day (D) and the 2 days before that (D-1 and D-2) as yellow water trap catches increased with rising temperatures and decreased when temperature remained constant over a 3-day period. Within the categories of maximum temperature, the flight indices for each parasitoid were further defined by use of the daily mean temperature at 2 m and sunshine hours. A flight index (FI) was assigned to each combination of weather parameters that permitted flight. A FI = 1 indicated that flight was possible but at a low level, a FI = 2 indicated conditions for flight were good, and an FI = 3 indicated that conditions were optimal. Wind speed at 2 m was used as a factor for exclusion of flight only; too high a wind speed (> 6.5 m/s) prohibits flight of all species of parasitoids even if all other weather conditions are optimal. In this way, flight tables were produced for each of the key parasitoids (Tables 15.6, 15.7, 15.8, and 15.9). The flight tables are based on a comparison of the data sets of the yellow water trap samples (see Table 15.5) with the weather data of the next weather station. This was not done with a statistical method but with expert knowledge analysing periods with and without migration and the related weather characteristics in these

Table 15.6 Combinations of weather parameters that allow migration of the parasitoid *Tersilochus microgaster* into winter oilseed rape

Weather parameters			
Maximum temperature (2 m) °C	Mean temperature (2 m) °C	Sunshine (h)	Flight index
>= 10	>= 2.5	>= 10	1
>=12	>=5.0	>=7	1
	>=5.0	>=10	2
	>=2.5	>=7	1
	>=2.5	>=10	2
	>=5.0	>=5	1
	>=5.0	>=7	2
	>=9.0	>=3	1
	>=9.0	>=5	2
	>=9.0	>=7	3
>=14	–	>=1	1
	–	>=3	2
	–	>=5	3

Conditions for migration are low when flight index (FI) = 1, good when FI = 2 and optimal when FI = 3.

Table 15.7 Combinations of weather parameters that allow migration of the parasitoids *Tersilochus obscurator* and *Phradis interstitialis* into winter oilseed rape

Weather parameters			
Maximum temperature (2 m) °C	Mean temperature (2 m) °C	Sunshine (h)	Flight index
>= 14	>= 7	>= 4	1
	>= 7	>= 8	2
	>=10	>= 2	1
	>=10	>= 4	2
>= 16	>= 7	>= 2	1
	>= 7	>= 4	2
	>= 7	>= 8	3
	>=10	>= 2	2
	>=10	>= 4	3
>= 19	>= 7	>= 1	1
	>= 7	>= 2	2
	>= 7	>= 4	3
	>=10	>= 1	2
	>=10	>= 2	3

Conditions for migration are low when flight index (FI) = 1, good when FI = 2 and optimal when FI = 3.

periods. Recurring periods without parasitoids captured in the traps were for example used to decide on the minimum temperature level to allow migration. After that, periods with low or no captured parasitoids and temperatures above that level were used to find other weather parameters that limit the migration even with sufficient temperatures such as sunshine hours or wind speed.

Table 15.8 Combinations of weather parameters that allow migration of the parasitoid *Tersilochus heterocerus* and *P. subuliformis* into winter oilseed rape

Weather parameters			
Maximum temperature (2 m) °C	Mean temperature (2 m) °C	Sunshine (h)	Flight index
>=14	>=10	>=4	1
	>=10	>=8	2
	>=12	>=2	1
	>=12	>=4	2
>=17	>=10	>=2	1
	>=10	>=4	2
	>=10	>=8	3
	>=12	>=1	1
	>=12	>=2	2
	>=12	>=4	3
>=19	>=10	>=1	1
	>=10	>=2	2
	>=10	>=4	3
	>=12	>=1	2
	>=12	>=2	3

Conditions for migration are low when flight index (FI) = 1, good when FI = 2 and optimal when FI = 3.

Table 15.9 Combinations of weather parameters that allow migration of the parasitoid *Trichomalus perfectus* into winter oilseed rape

Weather parameters			
Maximum temperature (2 m) °C	Mean temperature (2 m) °C	Sunshine (h)	Flight index
>=17	>=12	>=8	1
	>=14	>=4	1
>=19	>=12	>=4	1
	>=14	>=2	1
	>=14	>=4	2
>=23	>=14	>=2	2
	>=14	>=4	3
	>=20	>=1	2
	>=20	>=2	3

Conditions for migration are low when flight index (FI) = 1, good when FI = 2 and optimal when FI = 3.

For *T. microgaster* (Table 15.6), conditions for the start of migration are fulfilled when (i) the maximum temperature on Day D is 3°C higher than that on D-1 or D-2, or (ii) the maximum temperature on Day D is at least 1°C higher than that on D-1 or D-2 and the mean temperature on Day D is at least 5°C. However, even when the conditions for the start of migration are fulfilled, flight is excluded if (i) the maximum temperature on Day D < 10°C and (ii) the mean temperature of Day D < 2.5°C.

For *T. obscurator* and *P. interstitialis* (Table 15.7), conditions for the start of migration are fulfilled when (i) the maximum temperature on Day D is 3°C higher than that on D-1 or D-2, or (ii) the maximum temperature on Day D is at least 1°C higher than that on D-1 or D-2 and the mean temperature on Day D is at least 10°C. However, even when the conditions for the start of migration are fulfilled, flight is excluded if (i) the maximum temperature on Day D < 14°C and (ii) the mean temperature of Day D < 7°C.

For *T. heterocerus* and *P. subuliformis* (Table 15.8), conditions for the start of migration are fulfilled when (i) the maximum temperature on Day D is 3°C higher than that on D-1 or D-2, or (ii) the maximum temperature on Day D is at least 1°C higher than that on D-1 or D-2 and the mean temperature on Day D is at least 12°C. However, even when the conditions for the start of migration are fulfilled, flight is excluded if (i) the maximum temperature on Day D < 14°C and (ii) the mean temperature of Day D < 10°C.

For *T. perfectus* (Table 15.9), conditions for the start of migration are fulfilled when (i) the maximum temperature on Day D is 3°C higher than that on D-1 or D-2, or (ii) the maximum temperature on Day D is at least 1°C higher than that on D-1 or D-2 and the mean temperature on Day D is at least 14°C. However, even when the conditions for the start of migration are fulfilled, flight is excluded if (i) the maximum temperature on Day D < 17°C and (ii) the mean temperature of Day D < 12°C.

The phenological migration flight tables for parasitoids deduced by the analysis of the yellow water trap catches and the weather data were integrated into the proPlant DSS for validation. A comparison of parasitoid catches in yellow water traps and the prediction by proPlant DSS showed good compliance for all six species of parasitoid. Prediction of migration based on weather data can therefore be expected to help optimize insecticide application dates and to define spray windows, which allow the application of insecticides at times that minimize harm to parasitoids.

15.5 Spray Windows

Where treatment with insecticide is essential, it should be applied at a time that maximises pest control while minimising harm to their parasitoids (see also Thieme et al. Chapter 12 this volume and Ulber et al. Chapter 13 this volume). This is most effectively achieved when it takes account of the phenology of the pests as well as the phenology of their parasitoids rather than when made at a fixed date or crop growth stage.

The combined parasitoid complex in winter rape is active over a long period during the spring/summer from mid-March to the end of June (Fig. 15.7); individual species also have long periods of activity and several peaks of migration. To avoid killing any parasitoids, insecticide applications would need to be applied outside this period of activity which is impractical for control of the pests. A compromise must be sought to find spray windows that allow insecticide application consistent with good pest control but that minimize effects on parasitoids.

Studies of the phenology of the cabbage seed weevil and its key parasitoid *T. perfectus* have shown that there is dissociation between the immigration flights of pest and parasitoid, the former arriving before the latter (Murchie et al. 1997). This provides an opportunity for the temporal targeting of pyrethroids before the main migration period of the parasitoids. Avoidance of treatments post-flowering, when the parasitoids are most active, conserves natural populations of *T. perfectus* and has been shown to be effective and economically-viable as a method for controlling cabbage seed weevil populations in UK (Alford et al. 1996).

The use of phenological models can help find spray windows during this period which allow some parasitoid activity to continue. For example, a spray window that opens before the main period of parasitoid migration but closes before the end of parasitoid migration into the crop would allow parasitization by later-migrating parasitoids. In general, if treatment is carried out at the beginning of migration, it will have less effect on those parasitoids that migrate at a later date, because of the decreasing efficacy of the applied chemical.

In the proPlant DSS, the definition of spray windows with deduction of activity days based on weather data has been integrated for testing. Figure 15.8 shows an example for the cabbage stem weevil in Germany in 2003. Below the conditions for migration and egg laying of the cabbage stem weevil, the activity periods (or days) of the cabbage stem flea beetle parasitoid *T. microgaster* and of the cabbage stem weevil parasitoid *T. obscurator* are shown. In this example, the proPlant screen display indicates that insecticide against the cabbage stem weevil should be

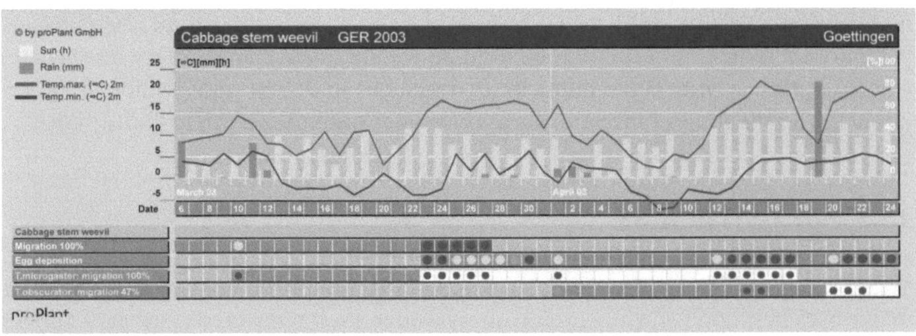

Fig. 15.8 proPlant computer screen display showing integration of phenological models for key parasitoids in winter oilseed rape and deduced spray windows. The upper graph displays weather data between 6 March and 24 April 2003: sunshine (h), *light grey bars*; rain (mm), *dark grey bars*; maximum temperature at 2 m (°C), *upper line* and minimum temperature at 2 m (°C), *lower line*. The *upper two horizontal bars* show weather-based proPlant predictions of flight activity (*upper*) and egg-laying (*lower*) by the cabbage stem weevil during these dates. The *lower two horizontal bars* show weather-based proPlant predictions of flight activity by *T. microgaster* (*upper*) and by *T. obscurator*. The *dots* indicate days when pest and parasitoid flight or pest egg-laying is possible/good (*lighter dot*) or optimal (*darker dot*). Days when migration was possible do not show in *greyscale*. The lighter parts of the parasitoid *bars* show time periods when no insecticide should be applied to conserve parasitoids, i.e., between 23 March and 15 April for *T. microgaster* and between 20 and 24 April for *T. obscurator*

applied early, before 22 March, to minimize the effects on parasitisation of cabbage stem flea beetle larvae by *T. microgaster.* A second application of insecticide against pollen beetle and/or flowering pests, could be made between 16 April (end of the *T. microgaster* spray window) and 19 April (start of the *T. obscurator* spray window). The effectiveness of different insecticide application dates, as recommended by the proPlant phenological models on pest control with parasitoid conservation, now need to be checked in field trials.

Greater understanding of the diel periodicities of pests and parasitoids within the crop may help define times of day when parasitoids are less vulnerable than pests to insecticides. Ferguson et al. (2010) found that, during flowering, peak flight activity of the pollen beetle and its parasitoid *P. interstitialis* and of *T. obscurator* (a parasitoid of the cabbage stem weevil) was around midday and that few insects were caught before 10.00 h. Further work on pest and parasitoid behaviour and vulnerability within the crop canopy is needed to determine whether application of insecticide in early morning (or late evening) would help conserve parasitoids without compromising control of the pests.

Thus, spray windows defined by weather-based phenological models and diel periodicities of pests and their parasitoids have potential to help farmers and advisors in the future to decide when to apply insecticide to control oil seed rape pests at times when they will have minimum effects on parasitoids thereby supporting the general trend towards reduced pesticide use and greater reliance on biological control by naturally-occurring parasitoids.

15.6 Implications for Biocontrol-Based Integrated Pest Management

The prophylactic application of a chemical insecticide to kill insect pests, especially when it applied to the entire rape crop, is wasteful of resources and harmful to the environment; it may also be counterproductive if it kills natural enemies of the pests. The use of economic thresholds, crop monitoring and computer-based decision support systems (DSSs) help decide on the need for pest control measures to be taken, allow more efficient targeting of insecticide in time and space, and help define spray windows that target the pests while conserving their parasitoids.

Use of the proPlant DSS by German farmers and consultants has shown that the increased use of pest thresholds in combination with the phenological models increases precision of insecticide application to selected crops during the predicted activity periods of the pests. It also reduces the number of insecticide treatments applied to crops, thereby reducing the time expended on and the cost of crop protection (Johnen and Meier 2000); current practice in Germany usually involves two or three applications of insecticide each season whereas using proPlant only one or two applications were recommended, as the system aims at delaying the first spring treatment until pests that migrate later have arrived.

The integration of economic thresholds, crop monitoring and computer-based DSSs into pest management strategies helps to make insecticide application more efficient and compatible with the conservation of natural enemies.

Acknowledgements The project MASTER (QLK5-CT-2001-01447) was part-funded by the EU under its Framework 5 Quality of Life and Management of Living Resources programme. We thank all who helped with sampling and insect identification for this work, particularly the following: Britt Ahman, Annely Kalle, Hassan Barari, Bertil Christensson, Andrew Ferguson, Pierre Fernandez, Kirsa Fischer, Dave Frearson, Ross Holdgate, Mariusz Kaczmarzyk, Reelika Kevvai, Beate Klander, Neil Mason, Helen Nuss, Ross Piper, Emma Smith, Tiiu Tarang, Eve Veromann, Nigel Watts and Rainer Wedemeyer.

References

Alford DV, Walters KFA, Williams IH, Murchie AK (1996) A commercially viable low-cost strategy for the management of seed weevil populations on winter oilseed rape in the UK. *Proc British Crop Prot Conf, Pests and Diseases, Brighton,* pp 609–614.

Bonnemaison L, Jourdheuil P (1954) L'altise d'hiver du colza (*Psylliodes chrysocephala* L.). *Ann Ēpiphyties* 5: 345–524.

Büchs W (1992) Insektizide gegen Rapserdfloh einsparen. *Der Pflanzenarzt* 45: 23–25.

Epke K, Michel M, Volk T (1996) Entwicklung des wissensbasierten Beratungssystems PRO_PLANT für den umweltschonenden Einsatz von Pflanzenschutzmitteln in der Landwirtschaft. *Forschung und Beratung Reihe C* 50: 154.

Ferguson AW, Holdgate R, Mason NS, Clark SJ, Williams IH (2010) Phenologies and diel periodicities of within-crop flight by pests and parasitoids in oilseed rape. *IOBC/wprs Bull* (in press).

Frahm J, Johnen A, Volk T (1996) Development of the PRO_PLANT decision support system for plant protection in cereals, sugar beet and rape. *EPPO Bull* 26: 609–622.

Hossfeld R (1993) Die Gelbschale als Entscheidungshilfe bei der Bekämpfung des Rapserdflohs (*Psylliodes chrysocephala* L.). *Gesunde Pflanzen* 45: 291–295.

Johnen A, Meier H (2000). A weather-based decision support system for managing oilseed rape pests. *Proc BCPC Conf-Pests & Diseases* 2000, 2: 793–800.

Johnen A, Williams IH, Ferguson AW, Büchs W, Klukowski Z, Luik A, Nilsson C, Ulber B (2006a) MASTER: Validation of existing phenological models of the proPlant DSS for key pests in winter oilseed rape in different climatic areas in Europe and prospects for IPM. *Proc Int Symp Integrated Pest Management in Oilseed Rape, 3–5 April 2006, Göttingen, Germany*: 140–142.

Johnen A, Williams IH, Ferguson AW, Büchs W, Klukowski Z, Luik A, Nilsson C, Ulber B (2006b) MASTER: Construction of phenological models of key parasitoids in Europe and prospects for spray windows compatible with their conservation in winter oilseed rape. *Proc Int Symp Integrated Pest Management in Oilseed Rape, 3–5 April 2006, Göttingen, Germany*: 137–139.

Lane A, Holliday JM, Walters KFA (1995) A rapid method for assessing infestations of cabbage stem flea beetle larvae in winter oilseed rape and implications for control. *IOBC/wprs Bull* 18: 127–133.

Meier U (ed.) (2001) Growth stages of mono- and dicotyledonous plants. http://www.bba.de/veroeff/bbch/bbcheng.

Murchie AK, Williams IH, Alford DV (1997) Effects of commercial insecticide treatments to winter oilseed rape on parasitism of *Ceutorhynchus assimilis* Paykull (Coleoptera: Curculionidae) by *Trichomalus perfectus* (Walker) (Hymenoptera: Pteromalidae). *Crop Prot* 16: 199–202.

Newe M, Meier H, Johnen A, Volk T (2003) proPlant expert.com – an online consultation system on crop protection in cereals, rape, potatoes and sugar beet. *EPPO Bull* 33: 443–449.

Nilsson C (2002) Strategies for the control of cabbage stem flea beetle on winter rape in Sweden. *IOBC/wprs Bull* 25(2): 133–139.
Williams IH (2006) Integrating biological control within IPM for winter oilseed rape in Europe: An overview of the MASTER project. *Proc Symp on Integrated Pest Management in Oilseed Rape, 3–5 April 2006, Göttingen, Germany.*
Williams IH, Büchs W, Hokkanen H, Menzler-Hokkanen I, Johnen A, Klukowski Z, Luik A, Nilsson C, Ulber B (2005) MASTER – Integrating biological control within IPM for winter oilseed rape across Europe. *Proc BCPC Int Cong, Crop Science & Technology, Glasgow, 31 October–2 November 2005*, 1: 301–308.

Chapter 16
Farming Systems, Integrated Crop Management and Winter Oilseed Rape Production

Christer Nilsson

Abstract Farming systems for arable cropping, should, in the future, combine sustainability with environmental acceptability, be high-yielding yet energy efficient, providing a good net return. Integrated crop management aims to achieve a sustainable farming system by using natural resources and regulating mechanisms to replace polluting inputs. In oilseed rape production, nitrogen inputs to the crop and machinery use are greater than for many other crops and a system that reduces these inputs is desirable. The crop is attacked by more pests than most other arable crops and consequently pesticide inputs are also high; increased use of natural control provided by parasitoids, predators and pathogens would improve sustainability and environmental acceptability. An Integrated Crop Management (ICM) System can be designed to be more sustainable and resource-efficient and to enhance biological control of pests, by reduced tillage with no ploughing, use of a seed mix to provide an internal trap crop to reduce pest damage, plant density to increase parasitisation, and insecticide application using control thresholds. This IPM strategy is exemplified by the EU project MASTER (QLK5-CT-2001-01447).

16.1 Farming Systems and Integrated Crop Management

Crop management systems for the production of arable crops in Europe should, in the future, combine sustainability with environmental acceptability to satisfy both social and economic demands. They should be high-yielding yet energy efficient and provide a good economic net return.

Sustainability in agricultural production of food, fodder and raw material for industrial use depends, at present, largely on the use of non-renewable energy and the environmental impact of the farming system used. The conventional farming systems of today are heavily dependent on fossil energy, in the form of diesel for soil tillage and for transport, and in the form of nitrogen fertilisers, each taking around

C. Nilsson (✉)
Swedish University of Agricultural Sciences, Alnarp SE-230 53, Sweden
e-mail: NCNilsson@telia.com

I.H. Williams (ed.), *Biocontrol-Based Integrated Management of Oilseed Rape Pests*, DOI 10.1007/978-90-481-3983-5_16, © Springer Science+Business Media B.V. 2010

a third of the total energy needed for a winter wheat or an oilseed rape crop (Rathke and Diepenbrock 2006). In return, at least five to ten times the total input energy is gained in the harvested products (Hülsberger et al. 2001, Kuesters and Lammel 1999, Zentner et al. 2004, Tzilivakis et al. 2005). With the present technology of producing renewable energy, a future scenario for a sustainable farming system would be very dependent on the acreage available for the supply of production energy (Bailey et al. 2003). Acreage will thus be one of the most important resources and pressure will be on producing as much as possible from every hectare. Pesticides and nitrogen fertilizers will have to continue to be used, at least to some extent. Such a farming system will minimize soil tillage to save energy. Crop rotation will be central to the potential to decrease the need for pesticides (Edwards and Stinner 1990, Jordan 1990, Jordan and Hutcheon 1995, Alford et al. 2003). Conserving and enhancing biocontrol will be an integral part of this arable farming system.

Integrated Crop Management (ICM) is the method that ultimately should lead to a farming system exemplified above, a farming system that is sustainable. There are many definitions of ICM, but that of the IOBC (El Titi et al. 1993) and further elaborated in Boller et al. (1997) and Boller et al. (2004) is the most comprehensive, where ICM is '... using natural resources and regulating mechanisms to replace polluting inputs and to secure sustainable farming.'

The study of the whole farming system has been a main focus within economics and sociology for a long time, but has gained importance in biology only during the last decade or two. Tools to study and quantify the performance of systems have been developed within areas of product control and certification of production processes, and have been adopted by experimental biology. The farming system experiment often aims at continuous improvement, using the approach of objectives – design – evaluation – changing design, in a repeated yearly process. It is important to use large plots, at least 1 ha in area is recommended, to give them an identity separate from the surrounding fields, to make it possible to use normal farm machinery, and to allow for the natural distribution of diseases, pests and their natural enemies within plots.

The tools used for evaluating system performance (Pervanchon et al. 2002) depend on the scale of the study (field, farm, region and country) and the problem focus (production efficiency in relation to resources, environmental questions, rural development and so on). The tools or indicators often used in the analyses of arable farming systems are summarised in Table 16.1. Many of these indicators have

Table 16.1 Indicators used to study and quantify the performance of arable farming systems

Economic and social aspects	Net return, labour hours, health, employment
Production	Quantity, quality, stability, acreage needed
Efficiency in utilization of external resources (per ha, per kg product)	Annual balance of nutrients, energy & carbon, energy account & fuel consumption, pesticide use
Abiotic environment	Emissions to water, air, soil pollution, pH
Biotic environment	Bio-diversity (farm, landscape), biocontrol, plant protection needs, pesticide effects, crop & crop stand, ecological infrastructure

been discussed and used in the EU-project AIR 3 CT920755 'Research network on integrated and ecological arable farming systems' (Vereijken 1994, 1997).

System economics should be calculated using a standard net return method. Yield revenue is compared to costs for seed, nitrogen, other fertilizers, pesticides, machinery and labour. Energy use and gain is a more universal way of analysing farming systems. Each item introduced into the net return calculation can also be given an energy value, enabling a comparison of the production that can be used over long periods of time without recalculation.

16.2 Integrated Crop Management in Oilseed Rape Production

16.2.1 Nitrogen

Nitrogen use is high in modern, high-yielding arable crop production, and in oilseed rape (*Brassica napus*; OSR) production, more nitrogen is used than in most other crops. It is an important yield-determining factor, determining the protein content of the seed and thus the value of the seed press residues for protein fodder for livestock. A high nitrogen utilization rate is important as the energy in nitrogen is a very important component of the energy balance of the crop (Zentner et al. 2004). One kilo of nitrogen fertilizer takes about the same energy to produce as is in one litre of diesel. As sustainability in the first place means replacing fossil energy with energy produced in other ways, e.g., through photosynthesis, nitrogen prices will be higher and nitrogen levels lower on sustainable farms than at present. Taken over many years, seed yields would decrease moderately, while nitrogen use would be substantially lower, as farmers, at present, usually fertilize for best yield and not for average attained yield. Oil takes more energy for the plant to produce than starch, the yield of winter OSR is usually a little more than half that of winter wheat, but produced with the same amount of nitrogen. Often half of the nitrogen applied is left on the field when the seed has been harvested, much more than is left after a wheat crop. The fate of this nitrogen is of great environmental importance, as nitrogen fertilisation is not changed very much in the successive crop, e.g., winter wheat. Nitrogen leaching out of the system would be avoided through catch crops in the rotation, actively improving soil structure and making soil tillage less important. Keeping the soil covered with plants for most of the year would probably also add to an increase in predator numbers in the crop (Büchs 2003).

The higher the yield at the same nitrogen supply, the better the utilisation of the nitrogen. Obtaining a high yield is thus one of the goals in all crop production. Plant protection is very important to achieve this. Losses in crop yield can often be related to losses of photosynthetic ability and pests, diseases and weeds (shadowing), can greatly reduce nitrogen efficiency.

16.2.2 Soil Tillage

Manufacture and use of machinery is the other important part of the energy needed to grow an OSR crop. The sustainable farm would thus minimise soil tillage and also aim to simplify tillage operations so as to use fewer machines. Reduced tillage is based upon the assumption that the main tillage operations can be replaced by biological and physical soil structure-forming processes of most clay-containing soils, having positive effects on nitrogen efficiency, root growth, yield formation, resilience to unfavourable weather conditions and probably also lower production of climatic gasses. This transformation takes time and the new structure is usually formed within 5 years. In crops that are sensitive to soil structure, such as sugar beet and winter OSR, it can take longer before the yields in reduced tillage fields are comparable to those under conventional tillage. Deep rooted cover crops will markedly improve the soil structure. In crop rotations with OSR, mustard should be avoided as, like OSR, it is a host of club-root (*Plasmodiophora*). Minimising soil tillage operations should lower production costs, by lowering use of machinery, fuel and labour costs. On a large farm, this could mean that fewer employees are needed, in turn leading to a substantial increase in profitability.

Conservation tillage also increases the number of predators on the soil surface (Stinner and House 1990, Büchs 2003). Many parasitoid and predator species overwinter in the soil of the rape field (see also Ulber et al. Chapter 2 and Williams et al. Chapter 4, this volume). A potential way of increasing their population densities is to minimize their overwintering mortality through the post-harvest soil cultivations of rape stubble to establish a new crop, often winter wheat. Ploughing and rotary harrowing, can reduce their survival, whereas non-inversion tillage, in particular direct drilling, is less harmful (Nilsson 1985, 2006). The benefits of reduced tillage to the overwinter survival of the parasitoids of oilseed rape pests are reviewed by Nilsson in Chapter 11 (this volume).

16.2.3 Plant Stand

Variety differences are not considered to be of relevance for insect pest populations at present. The OSR seed can be admixed with 2% turnip rape to produce an internal trap crop. Turnip rape develops more quickly in the spring than oilseed rape and is the more attractive to the bud-stage pests like the pollen beetle, thereby reducing damage to the main cultivar (Büchi 1990, 1995, Nilsson 2004, Cook et al. 2007).

Plant density, spatial distribution of plants and plant morphology can affect parasitism of the larvae of target stem-boring pests. The most common row spacing is 12.5 cm. Wider row spacing and higher plant density can increase parasitisation rates of stem-mining larvae (Nuss 2004, Neumann and Ulber 2006, Ulber and Fischer 2006), although Zaller et al. (2008) found more damage to plants in denser plant stands. Herbicide use can also be decreased by increasing row spacing and using a combination of row-sprayed herbicide and inter-row hoeing. Wider row spacing will sometimes give a lower yield.

A final plant density of around 30–40 plants m^{-2} is recommended to farmers (Schulz 1998, Nuss 2004). The seed rate sown is usually greater than this to compensate for winter losses and slug, bird and mouse damage, and thus plant densities achieved in practice vary considerably. Sometimes, when plant losses are low, a much higher plant density than intended is achieved. The discrepancy between seeds sown and final plant stand is one of the main obstacles to a better control of OSR husbandry.

16.2.4 Plant Protection

Oilseed rape harbours more pests than most other arable crops. Location, farming system and the use of pesticides determine which insect species become pests in a given crop. In Europe, OSR is grown from east to west in the northern regions, in both maritime and continental climates. Four insect species are common pests throughout the rape-growing area: cabbage stem weevil (*Ceutorhynchus pallidactylus*), cabbage seed weevil (*C. obstrictus* syn. *C. assimilis*), pollen beetle (*Meligethes* species, mainly *M. aeneus*) and brassica pod midge (*Dasineura brassicae*). The rape stem weevil (*C. napi*) is a serious pest only in central Europe and cabbage stem flea beetle (*Psylliodes chrysocephala*) is confined to maritime areas, i.e., northern and western Europe (Alford et al. 2003, Williams Chapter 1 this volume). Each pest species has its own specific key parasitoids but these are the same throughout Europe (Ulber et al. 2006, Ulber et al. Chapter 2 this volume). There are also great similarities in the species composition of polyphagous soil-surface dwelling predators (Büchs and Alford 2003, Büchs et al. 2006, Warner et al. 2008, Williams et al. Chapter 4 this volume). Species richness and activity-densities of predators have been shown to increase with more extensive management systems, reduced tillage and lower pesticide inputs in oilseed rape (Büchs et al. 1997, Büchs 2003, Luik et al. 2006) as in cereals (Holland and Luff 2000, Hance 2002, Holland et al. 2002). There is thus common ground for attempting to develop an European ICM strategy for protecting OSR crops against damage from these pests.

The economic importance of these six pests can be great, for example, yield losses of up to 50% have been reported from the pollen beetle (Nilsson 1987). The routine use of insecticides is therefore common practice throughout Europe: in the autumn once or twice against cabbage stem flea beetle, early in spring against rape stem weevil, later several times against pollen beetle and cabbage stem weevil and finally during flowering or early pod setting against cabbage seed weevil and brassica pod midge (Walters et al. 2003, Thieme et al. Chapter 12, and Williams Chapter 1 this volume).

One objective of an European strategy would be to decrease the use of pesticides and increase the natural control provided by local populations of parasitoids, predators and pathogens. This cannot be achieved by simply modifying pesticide use, but must also involve crop husbandry in general and can be better described as a part of ICM incorporating Integrated Pest Management (IPM) strategies in a sustainable and environmentally-acceptable farming system.

Decision support systems, such as ProPlant (Frahm et al. 1996, Johnen and Meier 2000, Johnen et al. 2006a, b, Johnen et al. Chapter 15 this volume) can be used to minimise the use and optimise the timing of pesticide application, so as to create a balance between yield loss, environmental effects and biological control.

Increasing natural regulation of pest populations is often a long term project, and without a long term strategy, pesticides are the only alternative. Parasitoids and predators are sensitive to soil tillage, pesticide use and timing (Johnen et al. 2006a, b, Ulber et al. Chapter 13 this volume), and to some extent also to other factors, but, in general, it is possible to increase natural regulation of pest populations through changes in crop management.

Economic pest threshold levels above which insecticide application is recommended are an important component of IPM for the crop with potential for minimising insecticide input. Local economic pest thresholds (Alford et al. 2003, Williams Chapter 1 this volume) are available for most pests in most countries but vary with country and even within a country depending on severity of a particular pest species. Recent surveys have revealed that crops of winter and spring rape commonly receive between one and four applications and that some receive more than five (Menzler-Hokkanen et al. 2006, Richardson 2008). In systems where thresholds for control of insects are used, insecticide application may be more than halved compared to schedule sprayed systems. More herbicides have to be used in the ICM system with reduced tillage, because of the need to kill volunteer plants of the preceding cereal crop normally ploughed under.

Reducing the dose rate of insecticide applied can conserve parasitoids more effectively than selecting active ingredients of higher selectivity (see Ulber et al. Chapter 13 this volume). It also reduces input costs and input of environmentally-undesirable pollutant.

Integrated disease management is primarily a function of crop rotation, and no other factor can offset the impact of crop rotation for the most important soil borne diseases *Plasmodiophora, Phoma, Sclerotinia* and *Verticillium* wilt incidence. Resistance breeding and a knowledge-based use of fungicides can often give satisfactory control of most of the other diseases.

16.3 The MASTER Experiment

Recently most of these component parts of integrated crop management in OSR were tested in the integrated crop management experiment of the EU project MASTER (QLK5-CT-2001-01447) (outlined in Williams et al. 2005) which was part-funded by the EU under its Framework 5 Quality of Life and Management of Living Resources programme. An ICM System for growing winter rape, within a cereal rotation was designed and compared to a Standard European Farming System aimed to depict a practical, modern way of growing winter rape in today's Europe, essentially French, German and UK husbandry (Cetiom, UFOP and HGCA websites, respectively). The ICM System was designed to be more sustainable and resource-efficient than the current modern system and to enhance biological control of pests, particularly parasitoids and predators, through modification of crop husbandry. Three joint field experiments in five EU countries: Estonia, Germany, Poland, Sweden and the UK were conducted over 4 years; each experiment was of 2

Table 16.2 Plot treatments in three collaborative European experiments comparing a standard system of growing winter oilseed rape (STN) with a system enhanced to conserve biological control agents (ICM)

						Insecticide applications	
Year	Plot	Soil tillage	Seed mix OSR:TR (%)	Row spacing (cm)	Seed/m^2	No.	Rate
2002/4	STNii	Plough	100:0	12.5	50	3	Full
	ICMie	Reduced	98:2	12.5	50	0	–
2003/5	STNii	Plough	100:0	12.5	60	3	Full
	STNie	Plough	100:0	12.5	60	0–3	Full
	ICMi0	Reduced	98:2	12.5	60	0	–
	ICMie	Reduced	98:2	12.5	60	0–3	Full
2004/6	STNii	Plough	100:0	12.5	60	3	Full
	STNie	Plough	100:0	12.5	60	0–3	Full
	ICMi0	Reduced	98:2	25.0	40	0	–
	ICMie	Reduced	98:2	25.0	40	0–3	Half

ii, insecticide applied prophylactically to schedule; ie, insecticide applied only when the local economic threshold for control was exceeded; i0, no insecticide applied. OSR, oilseed rape; TR, turnip rape (After Williams et al. 2005)

year's duration with pest and natural enemy monitoring in the winter rape crop and in the following winter wheat crop (Table 16.2).

The MASTER experiments demonstrated that a farming system based on ICM principles, with non-inversion soil tillage and the use of pest control thresholds to determine the need for insecticide application, was feasible and could be recommended to farmers as a strategy to actively-enhance natural enemy populations and thereby improve biological control of economically-important pests of oilseed rape, and, at the same time, use less resources and decrease environmental impact.

References

Alford DV, Nilsson C, Ulber B (2003) Insect pests of oilseed rape crops. In: Alford DV (ed.) *Biocontrol of oilseed rape pests*. Blackwell, Oxford, UK.

Bailey AP, Basford WD, Penlington N, Park JR, Keatinge JDH, Rehman T, Tranter RB, Yates CM (2003) A comparison of energy use in conventional and integrated arable farming systems in the UK. *Agr Ecosyst Environ* 97: 241–253.

Boller EF, Avilla J, Joerg E, Malavolta C, Wijnands F, Esbjerg P (2004) Guidelines for integrated production. Principles and technical guidelines. *IOBC/wprs Bull* 27(2): 49 pp.

Boller EF, Malavolta C, Jörg E (1997) Guidelines for integrated production of Arable Crops in Europe. Technical guidelines. *IOBC/wprs Bull* 20(5): 115 pp.

Büchi R (1990) Investigations on the use of turnip rape trap plants to control oilseed rape pests. *IOBC/wprs Bull* 13(4): 32–39.

Büchi R (1995) Combination of trap plants (*Brassica rapa* var. *silvestris*) and insecticide use to control rape pests. *IOBC/wprs Bull* 18: 102–121.

Büchs W (2003) Impact of on-farm landscape structures and farming systems on predators. In: Alford DV (ed.) *Biocontrol of oilseed rape pests*. Blackwell, Oxford, UK.

Büchs W, Alford DV (2003) Predators of oilseed rape pests. In: Alford DV (ed.) *Biocontrol of oilseed rape pests*. Blackwell, Oxford, UK.

Büchs W, Felsman DS, Klukowski Z, Luik A, Nilsson C, Williams IH (2006) Key predator species in oilseed rape crops – a review of literature and MASTER results. *CD-Rom Proc Int Symp Integrated Pest Management in Oilseed Rape, 3–5 April 2006, Göttingen, Germany.*

Büchs W, Harenberg A, Zimmerman J (1997) The invertebrate ecology of farmland as a mirror of the intensity of the impact of man? – an approach to interpreting results of field experiments carried out in different crop management intensities of a sugar beet and an oil seed rape rotation including set-aside. *Biol Agric Hort* 15: 83–108.

Cetiom homepage (http://www.cetiom.fr/).

Cook SM, Rasmussen HB, Birkett MA, Murray DA, Pye BJ, Watts NP, Williams IH (2007) Behavioural and chemical ecology underlying the success of turnip rape (*Brassica rapa*) trap crops in protecting oilseed rape (*Brassica napus*) from the pollen beetle (*Meligethes aeneus*). *Arthropod-Plant Interact* 1: 57–67.

Edwards CA, Stinner BR (1990) The use of innovative agricultural practices in a farming systems context for pest control in the 1990 s. *Proc BCPC Conf – Pests & Diseases 1990*: 679–684.

El Titi A, Boller EF, Gendrier JP (1993) Integrated production. Principles and technical guidelines. *IOBC/wprs Bull* 16(1): 96 pp.

Frahm J, Johnen A, Volk T (1996) Development of the PRO_PLANT decision support system for plant protection in cereals, sugar beet and rape. *EPPO Bull* 26: 609–622.

HGCA homepage (http://www.hgca.com/)

Hance T (2002) Impact of cultivation and crop husbandry practices. In: Holland JM (ed.) *The agroecology of carabid beetles*. Intercept, Andover, UK.

Holland JM, Frampton GK, Van der Brink P (2002) Carabids as indicators within temperate arable farming systems: Implications from scarab and link integrated farming systems projects. In: Holland JM (ed.) *The agroecology of carabid beetles*. Intercept, Andover, UK.

Holland JM, Luff ML (2000) The effects of agricultural practices on Carabidae in temperate ecosystems. *Integr Pest Manag Rev* 5: 109–129.

Hülsberger KJ, Feil B, Biermann S, Rathke GW, Kalk WD, Diepenbrock W (2001) A method of energy balancing in crop production and its application in long-term fertilizer trial. *Agr Ecosyst Environ* 86: 303–321.

Johnen A, Meier H (2000) A weather-based decision support system for managing oilseed rape pests. *Proc BCPC Conf – Pests & Diseases 2000*, 2: 793–800.

Johnen A, Williams IH, Ferguson AW, Büchs W, Klukowski Z, Luik A, Nilsson C, Ulber B (2006a) MASTER: Construction of phenological models of key parasitoids in Europe and prospects for spray windows compatible with their conservation in winter oilseed rape. *CD-Rom Proc Int Symp Integrated Pest Management in Oilseed Rape, 3–5 April 2006, Göttingen, Germany*

Johnen A, Williams IH, Ferguson AW, Büchs W, Klukowski Z, Luik A, Nilsson C, Ulber B (2006b) MASTER: Validation of existing phenological models of the proPlant DSS for key pests in winter oilseed rape in different climatic areas in Europe and prospects for IPM. *CD-Rom Proc Int SympIntegrated Pest Management in Oilseed Rape, 3–5 April 2006, Göttingen, Germany.*

Jordan VWL (1990) Conceptual reappraisal of UK arable farming practices for increased flexibility in crop protection options. *Proc BCPC Conf – Pests & Diseases 1990*: 1221–1230.

Jordan VWL, Hutcheon JA (1995) Less intensive farming and the environment: An integrated farming systems approach for UK arable crop production. In: Glen DM, Greaves MP, Anderson HM (eds.) *Ecology and integrated faming systems*. Wiley & Sons, Chichester, UK.

Kuesters J, Lammel J (1999) Investigations of the energy efficiency of the production of winter wheat and sugar beet in Europe. *European J Agron* 11: 35–43.

Luik A, Veromann E, Kevväi R, Kruus M (2006) A comparison of the pests, parasitoids and predators on winter and spring oilseed rape crops in Estonia. *CD-Rom Proc Int Symp Integrated Pest Management in Oilseed Rape, 3–5 April 2006, Göttingen, Germany.*

Menzler-Hokkanen I, Hokkanen HMT, Büchs W, Klukowski Z, Luik A, Nilsson C, Ulber B, Williams IH (2006) Insect problems in European oilseed rape cultivation, and how to deal with them: The OSR farmers' perspective. *IOBC/wprs Bull* 29(7): 91–94.

Neumann N, Ulber B (2006) Adult activity and larval abundance of stem weevils and their parasitoids at different crop densities of oilseed rape. *IOBC/wprs Bull* 29(7): 193–199.

Nilsson C (1985) Impact of ploughing on emergence of pollen beetle parasitoids after hibernation. *Z Angew Entomol* 100: 302–308.

Nilsson C (1987) Yield losses in summer rape caused by pollen beetles (*Meligethes* spp.). *Swedish J AgrRes* 17: 105–111.

Nilsson C (2004) Trap plants to avoid insecticide application against pollen beetles in oilseed rape. *IOBC/wprs Bull* 27(10): 217–224.

Nilsson C (2006) Effects of post-harvest soil tillage on the survival of key parasitoids of oilseed rape pests. *CD-Rom Proc Int Symp Integrated Pest Management in Oilseed Rape, 3–5 April 2006, Göttingen, Germany.*

Nuss H (2004) Effect of plant density and plant architecture on the abundance and within-plant distribution of stem borers in winter oilseed rape. *PhD Thesis, University of Göttingen* http://webdoc.sub.gwdg.de/diss/2004/nuss/nuss.pdf .

Pervanchon F, Bockstaller C, Girardin P (2002) Assessment of energy use in arable farming systems by means of an agro-ecological indicator: The energy indicator. *Agr Syst* 72: 149–172.

Rathke GW, Diepenbrock W (2006) Energy balance of winter oilseed rape (*Brassica napus* L.) cropping as related to nitrogen supply and preceding crop. *European J Agron* 24: 35–44.

Richardson DM (2008) Summary of findings from a participant country pollen beetle questionnaire. *EPPO Bull* 38: 68–72.

Schulz RR (1998) Möglichkeiten zur Vermeidung einer vorzeitigen Sproβstreckung. *Raps* 16(3): 98–101.

Stinner BR, House GJ (1990) Arthropods and other invertebrates in conservation-tillage agriculture. *Annu Rev Entomol* 35: 299–318.

Tzilivakis J, Warner DJ, May M, Lewis KA, Jaggard K (2005) An assessment of the energy input and greenhouse gas emissions in sugar beet (*Beta vulgaris*) production in the UK. *Agr Syst* 85: 101–119.

UFOP homepage (www.ufop.de).

Ulber B, Fischer K (2006) Effect of plant density and host plant architecture on the abundance and within-plant distribution of stem-mining pests and the level of parasitism. *Proc Int Symp Integrated Pest Management in Oilseed Rape, 3–5 April 2006, Göttingen, Germany.*

Ulber B, Fischer K, Klukowski Z, Luik A, Veromann E, Nilsson C, Ahman B, Williams IH, Ferguson AW, Piper R, Barari H (2006) Identity of parasitoids and their potential for biocontrol of oilseed rape pests in Europe. *CD-Rom Proc Int SympIntegrated Pest Management in Oilseed Rape, 3–5 April 2006, Göttingen, Germany.*

Vereijken P (1994). Research network on integrated and ecological arable farming systems. 1. Designing Prototypes. AB-DLO Wageningen, p 87.

Vereijken P (1997) A methodical way of prototyping integrated and ecological arable farming systems (I/EAFS) in interaction with pilot farms. *European J Agron* 7: 235–250.

Walters KFA, Young JEB, Kromp B, Cox PD (2003) Management of oilseed rape pests. In: Alford DV (ed.) *Biocontrol of oilseed rape pests*. Blackwell, Oxford, UK.

Warner DJ, Allen-Williams LJ, Warrington S, Ferguson AW, Williams IH (2008) Implications for conservation biocontrol of spatio-temporal relationships between carabid beetles and coleopterous pests in winter oilseed rape. *Agr Forest Entomol* 10: 375–387.

Williams IH, Büchs W, Hokkanen H, Menzler-Hokkanen I, Johnen A, Klukowski Z, Luik A, Nilsson C, Ulber B (2005) MASTER – Integrating biological control within IPM for winter oilseed rape across Europe. *Proc BCPC Int Cong, Crop Science & Technology, Glasgow, 31 Oct–2 Nov 2005,* 1: 301–308.

Zaller JG, Moser D, Drapela T, Schmöger C, Frank T (2008) Effect of within-field and landscape factors on insect damage in winter oilseed rape. *Agr Ecosyst Environ* 123: 233–238.

Zentner RP, Lafond GP, Derksen DA, Nagy CN, Wall DD, May WE (2004) Effect of tillage and crop rotation on non-renewable energy use efficiency for a thin Black Chernozem in the Canadian Prairies. *Soil Till Res* 77: 125–136.

Chapter 17
Integrating Crop and Landscape Management into New Crop Protection Strategies to Enhance Biological Control of Oilseed Rape Insect Pests

Adrien Rusch, Muriel Valantin-Morison, Jean Pierre Sarthou, and Jean Roger-Estrade

Abstract The development of sustainable cropping systems is a major challenge for agronomists and crop scientists in many regions of the world. The prophylactic uses of broad spectrum insecticides are actually the main solution for farmers to control insect pests. Therefore, there is a growing need to develop innovative crop protection strategies through an integrated approach which aims at favouring natural enemies and enhancing biological control in agroecosystems. The development of such strategies requires a thorough understanding of agroecosystem functioning. In this chapter, we present the effects of different elements, from the field scale to the landscape scale that are known to enhance biological control in agroecosystems and limit pest damage with particular reference to oilseed rape. Linking integrated pest management and landscape ecology brings a regional perspective to the management of pest populations. Available techniques are often added together rather than combined in an integrated way and are rarely evaluated through environmental and economical criteria. Therefore, we present a methodological framework to design and assess sustainable cropping systems, with a particular emphasis on complementariness between models, systemic trials and more analytical approaches. Implementation of sustainable cropping systems implies the development of new integrated pest management strategies and thereby an increased participation of the different stakeholders from farmers to policy makers.

17.1 Introduction

Conventional modern agriculture was developed to increase and maximize productivity in order to meet increasing demands for food. The pursuit of this goal has led to an oversimplification of crop diversity at the field, cropping system and landscape levels, with increasing reliance on agrochemicals. Indeed, a range of farmers' practices have been developed without foreseeing any of the long-term consequences on the environment and on the ecological dynamics of agroecosystems. However, in

A. Rusch (✉)
INRA-AgroParisTech, UMR211 Agronomie, Thiverval-Grignon, France
e-mail: adrien.rusch@grignon.inra.fr

I.H. Williams (ed.), *Biocontrol-Based Integrated Management of Oilseed Rape Pests*,
DOI 10.1007/978-90-481-3983-5_17,

today's socio-economic context, there is an increasing demand to reduce chemical inputs on arable crops and to develop more sustainable crop management strategies. Moreover, there is clear evidence that widespread resistance to pesticides resulting from the intensive use of broad spectrum insecticides necessitates the development of more sustainable and environmentally friendly insect pest management. Our aim in this chapter is to review the ways in which the cropping system, in its broadest definition, could be modified to enhance biological control in agroecosystems and limit pest damage. Consideration of different scales of analysis provides a good overview of how farmers' practices affect biological control and implications for integrated management of oilseed rape pests. Furthermore, multiple levels analysis permits to take into account the function of uncultivated areas on pest population dynamics and their biological control. In this scaling up approach we consider three different levels: the field level, the cropping system level and the landscape scale (from field edge management to spatial organization of crops and non-cultivated areas). These different levels include a crop management modification at the field scale, a diversification in crop sequences at the farm scale and spatial patterns at the landscape scale. At each level, we analyse which elements can enhance beneficial biological interactions and their impacts (direct or indirect) on pest regulation and pest damage with a particular emphasis on oilseed rape pests.

We will first present the effect of single practices at the field scale that play a role on crop attractiveness. Secondly, we will review the effect of crop management on the local habitat environment to promote biological control. Thirdly, we will summarize the work that has been carried out on the effect of crop diversification in space and over time at the field and the cropping system scale. Fourthly, we will discuss studies that have accounted for spatial pattern effect on pest population and pest regulation in agricultural landscapes. In conclusion, we will examine the perspective for designing new crop protection strategies and we will discuss the advantages of a complementary approach between modelling and on-farm diagnosis for integrated pest management.

17.2 Effect of Single Practices to Induce Pest Avoidance and Limit Pest Damage

There is a growing body of evidence that modifying some elements of crop management at the field scale can induce pest avoidance and limit pest damage. This strategy involves avoiding the simultaneous presence of crops and pests through different factors: the plant's ability to attract and repel pests (e.g., the 'push-pull' strategies), the plant's ability to recover from injury and the desynchronization between crop susceptibility periods, pest and predator life cycles.

17.2.1 Effect of Species, Cultivar and Crop Management on Crop Attractiveness for Pests at the Field Scale

We review here some important results concerning the effect of species, cultivar and crop management on crop location by oilseed rape pests. More information on the behavioural ecology associated with location of the oilseed rape crop by its major pests and their key hymenopterous parasitoids is given by Williams and Cook (Chapter 7 this volume).

The capacity of insects to identify a host plant suitable for its feeding and reproduction depends on the morphological and/or metabolic characteristics of the plant. The pollen beetle (*Meligethes aeneus*), an oilseed rape pest, locates its host plant through visual and olfactory signals (Evans and Allen-Williams 1989, Evans and Allen-Williams 1998). The beetles are principally attracted by the yellow colour of the flowers and by certain chemical signals released by the plant. It has been shown that degradation products of glucosinolates attract insects specialised on cruciferous host plants (Feeny et al. 1970, Finch 1978, Free and Williams 1978). In particular, several studies have demonstrated the particular importance of alkenyl glucosinolates (which release low amounts of the volatile isothiocyanates (ITCs) most attractive to pests) for host plant location and selection (Smart and Blight 2000, Cook et al. 2006).

Based on the hypothesis that the production of glucosinolates by cultivars of winter oilseed rape (WOSR) and other Brassicaceae may attract pollen beetles, many studies have focused on the effects of host plants on insect orientation and feeding (Bartlet et al. 2004), oviposition behaviour (Borg and Ekbom 1996), and egg production of the pollen beetle (Hopkins and Ekbom 1999). Turnip rape (*Brassica rapa*) has been found to attract more pollen beetles in both laboratory and field conditions (Hokkanen 1989, Cook et al. 2002, 2006, Valantin-Morison and Quéré 2006, Rusch and Valantin-Morison 2010). The same effect has been reported for other oilseed rape pests such as cabbage stem flea beetle (*Psylliodes chrysocephala*) (Büchi 1995, Barari et al. 2005) and cabbage seedpod weevil (Carcamo et al. 2007).

Many physiological properties play a role in determining the attractiveness or repulsiveness of certain plant species, and stage of development seems to be a major factor in host selection. In particular, Cook et al. (2006) have demonstrated that pollen beetles preferred turnip rape to oilseed rape when both species had closed buds, preferred flowering plants regardless of the given species and had no preference for a particular species when both species were flowering. The importance of the stage of crop development in the implementation of a catch crop strategy has been confirmed by Valantin-Morison and Quéré (2006) and Rusch and Valantin-Morison (2010). The earlier flowering of turnip rape than of WOSR proved to act in synergy with the differences in glucosinolate synthesis between species discussed above.

Turnip rape is thus often used in this particular situation as a so-called trap crop. Simulations using a spatially explicit individual-based model show that for herbivores that actively immigrate from a nearby source via the field edge, a surrounding border trap crop is the optimal arrangement (Potting et al. 2005). Moreover, a trap crop can be used for the capture, in July and August, of the second generation of insect pests, to reduce population sizes just before hibernation (Husberg et al. 1985). In such a situation, landscape organization of these trap crops should be taken into account.

Nitrogen supply affects the glucosinolate content of cruciferous crops. As described above, these metabolites are involved in the plant host identification processes of pollen beetles and other oilseed rape pests. Indeed, several authors have shown that the rate of nitrogen application had a highly significant effect on the glucosinolate content of the seeds (Milford and Evans 1991) or in the aerial parts of the plant (Markus et al. 1996). So nitrogen supply strategy (in terms of dates and rates) can play an indirect role on host plant attractiveness (Rusch and Valantin-Morison 2010). In the case of winter oilseed rape, Rusch and Valantin-Morison (2010) revealed an important effect of concordance between nitrogen supply and the number of stem weevil punctures. Indeed, plants which had received nitrogen earlier in the season (just before the first stem weevil flights) showed a greater stem elongation. As stem elongation is one of the most important criteria for stem weevil host selection, plants with greater elongation showed significantly higher numbers of punctures per main raceme and higher stem damage (Rusch and Valantin-Morison 2010).

Recent studies have shown that botanical extracts can function as insecticides, deterrents or repellents. Essential oils from plant species of the *Lamiaceae* have a broad spectrum of biological activity and lavender essential oil has been reported as interfering with orientation to, and selection of, host plants by a range of pest insects (Landolt et al. 1999, Mauchline et al. 2005, Cook et al. 2007). This lavender oil has been identified as a repellent and represents an interesting tool in push-pull pest control strategies developed against pollen beetle. Moreover, Cook et al. (2007) showed that parasitoids gave no significant responses to the odour of lavender oil in behavioural assays. This study suggests that lavender treated oilseed rape plants could be used as an element of an integrated pest management strategy as it would not reduce host habitat location by parasitoids of the given pest.

17.2.2 Effect of Species and Varietal Resistance on Pest Damage

Numerous studies have pointed out the important role of host plant resistance in arthropod pest management (Van Emden 1991, Kogan 1994, Gatehouse 2002, Sharma and Ortiz 2002). Kogan (1994) explained that the damage caused by an insect pest to a crop depends on the feeding habit of the pest species, the size of its population, and the capacity of the plant to withstand the type and amount of injury that results. One of the most important biotic factors that regulates pest population size is the adequacy of a plant as a host for the insect pest. Therefore, Kogan (1994) highlighted that resistance of a host plant to an insect pest is expressed through properties that enable the plant to restrain the growth of the pest population (antibiosis phenomenon) or the capacity of the plant to withstand attacks and recover from injury. In cruciferous plants, all three types of insect resistance modalities (antixenosis, antibiosis and tolerance) have been found in the case of various insect pests (Palaniswamy 1996). For example, Dosdall et al. (1994) compared the susceptibilities of species and cultivars of oilseed rape and mustard to infestation by root maggots (*Delia* spp.) and found that oilseed rape was the most susceptible. The authors found that differences in susceptibility occurred among and within oilseed rape and Indian mustard species, but generally the differences were greater between species than between cultivars within species. This study demonstrated that the mechanism of resistance by cruciferous species under study to infestation by *Delia* spp. is antixenosis. Moreover, Ellis and Farrell (1995) have compared the response of six *Brassica* accessions to aphid infestations and found both antixenotic and antibiotic resistance to cabbage, while Palaniswamy (1996) reported that both antixenosis and tolerance mechanisms were found in the resistance of cruciferous plants (e.g., *Sinapsis alba*, *Brassica juncea* and *Brassica napus*) to the flea beetle (*Phyllotreta striolata*). Lamb et al. (1993) identified a line derived from *Brassica rapa* that shows quantitative resistance to flea beetles (*P. striolata*). This resistance is attributed to a low level of antixenosis, rapid growth at the cotyledon stage, and tolerance to damage. The existence of natural variation in response to pest infestations among oilseed rape cultivars and *Brassica* species suggests the potential for breeding resistant *B. napus* cultivars. Further investigations are necessary in order to understand the underlying mechanisms and to develop more adapted cultivars.

17.2.3 Effect of Nitrogen Supply and Cultivar on Crop Condition and Crop Ability to Reduce Pest Damage

Numerous studies have pointed out that the oilseed rape crop can compensate for damage caused by different herbivores (Williams and Free 1979, Lerin 1988). Williams and Free (1979) showed that the removal of up to 60% of the buds caused no yield loss. Compensation mechanisms take place at different levels of the plant and through different yield components as a function of pedoclimatic conditions, timing of herbivore attacks and crop status. For example, pollen beetle attacks involve loss of apical dominance and result in a stunting of the attacked racemes and the outgrowth of side racemes (Nilsson 1994) with more pods per side raceme (Podlaska et al. 1996). Pollen beetle attacks reduce in most cases the number of seeds per pod due to the delay in pod formation. Thus compensation for seed loss takes place through an increase in seed weight. It is also known that oilseed rape crops can compensate for changes in plant density. Indeed single plants in low density populations grow larger and produce more pods and more seeds than those in high seeded density. As the ability of the crop to compensate for pest damage directly depends on crop status, nutritional resources and nitrogen availability, plants with sufficient available nitrogen in their direct environment will be more able to reduce yield losses due to herbivory. Indeed, Valantin-Morison et al. (2007) found a negative correlation between nitrogen availability in the soil and pollen beetle damage certainly due to nitrogen effect on plant vigour and compensation ability. The synchronization between pest attack and nitrogen supply appears to be very important to consider even if little is known about it.

Insects choose the most favourable host plant for the development of their eggs, as a function of their potential capacity to meet the needs of the larvae. Several studies have shown that pollen beetles adjust the number of eggs laid per bud and the amount of resources in the egg for larval development as a function of plant quality (Hopkins and Ekbom 1996, 1999). In addition, Nilsson (1994) showed that females selected flower buds as a function of size, showing a marked preference for buds between 2 and 3 mm long and refusing the suboptimal buds unless there was very strong competition for oviposition. Another pest of oilseed rape crops, the cabbage root fly has also been shown to select the most favourable plants for oviposition on the basis of stem diameter (Dosdall et al. 1996, Valantin-Morison et al. 2007). Finally, the architecture and height of the plant also seem to play an important role in host selection mechanisms (Kostal 1993, Rojas and Wyatt 1999, Finch and Collier 2003). The amount and timing of nitrogen supply may play a significant role in determining the potential attractiveness of the crop to pests by modifying plant quality and architecture. Indeed, high nitrogen amounts could induce the production of large flower buds which is more favourable for pollen beetle oviposition but also produce more secondary racemes, which allow better compensation for pest damage.

Even if some knowledge has been acquired, the general understanding of the effect of some crop practices, particularly nitrogen supply and cultivar, on crop response to pest infestations in oilseed rape, is not well established. Nevertheless, this understanding is the cornerstone of establishing injury and economic thresholds for oilseed rape pests which are still lacking in general or showing great variability between countries (as demonstrated by Nilsson 1994, see also Williams Chapter 1 this volume).

17.2.4 Effect of Sowing and Harvesting Dates on the Synchronization Between Crop, Pest, and Predator Life Cycles

One of the strategies underlying pest avoidance is the desynchronization between crop susceptibility periods and the biological cycle of pests. This effect is widely known in cereal and oilseed crops. Dosdall and Stevenson (2005) demonstrated that the sowing date of oilseed rape strongly affects flea beetle (*Phyllotreta cruciferae*) damage. Indeed, the damage was greater on spring-seeded oilseed rape than on plants seeded in the autumn. Flea beetle feeding damage to oilseed rape apical meristems can prevent a compensatory response, but by the time of greatest injury, winter oilseed rape had well-developed, enlarged apical meristems making them less susceptible to damage. Winter oilseed rape enabled plants to progress beyond the vulnerable cotyledon stage by the time that most flea beetle damage was inflicted, resulting in less crop damage.

The same effect was observed on the maize crop. Early-sown maize is less susceptible to maize earworm and stem borer, *Diatrae grandiosella* (Bajwa and Kogan 2004). This lower susceptibility results from the fact that *D. grandiosella* tends to lay fewer eggs on more mature plants, which have already passed their critical growth stage before most of the larvae begin to feed (Herzog and Funderburk 1985). Pest avoidance can also be achieved by modifying harvest date to limit the damage caused by various pests attacking the crop shortly before harvest (Flint and Gouveia 2001). For instance, for *D. grandiosella* on maize, in addition to the effects of early sowing described above, early sown maize can be harvested before fully grown pre-diapause larvae have girdled the plants and caused yield losses through lodging (Bajwa and Kogan 2004). But sowing date effects can be antagonistic when considering different pest populations. Indeed, Valantin-Morison et al. (2007) have shown that sowing oilseed rape early tended to increase cabbage root fly damage, whereas it was associated with a lower level of attack by cabbage stem flea beetle.

Effects of sowing date on the incidence of the barley yellow dwarf luteovirus have been reported (McKirdy and Jones 1997). The virus is transmitted by aphids and the main vectors are bird cherry-oat aphid (*Rhopalosiphum padi*) and grain aphid (*Sitobion avenae*). As the major flight of aphids peaks in September and October, delayed sowing reduced the number of aphids per plant and decreased disease incidence. Effects of sowing date on the severity of other crop diseases have also been reported (e.g., Krupinsky et al. 2002). In the case of oilseed rape, Aubertot et al. (2004) demonstrated that early sowing dates reduced damage caused by *Leptosphaeria maculans*, the causal agent of phoma stem canker, one of the most important oilseed rape diseases. In this pathosystem, damage at harvest is most severe if infection occurs soon after emergence (Brunin and Lacoste 1970). As the primary inoculum of *L. maculans* generally peaks between September and December (West et al. 2002), early crop sowing leads to a lower risk of infection just after emergence than other sowing dates.

Moreover, harvesting can have a major impact on field fauna. In fact, it produces a brutal perturbation of the agroecosystem involving microclimate changes that impact natural enemies at the field scale. According to Riechert and Lockley (1984) harvest effects have a greater impact on spider communities than does the use of pesticides. Harvest effects depend on the time at which harvest actually occurs. For spring crops (such as maize) harvest is sufficiently late for most of the predatory species to be at the end of their activity period and to have reached their overwintering sites. For winter crops (such as winter oilseed rape and most of the cereals) harvest dates generally coincide with the maximum abundance and activity of some predators (Büchs 2003). This demonstrates that sowing and harvest date can have an important effect on predator population and pest regulation. But possible antagonistic effects, highlighted above, imply that pest avoidance should not be seen as a simple technical operation that is easy and straightforward to apply. Instead, many interactions must be taken into account when designing pest management strategies. Overall, pest avoidance strategies should be designed within the conceptual framework of integrated crop management (as defined by Royer et al. 1999).

17.3 Crop Management Effect on Local Habitat Environment to Promote Biological Control of Insect Pests

Different aspects of crop management affect the local condition of the field and influence (directly or indirectly) pest dynamics, pest damage and natural enemies. We review the most important factors of crop management with particular reference to oilseed rape. The objective of this part is to present the main local elements that could be combined in an integrated pest management approach.

17.3.1 Direct Effects of Soil Tillage on Predator and Parasitoid Populations

Parasitoid populations are very susceptible to post-harvest soil tillage as the majority of the parasitoid species of oilseed rape pests overwinter in the soil of rape fields. Thus, it has been shown that soil cultivation techniques used to establish the crop following rape can greatly influence survival, emergence rates and then parasitization of oilseed rape pests of the following year (Nilsson 1985, 1994, Hokkanen et al. 1988). For further details on the effects of soil tillage on parasitoids of oilseed rape pests, see Nilsson (Chapter 11 this volume). Reduction in arthropod densities caused by different crop management activities (which cause direct mortality) varies from 25 to 60% according to the different taxa (Thorbek and Bilde 2004), and spiders seem to be more vulnerable to mechanical crop treatments than carabid and staphylinid beetles. Moreover, important and sometimes higher effects of habitat disruption are observed a few days after the crop management operation (Thorbek and Bilde 2004).

The timing of tillage procedures also has an important impact on predator populations. Ploughing in spring has a more harmful effect on ground beetle populations than autumn ploughing. It is known that tillage in late summer or in autumn is more beneficial to population development of some beetles than spring ploughing. It is assumed that later in the season most of the predators will have moved out to their overwintering sites, mostly in uncultivated areas, or will have moved deeper into the soil (Büchs et al. 1999).

17.3.2 Effects of Mulch on Generalist Predators and Biological Control

Physical disturbance of the soil caused by tillage and residue management is a crucial factor in determining soil biotic activity and species diversity in agroecosystems (Altieri 1999). Soil tillage affects organic matter characteristics and location, water regime and structure of the soil surface (Holland 2004). Reduced tillage (with surface placement of residues) creates a relatively more stable environment, encourages the development of more diverse species (in particular decomposer communities) and slower nutrient turnover. Several studies on different crops including oilseed rape have shown that increasing the structural complexity of the soil through the addition of organic matter almost always leads to an increase in the diversity of generalist predators (ground beetles, spiders and hoverflies) (Kromp 1999, Schmidt et al. 2004, Pullaro et al. 2006, Brevault et al. 2007). In most cases it leads to a decrease in the populations of certain crop pests (Zehnder and Hough-Goldstein 1989, Brust 1994, Schmidt et al. 2004, Pullaro et al. 2006, Zehnder et al. 2007).

Landis et al. (2000), in their review of habitat management to conserve natural enemies of arthropod pests, reported that manure and straw increased numbers of the carabid *Bembidion lampros*, an egg predator of the cabbage root fly, and increased total carabid populations in cabbage. This increase was apparent even into the year following the last application and was attributed to increased reproduction by the predators, which, in turn, may have resulted from the observed increase in alternative prey availability. Indeed, the presence of decomposing organic matter at the surface of the soil provides predators with alternative prey when crop pests are no longer present in the plot. Therefore, no-tillage systems, such as direct drilling, that leave crop residues on the soil surface, increase the populations and impacts of predatory carabids. Kendall et al. (1991) even showed that, in winter barley crops, the amount of straw present on the soil surface was positively correlated with the diversity of polyphagous predators and negatively correlated with barley yellow dwarf virus infection levels. Some studies have also reported that mulch causes disturbances in pest prospecting and approach behaviour. Indeed, pests are less efficient at host plant location, due to physical barriers and release of allelochemical substances from the decomposing (Mabbett 1991) or living mulch (Finch and Collier 2000). Büchs and Katzur (2004) confirmed that reduced-tillage favours natural enemies in the case of organic oilseed rape. In their study, they focused on the occurrence and control of rape pests by comparing three different treatments: a plough/comb harrowing treatment, a mulch/comb harrowing treatment and a mulch/hoeing treatment. They found that the mulch/comb harrowing treatment promoted the natural enemies of oilseed rape pests compared to the other treatments. However, higher risk of yield loss due to weed populations was found in the case of the mulch/comb harrowing treatment. The mulch/hoeing treatment appeared to reduce the weed population and yield loss risks but also reduced natural enemies and increased pest damage. Thus no consensus has been reached about the effects of mulch left by simplified soil cultivation, and indeed, in some cases, the number of pests increased. Mabbett (1991) has reported an enhancement of slug (*Deroceras reticulatum*) populations after mulch application to winter barley fields. Different hypotheses have been proposed to explain the enhancement effects of mulch on pest populations. First of all, a lack of natural enemies, including large ground beetles specialising in mollusc predation, due to deleterious insecticide treatments may be responsible for such effects (Chabert and Gandrey 2005). Secondly, deficiencies in the ecological infrastructure (Bohan et al. 2000, de la Peña et al. 2003, Holland et al. 2005), with slugs being favoured by the presence of organic matter can also be responsible for this enhancement. Finally, an overabundance of alternative prey, such as springtails, aphids, fly eggs and larvae, may favour pest populations (Mair and Port 2002, Symondson et al. 2006).

For many crops, the effects of mulch on pest activity and generalist predators appear to be well known but little is known about its effects on oilseed rape-specific pests. Furthermore, as mentioned in part 1 of this chapter, leaving mulch on the soil surface should not be thought of as the sole alternative to pesticide use, since many interactions between other practices and other pests must be taken into account.

17.3.3 Effect of Pesticide Use on Natural Enemy Populations

The impact of insecticides on parasitoid populations has been addressed by Ulber and Klukowski (Chapter 13 this volume). Many parasitoid species of different oilseed rape pests emerge 1–2 weeks before rape flowering and are particularly active in the crop during the flowering stage, searching for suitable hosts. Thus, they can be very affected by late insecticide spraying occurring around flowering (Nilsson 1994, Nitzsche and Ulber 1998, see also Johnen et al. Chapter 15 this volume).

Other effects of pesticides on natural enemies have also been reported on different crops including oilseed rape. In the literature, pesticides are considered to be one of the main causes for biodiversity loss (Ewald and Aebischer 2000). Indeed, numerous studies have illustrated the impact of pesticides on different communities of natural enemies (Chabert and Gandrey 2005, Koss et al. 2005, Tietjen and Cady 2007). Some have pointed out that the side effect of insecticide use on natural enemies can be the origin of important pest proliferations (Gordon and McEwen 1984). Even if it is admitted that species inhabiting higher parts of the field vegetation are more exposed to pesticide effects (Vickerman 1992), it has been shown that pesticide use decreased the number of arthropod species (hypogeous or epigeous species) at the field scale and decreased biological control by natural enemies (Burn 1988, Duffield 1991). Stark et al. (2004) have shown that life story attributes as well as population structure at the time of pesticide exposure both play a major role in population susceptibility to pesticides. Therefore, the authors have highlighted the need to explicitly consider differences in life history variables among species when calculating compatibility of pesticides and biological control agents as well as the population structure of beneficial species at the time of pesticide application. Furthermore, herbicides and fungicides have important indirect effects on arthropod communities (such as phytophagous or flower-visiting arthropods) as they generally suppress their resources (Landis et al. 2000). Büchs et al. (1991) also showed that insecticide seed treatments on oilseed rape may have seriously affected epigaeic predators such as ground beetles, because even predatory species are partly phytophagous.

17.4 Effect of Crop Diversification in Space and Over Time on Biological Control: Crop Succession and Within-Crop Diversification

Diversified cropping systems in space and over time, such as those based on intercropping, agroforestry, cover cropping and on more diversified crop rotations have been the target of much recent research. This interest is partly based on the emerging evidence that these systems are more stable and more resource conserving (Vandermeer 1995). Indeed, crop monocultures are environments in which it is difficult to induce efficient biological pest control because these systems lack adequate resources for effective performance of natural enemies. Therefore such systems are more susceptible to pest or disease outbreaks. Some elements at the field and farm scales can be used to increase crop diversity and induced biological control.

17.4.1 Diversifying Species in Crop Succession: Effect on Pests and Natural Enemy Communities

Rotation of annual crops has been empirically developed by farmers to reduce and control soil-borne pest and disease proliferation. In the middle of the last century, a well developed rotation consisted of six to eight different crops in sequence (Häni et al. 1998). An increase in economic pressure and food demand pushed farmers to spread more pesticides and to maximise land use. The rotation was reduced to a few species, leading to an increase in pest proliferation and a decrease in biodiversity of beneficial species.

A meta-analysis (Tonhasca and Bryne 1994) showed that crop diversification led, in 52–70% of cases (21 studies), to a decrease in pest density. Since this study, several publications have taken part in the debate over the consequences of crop diversification for pests (Coll and Bottrell 1995, Theunissen et al. 1995, Khan et al. 1997, Schellhorn and Sork 1997, Harmon et al. 2003, Hooks and Johnson 2003, Aquilino et al. 2005, Costamagna and Landis 2006, Khan et al. 2006, Bjözrkman et al. 2007). Eight of these publications have reported that crop diversification successfully decreased the populations of all or some pests and two studies found that crop diversification had no effect on pest populations.

Other studies showed that diversifying crop succession increased some natural enemies. Indeed, Büchs et al. (1997) studied the effects of different crop rotation intensities on the arthropod community in a sugar beet rotation and an oilseed rape rotation. They showed that certain pest species were enhanced by an increasing intensity of crop rotation, while some beneficial insects were not able to build up stable populations in arable crops in such intensive succession. The authors found that the number of individuals, species richness, body length, and reproductive rates of beneficial insects increased with progressive extensification of crop sequences and especially in set-aside areas with natural succession.

However, higher crop diversification within crop sequence does not always increase abundance or activity of pest antagonists. Indeed, different studies have compared carabid populations between different crop successions including monoculture and no significant differences were found (Holland et al. 1996, Winstone et al. 1996, Kromp 1999).

Some advantages of crop diversification are also illustrated by the use of break crops to control soil-borne diseases. The term ‘break crop’ refers to breaking the life cycle of a crop-specific pathogen by growing a non-host crop in the rotation. Winter oilseed rape and other cruciferous plants could be considered as a biofumigation crop (Kirkegaard et al. 2008).

17.4.2 Increasing Within-Field Diversity: Effect of Cover Cropping and Intercropping on Pests and Natural Enemy Communities

Cover cropping and intercropping increase the diversity within the field. Plant cover has many roles like improving the resources and growth conditions or decreasing the impact of pests, thereby increasing crop productivity. Cover crops can be separated in two main categories: annuals grown during an off-season that are killed before planting a cash crop, providing a mulch; and living mulches that grow at the same time as the cash crop for all or part of the growing season, resulting in an intercropping system.

An increasing body of literature demonstrates that pest proliferation decreases when crop specific diversity increases (Landis et al. 2000, Ferron and Deguine 2005). Many studies have shown that the introduction of a cover crop in perennial and annual crops may improve pest control (Wyss 1995, Pickett and Bugg 1998, Altieri and Nicholls 2004, Pfiffner and Wyss 2004, Broad et al. 2008). In fact, perennial crop systems are potentially more amenable to conservation biological control than are ephemeral annual systems because they are subject to lower levels of disturbance. Thus, resident populations of natural enemies may persist from year to year in perennial crops (Landis et al. 2000). However, cover crops may also aggravate pest damage or favour certain new pests when the plants used provide the pests with a key resource (Pfiffner and Wyss 2004). In 16 published cases reviewed by Russell (1989), natural enemies were more abundant in polycultures than in monocultures in 10 cases, in four cases no effect was reported, and in two cases predator numbers were lowered. Moreover, in a review, Andow (1991a) analysed the results of 209 studies on relations between intercropping and pest populations (287 pest species). Pests were significantly reduced in 52% of the cases compared to monocultures and were more important in 15%. These conclusions have been corroborated by different studies (Nickel 1973, Perrin 1977, Risch 1983) and are explained by different mechanisms.

First of all, the introduction of a cover crop produces indirect effects on pest populations involving host quality (Trenbath 1993) but also plant/crop architecture and crop microclimate (Altieri and Liebman 1986, Francis and Clegg 1990, Landis et al. 2000). Indeed, intercropping often leads to important changes in crop architecture and many authors have shown that insect pest behaviour (such as feeding, flight or reproduction) are very susceptible to general crop architecture. Cabbage moth (*Mamestra brassicae*) is affected by plant height (Rojas and Wyatt 1999) while the main factor governing the selection, during the landing, of the site most suitable for cabbage root fly was the conspicuousness of the plant (Kostal 1993).

Root (1973) incorporated the possible mechanisms that underlie the response of herbivore pest to habitat diversification into two hypotheses: the enemy hypothesis (first formulated by Pimentel (1961)) and the resource concentration hypothesis. The first one exposed indirect effects of the intercropping system on pests and states that the observed reduction of herbivores on intercropped systems is partly due to the attractiveness of the intercrop for more abundant and/or efficient predators and parasitoids, presumably because of the greater availability of resources and habitats as compared to the monoculture.

Intercropping also has more direct effects on the pest. The resource concentration hypothesis formulated by Root (1973) states that 'herbivores are more likely to find and remain on hosts that are growing in dense or nearly pure stands; the most specialized species frequently attain higher relative densities in simple environments; and, as a result, biomass tends to become concentrated in a few species, causing a decrease in the diversity of herbivores in pure stands'. According to this hypothesis the probability of pests finding their host plant, remaining on it and reproducing on it, is higher in monocultures than in a mixture of several species because the resource is diluted among other resources (Tahvanainen and Root 1972, Root 1973). This hypothesis therefore predicts a negative relationship between plant diversity and the level of invertebrate phytophagy (Root 1973), regardless of any interaction with the natural enemies of pest species. As emphasised in many studies exploring this hypothesis, the observed effects could have resulted from different mechanisms such as arthropod movement in approaching the crop and within it, immigration and emigration, and host finding and acceptance (Risch 1983). Other mechanisms such as barrier effects, as well as visual and olfactory effects on host location have also been reported (Perrin 1977, Altieri and Liebman 1986). Different studies (Risch 1983, Andow 1991a, b) have suggested that mechanisms of resource concentration rather than natural enemies contribute to lower herbivores numbers in polycultures.

Little is known about the specific effects of increasing within-crop diversity in the case of oilseed rape. Weiss et al. (1994) studied the influence of an intercropped agroecosystem of oilseed rape, canola-type (*Brassica napus*) and field pea (*Pisum sativum*) on the population level of the flea beetle (*Phyllotreta cruciferae)*, but no effect in reducing chrysomelid loads, nor increasing yield were recorded for the intercropped system. Butts et al. (2003) in their study on the effect of intercropping oilseed rape or pea with barley on assemblages of ground beetles (Coleoptera: Carabidae), found no effect of the intercropping system. Finally, Hokkanen (2008) reported that intercropping experiments revealed no differences in the number of pollen beetles per plant, percentage of parasitism, or overall predator activity densities between the monocrop and intercrop. However, the number of emerging new generation pollen beetles appeared drastically reduced in the intercrop as compared with the monocrop, which produced about five times as many F1 pollen beetles as the intercrop per surface area, or about two and half times as many per rapeseed plant. The author concluded that higher predator pressure (lower total number of prey, but equal number of predators) in the intercrop may be an explanation for such effects.

17.5 Influence of Spatial Context on Oilseed Rape Pests and Their Biological Control in Agricultural Landscapes

Modern agricultural landscapes generally consist of a changing mosaic of cultivated and uncultivated habitats. Agricultural intensification has led to a marked simplification and fragmentation of rural landscapes that are nowadays usually characterized by a high proportion of arable fields, an enlargement of field sizes and a high degree of fragmentation of small natural habitats. There is a growing body of evidence showing that species community structure and abundance and more generally biotic interactions have to be considered at a spatial scale much wider than a single patch of habitat. It is therefore important to link the biological characteristics and ecological processes of the given organism to spatial pattern. Indeed, spatial context does not influence all species in the same way and responses to landscape structure are species specific (i.e., species ability to disperse, species life cycle) (Dunning et al. 1992, Fahrig and Merriam 1994).

17.5.1 Effect of Semi-Natural Areas on Pests and Natural Enemy Populations

Even in situations of considerable crop diversity, all fields constitute ephemeral habitats which could be considered a hostile environment for many animal species, including natural enemies of phytophagous insects (Bianchi et al. 2006). On the contrary, non-cultivated habitats in rural landscapes, such as hedgerows, field margins, fallow land and woods, are relatively undisturbed, largely unchanging areas and provide both woody and herbaceous habitats. Non-crop habitats provide life support functions, maintaining the presence of alternative hosts and prey for predator and parasitoid populations (Sotherton 1984, Pickett et al. 2000, Denys and Tscharntke 2002). They can also provide sources of pollen and nectar, which are essential for many insects (Pickett and Bugg 1998). Finally, woody habitats often provide a more moderate microclimate than field centres, protecting parasitoids against extreme temperature variations (Rahim et al. 1991) and provide good habitats for overwintering of natural enemies and pests. According to Keller and Häni (2000), nine out of 10 auxiliary species need non-crop environments at some point in their life cycle, whereas this is the case for only one of two pest species. Therefore, most auxiliary species are heavily dependent on the resources provided by semi-natural environments requiring them to travel back and forth between uncultivated habitats and the given crop.

17.5.2 Effect of Landscape Context on Pests, Natural Enemies and Diseases

According to Pulliam (1988) and Pulliam and Danielson (1991) landscape mosaic can be conceived as functioning in terms of different sources and sinks for parasitoid populations. Non-crop habitats serve as the starting point for field colonization to various extents for many species beneficial, damaging or neutral to crops (Nentwig 1988, Dennis and Fry 1992, Thomas et al. 1992, Denys and Tscharntke 2002, Marshall 2004) and thus distance and spatial arrangement between fields and non-crop areas are important. It has been suggested that parasitoids and predators may generally act at smaller spatial scales than phytophagous pests (Roland and Taylor 1997, Zabel and Tscharntke 1998). This may account for the frequent observation that the intensity of the ecological pest control services, mediated by predators and parasitoids, decreases with increasing distance to the non-crop refuge zones (Altieri and Schmidt 1986, Klinger 1987, Collins et al. 2002). Thus, the spatial and temporal dynamics of many auxiliary species populations depend heavily on the characteristics of the landscape (Marino and Landis 1996, Elliot et al. 1998, Thies and Tscharntke 1999, Östman et al. 2001, Kruess 2003, Sarthou et al. 2005, Roschewitz et al. 2005), particularly in conventional production systems in which the presence of auxiliary species in crops is becoming increasingly dependent on the regular arrival of individuals from semi-natural habitats to re-colonize the field (Schmidt et al. 2005). The importance of local abundance of source habitats, refuges, and alternative hosts in the landscape on crop disease prevalence has also been demonstrated (Plantegenest et al. 2007). Landscape structure appeared to have an important effect on disease dynamics as pathogens disperse through different spatial dependent mechanisms, such as aerial dispersal to vector-borne viruses. Therefore some elements in the agricultural landscape, that influence air motion for example, may act as barriers limiting disease spread, or inversely as corridors facilitating the dispersal of the pathogen.

A literature review by Bianchi et al. (2006) analysed 28 studies focusing on pest pressure and/or on natural enemy populations in relation to landscape composition in the case of various crops. The authors found that pest pressure was reduced in complex landscapes in 45% of 10 studies reviewed. They also found that natural enemy activity was enhanced by complex landscape in 74% of the studies reviewed (24 publications). In 21% of the studies reviewed, no effect of landscape composition was reported, while in 5%, natural enemy activity was lower in complex than in simple landscapes. Even if the majority of the studies showed higher natural enemy activity, only 45% of them appeared to reduce pest pressure in more complex landscapes indicating important variability in organism responses to landscape structure.

For instance, in the case of oilseed rape pests, relations between pollen beetle, pod midge, stem weevil and landscape composition have been reported from two main studies (Thies et al. 2003, Zaller et al. 2008b). In these, pollen beetle responses to landscape complexity are contrasted. Thies et al. (2003) found that pollen beetle activity appeared to be negatively correlated to landscape complexity and that parasitism rates were positively correlated to landscape complexity. However, Zaller et al. (2008b) showed that the abundance of pollen beetles were negatively related to oilseed rape area and positively related to woody areas. Here the results suggest that complex landscapes enhanced pest populations either by supporting a greater variety of alternative host plants or by providing more suitable habitats for overwintering. This difference of response to landscape characteristics probably comes from the different landscape metrics chosen for quantifying landscape patterns, the landscapes and regions per se used in these studies, and also from fundamentally different methodological approaches. Effects of spatial context at the landscape scale on oilseed rape pests and their biological control are reviewed in more detail by Frank et al. (Chapter 10 this volume), and by Thies and Tscharntke (Chapter 9 this volume).

17.5.3 Effect of Field Margins on Pest Regulation

Vegetative buffers in agricultural landscapes can provide a range of important ecological services, including conservation of native flora and fauna, enhancement of biological pest control, and reduction of agrochemical drift. The characteristics of field margins, such as age and composition, contribute to the efficiency of these buffer zones on pest regulation.

First of all, it has been reported that the effects of these strips on the biodiversity of auxiliary species heavily depend on the plant species within the margin. Indeed, it is known that the provision of adequate floral resources favours the development and activity of predator and parasitoid populations (Rebek et al. 2006). Different studies have been interested in determining the species composition of flower strips most favourable to all sorts of auxiliary species important for crops (Nentwig et al. 1998, Chiverton 1999, Wäckers 2004). Moreover, many studies have been carried out on the effects on flower-dependent specialist predators and parasitoids of either single species flowering strips (Lővei et al. 1992, Hickman and Wratten 1996, Petanidou 2003, Pontin et al. 2006) or on flowering strips composed of several species (Klinger 1987, Sutherland et al. 2001, Rebek et al. 2006, Pontin et al. 2006). Cultivating a well-studied single flower species ensures the conservation of a particular target beneficial species and minimises the risk of non-target effects, such as inadvertently promoting populations of pests (Baggen et al. 1999), higher-order predators or hyperparasitoids (Stephens et al. 1998). In contrast, the use of flower mixtures diversifies the resources available, which caters for a greater diversity of pollinators and natural enemies because of selective feeding on the various different floral components. For the moment, no specific study has been reported on the effects of floral composition of field margins on biological control of oilseed rape pests.

Many studies have also shown that these flowering strips, favouring the development of flower-dependent specialist predators and parasitoids, also make very good refuges, after 2–3 years of ecological maturation, with the planting of several tussock grasses for ground-dwelling auxiliary species (Nentwig 1988, Frank and Nentwig 1995, Pfiffner and Luka 2000, Meek et al. 2002). Field margins, with a naturally diverse flora or with sown 'wild flowers', harbour the greatest abundance and diversity of arthropods (Lagerlöf and Wallin 1993). Thomas et al. (2002) also observed that the incorporation of wildflower seed into tussock grass mixtures sown for beetle banks provided resources for bumblebees, parasitoids, hoverflies and butterflies.

Secondly, many different studies have shown that the age of field margins plays an important role in biological control. Frank (1996) showed significant effects of sown weed strips of different age on the diversity of epigaeic predators (ground beetles and spiders) as well as of hover flies (Syrphidae) and solitary wasps (Sphecidae) and their tendency to disperse into adjacent crops. Büchi (2002) showed that parasitism rates of pollen beetle larvae by *Tersilochus heterocerus* were significantly higher in fields with wild flower strips than in fields with adjacent extensively-managed meadow. He also showed that, in both cases, pollen beetle larval density within the crop increased with increasing distance from field borders. Thies and

Tscharntke (1999) showed that age of field margin strips and fallow habitats had an important effect on parasitism rates of pollen beetle larvae. Old fallow field margins or fallow habitats exhibit higher pollen beetle mortality. These types of old habitats are less disrupted areas than cultivated ones and allow parasitoid populations to build up and to disperse into the crop. However, some studies have also shown that field margins did not always increase natural enemies and biological regulation of insect pests (Holland et al. 2008). Pfiffner et al. (2009) in their study of the impact of wildflower strips on biological control of cabbage Lepidoptera (*M. brassicae, Pieris rapae*) demonstrated that the provision of wildflower strips does not necessarily enhance biological control and suggested that site-specific environmental factors strongly affect their impact.

These results indicate that field margins are not systematically beneficial for pest control, since it depends on floral composition as well as on the age of the boundary and local conditions. Moreover, when designing low input cropping systems, it is necessary to consider field margin management in order to adapt cultural practices to the ecological services provide by the boundary.

17.5.4 Relevance and Robustness of Landscape Scale Approaches for Effective Biological Control

All the studies which have been carried out at the landscape scale provide evidence that the conservation of arthropods and the enhancement of biological control need a spatial context perspective. Furthermore, they allow us to understand how pest control could be strengthened by giving a basis for predicting how pest species and their natural enemies respond to landscape context. However, the effect of landscape characteristics in the biocontrol of pests remains a matter of debate. Firstly, the enhancement of natural enemy populations does not necessarily imply an effective pest control and the relationships between crop and non-crop habitats are complex and sometimes antagonistic (Thies and Tscharntke 1999, Valantin-Morison et al. 2007, Zaller et al. 2008a). Secondly, the effects of higher biological control on productivity are not well known and landscape effects on pest populations and crop damage have rarely been documented even though they are much more relevant than any effect on natural enemies (Bianchi et al. 2006). Many studies at the landscape scale suffer from the lack of information about the real effects of crop management and higher biodiversity on crop damage and yield losses. Indeed, Bianchi et al. (2006), in their review, found 10 studies dealing with pest densities and demonstrated that lowered pest pressure in complex landscape was found in 45% of the studies. In instances where agricultural biodiversity has enhanced biological control and reduced pest densities, the need for pesticide inputs may be lessened although linking higher biodiversity to higher yields through pest regulation is very difficult to demonstrate (Gurr and Wratten 2000). Thus, although the highly diverse auxiliary species communities present at the edge of the plot are capable of providing the crop with a certain level of protection (Klinger 1987, Hausammann 1996, Platt et al. 1999) without generally acting as a source of crop pests (Lagerlof and Wallin 1993,

Geiger et al. 2005), agronomists still have to demonstrate the beneficial effects on production and identify the key points relating to the cultivated field, neighbouring plots, and the surrounding area that might accentuate these effects.

In most of the studies previously cited, soil occupation within the landscape (in terms of crop or non-crop area, for example) was the only factor taken into account. Cropping systems in the farming landscape have generally been neglected (e.g., distribution of resistant varieties, ploughed and unploughed area, or frequency and timing of pesticide utilisation). A spatial consideration of the different cropping systems allocated throughout the landscape will allow a better comprehension of interactions occurring at a scale larger than the crop field. Agronomists designing crop protection strategies that maximise biological control at the landscape scale should integrate already existing ecological principles in order to build practical alternative systems that suit the specific needs of farmers and society. Because of the emerging importance of interactions between the field crop and the surrounding environment and the relative importance of spatial organization, agronomists have to reconsider the scale at which pest management strategies are being designed and have to use a less physicochemical approach to the environment than in the past.

17.6 Designing New Crop Protection Strategies: How Can We Use Crop Management and Landscape Effects to Improve Pest Management?

Integrated pest management strategies focus on long-term management of pest populations through a combination of techniques, including enhancing biological control, use of resistant varieties, chemical control, adoption of cultural practices such as crop rotation or sowing date to make habitat less convenient to pest population development, or physical control methods such as mechanical, pneumatic, thermal, or electromagnetic techniques for reducing pest populations. Pesticides are used only when careful monitoring indicates they are needed according to pre-established guidelines based on treatment and economic thresholds. More generally, these methods are the cornerstone of Integrated Crop Management strategies which are considered as a reasonable trade-off between profitability and environmental protection avoiding waste, enhancing energy efficiency and minimizing pollution (El Titi et al. 1993, see also Nilsson Chapter 16 this volume). The combinations of different techniques to achieve integrated pest management with more biological control involve profound modifications in the nature of the cropping system and different considerations. Firstly, the scale issue requires linking integrated pest management strategy to more ecological knowledge, especially landscape ecology. Secondly, designing innovative cropping systems needs a systemic approach that considers the entire agroecosystem and that particularly pays attention to farming techniques as well as economical and environmental impacts.

17.6.1 Linking Integrated Pest Management Against Insect Pests and Landscape Ecology

Designing integrated crop management strategies for farmers aims to achieve different objectives: (i) a production purpose (crop performance and quality of products), (ii) socio-economic imperatives (farm organization, farm income) and (iii) environmental issues (limitation of pesticides and nitrogen discharged into the environment, minimization of water and energy use). The integrated pest management paradigm holds that pests and their management exist at the cross roads of three multidimensional major fields of study: ecology, socio-economy and agriculture in hierarchical order, with ascending levels of complexity and expanding spatial scales. Based on these considerations, integrated pest management can be conceived as interactive systems with multiple levels of integration. Kogan (1988) proposed three different levels for insect pest management: (i) the integration of methods for the control of single species or species complexes (species/population level), (ii) the integration of impacts of multiple pest categories (insect, pathogen, and weeds) and the methods for their control (community level), and (iii) the integration of multiple pest impacts and the methods for their control at the total cropping system (agroecosystem level). A fourth level can be considered: the integration of social, political and legal constraints into integrated pest management (Kogan 1998).

Cumming and Spiesman (2006) have shown that, although integrated pest management was developed as a multiple level system and some area-wide pest management programs have known some success, integrated pest management has proceeded through an essay/error approach with few theoretical concerns (Kogan 1998) and especially without turning to theory developed in ecology. The integrated pest management paradigm would benefit from incorporating more ecological theory, particularly landscape ecology, into its foundation (Cumming and Spiesman 2006). They also argued that integrated pest management and habitat fragmentation control are two complementary aspects of the same problem. Crop protection strategy need no longer be a phytosanitary issue at the field scale and at a given moment, but could benefit from a more holistic approach at the farm and landscape levels. The link between landscape ecology theory and integrated pest management knowledge should enhance the effectiveness of integrated pest management especially management based on biological control. Landscape ecology theory brings a regional perspective to the integration of pest control strategies and it should lead us to a better understanding of the multi-scale relationships between the control of pest outbreaks and the larger landscape. This entails a real management of pest population in space and over time, taking into account cultivated and uncultivated habitat management (Ferron and Deguine 2005). Indeed, uncultivated habitats within the landscape have to be considered in integrated pest management strategies as they are key elements in providing ecological services. The development of an integrated pest management at a broader scale requires a thorough understanding of the cropping system effects on both pest and natural enemy populations and the spatial ecology of the given pest and beneficial species in order to maintain and optimise a more complete range of ecological functions and ecosystem services within farming systems accounting for farmers' economical imperatives.

17.6.2 Methods of the Agronomist, a Complementary Approach Between Models, Diagnosis and Systems Experiments

17.6.2.1 General Framework for the Design of Innovative Cropping Systems

The elementary control methods of integrated pest management can be combined to control pest communities through integrated pest management strategies (Dhaliwal et al. 2004). The literature reveals that a single cultural practice given the pests and the objectives can lead to antagonistic effects. Nevertheless, the assessment of the combination of different control methods into crop management has received little attention and there is currently a lack of pesticide-free crop protection strategies in the case of oilseed rape. We present here a general methodological framework for designing innovative cropping systems and report the implications of reconsidering the scale for such an approach.

As explained earlier, designing cropping systems is a multi-objective task that has to consider a range of imperatives such as crop yield, environmental impacts, and farmers' incomes. In order to face the challenge of multi-functional and sustainable cropping systems, agronomists have developed a methodological framework to evaluate and conceive new cropping systems. Meynard et al. (2001) have developed a holistic approach which takes into account limiting factors of a given situation and thereby consider possible antagonist effects for designing and evaluating cropping systems (Fig. 17.1). In the case of pest management strategy, this methodological framework takes into account all pest damage of a given region and does not consider only one type of pest. The first step, the regional diagnosis step, is based on

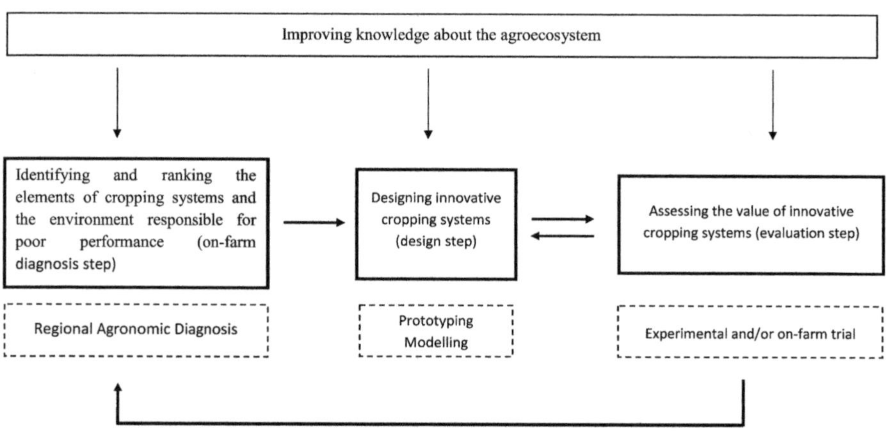

Fig. 17.1 Methodological framework for cropping system improvement (adapted from Meynard et al. 2001). The three iterative main steps of the general framework are in *bold lines*. The *dashed lines* indicate particular relevant examples of tools used at each step of the conception process

identifying and ranking the elements of the cropping systems and the environment responsible for poor performance. The second step consists in designing innovative cropping systems through different ways, such as prototyping or modelling. The last step is the evaluation step, which consists of assessing the value of the innovative cropping system.

17.6.2.2 The Regional Agronomic Diagnosis: A Pre-requisite for Designing Cropping System

The Regional Agronomic Diagnosis (RAD) proposed by Doré et al. (1997) aims to identify and rank the factors limiting crop yield on the regional scale, based on the study of a farmers' field network. The relevance of such a methodological framework has been demonstrated in various case studies (Doré et al. 2008). Weed biomass and nitrogen deficiencies linked to sowing date, soil management and the quantity of organic manure applied affecting organic winter oilseed rape (Valantin-Morison and Meynard 2008), nitrogen deficiencies linked to soil compaction affecting pea yield (Doré et al. 1997) and weed and nitrogen deficiencies linked to soil tillage, type of weed and crop density affecting yield performance (David et al. 2005), all provide good examples of major limiting factors identified by RAD. This type of approach is based on an analysis of the functioning of the agroecosystem and often helps to increase our knowledge about it. The RAD can be considered as a pre-requisite to the design of new cropping systems. Indeed, it allows us to rank the major pest problems and to identify the main crop practices, or cropping system elements that have to be considered. RAD thereafter focuses on the possible way to improve cropping systems and points out the knowledge that is required to achieve new crop management objectives. Nevertheless, considering that many environmental variables depend on processes operating at a scale larger than the field scale, Doré et al. (2008) in their review, suggested that the RAD requires adaptations, as the classical methodology does not take into account the surrounding environment in which a field is embedded. In fact, Valantin-Morison et al. (2007) in their study about the effect of cropping systems on pest damage to organic winter oilseed rape showed that surrounding environment variables can have an important explanatory role.

17.6.2.3 Prototyping and Modelling: Basis for Integrated Crop Management of Oilseed Rape Pests

Prototyping and modelling are two different and sometimes complementary ways for designing integrated crop management strategies. Vereijken (1997) formalized a methodical way of prototyping integrated and ecological arable farming systems founded on expert knowledge, which could be summarized by four main steps: (1) identifying the sets of constraints and establishing a hierarchy of objectives and criteria, (2) designing a theoretical prototype of the cropping system, based on a concerted action of an expert collective, that fits both with the multi-objective parameters imposed by the shortcomings of the current farming system in a given region, and with the multi-constraints imposed by the whole environment, (3) improving the prototype on a few experimental and on-farm trials that represent the different constraints of the given region, and evaluating its performance through different criteria (such as crop performance, environmental, economical, or social), (4) adapting this cropping system prototype according to the multi-criteria evaluation. Iteration cycles between the two last steps lead to a technical prototype that could be tested in a larger range of situations in on-farm trials. Even if it has been proved that this method is very useful for designing a new cropping system (Lançon et al. 2007) and for involving different stakeholders, such as the researcher and the farmer, this type of approach has limits. Indeed, pedoclimatic variability is very difficult to take into account and the rules drawn up by the expert are sometimes not very formalized and are tough to evaluate and criticize.

Prototyping methodology and the modelling approach can be complementary. In fact, farm system models provide a means to expand, refine and formalize expert knowledge, as well as to integrate these and scientific agro-ecological knowledge at the farm level (Sterk et al. 2007). Traditionally, empirical approaches were often used to quantify economic threshold levels in decision-support systems for pest management. These empirical damage functions are generally derived by regression analysis relating a measurement of pest severity at a given crop stage to yield loss. But one of the main limits of such approaches is that they ignore crop-pest interactions and their value is generally limited to the local conditions where the measurements have been taken (Kropff et al. 1995). Moreover, the possible antagonist effects of crop practices on several pests are not considered with such an approach. Since then, a range of models that take into account pest life cycle, genotype, environment and/or management practices in designing crop management processes have been developed but mainly at the field level. For example, in the case of oilseed rape, a bio-economic model (OMEGAsys) has been developed to represent the effect of crop management either on crop yield, weed biomass, and stem canker attacks. The first aim of this model is to help in the conception of environmentally-friendly crop management at the field scale (Valantin-Morison et al. 2010). The output variables that are used to rank the different crop management strategies are attainable yield, frequency of pesticides treatments and gross margin. Some recent studies have reported an interest in using models to design new cropping systems for a range of problematics at a larger scale (Colbach et al. 2001a, b, Souchere et al. 2005, Tixier et al. 2007). Taking into account, to a greater extent, biological interactions in agroecosystems and their related scales raises some methodological issues. Indeed, given the considered spatial and time scales and the number of technical operations that have to be considered, it is sometimes very difficult to assess new pest management strategies using traditional field experiments. On these bases, spatial modelling appears to be a relevant tool for designing innovative cropping systems in which pest populations and incidence could be managed at such a scale. Experimental trials are used here to evaluate the relevance of the cropping systems, improve their robustness, and their technical feasibility. The SIPPOM-WOSR model (Lô-Pelzer 2008) is a good example of the development of sustainable strategies to control a disease of the oilseed rape crop, phoma stem canker, at the regional scale. This model has been developed to evaluate the agronomic, economic, and environmental performances (through output variables such as disease severity indices and the associated yield losses, actual yields, gross margins, energetic costs of cultural practices and Treatment Frequency Indices) of spatially-distributed cropping systems that combine cultural, genetic, and chemical control. It also calculates the genetic structure of pathogen populations depending on evolutionary forces or genetic mechanisms: migration, selection and recombination. This model allows ranking integrated crop management strategies. It demonstrates that the spatial distribution of a specific resistant cultivar combined with other control methods (such as cultural and chemical) can enhance the durability of the specific resistance gene. For the moment no such modelling approach that integrates spatial distribution of cropping systems has been developed concerning insect pest management particularly those based on biocontrol.

However, a few models have been developed to investigate the interaction between crop and semi-natural habitat on natural enemy populations but with little concern on crop management effect. Halley et al. (1996) studied the role of landscape heterogeneity on linyphiid spiders and found that inclusion of a small amount of refuge areas in a cereal landscape increased the population size of spiders in fields, while pesticide use and crop rotation decreased population size. Bianchi et al. (2007) recently developed a spatially-explicit model that assesses the effect of land use (distribution of crop and semi-natural habitat, quality of crop habitat) on the population viability of the ladybeetle (*Coccinella septempunctata*) and the aphid population dynamics in the agricultural landscape. The authors found that the primary cause of the decline of ladybeetles in Czech landscapes is the decrease of aphid populations in alfalfa and cereal crops due to a major reduction in fertilizer input from 1978 to 2005. This model demonstrates that the population viability of the ladybeetle depends on the availability of aphid prey in crops (particularly cereal) distributed in the landscape.

It is important to highlight that designing cropping systems through modelling is only possible after preliminary studies on the effects of cropping systems and landscape elements on pest populations and their natural enemies. Knowledge about their biology and ecology is also required. For example, Gu et al. (2007) explain that the deployment of integrated pest management at a larger scale requires that particular attention be paid to: (i) winter breeding areas and the dynamics of overwintering populations, (ii) the pattern of spring migration and seasonal variations in the population distribution areas, (iii) the relationship between incidence of migration events and weather systems, and (iv) the contribution of migrant populations to local infestations. Models are able to integrate very different pieces of knowledge and are therefore useful tools for understanding complex agroecological interactions occurring in rural landscapes and for identifying crucial knowledge gaps.

17.7 Conclusion

The development of sustainable agricultural systems is now a major concern of many researchers, farmers, and policymakers worldwide. One of the key strategies in sustainable cropping systems is to restore and maximize ecosystem services in agricultural landscapes. Improving biological control appears to be the cornerstone of insect pest management. This review provides clear evidence that the development of integrated pest management needs to take into account various aspects, such as ecological, agronomical or socio-economical ones and different spatial or temporal scales. Linking integrated insect pest management and landscape ecology requires more studies to investigate the effects of landscape composition on interactions between natural enemies, pests, and crop productivity. It is clear that, although there are numerous (and rapidly increasing) studies showing that the biological control of pests is technically possible, via new crop management systems or habitat management, the economic efficiency of this biological control remains to be demonstrated. Moreover, consideration of relevant crop management effects at the landscape scale will certainly allow a better discrimination and identification of the semi-natural habitat effects and the cropping system effects on natural enemies and pest populations.

The development of area-wide pest management requires a more holistic approach that would integrate crop and landscape management effects. It would also fill gaps in knowledge about the ecology of insect pests (e.g., overwintering areas, pattern of migration) and quantify those effects in terms of environmental (e.g., energy use, pesticides use, nitrogen discharge) and economical consequences (e.g., crop damage, yield losses, cost/benefit). In order to meet such a scientific challenge, complementing on-farm trials that produce knowledge and improve cropping systems or scenarios with modelling approaches is likely to be a potentially productive approach.

Moreover, designing innovative cropping systems with more reliance on biological control requires that all stakeholders be consulted. Indeed, development of an integrated pest management at the landscape scale needs, on the one hand, solid scientific knowledge, and on the other hand, a good synergy between the different stakeholders, such as farmers, policy makers, and managers or private land owners. This synergy could be used to implement integrated pest management on a wider scale.

The review of studies on the integration of crop and landscape management into new crop protection strategies has led us to conclude that further investigations are required on different key fronts: (i) knowledge production on cropping system effect and ecosystem production functions and services, (ii) improving knowledge about pest and natural enemy ecology, (iii) designing innovative cropping systems, and (iv) implementing and adapting these in diverse biophysical and social contexts.

Acknowledgements We would like to thank the editor, Ingrid H. Williams, for useful suggestions and S. Tanis-Plant for helpful English corrections on an earlier draft of the manuscript.

References

Altieri MA (1999) The ecological role of biodiversity in agroecosystems. *Agr Ecosyst Environ* 74: 19–31.

Altieri MA, Liebman M (1986) Insect, weed and plant disease management in multiple cropping systems. In: Francis CA (ed.) *Multiple cropping systems*. MacMillan, New York.

Altieri MA, Nicholls CI (2004) An agroecological basis for biological control through conservation. *California conf on biological control IV*, Berkeley, California, USA, 13–15 July 2004, pp 28–39.

Altieri MA, Schmidt LL (1986) The dynamics of colonizing arthropod communities at the interface of abandoned, organic and commercial apple orchards and adjacent woodland habitats. *Agr Ecosyst Environ* 16: 29–43.

Andow DA (1991a) Vegetational diversity and arthropod population response. *Annu Rev Entomol* 36: 561–586.

Andow DA (1991b) Yield loss to arthropods in vegetationally diverse agroecosystems. *Environ Entomol* 20: 1228–1235.

Aquilino KM, Cardinale BJ, Ives AR (2005) Reciprocal effects of host plant and natural enemy diversity on herbivore suppression: An empirical study of a model tritrophic system. *Oikos* 108: 275–282.

Aubertot JN, Pinochet X, Dore T (2004) The effects of sowing date and nitrogen availability during vegetative stages on *Leptosphaeria maculans* development on winter oilseed rape. *Crop Prot* 23: 635–645.

Baggen LR, Gurr GM, Meats A (1999) Flowers in tri-trophic systems: Mechanisms allowing selective exploitation by insect natural enemies for conservation biological control. *Entomol Exp Appl* 91: 155–161.

Bajwa WI, Kogan M (2004) Cultural practices: Springboard to IPM. In: Koul O, Dhaliwal GS, Cuperus GW (eds.) *Integrated pest management: Potential, constraints and challenges*. CABI Publishing, UK.

Barari H, Cook SM, Clark SJ, Williams IH (2005) Effect of a turnip rape (*Brassica rapa*) trap crop on stem-mining pests and their parasitoids in winter oilseed rape (*Brassica napus*). *BioControl* 50: 69–86.

Bartlet E, Blight MM, Pickett JA, Smart LE, Turner G, Woodcock CM (2004) Orientation and feeding responses of the pollen beetle, *Meligethes aeneus*, to candytuft, *Iberis amara*. *J Chem Ecol* 30: 913–925.

Bianchi FJJA, Booij CJH, Tscharntke T (2006) Sustainable pest regulation in agricultural landscapes: A review on landscape composition, biodiversity and natural pest control. *P Roy Soc Lond* 273: 1715–1727.

Bianchi FJJA, Honek A, Werf Wvander (2007) Changes in agricultural land use can explain population decline in a ladybeetle species in the Czech Republic: Evidence from a process-based spatially explicit model. *Landscape Ecol* 22: 1541–1554.

Bjözrkman M, Hamback PA, Ramert B (2007) Neighbouring monocultures enhance the effect of intercropping on the turnip root fly (*Delia floralis*). *Entomol Exp Appl* 124: 319–326.

Bohan DA, Glen DM, Wiltshire CW, Hughes L (2000) Parametric intensity and the spatial arrangement of the terrestrial mollusc herbivores *Deroceras reticulatum* and *Arion intermedius*. *J Anim Ecol* 69: 1031–1046.

Borg A, Ekbom B (1996) Characteristics of oviposition behaviour of the pollen beetle, *Meligethes aeneus* on four different host plants. *Entomol Exp Appl* 81: 277–284.

Brevault T, Bikay S, Maldes JM, Naudin K (2007) Impact of a no-till with mulch soil management strategy on soil macrofauna communities in a cotton cropping system. *Soil Till Res* 97: 140–149.

Broad ST, Schellhorn NA, Lisson SN, Mendham NJ, Corkrey R (2008) Host location and parasitism of *Brevicoryne brassicae* in diversified broccoli cropping systems. *Entomol Exp Appl* 129: 166–171.

Brunin B, Lacoste L (1970) Recherche sur la maladie du colza due a *Leptosphaeria maculans*. *Ann Phytopathol* 2: 477.

Brust GE (1994) Natural enemies in straw-mulch reduce Colorado potato beetle populations and damage in potato. *Biol Control* 4: 163–169.

Büchi R (1995) Combination of trap plants (*Brassica rapa* var. *silvestris*) and insecticide use to control rape pests. *IOBC/wprs Bull* 18(4): 102–121.

Büchi R (2002) Mortality of pollen beetle (*Meligethes* spp.) larvae due to predators and parasitoids in rape fields and the effect of conservation strips. *Agr Ecosyst Environ* 90: 255–263.

Büchs W (2003) Impact of on-farm landscape structures and farming systems on predators. In: Alford DV (ed.) *Biocontrol of oilseed rape pests*. Blackwell, Oxford, UK.

Büchs W, Harenberg A, Prescher S, Weber G, Hattwig F (1999) Entwicklung von Evertebratenzönosen bei verschiedenen Formen der Flächenstillegung und Extensivierung. *Mitteilungen aus der Biologischen Bundesanstalt für Land-und Forstwirtschaft Berlin-Dahlem* 368: 9–38.

Büchs W, Harenberg A, Zimmermann J (1997) The invertebrate ecology of farmland as a mirror of the intensity of the impact of man? – An approach to interpreting results of field experiments carried out in different crop management intensities of a sugar beet and an oil seed rape rotation including set-aside. *Biol Agr Hort* 15: 83–107.

Büchs W, Heimbach U, Czarnecki E (1991) Effects of different rape seed dressings on carabid beetles: Laboratory and semifield tests. *Gesunde Pflanzen* 43: 299–306.

Büchs W, Katzur K (2004) Cultivation techniques as means to control pests in organic oilseed rape production. *IOBC/wprs Bull* 27(10): 225–236.

Burn AJ (1988) Assessment of the impact of pesticides on invertebrate predation in cereal crops. *Aspects Appl Biol* 17: 279–288.

Butts RA, Floate KD, David M, Blackshaw RE, Burnett PA (2003) Influence of intercropping canola or pea with barley on assemblages of ground beetles (*Coleoptera: Carabidae*). *Environ Entomol* 32: 535–541.

Carcamo HA, Dunn R, Dosdall LM, Olfert O (2007) Managing cabbage seedpod weevil in canola using a trap crop – a commercial field scale study in western Canada. *Crop Prot* 26: 1325–1334.

Chabert A, Gandrey J (2005) Impact of some insecticides on Carabidae and consequences for slug populations. *IOBC/wprs Bull* 28(6): 111–113.

Chiverton PA (1999) Buffer zones of different floral composition – effects on the beneficial arthropod fauna. *Aspects Appl Biol* 54: 307–314.

Colbach N, Clermont-Dauphin C, Meynard JM (2001a) GENESYS: A model of the influence of cropping system on gene escape from herbicide tolerant rapeseed crops to rape volunteers. I. Temporal evolution of a population of rapeseed volunteers in a field. *Agr Ecosyst Environ* 83: 235–253.

Colbach N, Clermont-Dauphin C, Meynard JM (2001b) GENESYS: A model of the influence of cropping system on gene escape from herbicide tolerant rapeseed crops to rape volunteers. II. Genetic exchanges among volunteer and cropped populations in a small region. *Agr Ecosyst Environ* 83: 255–270.

Coll M, Bottrell DG (1995) Predator-prey association in mono- and dicultures: Effect of maize and bean vegetation. *Agr Ecosyst Environ* 54: 115–125.

Collins KL, Boatman ND, Wilcox A, Holland JM, Chaney K (2002) Influence of beetle banks on cereal aphid predation in winter wheat. *Agr Ecosyst Environ* 93: 337–350.

Cook SM, Bartlet E, Murray DA, Williams IH (2002) The role of pollen odour in the attraction of pollen beetles to oilseed rape flowers. *Entomol Exp Appl* 104: 43–50.

Cook SM, Jonsson M, Skellern MP, Murray DA, Anderson P, Powell W (2007) Responses of *Phradis* parasitoids to volatiles of lavender, *Lavandula angustifolia* – a possible repellent for their host, *Meligethes aeneus*. *BioControl* 52: 591–598.

Cook SM, Smart LE, Martin JL, Murray DA, Watts NP, Williams IH (2006) Exploitation of host plant preferences in pest management strategies for oilseed rape (*Brassica napus*). *Entomol Exp Appl* 119: 221–229.

Costamagna AC, Landis DA (2006) Predators exert top-down control of soybean aphid across a gradient of agricultural management systems. *Ecol Appl* 16: 1619–1628.
Cumming GS, Spiesman BJ (2006) Regional problems need integrated solutions: Pest management and conservation biology in agroecosystems. *Biol Conserv* 131: 533–543.
David C, Jeuffroy MH, Henning J, Meynard JM (2005) Yield variation in organic winter wheat: A diagnostic study in the Southeast of France. *Agron Sustain Dev* 25: 213–223.
Dennis P, Fry GLA (1992) Field margins: Can they enhance natural enemy population densities and general arthropod diversity on farmland? *Agr Ecosyst Environ* 40: 95–115.
Denys C, Tscharntke T (2002) Plant-insect communities and predator-prey ratios in field margin strips, adjacent crop fields, and fallows. *Oecologia* 130: 315–324.
Dhaliwal GS, Opender K, Ramesh A (2004) Integrated pest management: Retrospect and prospect. In: Koul O, Dhaliwal GS, Cuperus GW (eds.) *Integrated pest management: Potential, constraints and challenges*. CABI Publishing, UK.
Doré T, Clermont-Dauphin C, Crozat Y, David C, Jeuffroy MH, Loyce C, Makowski D, Malezieux E, Meynard JM, Valantin-Morison M (2008) Methodological progress in on-farm regional agronomic diagnosis, a review. *Agron Sustain Dev* 28: 151–161.
Doré T, Sebillotte M, Meynard JM (1997) A diagnostic method for assessing regional variations in crop yield. *Agr Sys* 54: 169–188.
Dosdall LM, Herbut MJ, Cowle NT (1994) Susceptibilities of species and cultivars of canola and mustard to infestation by root maggots (*Delia* spp.) (*Diptera: Anthomyiidae*). *Can Entomol* 126: 251–260.
Dosdall LM, Herbut MJ, Cowle NT, Micklich TM (1996) The effect of seeding date and plant density on infestations of root maggots, *Delia* spp. (Diptera: Anthomyiidae), in canola. *Canad J Plant Sci* 76: 169–177.
Dosdall LM, Stevenson FC (2005) Managing flea beetles (*Phyllotreta* spp.) (Coleoptera: Chrysomelidae) in canola with seeding date, plant density, and seed treatment. *Agron J* 97: 1570–1578.
Duffield S (1991) Does the size of area you spray with insecticides influence the extent of side-effects on invertebrate populations? *Game Conservancy Rev* 22: 64–65.
Dunning JB, Danielson JB, Pulliam HR (1992) Ecological processes that affect populations in complex landscapes. *Oikos* 65: 169–175.
El Titi A, Boller EF, Gendrier JP (ed.)(1993) Integrated production – Principles and Technical Guidelines. *IOBC/wprs Bull* 16(1): 1–97.
Elliot NC, Kieckhefer RW, Lee J, French BW (1998) Influence of within-field and landscape factors on aphid predator populations in wheat. *Landscape Ecol* 14: 239–252.
Ellis PR, Farrell JA (1995) Resistance to cabbage aphid (*Brevicoryne brassicae*) in six brassica accessions in New Zealand. *NZ J Crop Hort Sci* 23: 25–29.
Evans KA, Allen Williams LJ (1989) The response of the cabbage seed weevil (*Ceutorhynchus assimilis* Payk.) and the brassica pod midge (*Dasineura brassicae* Winn.) to flower colour and volatiles of oilseed rape. *Aspects Appl Biol* 23: 347–353.
Evans KA, Allen-Williams LJ (1998) Response of cabbage seed weevil (*Ceutorhynchus assimilis*) to baits of extracted and synthetic host-plant odor. *J Chem Ecol* 24: 2101–2114.
Ewald JA, Aebischer NJ (2000) Trends in pesticide use and efficacy during 26 years of changing agriculture in southern England. *Environ Monit Assess* 64: 493–529.
Fahrig L, Merriam G (1994) Conservation of fragmented populations. *Conserv Biol* 8: 50–59.
Feeny P, Paauwe KL, Demong JN (1970) Flea beetle and mustard oils: Host plant specificity of *Phyllotreta cruciferae* and *P. striolata* adults (Coleopera: Chrysomelidae). *Ann Entomol Soc Am* 63: 832–841.
Ferron P, Deguine JP (2005) Crop protection, biological control, habitat management and integrated farming, a review. *Agron Sustain Dev* 25: 17–24.
Finch S (1978) Volatile plant chemicals and their effect on host plant finding by the cabbage root fly (*Delia brassicae*). *Entomol Exp Appl* 24: 350–359.

Finch S, Collier RH (2000) Host-plant selection by insects – a theory based on 'appropriate/inappropriate landings' by pest insects of cruciferous plants. *Entomol Exp Appl* 96: 91–102.

Finch S, Collier RH (2003) Host-plant selection by insects – the 'missing link'. *IOBC/wprs Bull* 26(3): 103–108.

Flint ML, Gouveia P (2001) *IPM in practice: Principles and methods of integrated pest management*. University of California Press, Oakland, CA.

Francis CA, Clegg MD (1990) Crop rotations in sustainable production systems. In: Edwards CA, Lal R, Madden P, Miller RH, House G (eds.) *Sustainable agricultural systems*. Soil and Water Conservation Society, Ankeny, IA.

Frank T (1996) Species diversity and activity densities of epigaeic and flower visiting arthropods in sown weed strips and adjacent fields. *IOBC/wprs Bull* 19(3): 101–105.

Frank T, Nentwig W (1995) Ground dwelling spiders (Araneae) in sown weed strips and adjacent fields. *Acta Oecologica* 16(2): 179–193.

Free JB, Williams IH (1978) The response of pollen beetle, *Meligethes aeneus*, and the seed weevil, *Ceutorhynchus assimilis*, to oilseed rape, *Brassica napus,* and other plants. *J Appl Ecol* 15: 761–774.

Gatehouse JA (2002) Plant resistance towards insect herbivores: A dynamic interaction. *New Phytologist* 156: 145–169.

Geiger F, Bianchi FJJA, Wäckers FL (2005) Winter ecology of the cabbage aphid *Brevicoryne brassicae* (L.) (Homo., *Aphididae*) and its parasitoid *Diaeretiella rapae* (McIntosh) (Hym., Braconidae: Aphidiidae). *J Appl Entomol* 129: 563–566.

Gordon PL, McEwen FL (1984) Insecticide-stimulated reproduction of *Myzus persicae*, the green peach aphid (Homoptera: Aphididae). *Can Entomol* 116: 783–784.

Gu H, Fitt GP, Baker GH (2007) Invertebrate pests of canola and their management in Australia: A review. *Aust J Entomol* 46: 231–243.

Gurr GM, Wratten SD (eds) (2000) *Biological control: Measures of success*. Kluwer Academic Press, Dordrecht.

Halley JM, Thomas CFG, Jepson PC (1996) A model for the spatial dynamics of linyphiid spiders in farmland. *J Appl Ecol* 33: 471–492.

Harmon JP, Hladilek EE, Hinton JL, Stodola TJ, Andow DA (2003) Herbivore response to vegetational diversity: Spatial interaction of resources and natural enemies. *Population Ecology* 45: 75–81.

Hausammann A (1996) Strip-management in rape crop: Is winter rape endangered by negative impacts of sown weed strips? *J Appl Entomol* 120: 505–512.

Herzog DC, Funderburk JE (1985) Host plant resistance and cultural control interactions with biological control. In: Hoy MA, Herzog DC (eds.) *Biological control in agricultural IPM systems*. Academic Press, New York, USA.

Hickman JM, Wratten SD (1996) Use of *Phacelia tanacetifolia* strips to enhance biological control of aphids by hoverfly larvae in cereal fields. *J Econ Entomol* 89: 832–840.

Hokkanen HMT (1989) Biological and agrotechnical control of the rape blossom beetle *Meligethes aeneus* (Coleoptera, Nitidulidae). *Acta Entomol Fenn* 53: 25–29.

Hokkanen HMT (2008) Biological control methods of pest insects in oilseed rape. *EPPO Bull* 38(1): 104–109.

Hokkanen HMT, Husberg GB, Söderblom M (1988) Natural enemy conservation for the integrated control of the rape blossom beetle *Meligethes aeneus* F. *Ann Agr Fenn* 27: 281–294.

Holland JM (2004) The environmental consequences of adopting conservation tillage in Europe: Reviewing the evidence. *Agr Ecosyst Environ* 103: 1–25.

Holland JM, Drysdale A, Hewitt MV, Turley D (1996) The LINK IFS project – the effect of crop rotations and cropping systems on Carabidae. *Aspects Appl Biol* 47: 119–126.

Holland JM, Oaten H, Southway S, Moreby S (2008) The effectiveness of field margin enhancement for cereal aphid control by different natural enemy guilds. *Biol Control* 47: 71–76.

Holland JM, Thomas CFG, Birkett T, Southway S, Oaten H (2005) Farm-scale spatiotemporal dynamics of predatory beetles in arable crops. *J Appl Ecol* 42: 1140–1152.

Hooks CRR, Johnson MW (2003) Impact of agricultural diversification on the insect community of cruciferous crops. *Crop Prot* 22: 223–238.

Hopkins RJ, Ekbom B (1996) Low oviposition stimuli reduce egg production in the pollen beetle *Meligethes aeneus*. *Physiol Entomol* 21: 118–122.

Hopkins RJ, Ekbom B (1999) The pollen beetle, *Meligethes aeneus*, changes egg production rate to match host quality. *Oecologia* 120: 274–278.

Husberg GB, Granlund H, Hokkanen H (1985) Control of rape beetles with the aid of trap crops. *Vaxtskyddsnotiser* 49: 98–101.

Häni FJ, Boller EF, Keller S (1998) Natural regulation at the farm level. In: Pickett CH, Bugg RL (eds.) *Enhancing biological control, habitat management to promote natural enemies of agricultural pests*. University of California Press, USA.

Keller S, Häni F (2000) Ansprüche von Nützlingen und Schadlingen an den Lebensraum. In: Nentwig W (ed.) *Streifenformige okologische Ausgleichsflächen in der Kulturlandschaft: Ackerkrautstreifen, Buntbrache, Feldrander*. Verlag Agrarokologie, Bern.

Kendall DA, Chinn NE, Smith BD, Tidboald C, Winstone L, Western NM (1991) Effects of straw disposal and tillage on spread of barley yellow dwarf virus in winter barley. *Ann Appl Biol* 119: 359–364.

Khan ZR, Ampong-Nyarko K, Chiliswa P, Hassanali A, Kimani S, Lwande W, Overholt WA, Pickett JA, Smart LE, Wadhams LJ, Woodcock CM (1997) Intercropping increases parasitism of pests. *Nature* 388: 631–632.

Khan ZR, Pickett JA, Wadhams LJ, Hassanali A, Midega CAO (2006) Combined control of *Striga hermonthica* and stemborers by maize-*Desmodium* spp. intercrops. *Crop Prot* 25: 989–995.

Kirkegaard J, Christen O, Krupinsky J, Layzell D (2008) Break crop benefits in temperate wheat production. *Field Crops Res* 107: 185–195.

Klinger K (1987) Effects of margin strips along a winter wheat field on the predatory arthropod fauna and cereal aphid infestation. *J Appl Entomol* 104: 47–58.

Kogan M (1988) Integrated pest management theory and practice. *Entomol Exp Appl* 49: 59–70.

Kogan M (ed.) (1994) *Areawide management of the codling moth: Implementation of a comprehensive IPM program for pome fruit crops in the Western U.S.* Integrated Plant Protection Center, Corvallis, Oregon.

Kogan M (1998) Integrated pest management: Historical perspectives and contemporary developments. *Annu Rev Entomol* 43: 243–270.

Koss AM, Jensen AS, Schreiber A, Pike KS, Snyder WE (2005) Comparison of predator and pest communities in Washington potato fields treated with broad-spectrum, selective, or organic insecticides. *Environ Entomol* 34: 87–95.

Kostal V (1993) Physical and chemical factors influencing landing and oviposition by the cabbage root fly on host-plant models. *Entomol Exp Appl* 66: 109–118.

Kromp B (1999) Carabid beetles in sustainable agriculture: A review on pest control efficacy, cultivation impacts and enhancement. *Agr Ecosyst Environ* 74: 187–228.

Kropff MJ, Teng PS, Rabbinge R (1995) The challenge of linking pest and crop models. *Agricultural Systems* 49: 413–434.

Kruess A (2003) Effects of landscape structure and habitat type on a plant-herbivore-parasitoid community. *Ecography* 26: 283–290.

Krupinsky JM, Bailey KL, McMullen MP, Gossen BD, Turkington TK (2002) Managing plant disease risk in diversified cropping systems. *Agron J* 94: 198–209.

Lagerlof J, Wallin H (1993) The abundance of arthropods along two field margins with different types of vegetation composition: An experimental study. *Agr Ecosyst Environ* 43: 141–154.

Lamb RJ, Palaniswamy P, Pivnick KA, Smith MAH (1993) A selection of oilseed rape, *Brassica rapa* L., with resistance to flea beetles, *Phyllotreta cruciferae* (Goeze) (Coleoptera: *Chrysomelidae*). *Can Entomol* 125: 703–713.

Landis DA, Wratten SD, Gurr GM (2000) Habitat management to conserve natural enemies of arthropod pests in agriculture. *Annu Rev Entomol* 45: 175–201.
Landolt PJ, Hofstetter RW, Biddick LL (1999) Plant essential oils as arrestants and repellents for neonate larvae of the codling moth (Lepidoptera: *Tortricidae*). *Environ Entomol* 28: 954–960.
Lançon J, Wery J, Rapidel B, Angokaye M, Gerardeaux E, Gaborel C, Ballo D, Fadegnon B (2007) An improved methodology for integrated crop management systems. *Agronomy for Sustainable Development* 27: 101–110.
Lerin J (1988) Yield losses associated with 2 successive pests (*Ceuthorrhynchus napi* Gyll. and *Meligethes aeneus* F.) on winter rape (cultivar Bienvenu). *Agronomie* 8: 251–256.
Lô-Pelzer E (2008) *Modelling the effects of cropping systems and their spatial distribution on phoma stem canker on Winter Oilseed Rape and the adaptation of fungal pathogen*. PhD Thesis, AgroParisTech.
Lővei GL, McDougall D, Bramley G, Hodgson DJ, Wratten SD (1992) Floral resources for natural enemies: The effect of *Phacelia tanacetifolia* (Hydrophyllaceae) on within-field distribution of hoverflies (Diptera: Syrphidae). *Proc 45th NZ Plant Prot Conf*, Wellington, NZ, 11–13 August 1992, pp 60–61.
Mabbett T (1991) Straw incorporation trials reveal arable slug damage will increase. *Agr Int* 43: 304–306.
Mair J, Port GR (2002) The influence of mucus production by the slug, *Deroceras reticulatum*, on predation by *Pterostichus madidus* and *Nebria brevicollis* (Coleoptera: Carabidae). *Biocontrol Sci Techn* 12: 325–335.
Marino PC, Landis DA (1996) Effect of landscape structure on parasitoid diversity and parasitism in agroecosystems. *Ecol Appl* 6: 276–284.
Markus J, Podlaska J, Dmoch J, Pietkiewicz S, Oboda T, Lewandowski M (1996) Compensation of the damage caused by pollen beetle (*Meligethes aeneus*) on winter oilseed rape under different plant density and fertilisation III. Chemical composition of winter oilseed rape cv. Leo. *Rosliny Oleiste* 17: 325–330.
Marshall EJP (2004) Agricultural landscapes: Field margin habitats and their interaction with crop production. *J Crop Improvement* 12: 365–404.
Mauchline AL, Osborne JL, Martin AP, Poppy GM, Powell W (2005) The effects of non-host plant essential oil volatiles on the behaviour of the pollen beetle *Meligethes aeneus*. *Entomol Exp Appl* 114: 181–188.
McKirdy SJ, Jones RAC (1997) Effect of sowing time on barley yellow dwarf virus infection in wheat: Virus incidence and grain yield losses. *Australian J Agr Res* 48: 199–206.
Meek B, Loxton D, Sparks T, Pywell R, Pickett H, Nowakowski M (2002) The effect of arable field margin composition on invertebrate biodiversity. *Biol Conserv* 106: 259–271.
Meynard JM, Doré T, Habib R (2001) Evaluation and conception of cropping systems for a sustainable agriculture. *Comptes Rendus Académie d'Agriculture de France* 87: 223–236.
Milford GFJ, Evans EJ (1991) Factors causing variation in glucosinolates in oilseed rape. *Outlook Agr* 20: 31–37.
Nentwig W (1988) Augmentation of beneficial arthropods by strip-management. 1. Succession of predacious arthropods and long-term change in the ratio of phytophagous and predacious arthropods in a meadow. *Oecologia* 76: 597–606.
Nentwig W, Frank T, Lethmayer C (1998) Sown weed strips: Artificial ecological compensation areas as an important tool in conservation biological control. In: Barbosa P (ed.) *Conservation biological control*. Academic Press, San Diego, USA.
Nickel JL (1973) Pest situation in changing agricultural systems – a review. *Bull Entomol Soc Amer* 19: 136–142.
Nilsson C (1985) Impact of ploughing on emergence of pollen beetle parasitoids after hibernation. *Z Angew Entomol* 100: 302–308.
Nilsson C (1994) *Pollen beetles (*Meligethes spp.*) in oil seed rape crops* (Brassica napus *L) biological interactions and crop losses*. Department of Plant Protection Sciences, SLU, Dissertations 1, 39 pp.

Nitzsche O, Ulber B (1998) Influence of different tillage treatments following the harvest of oilseed-rape on the mortality of pollen beetle (*Meligethes* spp.) parasitoids. *Z Pflanzenkr Pflschutz* 105: 417–421.

Östman Ö, Ekbom B, Bengtsson J (2001) Landscape heterogeneity and farming practice influence biological control. *Basic Appl Ecol* 2: 365–371.

Palaniswamy P (1996) Host plant resistance to insect pests of cruciferous crops with special reference to flea beetles feeding on canola – a review. *Acta Hort* 407: 469–481.

de la Peña NM, Butet A, Delettre Y, Morant P, Burel F (2003) Landscape context and carabid beetles (Coleoptera: Carabidae) communities of hedgerows in western France. *Agr Ecosyst Environ* 94: 59–72.

Perrin RM (1977) Pest management in multiple cropping systems. *Agron Ecosys* 3: 93–118.

Petanidou T (2003) Introducing plants for bee-keeping at any cost? – assessment of *Phacelia tanacetifolia* as nectar source plant under xeric Mediterranean conditions. *Plant Systematics and Evolution* 238: 155–168.

Pfiffner L, Luka H (2000) Overwintering of arthropods in soils of arable fields and adjacent semi-natural habitats. *Agr Ecosyst Environ* 78: 215–222.

Pfiffner L, Luka H, Schlatter C, Juen A, Traugott M (2009) Impact of wildflower strips on biological control of cabbage lepidopterans. *Agr Ecosyst Environ* 129: 310–314.

Pfiffner L, Wyss E (2004) Use of sown wildflower strips to enhance natural enemies of agricultural pests. In: Gurr GM, Wratten SD, Altieri MA (eds.) *Ecological engineering for pest management. Advances in habitat manipulation for arthropods*. CSIRO Publishing, Melbourne, Australia.

Pickett CH, Bugg RL (1998) *Enhancing biological control: Habitat management to promote natural enemies of agricultural pests*. University of California Press, Berkeley, USA.

Pickett CH, Roltsch WJ, Corbett A, Daane KM (2000) Habitat management for enhancing biological control: Benefits and pitfalls. *California Conf on Biological Control II*, Riverside, California, USA, 11–12 July, 2000, pp 81–85.

Pimentel D (1961) Species diversity and insect population outbreaks. *Ann Entomol Soc Am* 54: 76–86.

Plantegenest M, Le May C, Fabre F (2007) Landscape epidemiology of plant diseases. *J Roy Soc Lond* 4: 963–972.

Platt JO, Caldwell JS, Kok LT (1999) Effect of buckwheat as a flowering border on populations of cucumber beetles and their natural enemies in cucumber and squash. *Crop Prot* 18: 305–313.

Podlaska J, Markus J, Dmoch J, Loboda T (1996) Compensation of the damage caused by pollen beetle (*Meligethes aeneus*) on winter oilseed rape under different plant density and fertilisation. I – Some morphological characters. *Rosliny Oleiste* 17: 325–330.

Pontin DR, Wade MR, Kehrli P, Wratten SD (2006) Attractiveness of single and multiple species flower patches to beneficial insects in agroecosystems. *Ann Appl Biol* 148: 39–47.

Potting RPJ, Perry JN, Powell W (2005) Insect behavioural ecology and other factors affecting the control efficacy of agro-ecosystem diversification strategies. *Ecol Model* 182: 199–216.

Pullaro TC, Marino PC, Jackson DM, Harrison HF, Keinath AP (2006) Effects of killed cover crop mulch on weeds, weed seeds, and herbivores. *Agr Ecosyst Environ* 115: 97–104.

Pulliam HR (1988) Sources, sinks, and population regulation. *Am Nat* 132: 652–661.

Pulliam HR, Danielson JB (1991) Sources, sinks, and habitat selection: A landscape perspective on population dynamics. *Am Nat* 137: 50–66.

Rahim A, Hashmi A, Khan NA (1991) Effects of temperature and relative humidity on longevity and development of *Ooencyrtus papilionis* Ashmead (Hymenoptera: Eulophidae), a parasite of the sugarcane pest, *Pyrilla perpusilla* Walker (Homoptera: Cicadellidae). *Environ Entomol* 20: 774–775.

Rebek EJ, Sadof CS, Hanks LM (2006) Influence of floral resource plants on control of an armored scale pest by the parasitoid *Encarsia citrina* (Craw.) (Hymenoptera: Aphelinidae). *Biol Control* 37: 320–328.

Riechert SE, Lockley T (1984) Spiders as biological control agents. *Annu Rev Entomol* 29: 299 –320.

Risch SJ (1983) Intercropping as cultural pest control: Prospects and limitations. *Environ Manage* 7: 9–14.

Rojas JC, Wyatt TD (1999) Role of visual cues and interaction with host odour during the host-finding behaviour of the cabbage moth. *Entomol Exp Appl* 91: 59–65.

Roland J, Taylor PD (1997) Insect parasitoid species respond to forest structure at different spatial scales. *Nature* 386: 710–713.

Root RB (1973) Organization of a Plant-Arthropod Association in Simple and Diverse Habitats: The Fauna of Collards (*Brassica oleracea*). *Ecol Monog* 43: 95.

Roschewitz I, Hücker M, Tscharntke T, Thies C (2005) The influence of landscape context and farming practices on parasitism of cereal aphids. *Agr Ecosyst Environ* 108: 218–227.

Royer TA, Mulder PG, Cuperus GW (1999) Renaming (redefining) integrated pest management: Fumble, pass, or play? *Am Entomol* 45: 136–139.

Rusch A, Valantin-Morison M (2010) Effect of nitrogen fertilization, cultivar and species on attractiveness and nuisibility of two major pests of winter oilseed rape (*Brassica napus* L.): Pollen beetle (*Meligethes aeneus* F.) and stem weevil (*Ceutorhynchus napi* Gyl.). *IOBC/wprs Bull* (in press).

Russell EP (1989) Enemies hypothesis: A review of the effect of vegetational diversity on predatory insects and parasitoids. *Environ Entomol* 18: 590–599.

Sarthou JP, Ouin A, Arrignon F, Barreau G, Bouyjou B (2005) Landscape parameters explain the distribution and abundance of *Episyrphus balteatus* (Diptera: Syrphidae). *European J Entomol* 102: 539–545.

Schellhorn NA, Sork VL (1997) The impact of weed diversity on insect population dynamics and crop yield in collards, *Brassica oleracea* (Brassicaceae). *Oecologia* 111: 233–240.

Schmidt MH, Roschewitz I, Thies C, Tscharntke T (2005) Differential effects of landscape and management on diversity and density of ground-dwelling farmland spiders. *J Appl Ecol* 42: 281–287.

Schmidt MH, Thewes U, Thies C, Tscharntke T (2004) Aphid suppression by natural enemies in mulched cereals. *Entomol Exp Appl* 113: 87–93.

Sharma HC, Ortiz R (2002) Host plant resistance to insects: An eco-friendly approach for pest management and environment conservation. *J Environ Biol* 23: 111–135.

Smart LE, Blight MM (2000) Response of the pollen beetle, *Meligethes aeneus*, to traps baited with volatiles from oilseed rape, *Brassica napus*. *J Chem Ecol* 26: 1051–1064.

Sotherton NW (1984) The distribution and abundance of predatory arthropods overwintering on farmland. *Ann Appl Biol* 105: 423–429.

Souchere V, Cerdan O, Dubreuil N, le Bissonnais Y, King C (2005) Modelling the impact of agri-environmental scenarios on runoff in a cultivated catchment (Normandy, France). *Catena* 61: 229–240.

Stark JD, Banks JE, Acheampong S (2004) Estimating susceptibility of biological control agents to pesticides: Influence of life history strategies and population structure. *Biol Control* 29: 392–398.

Stephens MJ, France CM, Wratten SD, Frampton C (1998) Enhancing biological control of leafrollers (Lepidoptera: Tortricidae) by sowing buckwheat (*Fagopyrum esculentum*) in an orchard. *Biocontrol Sci Techn* 8: 547–558.

Sterk B, Ittersum MKvan, Leeuwis C, Wijnands FG (2007) Prototyping and farm system modelling – partners on the road towards more sustainable farm systems? *European J Agron* 26: 401–409.

Sutherland JP, Sullivan MS, Poppy GM (2001) Distribution and abundance of aphidophagous hoverflies (Diptera: Syrphidae) in wildflower patches and field margin habitats. *Agr Forest Entomol* 3: 57–64.

Symondson WOC, Cesarini S, Dodd PW, Harper GL, Bruford MW, Glen DM, Wiltshire CW, Harwood JD (2006) Biodiversity vs. biocontrol: Positive and negative effects of alternative prey on control of slugs by carabid beetles. *B Entomol Res* 96: 637–645.

Tahvanainen JO, Root RB (1972) The influence of vegetational diversity on the population ecology of a specialized herbivore, *Phyllotreta cruciferae* (Coleoptera: Chrysomelidae). *Oecologia* 10: 321.

Theunissen J, Booij CJH, Lotz LAP (1995) Effects of intercropping white cabbage with clovers on pest infestation and yield. *Entomol Exp Appl* 74: 7–16.

Thies C, Steffan-Dewenter I, Tscharntke T (2003) Effects of landscape context on herbivory and parasitism at different spatial scales. *Oikos* 101: 18–25.

Thies C, Tscharntke T (1999) Landscape structure and biological control in agroecosystems. *Science* 285: 893–895.

Thomas SR, Noordhuis R, Holland JM, Goulson D (2002) Botanical diversity of beetle banks: Effects of age and comparison with conventional arable field margins in southern UK. *Agr Ecosyst Environ* 93: 403–412.

Thomas CD, Thomas JA, Warren MS (1992) Distributions of occupied and vacant butterfly habitats in fragmented landscapes. *Oecologia* 92: 563–567.

Thorbek P, Bilde T (2004) Reduced numbers of generalist arthropod predators after crop management. *J Appl Ecol* 41: 526–538.

Tietjen WJ, Cady AB (2007) Sublethal exposure to a neurotoxic pesticide affects activity rhythms and patterns of four spider species. *J Arachnol* 35: 396–406.

Tixier P, Malezieux E, Dorel M, Bockstaller C, Girardin P (2007) Rpest – an indicator linked to a crop model to assess the dynamics of the risk of pesticide water pollution: Application to banana-based cropping systems. *European J Agron* 26: 71–81.

Tonhasca A, Bryne DN (1994) The effects of crop diversification on herbivorous insects: A meta-analysis approach. *Ecol Entomol* 19: 239–244.

Trenbath BR (1993) Intercropping for the management of pests and diseases. *Field Crops Res* 34: 381–405.

Valantin-Morison M, Lemarié S, Aubertot JN, Reau R, Lacroix A (2010) Utilization of a model to re-design integrated crop management for winter oilseed rape. *IOBC/wprs Bull* (in press).

Valantin-Morison M, Meynard JM (2008) Diagnosis of limiting factors of organic oilseed rape yield. A survey of farmers' fields. *Agronomy for Sustainable Development* 28: 527–539.

Valantin-Morison M, Meynard JM, Doré T (2007) Effects of crop management and surrounding field environment on insect incidence in organic winter oilseed rape (*Brassica napus* L.). *Crop Prot* 26: 1108–1120.

Valantin-Morison M, Quere L (2006) Effects of turnip rape trap crops on oilseed rape pests in a network of organic farmers' fields. *CD-Rom Proc Int Symp Integrated Pest Management in Oilseed Rape, 3–5 April 2006, Göttingen, Germany.*

Van Emden HF (1991) The role of host plant resistance in insect pest mis-management. *Bull Entomol Res* 81: 123–126.

Vandermeer J (1995) The ecological basis of alternative agriculture. *Annu Rev Ecol Syst* 26: 201–224.

Vereijken P (1997) A methodical way of prototyping integrated and ecological arable farming systems (I/EAFS) in interaction with pilot farms. *European J Agron* 7: 235–250.

Vickerman GP (1992) *The effects of different pesticide regimes on the invertebrate fauna of winter wheat. Pesticides, cereal farming and the environment: The boxworth project.* HMSO, London, UK.

Weiss MJ, Schatz BG, Gardner JC, Nead BA (1994) Flea beetle (Coleoptera: Chrysomelidae) populations and crop yield in field pea and oilseed rape intercrops. *Environ Entomol* 23: 654–658.

West JS, Fitt BDL, Leech PK, Biddulph JE, Huang YJ, Balesdent MH (2002) Effects of timing of *Leptosphaeria maculans* ascospore release and fungicide regime on phoma leaf spot and phoma stem canker development on winter oilseed rape (*Brassica napus*) in southern England. *Plant Pathol* 51: 454–463.

Williams IH, Free JB (1979) Compensation of oil seed rape (*Brassica napus* L.) plants after damage to their buds and pods. *J Agr Sci* 92: 53–59.

Winstone L, Iles DR, Kendall DA (1996) Effects of rotation and cultivation on polyphagous predators in conventional and integrated farming systems. *Aspects Appl Biol* 47: 111–118.

Wyss E (1995) The effects of weed strips on aphids and aphidophagous predators in an apple orchard. *Entomol Exp Appl* 75: 43–49.
Wäckers FL (2004) Assessing the suitability of flowering herbs as parasitoid food sources: Flower attractiveness and nectar accessibility. *Biol Control* 29: 307–314.
Zabel J, Tscharntke T (1998) Does fragmentation of *Urtica* habitats affect phytophagous and predatory insects differentially? *Oecologia* 116: 419–425.
Zaller JG, Moser D, Drapela T, Schmöger C, Frank T (2008a) Effect of within-field and landscape factors on insect damage in winter oilseed rape. *Agr Ecosyst Environ* 123: 233–238.
Zaller JG, Moser D, Drapela T, Schmöger C, Frank T (2008b) Insect pests in winter oilseed rape affected by field and landscape characteristics. *Basic Appl Ecol* 9: 682–690.
Zehnder G, Gurr GM, Kuhne S, Wade MR, Wratten SD, Wyss E (2007) Arthropod pest management in organic crops. *Annu Rev Entomol* 52: 57–80.
Zehnder GW, Hough-Goldstein J (1989) Colorado potato beetle (Coleoptera: Chrysomelidae) population development and effects on yield of potatoes with and without straw mulch. *J Econ Entomol* 83: 1982–1987.

Species Index

I.H. Williams (ed.), *Biocontrol-Based Integrated Management of Oilseed Rape Pests*, DOI 10.1007/978-90-481-3983-5, © Springer Science+Business Media B.V. 2010

Subject Index

MIX
Papier aus verantwortungsvollen Quellen
Paper from responsible sources
FSC® C105338
FSC
www.fsc.org

If you have any concerns about our products,
you can contact us on
ProductSafety@springernature.com

In case Publisher is established outside the EU,
the EU authorized representative is:
Springer Nature Customer Service Center GmbH
Europaplatz 3, 69115 Heidelberg, Germany

Printed by Libri Plureos GmbH
in Hamburg, Germany